T0145229

Advances in Intelligent Systems and Computing

Volume 815

Series editor

Janusz Kacprzyk, Polish Academy of Sciences, Warsaw, Poland
e-mail: kacprzyk@ibspan.waw.pl

The series "Advances in Intelligent Systems and Computing" contains publications on theory, applications, and design methods of Intelligent Systems and Intelligent Computing. Virtually all disciplines such as engineering, natural sciences, computer and information science, ICT, economics, business, e-commerce, environment, healthcare, life science are covered. The list of topics spans all the areas of modern intelligent systems and computing such as: computational intelligence, soft computing including neural networks, fuzzy systems, evolutionary computing and the fusion of these paradigms, social intelligence, ambient intelligence, computational neuroscience, artificial life, virtual worlds and society, cognitive science and systems, Perception and Vision, DNA and immune based systems, self-organizing and adaptive systems, e-Learning and teaching, human-centered and human-centric computing, recommender systems, intelligent control, robotics and mechatronics including human-machine teaming, knowledge-based paradigms, learning paradigms, machine ethics, intelligent data analysis, knowledge management, intelligent agents, intelligent decision making and support, intelligent network security, trust management, interactive entertainment, Web intelligence and multimedia.

The publications within "Advances in Intelligent Systems and Computing" are primarily proceedings of important conferences, symposia and congresses. They cover significant recent developments in the field, both of a foundational and applicable character. An important characteristic feature of the series is the short publication time and world-wide distribution. This permits a rapid and broad dissemination of research results.

Advisory Board

Chairman

Nikhil R. Pal, Indian Statistical Institute, Kolkata, India
e-mail: nikhil@isical.ac.in

Members

Rafael Bello Perez, Universidad Central "Marta Abreu" de Las Villas, Santa Clara, Cuba
e-mail: rbellop@uclv.edu.cu

Emilio S. Corchado, University of Salamanca, Salamanca, Spain
e-mail: escorchado@usal.es

Hani Hagras, University of Essex, Colchester, UK
e-mail: hani@essex.ac.uk

László T. Kóczy, Széchenyi István University, Győr, Hungary
e-mail: koczy@sze.hu

Vladik Kreinovich, University of Texas at El Paso, El Paso, USA
e-mail: vladik@utep.edu

Chin-Teng Lin, National Chiao Tung University, Hsinchu, Taiwan
e-mail: ctlin@mail.nctu.edu.tw

Jie Lu, University of Technology, Sydney, Australia
e-mail: Jie.Lu@uts.edu.au

Patricia Melin, Tijuana Institute of Technology, Tijuana, Mexico
e-mail: epmelin@hafsamx.org

Nadia Nedjah, State University of Rio de Janeiro, Rio de Janeiro, Brazil
e-mail: nadia@eng.uerj.br

Ngoc Thanh Nguyen, Wroclaw University of Technology, Wroclaw, Poland
e-mail: Ngoc-Thanh.Nguyen@pwr.edu.pl

Jun Wang, The Chinese University of Hong Kong, Shatin, Hong Kong
e-mail: jwang@mae.cuhk.edu.hk

More information about this series at http://www.springer.com/series/11156

Raju Surampudi Bapi · Koppula Srinivas Rao
Munaga V. N. K. Prasad
Editors

First International Conference on Artificial Intelligence and Cognitive Computing

AICC 2018

 Springer

Editors
Raju Surampudi Bapi
School of Computer
  and Information Sciences
University of Hyderabad
Hyderabad, Telangana, India

and

Cognitive Science Lab
International Institute
  of Information Technology
Hyderabad, Telangana, India

Koppula Srinivas Rao
Department Computer Science
  and Engineering
MLR Institute of Technology
Hyderabad, Telangana, India

Munaga V. N. K. Prasad
IDRBT
Hyderabad, Telangana, India

ISSN 2194-5357   ISSN 2194-5365   (electronic)
Advances in Intelligent Systems and Computing
ISBN 978-981-13-1579-4   ISBN 978-981-13-1580-0   (eBook)
https://doi.org/10.1007/978-981-13-1580-0

Library of Congress Control Number: 2018948127

© Springer Nature Singapore Pte Ltd. 2019
This work is subject to copyright. All rights are reserved by the Publisher, whether the whole or part of the material is concerned, specifically the rights of translation, reprinting, reuse of illustrations, recitation, broadcasting, reproduction on microfilms or in any other physical way, and transmission or information storage and retrieval, electronic adaptation, computer software, or by similar or dissimilar methodology now known or hereafter developed.
The use of general descriptive names, registered names, trademarks, service marks, etc. in this publication does not imply, even in the absence of a specific statement, that such names are exempt from the relevant protective laws and regulations and therefore free for general use.
The publisher, the authors and the editors are safe to assume that the advice and information in this book are believed to be true and accurate at the date of publication. Neither the publisher nor the authors or the editors give a warranty, express or implied, with respect to the material contained herein or for any errors or omissions that may have been made. The publisher remains neutral with regard to jurisdictional claims in published maps and institutional affiliations.

This Springer imprint is published by the registered company Springer Nature Singapore Pte Ltd.
The registered company address is: 152 Beach Road, #21-01/04 Gateway East, Singapore 189721, Singapore

Organizing Committee Chairs

Prof. K. Neeraja, MLR Institute of Technology, Hyderabad, India
Dr. G. Kiran Kumar, MLR Institute of Technology, Hyderabad, India
Prof. N. Chandra Sekhar Reddy, MLR Institute of Technology, Hyderabad, India

Special Session Chair

Dr. Mark Burgin, University of California, Los Angeles, USA

Programme Committee Chair

Dr. Suresh Chandra Satapathy, P. V. P. Siddhartha Institute of Technology, Vijayawada, India

Technical Programme Committee

Prof. K. L. Chugh
Dr. Satish Kumar, IIT Roorkee
Dr. Anil Kumar, IIT Patna
Dr. A. Nagaraju, Central University, Rajasthan
Dr. Damodar Reddy Edla, NIT Goa
Dr. Venkatanareshbabu Kuppili, NIT Goa
Dr. Vijaya Kumar, VIT University, Vellore

Steering Committee

Dr. Radhika Devi V., MLR Institute of Technology

Dr. M. Satyanarayana Gupta, MLR Institute of Technology
Dr. S. V. S. Prasad, MLR Institute of Technology
Dr. S. Madhu, MLR Institute of Technology
Dr. S. Shyam Kumar, MLR Institute of Technology

Organizing Committee

Mr. Ram Mohan Rao
Mr. T. Dharma Reddy, Professor, CSE
Dr. B. Rama, Professor, CSE
Dr. Susmithavalli G., Professor, CSE
Mr. Sk Khaja Shareef. Assoc. Professor, IT
Dr. Sheikh Gouse, Assoc. Professor, CSE
Dr. P. Amarnath Reddy, Assoc. Professor, IT
Mr. G. Prabhakar Reddy, Assoc. Professor, CSE
Mr. M. Srinivas Rao, Assoc. Professor, CSE
Mrs. L. Laxmi, Assoc. Professor, CSE
Mrs. P. Madhuravani, Assoc. Professor, CSE
Mrs. K. Archana, Assoc. Professor, CSE
Mr. K. Hemanath, Asst. Professor, IT
Mr. V. Krishna, Asst. Professor, IT
Mrs. B. DurgaSree, Asst. Professor, IT.
Mr. J. Pradeep Kumar, Asst. Professor, IT
Ms. M. Anila Rao, Asst. Professor, IT
Mr. A. Venkata Siva Rao, Asst. Professor, CSE
Mr. K. Sai Prasad, Asst. Professor, CSE
Ms. G. Divya Jyothi, Asst. Professor, CSE
Ms. K. Navya, Asst. Professor, CSE
Mrs. K. Nirosha, Asst. Professor, IT
Ms. B. Lakshmi, Asst. Professor, CSE
Mr. K. Rajesh, Asst. Professor, CSE
Mrs. V. Prashanthi, Asst. Professor, CSE
Mrs. N. Shirisha, Asst. Professor, CSE
Ms. Y. Prasanna, Asst. Professor, CSE
Ms. R. Anusha, Asst. Professor, CSE
Ms. K. Hepshiba Vijaya Kumari, Asst. Professor, CSE
Mr. K. Chandra Sekhar, Asst. Professor, CSE
Mr. E. Amarnatha Reddy, Asst. Professor, CSE
Mr. K. Srinivas Reddy, Asst. Professor, CSE
Ms. S. Poornima, Asst. Professor, CSE
Ms. Y. Dhana Lakshmi, Asst. Professor, CSE
Mr. G. Vijay Kanth, Asst. Professor, CSE
Mrs. A. Manusha Reddy, Asst. Professor, CSE
Ms. G. Roja, Asst. Professor, CSE

Mr. Anandkumar B., Asst. Professor, CSE
Mr. G. John Samuel Babu, Asst. Professor, CSE
Ms. D. Chaitrali Shrikant, Asst. Professor, CSE
Mr. Nayamtulla Khan, Asst. Professor, CSE
Ms. Madavi Deepala, Asst. Professor, CSE
Mr. M.Mahesh Kumar, Asst. Professor, CSE
Mr. B. S. S. Murali Krishna, Asst. Professor, CSE
Mrs. G. Sowmya, Asst. Professor, CSE
Mr. Rajesh Kumar Reddy, Asst. Professor, CSE
Ms. A. Gayatri, Asst. Professor, CSE
Ms. D. Kalavathi, Asst. Professor, CSE
Mr. P. Purushotham, Asst. Professor, CSE
Mr. D. Divakar, Asst. Professor, CSE
Ms. P. Padma, Asst. Professor, CSE
Mr. D. Mahesh Babu, Asst. Professor, CSE
Mr. Sheri Ramchandra Reddy, Asst. Professor, CSE
Mr. Yogender Nath, Asst. Professor, CSE
Mr. Rajinish J., Asst. Professor, CSE
Mr. T. Vijay Kumar, CSE
Mr. M. Srinivas, CSE
Mr. Hari Babau K., CSE
Mr. Syed Sajid, CSE

Preface

The first international conference on Artificial Intelligence and Cognitive Computing (AICC 2018) was successfully organized by MLR Institute of Technology, Hyderabad, India, from 2 to 3 February 2018. The objective of this international conference was to provide a platform for academicians, researchers, scientists, professionals and students to share their knowledge and expertise in the fields of artificial intelligence, soft computing, evolutionary algorithms, swarm intelligence, Internet of things, machine learning, etc., to address various issues to increase awareness of technological innovations and to identify challenges and opportunities to promote the development of multidisciplinary problem-solving techniques and applications. Research submissions in various advanced technology areas were received, and after a rigorous peer review process with the help of technical programme committee members, elite quality papers were accepted. The conference featured nine special sessions on various cutting-edge technologies which were chaired by eminent professors. Many distinguished researchers like Prof. D. K. Subramaniam; Tandava Popuri, Director, R&D Dell systems; Dr. Bapi Raju Surampudi; Dr. Suresh Chandra Satapathy; Dr. S. Rakesh; Dr. K. Venugopal; Mr. Nagarjun Malladi; Mr. Sriharsh Bhyravajjula, India, attended the conference. Plenary talk was given by Abhimanyu Aryan on artificial intelligence and virtual reality.

Our sincere thanks to all special session chairs Dr. N. Sandhya, VNRVJIET, Hyderabad; Dr. DVLN Somayajulu, NIT Warangal; Dr. Y. Rama Devi, CBIT, Hyderabad; Dr. B. Rama Abbidi, Kakatiya University, Warangal; and distinguished reviewers for their timely technical support. We would like to extend our special thanks to very competitive team members for successfully organizing the event.

Finally, we thank Dr. Suresh Chandra Satapathy, PVPSIT, Vijayawada, for his complete guidance being Publication Chair in bringing out a good volume in Springer AISC series.

Hyderabad, India

Dr. Koppula Srinivas Rao
Dr. Raju Surampudi Bapi
Dr. Munaga V. N. K. Prasad

Contents

About the Editors

Raju Surampudi Bapi is Professor at the School of Computer and Information Sciences, University of Hyderabad, India, and was Visiting Professor at the Cognitive Science Lab, International Institute of Information Technology, Hyderabad, during 2014–2017. He received his B.E. (electrical engineering) from Osmania University, Hyderabad, India, and his M.S. [biomedical engineering (BME)] and Ph.D. [Mathematical Sciences Computer Science (CS)] from the University of Texas at Arlington, USA. He was EPSRC (Engineering and Physical Sciences Research Council) Postdoctoral Fellow at the University of Plymouth, UK, and worked as Researcher in the Kawato Dynamic Brain Project, ATR Labs, Kyoto, Japan, before joining the University of Hyderabad in 1999. He has published over 100 papers in international journals and for international conferences. He served as the Coordinator of the Center for Neural and Cognitive Sciences, University of Hyderabad, India. His main research interests include biological and artificial neural networks, neural and cognitive modelling, machine learning, pattern recognition, neuroimaging and cognitive science. He is a senior member of IEEE and member of ACM, the Cognitive Science Society, USA, Association for Cognitive Science, India, and the Society for Neuroscience, USA.

Koppula Srinivas Rao is currently Professor and Head of the Department of Information Technology, MLR Institute of Technology, Telangana, India. He received his Ph.D. in computer science engineering from Anna University, Chennai, and his master's in computer science and engineering from Osmania University, Hyderabad. He has more than 20 years of teaching and research experience. His research interests include data mining, RFID data streams, natural language processing, and artificial intelligence studies and their applications to engineering. He has more than 19 publications in various reputed international journals and for conference proceedings to his credit and is a senior member of the Computer Society of India.

Munaga V. N. K. Prasad received his doctoral degree from the Institute of Technology, Banaras Hindu University, Varanasi, Uttar Pradesh, India. Currently, he is Associate Professor at the Institute for Development and Research in Banking/Technology, Hyderabad. He has published numerous papers in international and national journals. His research interests include biometrics, payment system technologies and digital watermarking. He is a senior member of IEEE and member ACM.

Automatic Retail Product Image Enhancement and Background Removal

Rajkumar Joseph, N. T. Naresh Babu, Ratan S. Murali
and Venugopal Gundimeda

Abstract Retailers need good-quality product images with clear background on their Web sites. Most of these product images captured have diverse backgrounds, posing a challenge to separate the foreground from the background along with the enhancement of the product image. Currently, most of these activities are done manually. Our study proposes a computer vision (CV)- and machine learning (ML)-based approach to separate foreground (FG) and background (BG) from retail product images and enhance them. This automated process of BG/FG extraction involves two steps. A neural network (NN) classifier to identify if the BG has a monocolor gradient or not, followed by the separation of FG from BG and enhancement applied on the FG from the input image. Our results show 91% accuracy for BG/FG extraction and identifying the product region of interest (ROI).

Keywords Background removal · Image matting · Image enhancement
Machine learning

R. Joseph (✉) · N. T. Naresh Babu · R. S. Murali
GTO-CDS-LAB, Cognizant Technology Solutions, SEZ Ave,
Elcot Sez, Sholinganallur, Chennai, Tamil Nadu 600119, India
e-mail: rajkumar.joseph@cognizant.com

N. T. Naresh Babu
e-mail: nareshbabu.nt@cognizant.com

R. S. Murali
e-mail: Ratan.SMurali@cognizant.com

V. Gundimeda
GTO-CDS-LAB, Cognizant Technology Solutions, Building 12A, Raheja
Mindspace, Hi-tech City Road, Hyderabad 500081, India
e-mail: Venugopal.Gundimeda@cognizant.com

© Springer Nature Singapore Pte Ltd. 2019   1
R. S. Bapi et al. (eds.), *First International Conference on Artificial Intelligence
and Cognitive Computing* , Advances in Intelligent Systems and Computing 815,
https://doi.org/10.1007/978-981-13-1580-0_1

1   Introduction

Retail product images generally are shot using various cameras with a contrasting background. In most of the cases, we have products being shot on a white background. White is generally not pure white when it is photographed and thus introduces various noises and some shades generally having gradient of gray and white whose variation changes depending on illumination and exposure settings of the camera. Generally, most algorithms require manual intervention to provide indicative markers to identify foreground and background regions. We propose a system that automatically removes background having monochrome gradients (not considering natural scene backgrounds) and enhances the foreground from input images. The proposed system has been optimized to work with retail product photography images only, images have monochrome-based background, and background color should be brighter than foreground object and with no outlier objects presented in camera view.

2   Related Work

The objective of background removal is to extract useful region/region of interest (ROI) from images without human assistance. Most of the BG vs FG separation techniques rely on color to determine the alpha matte and also consider few low-level features. Others like sampling-based methods rely on extracting colors by sampling the known FG and BG followed by a computation to determine the best FG/BG combination. Segmenting ROIs is a difficult problem in image processing, and it has been an active area of research for several decades. Different image segmentation techniques are shown in Fig. 1.

The thresholding methods like global, local, and adaptive techniques are used to extract ROIs [1–3]. The advantage of thresholding technique is to make threshold calculations faster and effective. Global algorithm is well suited only for the images with equal intensities. This method does not work well with variable illumination. The drawback of adaptive thresholding is computationally expensive, and therefore, it is not suitable for real-time applications. Zhu [4] proposed a new threshold-based edge detection and image segmentation algorithm. The threshold is computed for each pixel in the image on the basis of its neighboring pixels.

Yucheng [5] proposed a new fuzzy morphological-based fusion image segmentation algorithm. The algorithm uses morphological opening and closing operations to

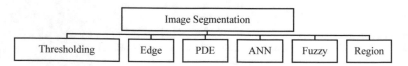

Fig. 1 Image segmentation techniques

smoothen the image and then perform the gradient operations on the resultant image [6, 7]. Khokher [8] presented a new method of image segmentation using fuzzy rule-based system and graph cuts [9]. The above fuzzy-based clustering technique has drawbacks, and it is sensitive to noise and computationally expensive, initialization condition of cluster number and cluster center. Also, the technique does not work well with non-globular clusters, and determination of fuzzy membership is not very easy.

Image matting is widely used to segment the target image from image/video. Popular matting techniques are blue screen matting, Bayesian matting, closed-form matting, geodesic matting, easy matting, graph cut, and deep matting. The different matting techniques are covered in [10–16]. In [13], authors used the geodesic framework to classify the pixels between foreground and background. However, this framework requires two manual inputs (select two lines, one on foreground region and other one on background region) to do automatic segmentation. This method exploited weights in the geodesic computation that depends on the pixel value distributions. The algorithm works best when these distributions do not significantly overlap. Human intervention is required to select foreground and background to achieve better accuracy. Also, it fails with transparent foreground images. Recently, various deep learning techniques have been proposed for image matting [15–18]. This area is still an active area of research due to complex nature of the FG/BG images.

3 The Proposed Work

The proposed system is designed to remove background and enhance the foreground object. The proposed framework is shown in Fig. 2.

To validate our prerequisite assumptions, the input product image is sent to automation classifier which identifies whether we have a monochrome background. Once we conclude that the input image has a monochrome background, we do a white/gray color classification check. Most of the retail product images are captured with white/gray background. For white/gray background images, fine-tuning method I is used. This method also works for non-contrasting BG/FG colors. For monochrome gradient (nonwhite/gray) background images with contrasting BG/FG colors, fine-tuning method II is used. If automation classifier is "No," manual selection is suggested.

3.1 Preprocessing

The input image is captured from camera device. Convert RGB image to L*a*b. Concentrate on "L" channel to further proceed. Apply histogram to "L-Channel."

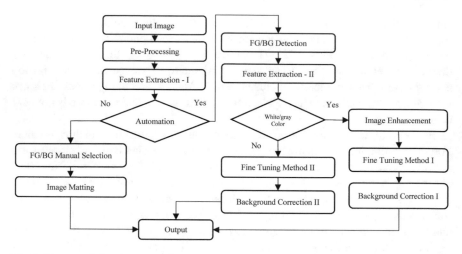

Fig. 2 Framework for the proposed system

Find highlighters' side peak. Stretch the image from 2% of highlighter. The stretched image (L*a*b) is converted back to RGB for further process.

3.2   Feature Extraction I and II

In this proposed system, we assume our scope has monochrome gradient background. So to identify monochrome background or not (texture pattern or natural scene, etc.), we introduced a classifier. Once classifier satisfies scope, then our proposed system will remove background automatically, else it will go to manual selection-based background removal. Histogram, vertical projection profile (VPP), and horizontal projection profile (HPP) are used as the features for the feature vector of the classifier, and the features are derived from grayscale image. VPP and HPP are calculated using Eq. (1). Number of histogram bins are 32, and they are normalized with respect to total numbers of pixels. Similarly, horizontal and vertical projection bins are 32, and they are normalized with respect to height/width and n-bit gray scale; it is derived from Eqs. (2) and (3). Therefore, total feature vector size is 96 (32X3).

$$\text{VPP} = \sum_{1 \leq x \leq m} f(x, y); \quad \text{HPP} = \sum_{1 \leq y \leq n} f(x, y) \tag{1}$$

$$\text{NVPP} = \text{VPP} / \left(\text{Height} * 2^{N\text{bitgrayscale}} \right) \tag{2}$$

$$\text{NHPP} = \text{HPP} / (\text{Width} * 2^{N\text{bitgrayscale}}) \tag{3}$$

Table 1 Parameter for MLP

Parameters	Number of perceptrons	Activation function	Training algorithm	No of epochs	Learning rate
Value	64	tanh	Levenberg–Marquardt, Back-propagation	100	0.1

where $f(x, y)$—input image, m/height—no of rows in the image, n/width—no of columns in the image, NVPP—normalized vertical projection profile, NHPP—normalized horizontal projection profile.

3.3 Classifier

Two classifiers are used in the proposed system. One is automation classifier, which is used to identify whether monochrome gradient background is present or not in the given input image. Another one is used to identity whether white/gray background is present or not in the given input image. We used multilayer perceptron (MLP), a type of neural network (NN) [19, 20] for classifications. Parameter for MLP is shown in Table 1.

3.4 Foreground Detection

Illumination correction is done using morphological operation. Convert input image into grayscale image, and apply morphological operation "open"; it is equivalent to opening (I_g). The illumination-corrected image is derived from Eq. (4). Diamond-shaped structuring element is used.

$$ICI = I_g - \text{Opening}(I_g) \tag{4}$$

where ICI—illumination-corrected image and I_g—grayscale image. Normalized vertical projection profile (NVPP) and normalized horizontal projection profile (NHPP) are calculated on illumination-corrected image. There will be significant change in intensity toward background to foreground. So differentiation is applied to identify starting point of foreground location. To suppress noise, normalization with respect to mean and moving average filter is used.

$$VPP = \left| \frac{\partial NVPP}{\partial X} \right|, \quad HPP = \left| \frac{\partial NHPP}{\partial y} \right| \tag{5}$$

$$NVPP = VPP/\mu_{VP}; \quad NHPP = HPP/\mu_{HP} \tag{6}$$

$$V(x) = \begin{cases} 1 \text{ if NVPP} \geq k * \sigma_{\text{NVPP}} \\ 0 \text{ if NVPP} < k * \sigma_{\text{NVPP}} \end{cases} \tag{7}$$

$$S_x = \min \arg (V(x) = 1); \quad E_x = \min \arg(V(x) = 1) \tag{8}$$

$$H(y) = \begin{cases} 1 \text{ if NHP} \geq k * \sigma_{\text{NHPP}} \\ 0 \text{ if NHP} < k * \sigma_{\text{NHPP}} \end{cases} \tag{9}$$

$$S_y = \min \arg (H(y) = 1); \quad E_y = \min \arg(H(y) = 1) \tag{10}$$

where S_x, E_x, S_y, E_y are the rectangle coordinates of ROI region.

3.5 Image Enhancement

Image enhancement is done only for white/gray background images using exposure setting. Exposure settings are derived mathematically using following Eqs. (11)–(17) and background RGB image has to convert into gray to apply histogram. In histogram graph, the darkest tone available is zero, and it is shown at the left-hand side of the graph. The lightest, whitest tone achievable is 255 on the scale, and it is shown on the extreme right of the graph. To get the best tonal range, and to avoid problems with underexposed shadows or overexposed highlights, the histogram should be vaguely bell-shaped. So, the histogram graph should move toward right-hand extremes with respect to $-2*(\sigma)$ from the peak histogram count.

$$\max H = \arg \max_{0 \leq i \leq n} \left(H_g(i) \right); \quad \lim_{i \to \text{optimal } H_g} H_g = 2 * \sigma_{H_g} \tag{11}$$

$$H_{\text{optimal}} = \begin{cases} \min \arg \left(\text{Optimal } H_g \right) \text{ if } H_g(\max H) \geq 2 * \sigma_{H_g} \\ \max H \quad\quad\quad\quad\quad\quad\quad \text{else} \end{cases} \tag{12}$$

$$E = 2^{n\text{bit}} / H_{\text{optimal}} \tag{13}$$

$$H = H_g * E \tag{14}$$

$$MI_g = H\left(I_g\right) \tag{15}$$

$$H_R = H_R * E; H_G = H_G * E; H_B = H_B * E; \tag{16}$$

$$I_R = H_R(I_R); I_G = H_G(I_G); I_B = H_B(I_B) \tag{17}$$

where $\max H$—location of peak histogram count, H_{optimal}—optimal location for image enhancement, E—exposure value of the given image, MI_g—modified grayscale image with respect to histogram stretched value, $I_R, I_{G,} I_B$—modified image, combination of histogram image H_R, H_G, H_B to red, green, blue channels, respectively.

3.6 Foreground Fine-Tuning Method I

This method is for white/gray gradient background images. Sometimes, product photographs are taken without contrasting color between background and foreground and some of input images do not have uniform background, and it is corrupted by noise, illumination, shadow lines, etc. So we used color and edge information combing together to fine-tune foreground detection to achieve better background subtraction [21]. Fine-tuning is done using sensitivity factor (SF). SF is purely based on probability-based metric foreground. Apply histogram to foreground image and background image. Here, the total number of histogram bins is 256. The SF is calculated using Eq. (20). From Eq. (21), FG_{mask0} is calculated. Local standard deviation $\overline{\sigma}_{local}$ of image is calculated using Eq. (23) with help of Eq. (22), FG_{mask0} is calculated from Eq. (24). Sensitivity factor for intensity of color information FG_{mask0} and sensitivity factor for local edge information $\overline{\sigma}_{local}$ both are fused by weighted average (weights are W_1, W_2). The FG_{mask} is calculated from Eq. (25). th_{SF}—threshold value for foreground region.

$$P_{FG}(i) = \frac{\sum H_{FG}(i)}{\sum H_{FG}(i) + \sum H_{BG}(i)} \tag{18}$$

$$P_{BG}(i) = \frac{\sum H_{BG}(i)}{\sum H_{BG}(i) + \sum H_{FG}(i)} \tag{19}$$

$$SF = P_{FG}/P_{BG} \tag{20}$$

$$FG_{mask0} = \begin{cases} 1 & \text{if } SF \geq th_{SF} \\ \dfrac{SF}{th_{SF}} & \text{if } SF < th_{SF} \end{cases} \tag{21}$$

$$\overline{X}(i, j) = \frac{1}{(2n+1)*(2m+1)} \sum_{i-n}^{i+n} \sum_{j-m}^{j+m} I_g(i, j) \tag{22}$$

$$\overline{\sigma}_{local}(i, j) = \sqrt{\frac{1}{(2n+1)*(2m+1)} \sum_{i-n}^{i+n} \sum_{j-m}^{j+m} \left[I_g(i, j) - \overline{X}(i, j)\right]^2} \tag{23}$$

$$FG_{mask1} = \begin{cases} 1 & \text{if } \overline{\sigma}_{local} \geq th_{\sigma} \\ 0 & \text{if } \overline{\sigma}_{local} < th_{\sigma} \end{cases} \tag{24}$$

$$FG_{mask} = (W_1 * FG_{mask0}) + (W_2 * FG_{mask1}) \tag{25}$$

$$W_1 + W_2 = 1; W_1, W_2 \geq 0; W_1, W_2 \leq 1 \tag{26}$$

3.7 Foreground Fine-Tuning Method II

In Sect. (3.6), we proposed foreground fine-tuning method I, which works better for gray/white background. This method will work for any monochrome gradient. We redefine SF which will play a major role for nonwhite/non-gray background. The multivariate Gaussian distribution for foreground (G_{FG}) and multivariate Gaussian distribution for background (G_{BG}) are derived from Eqs. (27) to (28), respectively. SF is calculated from Eq. (29), and it is known as foreground mask (FG_{mask}).

$$G_{FG} = \frac{1}{\sqrt{(2\pi)^k |\Sigma|}} \exp^{-\frac{1}{2}(X-\mu_{FG})^T \Sigma^{-1}(X-\mu_{FG})} \tag{27}$$

$$G_{BG} = \frac{1}{\sqrt{(2\pi)^k |\Sigma|}} \exp^{-\frac{1}{2}(X-\mu_{BG})^T \Sigma^{-1}(X-\mu_{BG})} \tag{28}$$

$$SF = \begin{cases} 1 & \text{if } I_R \geq \mu_R, I_G \geq \mu_G, I_B \geq \mu_B \\ 1 & \text{if } \frac{G_{FG}}{G_{BG}} \geq d \\ \frac{G_{FG}}{G_{BG*d}} & \text{if } \frac{G_{FG}}{G_{BG}} < d \end{cases} \tag{29}$$

$$FG_{mask} = SF \tag{30}$$

where Σ—covariance matrix, μ_{FG}—mean (foreground image), μ_{BG}—mean (background image), d—threshold, I_R, I_G, I_B—image pixel values red, green, blue, respectively.

3.8 Background Correction

Background correction is done by increasing brightness on background region only. To achieve this enhancement, separate each color component (I_R, I_G, I_B) and do following mathematical operation on foreground masked image (FG_{mask}) referring to Eq. (30). The background-corrected image is calculated using Eqs. (31)–(33).

$$Output_R = I_R + (1 - FG_{mask}) * (2^{nbit} - \mu_{BG_R}) \tag{31}$$

$$Output_G = I_G + (1 - FG_{mask}) * (2^{nbit} - \mu_{BG_G}) \tag{32}$$

$$Output_B = I_B + (1 - FG_{mask}) * (2^{nbit} - \mu_{BG_B}) \tag{33}$$

where FG_{mask}—foreground masked image, μ_{BGR}, μ_{BGG}, μ_{BGB}—mean of red, green, blue channels to background image, respectively. Thresholding is done to output image to recompute, FG_{mask} along with surface removal has been done using median filter, morphological to the processed image (output) and the mask have been updated.

$$SFG_{updatedmask} = \begin{cases} 1 \text{ if } Output_{grayscale} \geq th_{recompute} \\ 0 \text{ if } Output_{grayscale} < th_{recompute} \end{cases} \tag{34}$$

3.9 Image Matting

Moving forward, after identifying foreground detection, image matting technique is applied. We used Bayesian matting. The purpose of matting is to extract foreground objects from background, often for the purpose of compositing with new environments. A foreground object is extracted from the background by estimating the color and opacity, or alpha channel, for the foreground elements at each pixel. The color value can then be expressed by the composition Eq. (35):

$$SC = \alpha F + (1 - \alpha)B \tag{35}$$

where F and B are the foreground and background colors, alpha (α) is the opacity map, and C is the resulting color. Therefore, matting can be considered as the inverse process of composition, where we start from a composite image and attempt to extract the foreground and alpha images. This process is typically guided by a user indicating the location of foreground objects.

This proposed system implements the technique described in [13], where the matting problem is formulated in Bayesian framework and solved using maximum a posteriori (MAP) optimization. In this approach, we search for the most likely estimates of foreground (F), background (B), and alpha (α) given C, the observed color. More formally, Eq. (36) is written as

$$\arg \max_{F,B,\alpha} P(F, B, \alpha | C) \tag{36}$$

Applying Bayesian rule, taking the logarithm, and neglecting some terms, now the above equation becomes Eq. (37)

$$\arg \max_{F,B,\alpha} L(C|F, B, \alpha) + L(F) + L(B) \tag{37}$$

where $L (\bullet)$ denotes the log-probability. Chuang [13] explained model each of these terms by means of Gaussian distributions (isotropic for the first term, and unisotropic for the second and third), reflecting the spatial distribution of foreground and background colors in the image. As their resulting likelihood equation is not quadratic, it is solved using alternating iterations, until convergence. To guide the algorithm, a trimap M is required to be given by the user. This map indicates the background regions, foreground regions, and unknown regions. The pixels marked as foreground and background are automatically assigned alpha values 1 and 0, respectively, while the unknown pixels are processed based on the foreground and background information as described above.

(a)   (b)   (c)   (d)

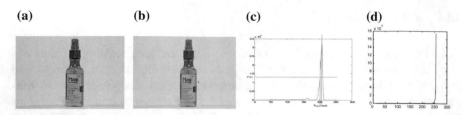

Fig. 3 **a** Input image, **b** gray image, **c** histogram plot, **d** histogram exposure up to 2%

Table 2 Data set for automation classifier

Number of sample for training	208
Number of sample for validation	44
Number of sample for testing	1408

Table 3 Training for MLP

Actual/prediction	0	0	Precision (%)
0	96	0	100
1	0	112	100
Recall (%)	100	100	
Accuracy (%)	100		

Table 4 Validation for MLP

Actual/prediction	0	1	Precision (%)
0	20	0	100
1	0	24	100
Recall (%)	100		
Accuracy (%)			100

4 Results and Discussion

The product input image has been captured using digital camera. After the conversion of RGB image to gray, apply histogram technique. Figure 3a–d shows input image, grayscale image, and histogram plot and histogram exposure up to 2%, respectively.

Table 2 describes data sets for automation classifiers. Confusion matrix has been given in Tables 3, 4, and 5 for training, validation, and testing, respectively.

Table 6 describes data sets for white/gray background classifiers. Confusion matrix has been given in Tables 7, 8 and 9 for training, validation, and testing, respectively.

Figure 4a shows the initial contour detection done using horizontal and vertical projections. Figure 4b shows results from horizontal and vertical projection techniques.

Table 5 Testing for MLP

Actual/prediction	0	1	Precision (%)
0	450	3	99.34
1	0	955	100
Recall (%)	100	99.69	
Accuracy (%)	99.79		

Table 6 Data set for white/gray background classifier

Number of sample for training	112
Number of sample for validation	24
Number of sample for testing	955

Table 7 Training for MLP

Actual/prediction	0	1	Precision (%)
0	33	0	100
1	0	79	100
Recall (%)	100	100	
Accuracy (%)			100

Table 8 Validation for MLP

Actual/prediction	0	1	Precision (%)
0	8	0	100
1	1	15	93.75
Recall (%)	88.89	100	
Accuracy (%)			95.83

Figure 4c shows background as black and foreground as white pixels. If we look at Fig. 4c, surface noise is presented at bottom of the image. To remove such a noise, we introduced foreground fine-tuning techniques which involve surface removal, median filer, morphological operations. Figure 4d shows clear mask image which does not have any surface noise. Figure 5a and b shows input image and the corresponding output produced by our proposed system, respectively, and background is

Table 9 Testing for MLP

Actual/prediction	0	1	Precision (%)
0	191	6	96.95
1	39	719	94.85
Recall (%)	39	99.17	
Accuracy (%)			95.29

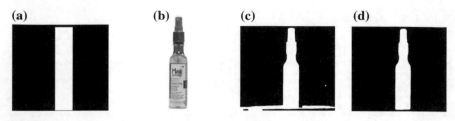

Fig. 4 **a** Initial contour, **b** projection output, **c** foreground detection, **d** foreground fine-tuning

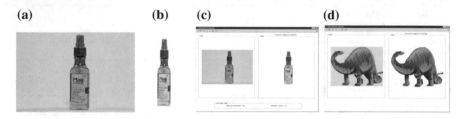

Fig. 5 **a** Raw input image, **b** output automation, **c** GUI output white/gray background, **d** GUI output different background

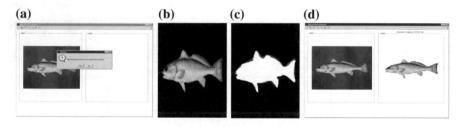

Fig. 6 **a** Different background image, **b** background selection, **c** foreground selection, **d** GUI semi-automation result

filled with white pixels (clear background). Figure 5c shows GUI representation of input sample images and desired output image with full automation. The above process will be applicable to white/gray color background and as well as monochrome gradient background images. Figure 5d shows the monochrome gradient background-removed results.

Let us consider the input image has natural scene as background and as well as textures variations, and it is shown in Fig. 6. Now, we introduced a semi-automation-based background removal.

In this case, user has to select both background regions. In Fig. 6b, foreground region (filled with black pixels) is selected. In Fig. 6c, background region is selected (filled with white pixels). The desired output is shown in Fig. 6d. Some of the challenged input images and corresponding results are shown in Fig. 7.

The proposed system has been tested and validated in different conditions. Fine-tuning method I works better for white/gray gradient background product images

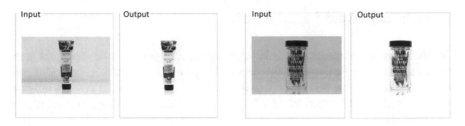

Fig. 7 Sample outcomes

Table 10 Consolidated results

S. No	Task type	Method	Total number of images	Error	Accuracy (%)
1	Foreground detection using rectangle ROI		955	11	98.85
2	Foreground extraction for white/gray gradient color	Fine-tuning method I	758	43	94.33
3	Foreground extraction for white/gray gradient color	Fine-tuning method II	758	128	83.11
4	Foreground extraction for monochrome gradient	Fine-tuning method I	197	47	76.14
5	Foreground extraction for monochrome gradient	Fine-tuning method II	197	26	86.80
6	Overall accuracy (2–3 combined)		955	85	91.10

as well as without contrasting color images. Fine-tuning method II works better for monochrome gradient (nonwhite/gray) background product images. The proposed system achieves better accuracy for monochrome background product images while fusing method I and II, the error rate and accuracy as shown in Table 10.

5 Conclusion and Future Work

Computer vision and machine learning shows great promise toward solving problems related to BG/FG separation and foreground image enhancement. The proposed system works better, even for non-contrasting background/foreground product images. This area can be further explored for improving accuracy over images with complex backgrounds.

References

1. S.U. Lee, S.Y. Chung, and R.H. Park: A Comparative Performance Study of Several Global Thresholding Techniques for Segmentation, Computer Vision, Graphics, and Image Processing, Vol. 52, pp. 171–190, 1990.
2. P.K. Sahoo, S. Soltani, A.K.C. Wong, and Y.C. Chen: A Survey of Thresholding techniques, Computer Vision, Graphics, and Image Processing, Vol. 41, pp. 233–260, 1988.
3. J.S. Weszka: A Survey of Threshold Selection Techniques, Computer Graphics and Image Processing, Vol. 7, pp. 259–265, 1978.
4. S. Zhu, X. Xia, Q. Zhang, and K. Belloulata, "An image segmentation algorithm in image processing based on threshold segmentation," in Proc. Third International IEEE Conference on Signal-Image Technologies and Internet-Based System, SITIS'0., pp. 673–678, 2007.
5. L. Yucheng and L. Yubin, "An algorithm of image segmentation based on fuzzy mathematical morphology," in International Forum on Information Technology and Applications, IFITA'09, pp. 517–520, 2009.
6. F. C. Monteiro and A. Campilho, "Watershed framework to region-based image segmentation," in Proc. International Conference on Pattern Recognition, ICPR 19th, pp. 1–4, 2008.
7. W. Haider, M. Sharif, and M. Raza, "Achieving accuracy in early stage tumor identification systems based on image segmentation and 3D structure analysis," Computer Engineering and Intelligent Systems, Vol. 2, pp. 96–102, 2011.
8. M. R. Khokher, A. Ghafoor, and A. M. Siddiqui, "Image segmentation using fuzzy rule based system and graph cuts," in Proc. 12th International Conference on Control Automation Robotics & Vision (ICARCV), pp. 1148–1153, 2012.
9. W. Cui and Y. Zhang, "Graph based multispectral high resolution image segmentation," in Proc. International Conference on Multimedia Technology (ICMT), pp. 1–5, 2010.
10. Chen, Tao, et al. "Sketch2photo: Internet image montage." ACM Transactions on Graphics (TOG) 28.5 (2009): 124.
11. Chen, Xiaowu, et al. "Image matting with local and nonlocal smooth priors." Proceedings of the IEEE Conference on Computer Vision and Pattern Recognition. 2013.
12. Bai, Xue, and Guillermo Sapiro. A Geodesic Framework for Fast Interactive Image and Video Segmentation and Matting (Preprint). No. Image preprint series 2171. Minnesota University Minneapolis institute for mathematics and its applications, 2007.
13. Yung-Yu Chuang, Brian Curless, David H. Salesin, and Richard Szeliski. A Bayesian Approach to Digital Matting. In Proceedings of IEEE Computer Vision and Pattern Recognition (CVPR 2001), Vol. II, 264–271, December 2001.
14. http://alphamatting.com/eval_25.php.
15. Ning Xu, Brian Price, Scott Cohen, Thomas Huang, "Deep image matting," in Proc. Computer Vision and Pattern Recognition 2017.
16. Varnousfaderani, Ehsan Shahrian, and Deepu Rajan. "Weighted color and texture sample selection for image matting." IEEE Transactions on Image Processing 22, no. 11 (2013): 4260–4270.

17. Huang, Haozhi, Xiaonan Fang, Yufei Ye, Songhai Zhang, and Paul L. Rosin. "Practical automatic background substitution for live video." Computational Visual Media 3, no. 3 (2017): 273–284.
18. L. Chen, J. T. Barron, G. Papandreou, K. Murphy, and A. L. Yuille. Semantic image segmentation with task-specific edge detection using CNN and a discriminatively trained domain transform. In CVPR, pages 4545–4554, 2016.
19. W. Zhao, J. Zhang, P. Li, and Y. Li, "Study of image segmentation algorithm based on textural features and neural network," in International Conference on Intelligent Computing and Cognitive Informatics (ICICCI), pp. 300–303, 2010.
20. Y. Chen, J. Wang, and H. Lu. Learning sharable models for robust background subtraction. In 2015 IEEE International Conference on Multimedia and Expo (ICME), pages 1–6. IEEE, 2015.
21. Y. Zhou, W.-R. Shi, W. Chen et al., "Active contours driven by localizing region and edge-based intensity fitting energy with application to segmentation of the left ventricle in cardiac CT images," Neurocomputing, Vol. 156, pp. 199–210, 2015. View at Publisher · View at Google Scholar · View at Scopus.

Acquiring Best Rules that Represent Datasets

L. M. R. J. Lobo and R. S. Bichkar

Abstract Finding interesting and usually unexpected optimistic rules which are used to normally represent a dataset becomes a question of concern when dealing with data repositories. Association rule mining can cater to generation and classification of these rules. They find all possible rules. This problem turns out to be NP-hard. Genetic algorithms can now be used since they perform well in a global search environment. Genetic algorithms will generate best rules from the population of association rules generated for a dataset. In this paper, an NP-hard association rule mining architecture is presented along with its genetic algorithm component. The popular Apriori algorithm is used to generate the primary association rules, and genetic algorithm, with modification in its mutation operator, is used to extract interesting rules from a transition database created from the item sets of the data repositories. The system developed shows comparative good accuracy of rules generated.

Keywords Association rules · Genetic algorithms · Apriori · RAGA OLEX-GA

1 Introduction

There is a great increase in data in the digital form in the present world. This requires solutions of fundamentally new research. This mining is performed on new types of data, formulating protocols and infrastructures to mine data distributed all over the

L. M. R. J. Lobo (✉)
Department of C.S.E, SGGS Institute of Engineering and Techology, Nanded, India
e-mail: lmrjlobo@rediffmail.com

L. M. R. J. Lobo
Department of C.S.E, Walchand Institute of Technology, Solapur, India

R. S. Bichkar
Department of E&TC, G.H. Raisoni College of Engineering and Management, Pune, India
e-mail: bichkar@yahoo.com

© Springer Nature Singapore Pte Ltd. 2019   17
R. S. Bapi et al. (eds.), *First International Conference on Artificial Intelligence and Cognitive Computing* , Advances in Intelligent Systems and Computing 815,
https://doi.org/10.1007/978-981-13-1580-0_2

network [1, 2]. The new trends, in data mining [3], rest solidly on a variety of research achievements. The basic research challenges involving findings lie in extending data mining algorithms [4] to newly framed data types, developing algorithms that work and enable their ease of use. Machine learning [5] has developed as a new paradigm where an algorithm is designed to address domains where a good theoretical model does not exist but where empirical observations can be made. For association learning, the output is in terms of generated rules with support and confidence attached to them. A genetic algorithm is a search technique used in computing to find exact or approximate solutions to optimization and search problems. They are inspired by evolutionary biology and find applications in computational science.

The present paper deals with the implementation and functioning of rule acquisition and betterment using a genetic algorithm. An evolutionary classifier [6] is used for extracting comprehensible rules in data mining.

It is shown that the prediction accuracy is robust. Comparison shows that such an evolutionary classifier produces good rules [7, 8] for the datasets in terms of number [9] and efficiency. This system is compared to RAGA [10], a system developed by Indira and Kanmani [11], and a standard genetic algorithm plug-in OLEX-GA to WEKA [12] using the same dataset of grilled mushrooms in the Agaricus and Lepiota family drawn from Lincoff 81 and the WEATHER dataset available in the WEKA datasets and also available at http://archive.ics.uci.edu/ml/. It is found that accuracy is increased to a great extent for same threshold support, confidence, and GA parameters.

2 Related Work

Association rule mining [13] is a technique of finding interesting association and correlation relationships among large sets of such data items. Attribute value conditions that occur frequently together in a given dataset are shown by rules.

The discovered knowledge is then represented in the form of IF-THEN reduction rules. A typical form

$$\text{If } X_1 \text{ and } X_2 \text{ and} \ldots X_n, \text{ then } Y. \tag{1}$$

where X_i, $\forall i \in \{1, 2, \ldots, n\}$ is the antecedent that leads to the prediction of the consequent Y [10]. Unlike if-then rules of logic, association rules are considered to be probabilistic in nature. There are two numbers that express the degree of the uncertainty about the rule. The first number is called support of the rule, which simply is the number of transactions that include all items in antecedent and consequent parts of the rule. It is expressed as a percentage of the total number of records in the database.

$$S = \frac{N_{a,c}}{N} \tag{2}$$

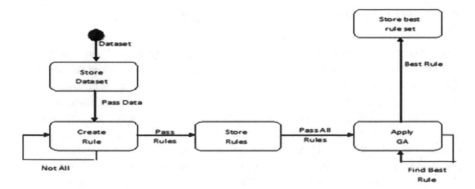

Fig. 1 System framework

where S is support, $N_{a,c}$ is number of transactions including all items in antecedent and consequent of a rule, and N is total number of transactions. The other number, known as confidence of the rule, is the ratio of the number of transactions that include all items in the consequent as well as the antecedent to number of transactions that include all items in the antecedent.

$$C = \frac{N_{c,a}}{N_a} \tag{3}$$

where C is confidence, $N_{c,a}$ is number of transactions including all items in consequent and antecedent of a rule, and N_a is number of transactions that include all items in antecedent.

The classic algorithm for generating these rules is the Apriori algorithm. The method followed begins by generating frequent item sets with just one item and to recursively generate frequent item sets with two items, then three item sets, and so on, until we have generated frequent item sets of all sizes. Cattral and Deugo [10] have implemented rule acquisition using genetic algorithms (RAGA).

Wakabi-Waiswa and Baryamureeba [7] have conducted experiments using real-world zoo dataset. The zoo database contains 101 instances corresponding to animal and 18 attributes. The attribute corresponding to the name of the animal was not considered in the evaluation of the algorithm.

3 Methodology

The framework used for experimentation is shown in Fig. 1. Finding interesting rules that are optimized is our goal.

3.1   Rule Generation Using Apriori Algorithm

The input dataset is analyzed and saved into a proprietary format. Then, the transaction database from the input dataset is created. The user is then asked to input [14] the values of threshold support and confidence. This is followed by construction of the items' database. Apriori algorithm is then applied for generating rules [15].

The method followed begins by generating frequent item sets with just one item and to recursively generate frequent item sets with two items, then 3 item sets, and so on, until we have generated frequent item sets of all sizes. The support and confidence for each set is then calculated. Now depending on the declared threshold support by the user, we select only those rules that can be built supporting this support value; then, we refer to threshold confidence value given by the user; we select only those rules that fit into this confidence condition. These rules are then saved in the rule database.

3.2   Application of Genetic Algorithm

After rule acquisition, genetic algorithm can be applied. The method used differs from a standard genetic algorithm in several crucial respects, including handling variable-length chromosomes. That is association rules, which contain n-place predicates. GA parameters are taken from the user. In order to efficiently explore the space of association rules, it uses macro-mutations as a generalization and specialization operators. The output from this work is the generation of the best association rules representing the complete dataset to its optimal level.

Every individual (perspective solution) used in the genetic algorithm is a chromosome. Association rules generated by Apriori algorithm are the chromosomes, whereas the antecedent and consequent in the rule form the genes in the chromosomes. Attributes of the dataset in a rule represent genes.

In the rule population, chromosomes are encoded to represent a single rule that is variable in length having symbolic structures, i.e., association rules that may contain n-place predicates ($n > 0$). Binary encoding using Michigan approach, where each rule is encoded as a single chromosome, is used. Two bits are used to encode the attribute. The antecedent part is denoted by 00, consequent part by 11, whereas 01 or 10 represent the absence of any attributes in the rule.

For example, consider a dataset having attributes {bread, butter, jam, milk, beer} and the rule

R1: if bread ˆ jam → milk
R1 is encoded as 00 01 00 11 01.

The mating pool is formed by selecting parents from the rule population using roulette wheel selection. Here, we find the sum of fitness of all chromosomes S. A random number is then generated from the interval 0 to S; we call this as r. The whole population is then gone through, and sum of fitness 0 to S is assigned. When sum S is greater than r, stop and return the chromosome at the point it is in. The sum is calculated only once for each population.

The genetic operators, such as crossover and mutation, are then applied to reproduce offspring for the new main population. The selected parents can be crossed over with user-specified crossover rate. Single-point crossover is used, and the crossover point is generated randomly. Single-bit flip mutation with user-specified mutation rate is used.

The evolution of the rule population continues until all the rule set populations have finished their evolutions.

Rule fitness depends primarily on confidence and support, and optionally on parsimony reward based on data coverage. For each proposed rule, confidence and support are calculated based on the available confidence $= 0$ or if it covers new data, it gets Fitness $= 0$, otherwise the distance of its confidence and support from the target values is determined, using a quasi-Euclidian distance.

$$Fitness = 100 - \sqrt{(C - T_c)^2 + (S - T_s)^2} \tag{4}$$

where C is the confidence, T_c is threshold confidence s is the support and T_s is the threshold support. The resulting value is normalized as a percentage, it is adjusted by rewarding by a small constant for each conjunct by which the maximum antecedent size exceeds the rules antecedent size so that

$$Fitness = Fitness - (max\ A_s - A_s) * R \tag{5}$$

where A_s is antecedent size and R is the reward. In classification tasks, fitness value is further adjusted by taking the rule's data coverage into account. The rules are maintained in descending order of fitness.

The initial and overall maximum consequent sizes for the classification are both 1. The crossover and mutation rates are set to 0.8. Macro-mutations operate within conjuncts, and macro-mutations achieve generalizations and specialization by deleting and adding conjuncts. The rate of mutations varies from 0.06 to 0.75. The processing stops after a specific number of generations or when full data coverage is achieved.

3.3 Method to Compare Results Achieved with RAGA

Using the same dataset of grilled mushrooms, used by the authors Cateral et al for demonstrating the RAGA environment (where they used nine test runs that produced between 14 and 25 rules yielding 100 % accuracy for the entire set) the experiments are designed. In the first experiment, test runs are created for the developed environment varying parameters of the Apriori algorithm, namely threshold support and threshold confidence. The parameters of Genetic algorithm namely crossover rate, mutation rate and maximum number of generations are then varied. The observed number of best rules are then recorded.

In the second experiment, variations are taken in minimum support and confidence for a constant crossover rate of 75% and mutation rate of 1% and observations are taken to observe changes in accuracy of the rules and fitness values.

3.4 Method to Compare Results Achieved with Work of Indira and Kanmani

The objective of the study of Indira and Kanmani was to compare the accuracy achieved in datasets by varying GA parameters. The accuracy and convergence of fitness values by controlling the GA parameters are recorded.

To compare with our implementation, we vary GA parameters of crossover rate, mutation rate, and number of generations and observe number of best rules, confidence, and fitness values of each rule.

3.5 Method to Compare Results Achieved with WEKA

To perform the task of working with WEKA tool plug-in OLEX-GA, the weather dataset, which is already processed in the proprietary form attribute-relation file format (.arff), is taken into consideration as input. After preprocessing, this dataset is fed as input to the OLEX-GA plug-in in classification algorithms available in WEKA. We run this algorithm which takes some time to build an output as a confusion matrix and gives correctly and incorrectly instances along with accuracy rate. We then perform analysis of our implementation on the dataset to calculate accuracy.

4 Experimental Setup and Results

The input used for the development of the system uses two datasets: the first dataset containing 8124 sample descriptions of 23 species of grilled mushrooms in the Agaracus and Lepiota Family. The dataset uses 22 attributes and classifies each mushroom as either edible or poisonous. The second was the weather dataset from UCI machine learning repository available at http://archive.ics.uci.edu/ml/. This dataset has five attributes, namely outlook, temperature, humidity, windy, and play; play attribute is used as class attribute in classification.

The association rule acquisition begins with loading the dataset from UCI machine repository, followed by applying Apriori algorithm to generate the association rules. Using these association rules, the initial population for genetic algorithm was formed.

Table 1 Test runs for developed environment

Thresh support	Thresh confidence	Crossover rate	Mut. rate	Max. num of generations	Num. of best rules
60	60	80	1	5	25
55	55	75	1	10	41
50	80	75	1	10	36

Table 2 Variations in support and confidence for crossover rate of 75% and mutation rate of 1%

Min. support	Min. confidence	Initial population	Num. of best rules	Num. of generations	Fitness best rule
50	50	1148	54	38	99
75	75	180	14	31	97
80	80	88	05	39	98

On these rules, the genetic algorithm technique was applied. Genetic algorithms work in generations and produce the best rules which represented the dataset most optimistically. For implementing the idea given, the programming language used was C. (VC++). The final results were stored in notepad.

4.1 Comparing Results with RAGA

The results for the developed system were compared with RAGA [10], and following were the observations. The four test run of the developed approach on mushroom dataset produced shows more number of rules than RAGA. This is shown in Table 1.

The test runs on RAGA produce between 14 and 25 rules only. A summary of the results varying minimum support and confidence is as shown in Table 2.

The table shows that variation in support and confidence brings changes in accuracy. Variations in minimum support and confidence bring changes in fitness of best rule. The optimum value of minimum support and confidence is based on the support and confidence values of the attributes in the dataset. These results are comparable with RAGA where each rule set yielded 100% accuracy for the entire set.

4.2 Comparing Results with Implementation of Indira and Kanmani

These results are comparable with those achieved by Indira and Kanmani in [15]. They have achieved an accuracy of up to 80% using different datasets. The fitness of the best rule is a measure of accuracy since accuracy of association rules is indicated

Table 3 Variations in GA parameters

Thresh support	Thresh confidence	Crossover rate	Mut. rate	Max. num of generations	Num. of best rules	Fitness best rule
50	80	80	1	10	7	99
60	60	80	1	5	10	89
55	55	75	1	10	11	84

Table 4 Confusion matrix

a	B	
7	2	$a = $ yes
2	3	$b = $ no

in terms of support and confidence measures. A summary of the results varying GA parameters is as shown in Table 3.

4.3 Comparing Results with WEKA

In the second experiment, results of rule analysis using the developed system were compared with OLEX-GA plug-in introduced in WEKA for rule induction.

Weather dataset was used as the input dataset. Then, the same dataset was applied on the developed system and OLEX-GA. The confusion matrix obtained from OLEX-GA is compared with that achieved from the developed environment.

For same Apriori parameters: threshold support: 30 and threshold confidence: 80, and for genetic algorithm input parameters: crossover rate: 70%, mutation rate: 1% and number of generations = 100 OLEX-GA gave 10 correctly classified instances and 4 incorrectly classified instances and a confusion matrix as shown in Table 4.

Accuracy rate = 71.4286%
The developed system gives three most optimized rules:
0001: humidity = normal, play = yes Fitness: 81%
0002: humidity = normal, windy = TRUE, play = yes Fitness: 55%
0003: humidity = normal, outlook = rainy, play = yes Fitness: 55%
Total cases in dataset containing normal = 7, total yes in these cases = 6, and total no in these cases = 1.
Correct predictions: 06
Incorrect predictions: 01

The following can be computed from this confusion matrix:
The model made six correct predictions (6 + 0), The model made 1 incorrect prediction (1 + 0), The model scored 7 cases (6 + 1), the error rate is $1/7 = 0.1428$, and the accuracy rate is $6/7 = 0.8571$.

Thus, accuracy rate for our system is higher as compared to that achieved using WEKA for similar analysis and same dataset.

5 Conclusion

We conclude here that the developed system architecture performed rule analysis on large data repositories using genetic algorithm and produced comparative results to RAGA [10] and the system developed by Indira and Kanmani in [11].

Large data repositories made the problem NP-hard. Genetic algorithms were chosen to find a solution. It is well known that genetic algorithms work in generations and cater to the reducibility of the problem. The accuracy of the rules produced by the developed system was compared to OLEX-GA plug-into WEKA, and it was found that the developed environment produced comparative accuracy rate for same Apriori algorithm and genetic algorithm parameters.

Acknowledgements The authors would like to thank their affiliated institutions for all the support given to them during their research and preparing of this publication.

References

1. Ba-alwi, F. M., : Knowledge acquisition tool for classification rules using genetic algorithm approach, In: International Journal of Computer Applications vol. 60, No. 1, pp. 29–34. (2012)
2. Freitas, A. A., : A review of evolutionary algorithms for data mining, In: Soft Computing for Knowledge Discovery and Data Mining, pp. 61–93 (2007)
3. Agrawal, R., Imielinski, T., Swami, A.: Mining association rules between sets of items in large databases, In: Proceedings of the 1993 ACM SIGMOD International Conference On Management Of Data, Washington DC (USA), pp. 207–216 (1993)
4. Kriegel, H.-P., Borgwardt, K. M., Krger, P., Matthias, A. P., Z imek, S. A., Al, H. P. K. E., : Future trends in data mining. In: Springer Science+Business Media, (2007)
5. Han, J., Kamber, M., : Data mining: Concepts and techniques, In :Data Min Knowl Disc 15:8797. Morgan Kaufmann, (2007)
6. Tan, Q. Y. K. C., Lee, T. H.,.: A distributed evolutionary classifier for knowledge discovery in Data mining, In: IEEE Transactions on Systems. Man and Cybernetics- Part C: Application and Reviews, vol. 35, No. 2, pp. 131–142 (2005)
7. Wakabi-waiswa, P. P., Baryamureeba, V., : Extraction of interesting association rules using genetic algorithms, In: International Journal of Computing and ICT Research, vol. 2 No. 1, pp. 26–33, (2008)
8. Yan, X., Zhang, C., Zhang, S., : Armga: Identifying interesting association rules with genetic algorithms, In : Journal of Applied Artificial Intelligence, 19:677689, pp. 677–689, (2005)
9. Salleb-aouissi, A., Vrain, C., Nortet, C., : Quantminer: A genetic algorithm for mining quantitative association rules, In: IJCAI,, pp. 1035–1040, (2007)
10. Cattral F. O. R., Deugo, D., : Rule acquisition with a genetic algorithm, In: Proc IEEE Congr. Evolutionary Computation, vol 1, pp. 125–129, (1999)
11. Indira, K., Kanmani, S.: Evaluating the rule acquisition in data mining using genetic algorithm, In: Journal of Computing, vol 4, Issue 5, pp. 128–133 (2012)

12. Witten, I. H., Frank, E., Trigg, L., Hall, M., Holmes, G., Cunningham, S. J.,: Weka: Practical machine learning tools and techniques with java implementations. In: Morgan Kaufmann, (1999)
13. Kotsiantis, S., Kanellopoulos, D., : Association rules mining: A recent overview, In: GESTS International Transactions on Computer Science and Engineering, vol. 32 (1), pp. 71–82 (2006)
14. Carvalho, . D. R., Freitas, A. A., Ebecken, N., : Evaluating the correlation between objective rule interestingness measures and real human interest, In: Proc. PKDD-2005, LNAI 3721. Springer, pp. 453–461, (2005)
15. Liu, J. J., yau Kwok, J. T.,: A concept learning method based on a hybrid genetic algorithm." in Proc. Science in China (Ser. E), 28(4), pp. 488–495, (1998)

Morphological-Based Localization of an Iris Image

S. G. Gino Sophia and V. Ceronmani Sharmila

Abstract Iris recognition is one of the reliable biometric techniques used for human identification purpose. It provides the unique information about a person with natural features such as both the left and right eye irises of a person is different and stable with the age and also the quality of the iris is not affected by contact lenses and eyeglasses. The authors suggested that iris recognition fails due to the tedious process involved during localization. The failure rate can be decreased by performing edge detection with a suitable localization algorithm. The authors proved that histogram equalization is one of the best image enhancement techniques to process an image with probability density function of different gray-level values. The edges of an image are identified using an edge detection algorithm using mean value and threshold values, and the localization of an image is rectified by the neighbors of a pixel and structuring element morphological operations. Compare the performance of the algorithms and prove that the localization of an edge using the structuring element of the morphological operation produces the best results compared with other morphological operations using the neighbors of a pixel.

Keywords Image localization · Morphological operations · Edge detection

1 Introduction

Image processing is to study and analyze the digital image by various steps such as preprocessing, enhancement, compression, segmentation, and feature extraction.

S. G. Gino Sophia (✉)
Department of Computer Science and Engineering, Hindustan Institute
of Technology and Science, Chennai, India
e-mail: sgsophia@hindustanuniv.ac.in

V. Ceronmani Sharmila
Department of Information Technology, Hindustan Institute of
Technology and Science, Chennai, India
e-mail: csharmila@hindustanuniv.ac.in

© Springer Nature Singapore Pte Ltd. 2019
R. S. Bapi et al. (eds.), *First International Conference on Artificial Intelligence
and Cognitive Computing* , Advances in Intelligent Systems and Computing 815,
https://doi.org/10.1007/978-981-13-1580-0_3

27

It finds its application in many areas such as medical field, remote sensing, human identification, color processing, satellite image processing. Among this, human identification is one of the areas to identify the human by various biometric resources such as fingerprint, eye, retina, face, and iris. The authors Ali et al. [6], Verma et al. [17], and Daugman [7] suggested that the iris recognition is one of the consistent biometric techniques and it is one of the natural passwords for human identification. Genetically identical people such as monozygotic and dizygotic also have independent irises. From the third month of gestation, the iris were formed and grows up to the eighth month by Daugman [5, 7]. The iris consists of a number of complex patterns with a high amount of melanin. But the blue iris lacks the presence of melanin. The innermost part of the iris is called the pupil, and the iris is covered by the sclera. The iris can be processed based on the features of shape, color, and texture as proposed by Hamouchene [8]. The pupil is distorted by the dilation and stretching that reduces the level of iris identification due to the wrong localization of an edge. Identification of a person using an iris was first proposed by Floam and Safir in 1987. Tsai [9] used the possibility of fuzzy matching strategy for finding the iris feature points. The similarity score table was used for comparing the feature points in matching algorithms.

1.1   Related Work

Daugman [1] suggested that the iris images are treated as iris codes with the logarithmic value of the probability value that is an information I of an image and the entropy H which is calculated using the probability value. The joint entropy $H(a, b)$ and conditional entropy $H(a/b)$ are to represent the iris codes. Daugman [2, 12] and Verma [17] illustrated the similarity of iris patterns for genetically identical iris and also using the phase information for iris patterns with Gabor wavelets and also comparing the irises using the EX-OR operation for finding out the amount of changes in patterns of texture. Hamouchene [8] proposed the texture feature extraction method as neighbors-based binary pattern for person identification in a Daugman's normalized iris image. Ali [6] calculated the iris key points using the convolution of images with the Gaussian filters for feature matching in scale-invariant feature transform (SIFT), and the author suggested that the best enhancement technique contrast-limited adaptive histogram equalization is used to provide the uniformity of the intensity levels and also removes the noise and detecting the boundary of the images using the Canny edge detector. Canny [3] summarized the detection of edge points using certain criteria such as threshold values, gradient vectors, fixing the edges by comparing the pixel values among the neighbors on the particular direction. The main motive of the paper [3] is to reduce the amount of data to be analyzed and maintain the similarity of the boundary data. Xin [4] suggested the new Canny edge detection algorithm for color images with the parameters such as intensity value and the distance between the pixel values and also sets the window size with the number of pixels. This algorithm was used to remove the Gaussian noise from an image. The rows and columns of Sobel operators and the neighboring intensity values with the number of pixels were used for finding out the response of an image. Verma [17] suggested that the extraction of features of the irises was in terms of codes. Wei [20] did the research for the identification of fake iris and also did the research for increasing the sharpness of the iris images. Chirchi [13] illustrated the concept of finding the center of the pupil using left, right, lower, and upper pixel values. Tsai [9] proposed that the extraction of features and the edges is detected using the Canny edge detector algorithms and circular Hough transform.

2 Proposed Method

The proposed method is used to solve the iris localization problem using improved morphological-based operations. The proposed system consists of the following steps, Image pre-processing, Image Edge Detection and Image Segmentation. The iris localization is done during edge detection to correctly segment the boundaries of an iris. Chandwadkar [15] proposed that the edge detection is the common approach for detecting discontinuities in the gray level. An edge is the set of connected pixels that lie on the boundary between two regions which is mainly used for subimage selection and retrieval. Xin [4] found the color image edges using the Sobel operators using row R and column C.

Mahlouji1 [16] and Birgale [18] proposed that the segmentation divides an image into regions or objects. The segmentation is based on two properties of intensity value (i.e., discontinuity and similarity). In the discontinuity, an image is partially based on abrupt changes in intensity, mainly on the edges of an image. Tsai [9] and Hao [11] proposed that the fuzzy curve tracing (FCT) algorithm was to detect a curve in a binary image using Eq. (1)

$$\text{FCT} = \sum_{I=1}^{N} \sum_{K=1}^{C} \cup ||X - Y|| + \alpha \sum_{K=1}^{C} ||X - Y - 1||^2 \tag{1}$$

where U are the membership values and α are the number of clusters. The author used the lower and higher threshold values that were selected automatically during the experimental process.

Canny [3], Xin [4], and Chandwadkar [15] suggested that among the number of edge detection algorithms, the Canny edge detector is used to detect the edges using the parameters such as gradient strength and direction at the location (x, y) of an image. The detection algorithm proposed by Mahlouji and Noruzi [16] uses Gaussian filter of the linear filtering, which can be processed using the convolution operator. Here, the output pixel value is calculated using multiplication of the weighted sum of the neighborhood pixel value and the Gaussian filter is calculated using Eq. (4). Then, find the gradient of an image; it is the changes in the intensity of an image on particular x and y directions. The gradient-based edge detection is also treated as the first-order and second-order derivatives of each intensity value on their directions using Eqs. (2) and (3).

$$\partial f = \frac{\partial f}{\partial x} + \frac{\partial f}{\partial y} \tag{2}$$

$$\partial^2 f = \frac{\partial^2 f}{\partial^2 x} + \frac{\partial^2 f}{\partial^2 y} \tag{3}$$

$$h = \sum_{i=1}^{n} \sum_{j=1}^{n} A_i K_j^T \tag{4}$$

Procedure edge detection (Gradient Strength G, Direction D, upper threshold U, lower threshold L) {

Loads the Source Image

Create the matrix with the intensity value and fix the size of the image. Convert the RGB image to grayscale image.

Determine the connected components of the binary image. Fix the upper threshold value U and lower threshold value L.

Find out the response of an image using edge detection masking filters. Find out the intensity gradient (Gx and Gy) of an image in both x and y directions.

Gradient strength (G) on a particular direction (D) in the x- and y-axis is calculated using Eqs. (5) and (6)

$$G = \sqrt{(Gx + Gy)} \tag{5}$$

$$D = \tan^{-1} \frac{Gy}{Gx} \tag{6}$$

If $G > U$, edge of the pixel is not selected.

If $G < L$, edge of the pixel is selected.

Apply the morphological operation of an image using the structuring element and the neighbors of a pixel. Compare the results.

2.1 Localization

Chirchi [13] and Cho [14] suggested that the localization of image is one of the tedious processes to find out the subimage from the original image and also locate them. Canny [3], Xin [4], and Daugman [2, 10] proved that localizing an image is complicated with the edge detection algorithm with the parameters such as the strength of the gradient value at a particular direction on (x, y) and their threshold values. So an algorithmic technique is necessary for image localization with the different parameters such as the automatic selection of the threshold values and the features of the morphological operations. Morphological processing is to extract the image components that are useful in image representation and description of shape.

The upper and lower threshold values are selected based on the intensity values of the histograms. Set the leftmost intensity point value on x-axis, and find their related number of pixels on the y-axis and continuously moving the intensity values on the x-axis with their pixel values on y-axis. Scanning the intensity value on the x-axis with the minimum intensity value on the left end refers to the lower value (LV) that refers to the dark pixel value and the right end on the x-axis refers to the higher value (HV) that is equivalent to the gray value. Then, fix the upper pixel point (UP) on the y-axis with the maximum number of pixels and lower pixel point (LP) on the y-axis with the minimum number of pixels. The center of the pupil is calculated using the average of the pixel intensity on the x-axis, and their related number of pixels on y-axis using Eq. (7) is shown in Fig. 1.

$$CP = \frac{LV + HV}{2}, \frac{UP + LP}{2} \qquad (7)$$

Where

$Q1 = (HV - CP)$
$Q2 = (CP - LV)$
$Q3 = (CP - UP)$
$Q4 = (LP - CP)$

Then, find the radius of the pixel value that is the maximum {Radius_Array ($Q1$, $Q2$, $Q3$, $Q4$)} using Eq. 7. The eight neighbors of a pixel (P, Q) are ($P - 1, Q - 1$), ($P - 1, Q$), ($P - 1, Q + 1$), ($P, Q - 1$), ($P, Q + 1$), ($P + 1, Q + 1$), ($P + 1, Q$), and ($P + 1, Q + 1$) which are shown in Fig. 2. The neighbors of 166 are 98, 110, 123, 145, 152, 171,178, and the median value is 145 as shown in Fig. 2b.

Morphological operation is the collection of operations by shape and features of a binary image. Birgale [18] constructed the number of connected component groups; each connected group satisfies the property of similarity. The structuring element is a small region or subimage with the values of the area, centroid of the structuring element, and the bounding box which is shown in Fig. 3b. If all the elements of the

Fig. 1 Finding the radius of an image

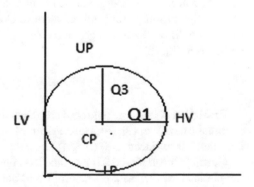

P-1,Q-1	P-1,Q	P-1,Q+1
P,Q-1	P,Q	P,Q+1
P+1,Q-1	P+1,Q	P+1,Q+1

98	17	178
110	16	152
98	14	123

(a)   (b)

Fig. 2 **a** Neighbors of a pixel and **b** example of neighbors of a pixel

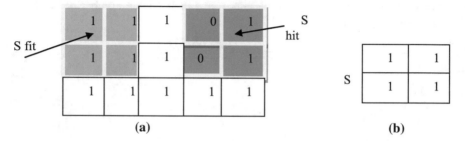

(a)   (b)

Fig. 3 **a** Example of morphological operations and **b** example of structuring element (SE)

Fig. 4 Example of erosion

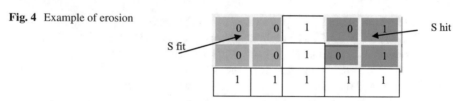

Structuring element are correctly matched to the binary pixel value is the fit and any one of the pixel values is correctly matched is hit as shown in Fig. 3a.

Suppose the SE is running over the original image, if the origin of SE visits every pixel of the original image, so the SE is contained in the original image or not by forming a new set of values being a member of new set or not. The erosion is applied to the original image for pupil extraction. Consider the two sets of values X and Y in the two-dimensional images in Z; the erosion is the subtraction of two images X and Y as in Eq. (8).

$$X \emptyset Y = ((Z|(Y)Z) \leq X) \tag{8}$$

Erosion of X and Y is the set of all points in Z, such that Y is translated by Z, that is contained in X in Eq. (8). Here, assume Y is SE. If the pixel values are fit with the SE, then 1 is replaced by 0 (Fig. 4).

Consider the input RGB image, and it is converted into a grayscale image. Initialize the parameters such as area, centroid, and bounding box for structuring element.

FALSE	FALSE	FALSE	FALSE
FALSE	FALSE	FALSE	FALSE
FALSE	TRUE	TRUE	TRUE
FALSE	TRUE	TRUE	TRUE

Erosion

FALSE	FALSE	FALSE	FALSE
FALSE	FALSE	FALSE	FALSE
FALSE	FALSE	FALSE	FALSE
FALSE	FALSE	TRUE	TRUE

(a)   (b)

Fig. 5 **a** Morphological operation output and **b** erosion output

Smooth the image using the filters, and then apply the convolution operations. The strength of the gradient value at x and y directions is calculated.

Find the center of the pupil using the minimum and maximum intensities and mean and median values of the images.

Formation of boundary of the pupil is done the technique used by the author Verma [17] as shown in the Fig. 3.

Kalsoom [19] is to localize a pupil using morphological operations using neighbors of a pixel.

2.2 Experimental Setup

Used the IIITD database of an iris image in Image analysis and Biometrics Lab. Using the Matlab R2017b for implementing the proposed method for solving the localization problem during edge detection. Consider the input grayscale image as shown in Fig. 6a. The erosion of the morphological operation using the structuring element for binary images and neighbors of a pixel in an edge detection image is shown in Fig. 6. An example of an input image with the size of $28 \times 146 = 4088$ consists of 28 row values and 146 column values. Here, P represents the subscripts of nonzero row values and Q represents the subscripts of nonzero column values. Here, the P consists of 4088 values with the minimum value of 1–28 and Q consists of 4088 values with the minimum value of 1–146 and every value is 28 times the subscripts of column values. The $(P - 1, Q)$ is above (P, Q) that consists of $4088 - 146 \times 2 = 3942 \times 2$. The neighbors of (P, Q) with the index values are shown in Table 1.

Figure 5a shows the output of the morphological operation, and then apply the erosion. The output pixel of the erosion is the lowest value of the neighbors as shown in Fig. 5b.

Consider the different types of image as shown in Fig. 6a–f.

If the morphological operation is applied to the images without applying edge detection, the localization problem is not solved correctly and the edges are also not detected as shown in Fig. 7.

Table 1 Neighbors of pixel (P, Q)

Rows	$P-1, Q-1$		$P-1, Q$		$P-1, Q+1$		$P, Q-1$		$P, Q+1$		$P+1, Q-1$		$P+1, Q$		$P+1, Q+1$	
126	18	5	18	5	18	6	14	5	14	6	19	5	15	5	19	6
127	19	5	19	5	19	6	15	5	15	6	20	5	16	5	20	6
127	20	5	20	5	20	6	16	5	16	6	21	5	17	5	21	6
128	21	5	21	5	21	6	17	5	17	6	22	5	18	5	22	6
129	22	5	22	5	22	6	18	5	18	6	23	5	19	5	23	6
130	23	5	23	5	23	6	19	5	19	6	24	5	20	5	24	6

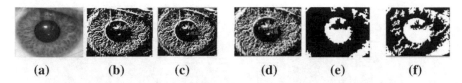

(a)   (b)   (c)   (d)   (e)   (f)

Fig. 6 **a** Input gray-level image, **b** edge detection output, **c** morphological localization of sclera, **d** morphological localization of pupil, **e** morphological erosion of iris using eight-connectivity, and **f** morphological erosion of iris using structural element

(a)   (b)

Fig. 7 Localization and extraction of pupil before applying edge detection

Table 2 Comparative study among the parameters used in the algorithms

Algorithm name for iris localization	MSE	PSNR
Gradient-based existing algorithms	0.25255	54.14
Morphological-based localization algorithm does not use gradient value of x and y directions	6.07E−04	80.3
Morphological-based localization proposed algorithm uses gradient value of x and y directions	0.00571	70.59

3   Results and Discussion

Comparing the algorithms with different parameters such as the gradient value, morphological operations using the structuring element and neighbors of pixels. The results of mean square error (MSE) and peak signal-to-noise ratio (PSNR) are tabulated in Table 2. To find out the performance of the gradient based algorithms by Mean Square Error and Peak Signal to Noise ratio in Fig. 8a, b using the values of Table 2. From the comparison of the values, the gradient-based parameter does not produce the best output.

Table 3 refers to the values of an morphological based localized image without applying edge detection. This is the value of the localization of an iris of finding the mean square error and peak signal-to-noise ratio as shown in Fig. 9a, b.

Table 4 refers to the values of an morphological based localized image with applying edge detection. It directly performs the localization of an iris for finding the

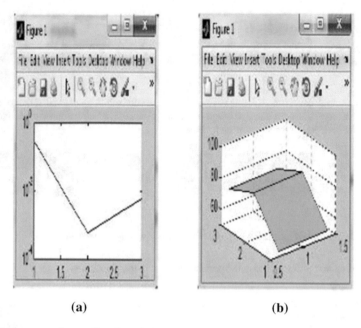

(a)   **(b)**

Fig. 8 Performance for gradient based algorithms: **a** using mean square error and **b** using peak signal-to-noise ratio

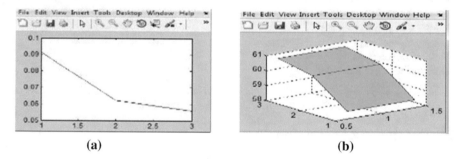

(a)   **(b)**

Fig. 9 Performance for morphological based localization before edge detection: **a** using mean square error and **b** using peak signal-to-noise ratio

mean square error and peak signal-to-noise ratio after iris localization and is shown in Fig. 10a, b. Comparing the differenet algorithms using the values of Tables 2, 3 and 4 and shows that the morphological based localization after the edge detection produces good results compared to before edge detection.

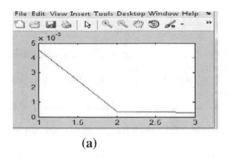
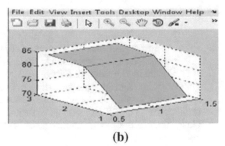

(a)   (b)

Fig. 10 Performance for morphological based localization after edge detection: **a** using mean square error and **b** using peak signal-to-noise ratio

Table 3 Values for morphological based localization applying before detection

Image	MSE	PSNR
Image1	0.0913	58.55
Image2	0.0621	60.22
Image3	0.0558	60.69

Table 4 Values for morphological based localization after edge detection

Image	MSE	PSNR
Image1	0.004524	71.60
Image2	3.54E−04	82.67
Image3	2.93E−04	83.49

4 Conclusion

The improved morphological operations on image localization technique with several techniques were studied and analyzed. The authors suggested that the histogram equalization of a normalized image results in perfect iris detection. But localization of iris image is the only risk factor involved in the process of iris recognition. To avoid such risk factors and to improve the results, better algorithm is developed with the improved morphological operations. The detection and localization of iris edges using the threshold values and the morphological operations produce better results in a better performance compared with the previous algorithm with the gradient value. Edge detection is also necessary before localization of an image and in future creates the simulation model for improving the performance.

References

1. John Daugman, "Information Theory and Iris Codes", 2016, IEEE Transactions on Information Forensics and Security, Volume 11, No 2, February.
2. John Daugman and Cathryn Downing, 2001, "Epigenetic Randomness, Complexity and Singularity of human iris patterns", The Royal Society.
3. John Canny, 1986, "Computational approach to edge detection. Pattern Analysis and Machine Intelligence", IEEE Transactions: 679–698.
4. Geng Xin, Chen Ke, Hu Xiaoguang, 2012, "An improved Canny edge detection algorithm for color image", IEEE.
5. John Daugman and Cathryn Downing, April 2001," Epigenetic randomness, complexity and singularity of human iris patterns", The Royal Society.
6. Han S. Ali, Asmaa I. Ismail, Fathi. A. Farag, 2016 "Speedup robust features for efficient iris recognition" Springer-Verlag, London 2016, SIViP 10:1385–1391, https://doi.org/10.1007/s1 1760-016-0903-8.
7. John Daugman, 2004, "How Iris Recognition works", IEEE transactions on Circuits and systems for video technology, Vol 14, No 1.
8. Izem Hamouchene, Saliha Aouat, 2016, "Efficient approach for iris recognition", Springer-Verlag, London, https://doi.org/10.1007/s11760-016-0900-y.
9. Chun- Chih Tsai, Heng –Yi Lin, Jinshiuh Taur, 2012, "Iris Recognition Using Possibilistic Fuzzy Matching on Local Featues", IEEE Transactions on Systems, Man and Cybernetics, Vol 42, No. 1.
10. John Daugman, 2003, "The importance of being random: statistical principles of iris recognition", Pattern Recognition 36 279 – 291.
11. Hao F, Daugman J, Zielinski P, June 2008, "A fast search algorithm for a large fuzzy database." IEEE Trans. Information Forensics and Security, Volume 3 and No 2, pp 203–212.
12. Daugman J, October 2007, "New methods in iris recognition." IEEE Trans. Systems, Man, Cybernetics, part- B volume 37 and No 5, pp 1167–1175.
13. Vanaja Roselin E. Chirchi and L. M. Waghmare, Research Scholar, JNTUH, Kukatpally Hyderabad-500085(AP), India, 2013, "Feature Extraction and Pupil Detection Algorithm Used for Iris Biometric Authentication System", International Journal of Signal Processing, Image Processing and Pattern Recognition, Vol. 6, No. 6, pp 141–160, http://dx.doi.org/10.14257/ij sip.2013.6.6.14.
14. Eun-suk Cho, Ronnie D. Caytiles, Seok-soo Kim, 2011, "New Algorithm Biometric-Based Iris Pattern Recognition System: Basis of Identity Authentication and Verification", (Journal of Security Engineering).
15. Radhika Chandwadkar, Saurabh Dhole, Vaibhav Gadewar, Deepika Raut, Prof. S. A. Tiwaskar ,October 2013, " Comparison of Edge Detection Techniques", Proceedings of Sixth IRAJ International Conference, Pune, India. http://dx.doi.org/10.13140/RG.2.1.5036.7123.
16. Mahmoud Mahlouji and Ali Noruzi, January 2012, "Human Iris Segmentation for Iris Recognition in Unconstrained Environments", IJCSI International Journal of Computer Science Issues, Vol. 9, Issue 1, No 3, ISSN (Online): 1694-0814.
17. Prateek Verma, Maheedhar Dubey, Praveen Verma, Somak Basu, June 2012, "Daughmans Algorithm Method For Iris Recognition-A Biometric Approach", International Journal of Emerging Technology and Advanced Engineering, Volume 2, Issue 6.
18. Lenina Birgale and M. Kokare, 2010, "Iris Recognition Without Iris Normalization", Journal of Computer Science 6 (9): 1042–1047, 2010 ISSN 1549-3636, Science Publications.
19. Sajida Kalsoom and Sheikh Ziauddin, 2012 "Iris Recognition: Existing Methods and Open Issues", PATTERNS: The Fourth International Conferences on Pervasive Patterns and Applications.
20. Zhuoshi Wei, Xianchao Qiu, Zhenan Sun and Tieniu Tan, 2008, "Counterfeit Iris Detection Based on Texture Analysis", 978-1-4244-2175-6/08/$25.00, IEEE.

A New Document Representation Approach for Gender Prediction Using Author Profiles

T. Raghunadha Reddy, M. Lakshminarayana, B. Vishnu Vardhan,
K. Sai Prasad and E. Amarnath Reddy

Abstract Author Profiling is used to predict the demographic characteristics like age, gender, country, nativity language, and educational background of anonymous text by analyzing their style of writing. Several researchers proposed different types of features like lexical, character based, content based, syntactic, topic specific, structural features, and readability features to discriminate the style of writing of the authors for Author Profiling. The representation of a document with extracted features is one of the important tasks in Author Profiling. In Author Profiling approaches, most of the researchers used the bag-of-words model for document representation. This paper concentrates on the alternative document representation to increase the performance of Author Profiling system. In this work, a new document representation model is proposed and compared the proposed model with existing document representation models like BOW and SOA. The proposed model is evaluated on the reviews dataset for predicting the gender of the authors using various machine learning classifiers. The proposed approach results were promising than most of the existing approaches for Author Profiling.

T. Raghunadha Reddy (✉)
Department of IT, Vardhaman College of Engineering, Hyderabad, India
e-mail: raghu.sas@gmail.com

M. Lakshminarayana
Department of CSE, Swarnandhra College of Engineering and Technology, Narsapuram, AP, India
e-mail: lachi9866516918@gmail.com

B. Vishnu Vardhan
Department of CSE, JNTUH Jagtiyal, Karimnagar, India
e-mail: mailvishnu@jntuh.ac.in

K. Sai Prasad · E. Amarnath Reddy
Department of Computer Science and Engineering, MLR Institute of Technology,
Hyderabad, India
e-mail: saiprasad.kashi@gmail.com

E. Amarnath Reddy
e-mail: amar.enumula@gmail.com

© Springer Nature Singapore Pte Ltd. 2019
R. S. Bapi et al. (eds.), *First International Conference on Artificial Intelligence and Cognitive Computing* , Advances in Intelligent Systems and Computing 815,
https://doi.org/10.1007/978-981-13-1580-0_4

Keywords Author Profiling · Gender prediction · Term weight measure
Document weight measure · PDW approach

1 Introduction

In the present era, the Internet is growing continuously with a massive amount of text mainly through reviews, twitter, blogs, and other social media. The availability of text challenges the information analysts and researchers to develop automated tools for extracting information from the text. In this context, Authorship Analysis is one of the interesting areas concentrated by various researchers to know the details of unknown text. Authorship Analysis is performed in three different ways such as Authorship Verification, Authorship Attribution, and Author Profiling.

Authorship verification finds whether the given document is written by a particular author or not by analyzing the documents of a single author [1]. Authorship Attribution is used to predict the given document author by analyzing multiple authors' documents [2]. Author Profiling is a type of text classification technique, which predicts the demographic characteristics of the authors by analyzing their writing styles [3].

Author Profiling is popular in various fields like marketing, literary research, forensic analysis. From the marketing point of view, most of the people post their reviews on various products without specifying their details. Author Profiling techniques were helpful to business experts to take strategic decisions on the business by finding the details of anonymous reviews. In literary research, Author Profiling helps in resolving the disputes of unknown documents authorship. In forensic analysis, Author Profiling is used to find the details of the perpetrator of harassing messages or threatening messages.

This paper is structured in six sections. Section 2 explains the overview of existing work carried in Author Profiling research domain. The proposed PDW approach is explained in Sect. 3. In Sect. 4, dataset characteristics, evaluation measures, and the empirical evaluations of existing approaches and the proposed approach were presented. In Sect. 5, an analysis is made on experimental results by comparing the proposed approach with existing approaches. The summary of the proposed work and possible extensions on the proposed work were addressed in Sect. 6.

2 Related Work

In Author Profiling, most of the researchers concentrated on the extraction of different types of textual features from the text to differentiate the style of writing of the authors. Various researchers proposed different types of stylistic features like word-based, character-based, syntactic, content-based, topic-specific, structural, and readability features to discriminate the style of writing of authors [4]. The document vectors are represented with these stylistic features. Different classifiers were used by the researchers to generate the classification model for Author Profiling system. The number of features and type of features places a major role in the representation of a document. One researcher in his various experiments used [5] most frequent 3000 terms as features and incremented up to 50,000 terms to represent the document vectors. Upendra Sapkota et al. used [6] 5000 most frequent terms for gender prediction and achieved good results for Spanish language, but the results were poor for English language.

Classification algorithms play a significant role for identification of Author profiles when the number of features was more. For classification, high-dimensional space leads to the overfitting problem. Principle component analysis (PCA) is a technique that transforms high-dimensional feature space into low-dimensional space. Wee-Yong Lim et al. used [7] PCA technique to diminish the feature set size. Delia-Irazu Hernandez et al. applied [8] Jaccard similarity coefficient on the texts to find informative words rather than only frequent words. Lucie Flekova et al. used [9] information gain feature selection approach to reduce the number of features from 1500 most frequent features.

Koppel, M. et al. considered [3] 566 documents from the British National Corpus (BNC). They achieved an accuracy of 77.3% for gender prediction using 1081 features that include part of speech (POS) N-Grams and function words. Argamon, S. et al. used [10] blog posts of 19,320 blog authors. The result of 76.1% accuracy is achieved for gender prediction by using both content-based and stylistic features. It was observed that the style-based features such as prepositions, pronouns, and determiners were most useful to predict the gender profile. Schler, J. et al., achieved [11] better accuracy of 80.1% for gender prediction using 1502 content-based and stylistic features on 37,478 blogs.

3 The Proposed Profile Specific Document Weighted Approach

In this approach, a new document representation model, namely Profile specific Document Weighted (PDW) model, was proposed. The PDW model is represented in Fig. 1. In this model, $\{D_1, D_2, ..., D_m\}$ is a set of documents in the dataset, and $\{T_1, T_2, ..., T_n\}$ is the collection of most frequent terms. TWF and TWM are the weights of the term in female and male dataset, respectively. DWM and DWF are the weights of the document specific to male profile and female profile, respectively.

In this model, a new document representation technique is proposed to predict the gender of anonymous text in the reviews' domain. The PDW model deals the challenging conditions of the Author Profiling task and also overcomes the drawbacks of BOW approach. In this proposal, the documents are represented in low-dimensional vector space and this representation captures the relationship between terms. The term weights are computed by capturing the discriminative information among the

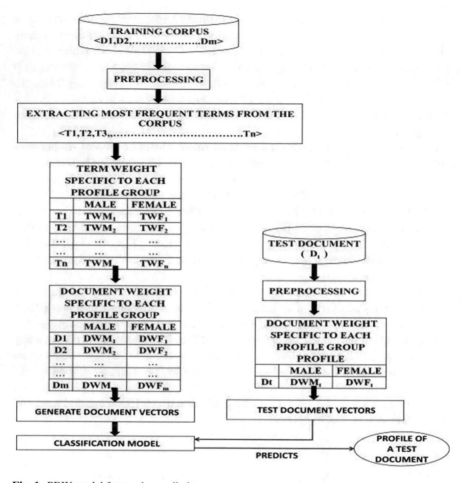

Fig. 1 PDW model for gender prediction

profiles. The weights of the documents were computed specific to profile groups by aggregating the term weights. This approach is termed as Profile specific Document Weight (PDW).

In this model, two steps majorly influence the accuracy of gender prediction. In the first step, finding suitable weight measure to compute the term weights for each profile group. In the second step, the document weight measure for computing the weights of documents specific to each profile group.

Section 3.1 describes the term weight measure used to compute the term weights specific to author profiles. The document weight measure specific to author profiles is explained in Sect. 3.2.

3.1 Term Weight Measure Specific to Profile

Term weight measures assign suitable weights to the terms based on their importance in the text. In this model, a term weight measure is used proposed by [12] to compute the term weight specific to the profile groups. This term weight measure gives more weight to the terms that are not distributed uniformly across the documents [12]. Equation (1) shows the term weight measure used in PDW model.

$$W_{tij} \quad w\, t_i, p_j \quad \log \text{TOTF}_{ti} \frac{tf(t_i, d_k)_{k1}^m}{\text{TOTF}_{ti}} \quad \log_1 \frac{tf(t_i, d_k)}{\text{TOTF}_{ti}} \tag{1}$$

where $tf(t_i, d_k)_{k1}^{m\,\text{TOTF}_{ti}}$.

where TOTFti is the total occurrence of the term t_i in all the documents labeled with profile p_j, $tf(t_i, d_k)$ is the term frequency of t_i in the document d_k.

The term weight measure calculates the term weights in individual documents of specific profile group. Then, these term weights are added to get the total weight of the term specific to one profile group. The term weights were calculated for every considered profile group. For gender dimension, two profile groups are considered such as male and female. The term weights are calculated in male dataset and female dataset.

3.2 Document Weight Measure Specific to Profiles

The document weight is computed based on the semantic relationship among the features in the document. In this representation, a document weight measure is used from information retrieval domain to compute the document weights. The document weights are computed specific to profile group by aggregating the term weights specific to that document. Equation (2) is used to determine the weight of a document specific to a profile group. Term frequency and inverse document frequency (TFIDF) measure as in Eq. (3) assigns the weight to a term based on the term frequency and document frequency in a total dataset of documents.

$$W_{dkj} \quad \text{TFIDF}(t_i, d_k) W_{tij}$$
$$t_i\, d_k, d_k\, p_j \tag{2}$$

$$\text{TFIDF}(t_i, d_k)\, tf(t_i, d_k)\, \log \frac{|D|}{|1DF_{ti}|} \tag{3}$$

where W_{dkj} is the document d_k weight in the profile group p_j, DF_{ti} is the number of documents contains the term t_i in the documents of profile group p_j, |D| is the total number of documents in the specific profile group. The document weights were used to represent the document vectors. The machine learning classifiers generate a classification model by using these document vectors.

Equation (4) shows the representation of a document vectors

$$Z_{d_k} \bigcup_{p_j} z_k, c_j \tag{4}$$

where z_k $W_{dk\,1}, W_{dk\,2}, \ldots, W_{dkq}$ and c_j is a class label of a profile p_j. z_k is a document vector of d_k.

4 Empirical Evaluations

4.1 Dataset Characteristics

The dataset was collected from TripAdvisor.com, and it consists of 4000 reviews of various hotels. The dataset is balanced in terms of gender dimension, and both male and female profile groups contain 2000 reviews of each.

4.2 Evaluation Measures

In Author Profiling approaches, several measures were used by the researchers to evaluate the performance of their system. In this work, an accuracy measure was used to evaluate the performance of our system. Accuracy is the ratio of number of documents correctly predicted their gender to the total number of test documents considered.

4.3 Bag-of-Words (BOW) Approach

Several researchers used the combination of content and stylistic features as attributes in their work on Author Profiling [3, 11]. The bag-of-words (BOW) is the most used representation for Author Profiling in different domains. In this BOW approach, the documents are represented with most frequent words as features.

Table 1 shows the accuracy of gender prediction for various classifiers using BOW approach. The simple logistic classifier obtained a good accuracy of 69.25% for gender prediction using the most frequent 7000 words when compared to other classifiers. It was observed that most of the classifiers show good accuracy when the number of features is increased. The experimentation was carried out with six machine learning classifiers such as random forest (RF), Naïve Bayes multinomial (NBM), simple logistic (SLOG), logistic (LOG), IBK, and bagging (BAG).

Table 1 Accuracy percentages of gender prediction in BOW approach

Classifier/number of terms	NBM	SLOG	LOG	IBK	BAG	RF
1000	65.95	68.80	68.55	57.65	62.90	59.25
2000	65.70	69.15	68.90	58.45	63.50	60.00
3000	65.85	69.05	69.00	58.65	63.40	60.05
4000	65.85	69.00	69.10	58.40	63.65	59.20
5000	66.00	69.10	69.10	58.60	63.55	59.45
6000	66.15	69.25	69.15	58.55	63.25	59.50
7000	66.05	69.25	69.15	58.65	63.35	59.85
8000	66.00	69.20	69.20	58.65	63.60	60.10

Table 2 Accuracy percentages of SOA approach for gender prediction

Classifier/number of terms used	NBM	SLOG	LOG	IBK	BAG	RF
1000	73.15	69.60	63.55	53.35	70.60	62.40
2000	73.35	69.10	56.10	53.30	71.70	61.05
3000	73.20	68.85	62.85	54.10	71.05	60.60
4000	72.75	69.20	63.70	54.25	70.75	62.10
5000	71.90	69.60	64.80	51.80	70.70	58.40
6000	71.95	69.85	65.20	50.45	70.60	58.90
7000	72.35	69.35	65.35	50.40	70.60	60.25
8000	72.30	69.95	65.10	51.00	70.65	57.85

4.4 Second-Order Attribute (SOA) Approach

A. Pastor López-Monroy [13] proposed an approach of second-order attribute (SOA) representation. The main purpose of this approach is to find the second-order attributes to represent the document. The accuracy values for gender prediction in SOA approach were given in Table 2. It is observed that the NBM classifier achieved a good accuracy of 73.35% for the 2000 words compared to other classifiers.

4.5 PDW Approach

The accuracy values of PDW approach for gender prediction is presented in Table 3. With the PDW approach, the Naive Bayes multinomial classifier obtained an accuracy of 90.45% when 8000 terms are used to compute the document weight. It was observed that the classifiers' accuracy for gender prediction is increased when the number of terms is increased for computing the document weight.

Table 3 Accuracy percentages of PDW approach for gender prediction in reviews' dataset

Classifier/number of terms used	NBM	SLOG	LOG	IBK	BAG	RF
1000	79.40	78.15	79.50	70.55	74.85	73.20
2000	82.20	80.95	81.80	73.90	74.20	73.35
3000	84.35	82.05	83.70	74.50	77.05	75.30
4000	85.55	84.75	85.55	78.90	79.15	78.80
5000	87.00	85.25	86.85	79.25	80.20	79.70
6000	88.40	87.30	88.30	82.15	81.80	81.65
7000	89.65	88.05	89.00	83.85	83.30	82.75
8000	90.45	89.60	90.30	84.00	84.65	84.75

Table 4 Comparison of BOW, SOA, and PDW approaches for 8000 most frequent terms

Classifier/approach	NBM	SLOG	LOG	IBK	BAG	RF
BOW	66.00	69.20	69.20	58.65	63.60	60.10
SOA	72.30	69.95	65.10	51.00	70.65	57.85
PDW	90.45	89.60	90.30	84.00	84.65	84.75

5 Discussion

The proposed approach used the term weight measure which computes the term weight in all the documents of specific profile group and document weight measure which computes the document weight by using all considered term weights. The document weight computation considers the semantic relationship between the terms in a document. The capturing of semantic relationship between the features helps to increase the accuracy of the gender prediction.

Table 4 shows the gender prediction accuracy results of the BOW, SOA, and PDW approaches for most frequent 8000 terms on different classifiers. It was also notable that the PDW approach outperformed than BOW and SOA approaches for all classifiers.

6 Conclusions and Future Scope

In this paper, the proposed PDW approach obtained an overall accuracy of 90.45% for gender prediction when 8000 most frequent terms are used. The obtained accuracy is much better when compared with the accuracies of existing approaches for the problem of Author Profiling. The proposed model can be extended for the identification of demographic characteristics of authors like age, location, native language, educational background, and personality traits with various combinations of term weight measures and document weight measures.

References

1. Koppel, M., Schler, J., Bonchek-Dokow, E.: Measuring differentiability: Unmasking pseudony-mous authors. J. Mach. Learn. Res. 8, 1261–1276 (Dec 2007).
2. M. Sudheep Elayidom, Chinchu Jose, Anitta Puthussery, Neenu K Sasi, "TEXT CLASSI-FICATION FOR AUTHORSHIP ATTRIBUTION ANALYSIS", Advanced Computing: An International Journal (ACIJ), Vol. 4, No. 5, September 2013.
3. Koppel, M., Argamon, S., Shimoni, A.R.: Automatically categorizing written texts by author gender. Literary and Linguistic Computing 17(4), pp. 401–412 (2002).
4. T. Raghunadha Reddy, B. VishnuVardhan, and P. Vijaypal Reddy, "A Survey on Authorship Profiling Techniques", International Journal of Applied Engineering Research, Volume 11, Number 5 (2016), pp 3092–3102.
5. Seifeddine Mechti, Maher Jaoua, Lamia Hadrich Belguith, "Author Profiling Using Style-based Features", Proceedings of CLEF 2013 Evaluation Labs, 2013.
6. Upendra Sapkota, Thamar Solorio, Manuel Montes-y-Gómez, and Gabriela Ramírez-de-la-Rosa, "Author Profiling for English and Spanish Text", Proceedings of CLEF 2013.
7. Wee-Yong Lim, Jonathan Goh and Vrizlynn L. L. Thing, "Content-centric age and gender profiling", Proceedings of CLEF 2013 Evaluation Labs, 2013.
8. Delia-Irazu Hernandez, Rafael Guzmán-Cabrera, Antonio Reyes, and Martha Alicia Rocha, "Semantic-based Features for Author Profiling Identification: First insights", Proceedings of CLEF 2013 Evaluation Labs, 2013.
9. Lucie Flekova and Iryna Gurevych, "Can We Hide in the Web? Large Scale Simultaneous Age and Gender Author Profiling in Social Media", Proceedings of CLEF 2013, 2013.
10. Argamon, S., Koppel, M., Pennebaker, J. W., and Schler, J. "Automatically profiling the author of an anonymous text". Communications of the ACM, 52(2):119. (2009).
11. Schler, J., Koppel, M., Argamon, S., and Penebaker, J. "Effects of Age and Gender on Blog-ging". AAAI Spring Symposium on Computational Approaches to Analysing Weblogs (AAAI-CAAW), AAAI Technical report SS-06-03, (2006).
12. Dennis, S.F., "The Design and Testing of a Fully Automated Indexing-Searching System for Documents Consisting of Expository Text", in Informational Retrieval: A Critical review, Thompson Book Company, Washington D.C., 1967, pages 67–94.

Local Edge Patterns for Color Images: An Approach for Image Indexing and Retrieval

A. Hariprasad Reddy and N. Subhash Chandra

Abstract Local Edge Patterns for Color Images (LEPCI) for image indexing and retrieval are a novel feature extraction technique supplied with this paper. The image converted into RGB and LEPCI encodes the one of a kind OR (XoR) operation between the middle pixel of each coloration plane and its surrounding associates of quantized orientation and gradient values. While neighborhood binary styles (LBP) and local orientation and gradient XoR styles (LOGXoRP) encode the relationship among the gray values of center pixel and its associates, we display that the LEPCI can extract effective texture (facet) features in comparison with LBP and LOGXoRP for color images. The overall performance of the proposed technique is tested by engaging in experiments on Corel-10K databases. The impacts of proposed procedure show advancement as far as their evaluation measures in contrast with LBP, LOGXoRP, and other present systems with particular databases.

Keywords Feature extraction · Local binary patterns (LBP) · Local orientation and gradient XoR patterns (LOGXoRP)
Content-based image retrieval (CBIR) · Texture

1 Introduction

The want of better and speedier recovery systems has constantly fueled to the investigations in content-based image retrieval (CBIR). The component taken out plays a vital capacity in CBIR whose adequacy relies on the strategy that took after for removing highlights from given images. The noticeable substance and material descriptors are both global or neighborhoods. A global identifier speaks to the unmistakable abil-

A. H. Reddy (✉)
Department of Computer Science and Engineering, JNTU Hyderabad, Hyderabad, India
e-mail: ankireddypalli.reddy@gmail.com

N. S. Chandra (✉)
Department of CSE, CVR College of Engineering, Hyderabad, India
e-mail: subhashchandrancse@gmail.com

© Springer Nature Singapore Pte Ltd. 2019
R. S. Bapi et al. (eds.), *First International Conference on Artificial Intelligence and Cognitive Computing* , Advances in Intelligent Systems and Computing 815,
https://doi.org/10.1007/978-981-13-1580-0_5

ities of the entire image, while a local identifier speaks to the obvious capacities of ranges or protests depict the images. Those are orchestrated as complex capacity vectors and amass the capacity database. For likeness, separate estimation of numerous techniques was created like Euclidean separation (L2), L1 distance, and numerous others. Choice of capacity descriptors and comparability separate measures influence recovery exhibitions of an image retrieval system eminently. Then some time ago accessible writing on CBIR is given in [1–6].

Mahmoudi et al. [7] have proposed the shape principally based component which orders image edges in view of the elements: their introductions and connection between neighboring edges. They demonstrated that the proposed plot is successful and heartily endures interpretation, scaling, shade, brightening, and seeing part varieties. So also they display a shiny new element coordinating way to deal with process the divergence esteem among the capacity vectors separated from images.

Local binary patterns (LBP) were proposed by Ojala et al. [8] that may demonstrate higher execution notwithstanding less computational multifaceted design for texture arrangement. Accomplishment of LBP varieties in expressions of speed and general execution is articulated in heaps of research regions which incorporates texture class [8, 9], face popularity [10, 11], object tracking, photo retrieval, and fingerprint matching and hobby factor detection.

Jun and Kim [10] have done work on the local gradient patterns (LGP) for face detection. LGP delineation is uncaring to global force varieties simply like alternate portrayals comprehensive of local binary patterns (LBP) and to neighborhood profundity varieties along the edge added substances.

Xie et al. [11] proposed LGXP for face recognition, LGP in [10] and local orientation gradient XoR patterns (LOGXoRP) [13] are encouraged us to endorse the local edge styles the use of color images (LEPCI) for image indexing and retrieval. The primary contributions of the LEPCI are given as follows: (a) the quantization of Gabor responses for LGXP operator could be very difficult because the variety of Gabor coefficients are exceptional for every image, whereas the LEPCI makes use of the quantized gradient and orientation responses whose variety (Orientation: $0°$ to $360°$ and Gradient: zero:255) is usually constant. (b) We prove that the LEPCI can extract powerful texture (facet) capabilities as compared to LBP and LGP. The execution of the LEPCI is tested by way of carrying out experiments on Corel-10K databases.

2 Local Patterns

2.1 Local Gradient Patterns (LGPs)

Jun and Kim [10] have proposed the LGP for face detection. The LGP value is calculated using Eqs. (1) and (2).

$$\text{LGP}_{P,R} = \sum_{i=1}^{P} 2^{(i-1)} \times f_1(|g_i - g_c| - T_h) \qquad (1)$$

$$T_h = \frac{1}{P} \sum_{i=1}^{P} |g_i - g_c| \qquad (2)$$

More details of LGP can be found in [10].

The proposed LGP is almost similar to the completed LBP magnitude (CLBP_M) [11]. The only difference between these two features is that the LGP calculates the threshold (T_h) from the mean/average of local difference operator (LDO) for a given pattern, whereas CLBP_M calculates from the mean/average of entire image LDO.

2.2 Proposed Method Local Edge Patterns for Color Images (LEPCI)

The idea of LBP [8], LGP [10], and LGXoRP [13] has been adopted to define the Local Edge Patterns for Color Images (LEPCI). A center pixel in an image and the gradients ($P = 8$) are calculated as:

$$I_{c1}|_{g_c}^{h} = g_1 - g_5$$
$$I_{c1}|_{g_c}^{v} = g_3 - g_7 \qquad (3)$$
$$c1 \forall \{R, G, B\} \qquad (4)$$

where $c1$ for every $\{R, G, B\}$ and $|\{g_1, g_2, g_3, g_4, g_5, g_6, g_7, g_8\}|_{P=8}$ are the pixel values of neighbors for a given center pixel, g_c.

The orientation and gradient values are calculated as follows:

$$I_{c1}|_{g_c}^{G} = \sqrt{\frac{(I_{g_c}^{h})^2 + (I_{g_c}^{v})^2}{2}} \qquad (5)$$

$$\Theta_{g_c} = \tan^{-1}\left(\frac{I_{c1}|_{g_c}^{v}}{I_{c1}|_{g_c}^{h}}\right) \qquad (6)$$

$$I_{g_c}^{O} = \begin{cases} 0° + \theta_{g_c} & I_{g_c}^{h} \geq 0 \text{ and } I_{g_c}^{v} \geq 0 \\ 180° - \theta_{g_c} & I_{g_c}^{h} < 0 \text{ and } I_{g_c}^{v} \geq 0 \\ 180° + \theta_{g_c} & I_{g_c}^{h} < 0 \text{ and } I_{g_c}^{v} < 0 \\ 360° - \theta_{g_c} & I_{g_c}^{h} \geq 0 \text{ and } I_{g_c}^{v} < 0 \end{cases} \qquad (7)$$

The gradient XoR patterns (LGXoRP) and orientation XoR patterns (LOXoRP) are calculated as:

$$\text{LECPI} = \begin{bmatrix} \{Q(I_{g_1}^G) \otimes Q(I_{g_c}^G)\}, \\ \{Q(I_{g_2}^G) \otimes Q(I_{g_c}^G)\}, \\ \cdots\cdots\cdots\cdots\cdots\cdots\cdots, \\ \{Q(I_{g_1}^G) \otimes Q(I_{g_c}^G)\}, \end{bmatrix} \tag{8}$$

$$\text{LECPI} = \begin{bmatrix} \{Q(I_{g_1}^O) \otimes Q(I_{g_c}^O)\}, \\ \{Q(I_{g_2}^O) \otimes Q(I_{g_c}^O)\}, \\ \cdots\cdots\cdots\cdots\cdots\cdots\cdots, \\ \{Q(I_{g_1}^O) \otimes Q(I_{g_c}^O)\}, \end{bmatrix} \tag{9}$$

where $Q(x)$ denotes the quantized value of X and $\otimes$ represents the exclusive OR (XoR) operation. After identifying the local pattern, construct a histogram using Eq. (10)

$$H_s(l) = \frac{1}{N_1 \times N_2} \sum_{j=1}^{N_1} \sum_{k=1}^{N_2} f_2(PTN(j,k), l); l \in [0, P(P-1)+2] \tag{10}$$

$$f_2(x, y) = \begin{cases} 1 & \text{if } x = y \\ 0 & \text{else} \end{cases} \tag{11}$$

where the size of the input image is $N1 \times N2$.

3   Feature Extraction and Analysis

Algorithm:
   Input: Image;   Output: Feature vector

1. Convert image into RGB planes.
2. Calculate the gradient and orientation values for each plane.
3. Quantize the gradient and orientation values.
4. Calculate the XoR patterns for gradients and orientations.
5. Histogram calculation.
6. Construct the feature vector by concatenating the histograms (Fig. 1).

3.1   Similarity Measure

Feature vector is constructed for query image Q as $f_Q = (f_{Q1}, f_{Q2}, \dots f_{Lg})$ similarly for the entire database also construct the feature vector, i.e., $f_{\text{DB}_j} =$

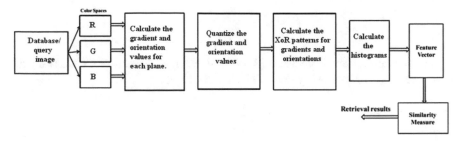

Fig. 1 Flowchart of proposed method

$(f_{DB_{j1}}, f_{DB_{j2}}, \ldots f_{DB_j Lg})$; $\quad j = 1, 2, \ldots, |DB|$. The goal is to select the best match over the query image. This can happen just by choosing 'n' top coordinated images by measuring the separation among the query image and images in the database $|DB|$. In this work, this should possibly by $d1$ similarity measure utilizing Eq. (12).

$$D(Q, DB) = \sum_{i=1}^{L_g} \left| \frac{f_{DB_{ji}} - f_{Q,i}}{1 + f_{DB_{ji}} + f_{Q,i}} \right| \tag{12}$$

where f_{DB_j} is jth database image and its ith feature.

3.2 Evaluation Measures

Proposed method LEPCI performance is measured based on average retrieval precision (ARP) and average retrieval rate (ARR) as shown below:

Precision for query image I_q is:

$$P(I_q, n) = \frac{1}{n} \sum_{i=1}^{|DB|} \left| \delta(\Phi(I_i), \Phi(I_q)) | \text{Rank}(I_q, I_q) \le n \right| \tag{13}$$

$$\delta(\Phi(I_i), \Phi(I_q)) = \begin{cases} 1 \ \Phi(I_i) = \Phi(I_q) \\ 0 \ \text{else} \end{cases} \tag{14}$$

Recall is defined as:

$$R(I_q, n) = \frac{1}{N_G} \sum_{i=1}^{|DB|} \left(\Phi(I_i), \Phi(I_q) \right) | \text{Rank}(I_i, I_q) \le n | \tag{15}$$

N_G—the number of relevant images in the database. The average precision is computed using Eq. (16)

$$P_{avg}^j(n) = \frac{1}{N_G} \sum_{i \in G} P(I_i, n) \tag{16}$$

The average retrieval precision (ARP) and average retrieval rate (ARR) for the database are computed using Eqs. (17) and (18), respectively.

$$ARP = \frac{1}{|DB|} \sum_{i=1}^{|DB|} P(I_i, n) \tag{17}$$

$$ARR = \frac{1}{|DB|} \sum_{i=1}^{|DB|} R(I_i, n) \Big|_{n \le 100} \tag{18}$$

4 Experimental Results and Discussions

In the work, the database called Corel is utilized for experimentation. Corel database [12] images are pre-grouped into selective classifications of length 100, by means of area experts. A few analysts feel that Corel database meets the greater part of the necessities to assess an image retrieval system. For Corel-10K database, 10,000 images are collected all these have sizes, both 126×187 and 187×126. The general execution of the proposed approach is measured as far as ARP and ARR. The outcomes are thought to be better, if regular estimations of exactness and consider are extreme (Fig. 2).

From Table 1, it has been observed that the proposed LEPCI experimentation values on precision and recall are compared with various existing method and there is an improvement in performance.

Table 2 tells that the proposed LEPCI is performing well to quantization level 4. From Fig. 3, we can observe that by giving an image as input from the Corel-10K database the proposed LEPCI is retrieving more relevant images.

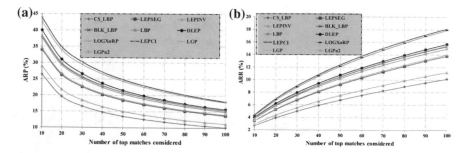

Fig. 2 **a, b** Comparison of ARP and ARR values of proposed LEPCI with various existing methods on the Corel-10K database

Table 1 Precision and recall values of various methods on Corel-10K database

Database	Performance Method		CS_LBP	LEPSEG	LEPINV	BLK_LBP	LBP	DLEP	LGP	LGPu2	LEPCI (proposed method)	LOGXoRP
Corel-10K	Precision (%)		26.4	34	28.9	38.1	37.6	40	38.4	34.4	43.8	42.6
	Recall (%)		10.1	13.8	11.2	15.3	14.9	15.7	15.4	14	18.5	18.1

Table 2 Results (precision and recall) of LEPCI on various quantization levels on Corel-10K

	Method	Quantization levels of orientation and gradients								
		18	12	9	8	7	6	5	4
Precision (%)	LOGXoPR	42.1	42.8	43.2	43.3	43.3	43.3	43.4	43.8
	PM(LEPCI)	42.7	43.2	43.6	43.7	43.7	43.8	43.9	44.2
Recall (%)	LOGXoRP	17.2	17.6	17.7	17.8	17.9	17.9	17.8	18.1
	PM(LEPCI)	17.3	17.7	17.9	18	18	18.1	18.2	18.5

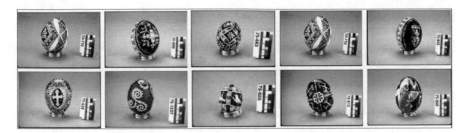

Fig. 3 LEPCI retrieved images on the Corel-10K database (top left image is the input image)

5 Conclusion

In this work, a fresh out of the box new image ordering and recovery calculation is proposed utilizing local aspect styles utilizing shading previews (LEPCI). The LEPCI encodes the color images principally in light of the XoR operation among the center pixel and its associates of quantized introduction and slope reactions.

The general execution improvement of the proposed technique has been as contrasted, and the LOGXoRP on tinge images, the basic exactness, and the normal recollect has advanced.

References

1. ML. Kherfi, D. Ziou, A. Bernardi 2004 Image Retrieval from the World Wide Web: Issues, Techniques and Systems. ACM Computing Surveys, 36 35–67.
2. Ke Lu and JidongZhao, Neighborhood preserving regression for image retrieval. Neurocomputing, 74 1467–1473 (2011).
3. Tong Zhaoa, Lilian H. Tang, Horace H.S. Ip, Feihu Qi, On relevance feedback and similarity measure for image retrieval with synergetic neural nets. Neurocomputing, 51 105–124(2003).
4. Kazuhiro Kuroda, Masafumi Hagiwara, An image retrieval system by impression words and specific object names-IRIS. Neurocomputing, 43 259–276 (2002).
5. Jing Li, Nigel M. Allinson, A comprehensive review of current local features for computer vision. Neurocomputing, 71 1771–1787 (2008).
6. Guang-Hai Liu, Lei Zhang, Ying-Kun Hou, Zuo-Yong Li and Jing-Yu Yang, Image retrieval based on multi-texton histogram, Pattern Recognition 43 2380–2389 (2010).
7. Fariborz Mahmoudia, Jamshid Shanbehzadeh, Amir-Masoud Eftekhari-Moghadam and Hamid Soltanian-Zadeh, Image retrieval based on shape similarity by edge orientation autocorrelogram, Pattern Recognition 36 1725–1736 (2003).
8. T. Ojala, M. Pietikainen, D. Harwood, A comparative study of texture measures with classification based on feature distributions, Pattern Recognition, 29 51–59 (1996).
9. Zhenhua Guo, Lei Zhang, and David Zhang, A Completed Modeling of Local Binary Pattern Operator for Texture Classification, IEEE Tans. Image Proc., 19 (6) 1657–1663 (2010).
10. Bongjin Jun and Daijin Kim, Robust face detection using local gradient patterns and evidence accumulation, Pattern Recognition 45 3304–3316 (2012).
11. ShufuXie, Shiguang Shan, Xilin Chenand Jie Chen, Fusing Local Patterns of Gabor Magnitude and Phase for Face Recognition, 19 (5) 1349–1361 (2010).

12. Corel-10K image database. [Online]. Available: http://www.ci.gxnu.edu.cn/cbir/Dataset.aspx.
13. A. Hariprasd Reddy, N, Subhash Chandra Local Orientation Gradient Xor Patterns: A New Feature Descriptor For Image Indexing And Retrieval, i-manager's Journal on Pattern Recognition, vol 2 No 4 1–10 (2016).

Secure Privacy Preserving of Personal Health Records Using Attribute-Based Encryption in Cloud Computing

R. China Appala Naidu, A. Srujan, K. Meghana, K. Srinivas Rao
and B. Madhuravani

Abstract Personal health information is outsourced to be stored in cloud to assure the patients' control on their information, and the data should be encrypted and stored in the database. The personal information should not be viewed by the doctor, and the health information should not be viewed by the insurance agents; such a way, the system is designed and protected the data in the database. In this paper, we used SHA 512 algorithm for attribute-based encryption and decryption. Based on the personal information, the system will generate the suitable medicine for the patient. An access policy is used to distribute the secret key of the patient to other parties to view the data of the patient. Using IBM-Bluemix load balancers, the system achieved good response time.

Keywords SHA 512 · Access policy · Bluemix

R. China Appala Naidu
CSE Department, St. Martins Engineering College, Secunderabad 500014,
Telangana, India
e-mail: chanr789@gmail.com

A. Srujan (✉) · K. Meghana
IT Department, St. Martins Engineering College, Secunderabad 500014,
Telangana, India
e-mail: srujanatluri@gmail.com

K. Meghana
e-mail: kmeghana789@gmail.com

K. Srinivas Rao · B. Madhuravani
Department of Computer Science and Engineering, MLR Institute of Technology,
Hyderabad, India
e-mail: ksrao2017@gmail.com

B. Madhuravani
e-mail: madhuravani.peddi19@gmail.com

© Springer Nature Singapore Pte Ltd. 2019   59
R. S. Bapi et al. (eds.), *First International Conference on Artificial Intelligence
and Cognitive Computing* , Advances in Intelligent Systems and Computing 815,
https://doi.org/10.1007/978-981-13-1580-0_6

1 Introduction

To achieve the privacy, we have different types of algorithms that are existed. One of the algorithms follows the encryption of the data and gives access to the data with the help of sharing-up of keys between users by a third-party involvement called a Trusted Authority (TA). In this case, there will be a unique ID that will be generated for the users whenever a user gets registered. The unique ID of the users will be encrypted and given keys to share their data with other users with the help of Trusted Authority (TA). No other data of the user will get encrypted. This method is called identity-based encryption (IBE) [1].

In this method, Trusted Authority and IBE are risk factors. The Trusted Authority staff can decrypt and see the data of the user whenever they need to. Due to only encryption of the identities, hackers can easily predict the data of the users with the help of other attributes like name, age.

There will be different parties like cloud server, patient, doctor, insurance company, friend. The data of the patient is accessed by other parties by sharing the key with the patient. There is a branching program for systolic BP which was implemented in this system [2, 3]. By providing sensors to the patient body, with the help of the branching program, the system gets activated and displays the medicine that should be taken by the patient in his respective smart phone. In this way, the system is able to protect the patient but still failed to gain privacy.

The next step is encryption of the data and storing the data. The encryption is performed to the attributes of the users. There will be no unique ID which will be created when a user gets registered. The data of the users will be taken in the form of attributes [4, 5]. Name, age, sex, phone number, etc., which are specified as patient details will be taken as attributes.

2 Load Balancing

Load balancing is used to control the impact that is being created on the system. Whenever the users who are accessing the system get increased, the load on the system will also increase and this works in vice versa. In order to make the system available even in the critical traffic times also, load balancing is used. It distributes the load between multiple servers to eradicate the server crashes.

There are different types of load balancers that are offered by IBM-Bluemix. Some of them are four-layer and seven-layer load balancers. Seven-layer load balancer works more efficient than the four-layer. There are different types of packages in load balancers which can be activated by pay-as-you-go to activate a particular package according to the load that is being created on the system [7].

3 Proposed Research Methodology

The proposed system is based on storing the data in an encrypted format and sharing the data with the users in a decrypted format by preserving the privacy. We have modified the branching program completely by adding temperature and sugar levels of the patient. The working of the branching program is based on

$$\text{if } h = L(1) \wedge h = M(1) \wedge h = R(1)$$
$$\text{then(“Display the tablet name”)}$$

In this case, the patient may get 27 different types of answers for his medication for each and every different criteria. The blood pressure, sugar levels and temperature of the patient are each sub-divided into three different criteria like if we take blood pressure, it may be below 100, 100–120 and above 120. Temperature and sugar levels also follow the same way.

This is the basic overview of the functioning of the system. When it comes to privacy preserving, we have used SHA 512 algorithm in order to encrypt the data of the users with the help of k-anonymity. There are two types in K-anonymity. First one is suppression, and the second one is generalization. We have used generalization method. The data of the users comes under attributes which is known as attribute mechanism.

In an attribute-based encryption, the secret key of a user and the cipher text is attribute dependent. In this type of system, the decryption of a cipher text is possible only if the attributes of the user key matches the attributes of the cipher text. The major role of attribute-based encryption is collision resistance. A system that holds multiple keys should only be able to access data individually through the access policy.

Now let us get into the framework of the work. Each and every user need to be registered into the application like patient, friend, insurance company or doctor. While being registered, the attributes of each user will be encrypted in the form of anonymization. In order to obtain security between users, an access policy has been introduced through which the details of the patient will be shared with the help of a key to which user, and the patient gives access to.

The patient's data is divided into two parts: medical history and personal profile. Doctor need not to know about the patient personal profile, and the insurance company need not to know about the medical history of the patient. Friend can know the personal and medical details of the patient. In this way, the patient can specify whose his friend, doctor and insurance company and can give them access to their respective fields. In this way, we can eradicate privacy concerns. Doctor is able to access to the medical history only, and insurance company can access only personal profile.

The attributes get decrypted only after the patient gives access to other users through access policy. In the key policy, the format is restricted to conjunctions among different attribute authority.

$$K := K1 \wedge K2 \wedge Kn$$

The set-up for the key generation, encryption, decryption will be explained below.

- Setup(1^n) The same as Setup from the multiple authority can be involved by setting up multiple key generations, except that each BB_n ($n = \{1,\dots, M-1\}$) defines an additional dummy attribute $B * n$ with its corresponding public key(Puk1) and master key components(Mak1), and each BB(Authority) initializes a version number ver $= 1$. The BBs publish (ver, Puk1), while (ver, Makn) is held by BB_n [6].
- KeyIssue(B^u, Mak1, Puk1) The same as KeyIssue from a multiple authority to the solution in which multiple trusted authorities can be involved by generating keys for the users in order to eradicate collusion, except the key policy Bu of each user must be end with $B * 1,\dots, B * M - 1$. The user receives (ver, SeK), wherever is the current version number and SeK is a secret key [6].
- Encrypt(M, B^u_{PUD}, Puk1) The same as Encryption solution, except that $B * n$ must be part of B^u_{BBn} ($\forall n \in \{1,\dots, M-1\}$). It outputs UJ $=$ ver, D0 $= X \cdot Y$ r, D1 $=$ qw 2, $\{Cn,i = Js \ i,f\}$ f $\in B^u_{PUD}$, k $\in \{1,\dots,N\}$. The encryptor stores the random numbers used to compute UJ.
- Decrypt(UJ, Puk1, SeK) The same as Decryption in [6], except it uses Puk1 and SeK with the same ver as in UJ.

4   Results

IBM-Bluemix is an IBM cloud which provides PaaS, IaaS, SaaS for clients to maintain their projects in cloud at lower prices. IBM provides Liberty for Java to deploy Java Web applications. There are different types of services offered by the Bluemix for maintaining security and maximum uptime for the application. We have used ClearDb as database services for storing the data of the users (Fig. 1).

IBM-Bluemix offers different types of servers, load balancers and other features like authentications, availability monitoring of the application.

In the above-drawn diagram (Fig. 2), there is an access policy which is being specified by the patient. With this access policy, the patient can provide the selected users to access his data.

In the above diagram (Fig. 3), there is an output which has been produced when the branching program is used and is stored in the database with an encrypted format. The first column is patients' name, second one is the physician who can access the patients' profile, next one is the prescription which has been generated with the help of sensor data and the last one is the date of prescription generated.

In Fig. 4, the data of the newly registered users is being stored in the database in an encrypted format. Username, password, user type and address are taken as attributes in this section. With the help of the ABE mechanism, we have encrypted all the attributes.

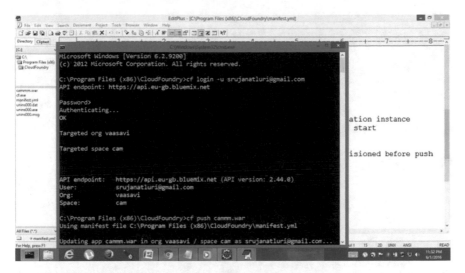

Fig. 1 Publishing into IBM-Bluemix

Fig. 2 Access policy

pname	physician	prescription	pre_date
cGF0aWVudDE=	cGh5c2ljaaWFu	eHI6enp6	2016-06-10

Fig. 3 Prescription data

username	password	usertype	contactno	address
srujan atluri	srujan	PHR Owner	8179293239	vijayawada
Krishnaiah	kitti	Physician	0987654321	warangal
narender	narender	Insurance Company	7890654321	hyderabad
cGh5c2ljaWFu	cGh5c2ljaWFu	UGh5c2ljaWFu	1234567890	aHlkZXJhYmFk
aW5zdXJhbmNlMQ==	aW5zdXJhbmNlMQ==	SW5zdXJhbmNlIENvbXBhbnk=	9876543210	aHlkZXJhYmFk
cGF0aWVudDE=	cGF0aWVudDE=	UEhSIE93bmVy	9876543210	aHlk
ZnJpZW5kMQ==	ZnJpZW5kMQ==	UEhSIFVzZXI=	7842627545	aHlkZXJhYmFk

Fig. 4 New users registered

Fig. 5 Sensor data

pid	temp	bp	sl
srujan atluri	MTEx	MjIy	MzMz
cGF0aWVudDE=	MTIy	MTEx	NTU=

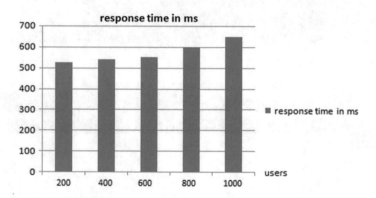

Fig. 6 Response time in milliseconds

Figure 5 describes the data which is being received from the sensors which are attached to the human body and is stored in the database in an encrypted format. There are three types of data that is being collected called temperature, blood pressure and sugar level. This sensor data is transferred to the branching program to produce a prescription (Fig. 6).

5 Test Analysis

The test analysis is performed on the system to calculate the response time and load balancing of the system. The response time is calculated in a way such that in what way does the response time of the system is being increased or decreased according to the number of users accessing the application (Table 1).

In the above figures and the response time of the proposed and existing system are being represented. The *y*-axis is taken as the response time of the system in the

Table 1 Response time of the proposed application

S.No.	Number of users	Response time in milliseconds
1	200	520
2	400	540
3	600	550
4	800	600
5	1000	650

Table 2 Response time of the existing application

S.No.	Number of users	Response time in milliseconds
1	200	100,000
2	400	110,000
3	600	118,000
4	800	120,000
5	1000	122,000

form of milliseconds. The x-axis is taken as the number of users using the application currently. When the users are being increased in existing system, the response is also being increased but in the proposed system, and there is no drastic difference in the response time for 200 users and thousand users due to load balancing. So in this way, we can say that our system more efficient (Table 2).

6 Conclusion

In this paper, we have designed a cloud system cloud-assisted privacy-preserving mobile health monitoring system (CAM) for the purpose of ensuring privacy of the clients' data. Moreover, this cloud system also helps in protecting the intellectual property of the organization named mHealth service providers. The privacy protection is achieved by applying anonymous attribute-based encryption (ABE) in medical diagnostic branching programs. We apply advanced encryption standard decryption outsourcing with privacy protection to shift client's pairing computation to the cloud server so that the decryption complexity can be reduced (Fig. 7).

We apply the random permutation technique to protect mHealth service providers programs. The decision thresholds at the branching nodes were randomized. And also to enable resource-constrained small firms to take part in mHealth business, our CAM design helps them to shift the computational burden to the cloud. Hence, our cloud-based system CAM has met the desired objective.

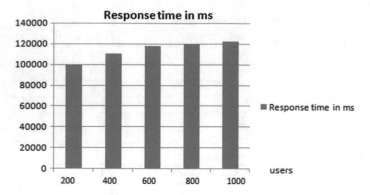

Fig. 7 Response time in milliseconds

References

1. M. Green, S. Hohenberger, and B. Waters, "Outsourcing the decryption of abe ciphertexts," in Usenix Security, 2011.
2. P. Mohan, D. Marin, S. Sultan, and A. Deen, "Medinet: personalizing the self-care process for patients with diabetes and cardiovascular disease using mobile telephony," in Engineering in Medicine and Biology Society, 2008. EMBS 2008. 30th Annual International Conference of the IEEE. IEEE, 2008, pp. 755–758.
3. Farmer, O. Gibson, P. Hayton, K. Bryden, C. Dudley, A. Neil, and L. Tarassenko, "A real-time, mobile phone-based telemedicine system to support young adults with type 1 diabetes," Informatics in primary care, vol. 13, no. 3, pp. 171–178, 2005.
4. J. Hur and D.K. Noh, "Attribute-Based Access Control with Efficient Revocation in Data Out-sourcing Systems," IEEE Trans. Parallel and Distributed Systems, vol. 22, no. 7, pp. 1214–1221, July 2011.
5. S. Jahid, P. Mittal, and N. Borisov, "Easier: Encryption-Based Access Control in Social Networks with Efficient Revocation," Proc. ACM Symp. Information, Computer and Comm. Security (ASIACCS), Mar. 2011.
6. M. Chase and S.S. Chow, "Improving Privacy and Security in Multi-Authority Attribute-Based Encryption," Proc. 16th ACM Conf. Computer and Comm. Security (CCS '09), pp. 121–130, 2009.
7. IBM Bluemix catalog.

DetectStress: A Novel Stress Detection System Based on Smartphone and Wireless Physical Activity Tracker

B. Padmaja, V. V. Rama Prasad, K. V. N. Sunitha, N. Chandra Sekhar Reddy and C. H. Anil

Abstract Stress has become an inevitable part of human lives and a major concern for public health. Especially, in today's highly competitive world, stress levels of individuals have increased and have pervaded their work life. In the workplace, a professional has to relentlessly confront a plethora of situations and issues such as work pressure, deadlines, disaster management, adapting to new changes. The incessant stress can cause many health issues, such as high blood pressure, insomnia, vulnerability to infections, and heart diseases. Our aim is to design a cognitive stress-level detection system (DetectStress) which unobtrusively assess an individual's stress levels based on smartphone daily activity data and wireless physical activity tracker (FITBIT) data. The device FITBIT records an individual's daily logs of food, weight, sleep patterns, heart rate, and physical activities. Individual stress was also measured using the most preferred psychological instrument perceived stress scale (PSS) questionnaire on monthly basis. The data was gathered using an online form which consists of ten questions in several categories, and these questions ask about their feelings and thoughts during the last month. For this study, data was collected from 35 young adults for a period of 2 months. The data includes their social behavior and other routine activities collected from smartphone. Our system uses machine learning approach for stress detection along with perceived stress scale

B. Padmaja (✉)
Department of CSE, IARE, Hyderabad, Telangana, India
e-mail: b.padmaja@gmail.com

V. V. R. Prasad
Department of CSE, SVEC, Tirupati, Andhra Pradesh, India
e-mail: vvramaprasad@gmail.com

K. V. N. Sunitha
Department of CSE, BVRITHCEW, Hyderabad, Telangana, India
e-mail: k.v.n.sunitha@gmail.com

N. C. S. Reddy · C. H. Anil
Department of CSE, MLRIT, Hyderabad, Telangana, India
e-mail: naguchinni@gmail.com

C. H. Anil
e-mail: anilchintalapudi5@gmail.com

© Springer Nature Singapore Pte Ltd. 2019
R. S. Bapi et al. (eds.), *First International Conference on Artificial Intelligence and Cognitive Computing* , Advances in Intelligent Systems and Computing 815,
https://doi.org/10.1007/978-981-13-1580-0_7

questionnaire score (PSS). The model is evaluated using two classifiers such as Naïve Bayes and Decision Tree and compared against a baseline classifier random classifier. Naïve Bayes classifier has increased in performance from 55 to 72% in terms of accuracy than other models in detecting stress levels (low < med < high). This paper gives the scope for stress detection more accurately using smartphone sensor technology rather than clinical experimentations. This system is assessed in real time with young college students in India. This paper also proposes the architecture of stress based on both physiological and behavioral response of individuals. The uniqueness of this work lies in the simplification of the stress detection process.

Keywords Social behavior · Smartphone · Wireless physical activity tracker Sleep patterns · Heart rate · Stress detection

1 Introduction

The dynamic and highly competitive workplaces, the utilization of advanced information and communication technologies, and extended office hours have given rise to stress among professionals. Many times employees face psychological problems due to job-related stress. Interim stress responses are sometimes advantageous, but long-term exposure to stress causes a number of diseases like hypertension and coronary artery disease [1, 2], as well as increases the possibility of increasing viral infections [3]. Furthermore, long-term exposure to stress can lead to mental illnesses causing depression, anxiety disorders, and suffer exhaustion [4]. Therefore, it is important that people are conscious of stressful situations so that they can take necessary actions to deal with them. Continuous exposure to stress is harmful to mental and physical health, but to fight with stress, one should first detect it.

The number of smartphone users in the world is expected to cross 5.2 billion, with more than a million new subscribers everyday. Smartphones provide low profile and economical access to big volumes of previously difficult-to-get data pertaining to human social behavior [5]. These devices can sense rich set of behavior data such as (i) call log and SMS data, (ii) location, (iii) nearby devices in physical proximity through Bluetooth. Due to availability of abundant stream of personal data related to daily activities, social interactions and daily routines [5, 6], it gives an exclusive opportunity to researchers to resolve several problems in our day-to-day life in a range of fields, such as mood [7], privacy attitudes [8], healthy living [9], productivity [10], spending behavior [11], and financial well-being [12]. Several researchers have also used smartphones to comprehend the links between sleep, mood, and sociability [13, 14].

This work is focused on one of the most common and debilitating problems in our society: stress. Stress is an omnipresent situation and an integral part of our lives, and studies have revealed that the increasing amount of stress plays a pivotal role in a number of physical, psychological, and behavioral conditions, such as anxiety, low self-worth, depression, lack of social interactions, cognitive impairments, insomnia, weak immunity, neurodegenerative diseases, and other medical conditions [15]. Besides it also escalates the healthcare costs. Hence, it is an important challenge to measure stress in daily life [16, 17].

There are two key stress hormones: cortisol and catecholamines in humans. Under normal (non-stressed) conditions, cortisol concentrations are high in the morning and then it progressively declines from late evening to early morning periods and shows sudden rise during the first few hours of sleep. According to Mason, the study of stress is psychological in nature and a range of questionnaires have been designed to evaluate the psychological factors that are linked to stress in humans [18]. A standard method of measuring stress is to enquire people regarding their current state of mind or feelings using questionnaires and perceived stress scale questionnaire (PSS) [19]. Survey questionnaires in the PSS ascertain to what degree a person feels stressed in a given situation.

Common indicators of stress comprise depression, sleep disorder, anxiety, nervousness, work mistakes, poor attention, and apathy, among many others. And sleep is an important component of maintaining internal stability in humans. Stress can be figured out by checking physical conditions and daily activities. Healthy eating, exercise, heart rate, and adequate amount of sleep lower stress levels [15, 20]. Sleep and mood are closely connected, where an inadequate amount of sleep has a significant effect on mood and causes irritation and stress. Insufficient sleep causes more stress and when sleep is resumed to normal, it shows a dramatic improvement in mood [21].

In this work, a wireless physical activity tracker called Fitbit tracker is used which uses a 3-axis accelerometer to recognize human motion. The accelerometer converts human movements into digital measurements when it comes in contact with a human body. By analyzing these data, Fitbit tracker gives comprehensive information about frequency, duration, intensity, and patterns of movement to tell step count, distance traveled, calories burned, and quality of sleep [22]. Therefore, Fitbit tracker can be used as a powerful tool to monitor stress.

In this paper, data regarding heart rate variability (HRV), sleep patterns, social life, and physical activity is used to detect the presence of stress in human lives. DetectStress employs this technology to identify stress in humans using data from various sensors.

The successive sections of this paper are organized as follows. In Sect. 2, DetectStress system architecture is described in detail. Experimental study and results are discussed in Sect. 3. Conclusion and the future scope of the work on DetectStress are stated in Sect. 4.

1.1 DetectStress System Architecture Overview

DetectStress follows machine learning (ML) approach which works independently to compute the stress level of the users. There are several links between social behavior of a person and his stress level, but DetectStress is a realistic approach and it develops

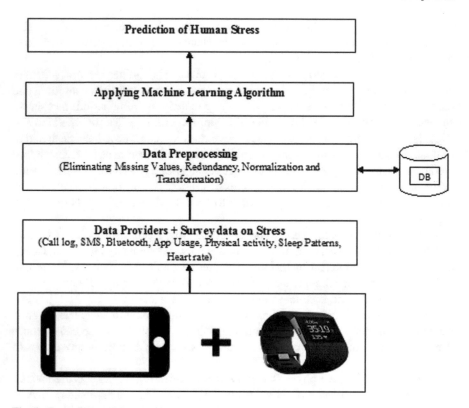

Fig. 1 DetectStress system architecture

a primary stress detection system which can be expanded and made even further advanced. Figure 1 shows the common architecture of DetectStress system which aims to find behavioral indicators for stress.

1.1.1 Data Providers

(a) Smartphone Data: This layer gathers and sends data to upper layers. This layer collects the following type of data in tables: Call log (source, destination, duration of call), SMS (message size), Bluetooth proximity (Bluetooth id, MAC, class and name of detected device), location (cellid, location area code), battery level, the screen on/off intervals, Wi-fi (base station id, MAC), and app usage.

(b) Fitbit Data: This module is responsible for tracking all-day activity such as physical activities (type of activity, step count, distance in miles, duration, calories burned), heart rate, and quality of sleep and stores them in CSV format. Heart rate (HR) is measured based on expansion and contraction of capillaries based on a change in blood volumes. Purepulse LED lights detect the blood volume changes and apply algorithms to measure heart rates automatically. There are three heart rate zones: peak zone (HR > 85% of maximum), cardio zone (HR > 70% and HR < 84% of maximum), fat burn zone (HR > 50% and HR < 69% of maximum). Sleep patterns are divided into three zones depending on the sleep quality: asleep, restless state, and awake. Sleep efficiency is calculated as

- Sleep efficiency = 100 * Time spent in sleeping/(Time spent in sleeping + Time spent in restlessness + Time of waking up during sleep)
- Sleep inefficiency = (1 − sleep efficiency).

1.1.2 Data Preprocessing

This module is responsible for cleaning the data collected from smartphone and wireless physical activity tracker Fitbit. Then normalization and transformation techniques are applied and data is stored in a database. Further, a suitable machine learning approach is selected to detect stress for each user which is elaborated in the later section.

1.1.3 Stress Detection

This section involves extracting necessary information and patterns from raw data and then combines the outcomes and computes stress levels of individuals.

2 Stress Detection

A stress detection system is developed with a better understanding of stress process. Stress is not necessarily a negative process, but continuous stress has adverse effects on health such as rise in heart rate, insomnia, change in mood, and social behavior. When humans undergo forceful events such as an exam, severe training, project deadlines, then body undergoes large physical and psychological stress, which increases metabolic rates by initiating a response to the sympathetic nervous system [23]. After this kind of vital events, the sympathetic nervous system readjusts itself and decelerates to reduce the stress, and the parasympathetic nervous system starts the repair processes [24]. Normally, these two nervous systems should maintain equilibrium with each other. If the sympathetic–parasympathetic balance is not maintained, then human body experiences stress too often. And if the sympathetic response continues to soar, then chronic stress is triggered. Figure 2 depicts the suggested method for stress based on both physiological and behavioral response of individuals. The extracted features are analyzed using machine learning approach for stress detection.

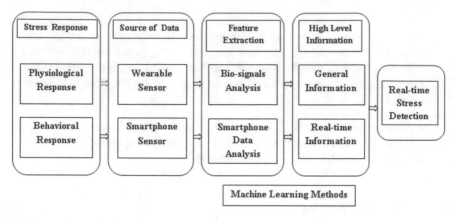

Fig. 2 Proposed approach for stress detection

Table 1 Parameters for measuring stress

Heart rate variability	Sleep pattern	Social interaction	Physical activity
Time interval between consecutive heartbeats	No of hours of sleep per day	Call, message, and presences	No of steps/day

In this work, smartphone data was gathered from 35 young adults for about 2 months. The data included their calls, SMS, Bluetooth presence, app usage, and Wi-fi. And a Web application is used to collect the perceived stress scale (PSS) data from each user. PSS is a popular stress assessment tool which helps us to understand how various parameters affect human stress levels. Several questions in PSS scale are about human feelings and thoughts during the last month. If individuals score high on the PSS, it indicates higher perceived stress. Based on scores, the baseline data is divided into three categories: low stress (PSS scores ranging from 0 to 13), medium stress (PSS scores ranging from 14 to 26), and high stress (PSS scores ranging from 27 to 40). But for experimental, task moderate and high stress are merged; so, there are two degrees of stress: low and high.

In this work, dissimilar dimensions of people's well-being are used to compute the human stress level, thus improving the quality of life. DetectStress considers four main dimensions of well-being: heart rate, sleeping patterns, social interaction, and their physical activity as shown in Table 1.

2.1 Heart Rate Variability (HRV)

The heartbeat frequency is normally computed as beats per minute (bpm) and R-R interval is calculated as the time between consecutive heartbeats in milliseconds. And heart rate variability (HRV) is beat to beat variation in the time between consecutive

Fig. 3 Shows average
statistic of last five sessions
HRV, stress and HR

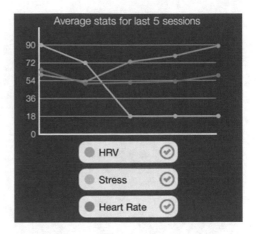

R-R intervals. When HR drops, HRV usually grows. And when HRV levels rise, an individual experiences less stress and vice versa. Figure 3 shows the association between HRV and stress levels of a person.

2.2 Sleeping Pattern

In this module, number of hours of sleep is taken into account and the number of hours for normal sleep is set to 8 with lower and upper threshold as 7 and 9 sleeping hours/day [25]. Figure 4 shows quality of sleep and total duration of sleep in minutes. Quality of sleep is indicated using no of times awake, minutes first awake, restless in minutes, and hours of sleep.

From Fig. 5, it is noticed that sleep inefficiency tends to go down toward the weekend, and increases again on Monday. Lack of sleep is stressful to the body, and it can cause high blood pressure and heart rate.

2.3 Social Interaction

Day-to-day social interactions of people have a severe impact on several dimensions of well-being [26]. Many feel stressed due to hectic work schedule, lack of social hold, and occasionally due to fear of social situations. People, who have many social networks, are likely to have sound mental and physical health. They have a tendency to resist stress and are frequently able to manage even chronic illness. In the digital era, many people are so used to receiving/making calls, messages, emails that the moment they do not receive one, their concern level increases. People constantly check their phones and when there is no new call or message, they feel stressed.

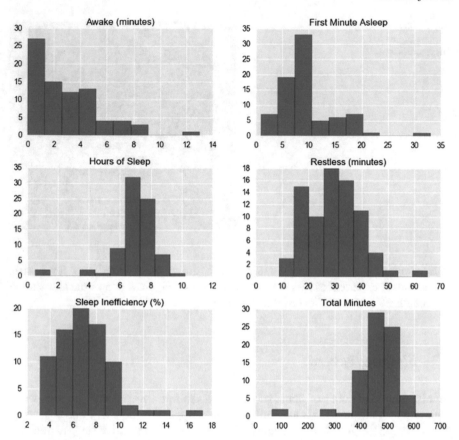

Fig. 4 Shows the distribution of each sleep type using histogram of bin size ten. It also shows sleep inefficiency and total minutes in bed

Repetitive checking of smartphone comes under compulsive behavior [27], and we have identified these features for detecting the stress level of a user. In DetectStress system, the number of calls, duration, no of SMS, the number of times the screen was turned on and off, and the app usage information.

Figure 6 shows a comparative study of calls, SMS, and Bluetooth presence on weekdays and at weekends. It shows that the number of calls, messages, and presence are more at weekends than on weekdays.

2.4 Physical Activity

Physical activity lowers overall stress levels and improves depression, self-esteem, and quality of life [28, 29]. In this work, Fitbit activity tracker monitors activities, such

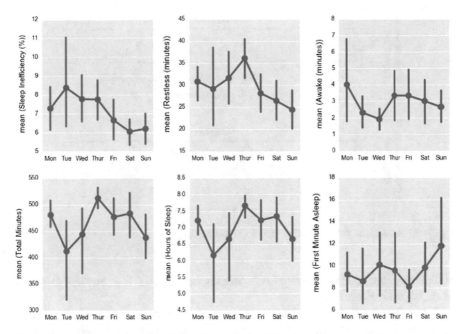

Fig. 5 Shows different measures of sleep by day of week, showing the estimated mean and confidence intervals

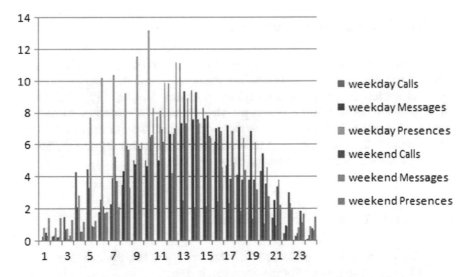

Fig. 6 Shows weekday versus weekend comparative study of calls, messages, and presences

as walking, jogging, warm up, cycling by implementing a step counter which keeps a count of steps taken by a person each day. According to American heart association,

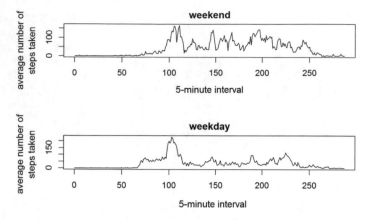

Fig. 7 Shows activity patterns during weekdays and weekends

10,000 steps are used as a metric for improving health and it is equivalent to 30 min of daily physical activity.

Figure 7 shows that the weekday's activity patterns seem to hit peaks during certain intervals while the activity patterns during weekends appear to be fairly consistent.

3 Experimental Study and Results

In this section, a pragmatic model is followed for evaluation of our data. The behavior of a person is observed in four dimensions (heart rate variability, sleeping pattern, social interaction, and physical activity) by collecting data for 5 months. All the participants are divided into two groups: high stress (PSS score $>= 14$) and low stress (PSS score $<= 13$). The data we analyzed in this study consist of call logs (13,035 voice calls), SMS logs (83,542 message records), proximity data (5,292,103 presence records Bluetooth scans made in every 5 min).

3.1 Social Interaction Feature Space

Basic smartphone features are categorized into two types: basic phone features and derived features. The lists of features are given in Table 2.

Table 2 List of smartphone features

Indegree of calls	n(Incalls)
Outdegree of calls	n(Outcalls)
Indegree of messages	n(SMS_rec)
Outdegree of messages	n(SMS_sent)
Effective size of outgoing calls	n(Outcalls)—redundancy
Effective size of incoming calls	n(Incalls)—redundancy
Effective size of incoming SMS	n(SMS_rec)—redundancy
Effective size of outgoing SMS	n(SMS_sent)—redundancy
Entropy of call contacts and SMS	Entropy $H (i - j) = -\sum_j f_j \log f_j$
No of Bluetooth IDs	Total(BID)
Times most common Bluetooth ID is seen	P(BID) [probability of BID]
Percentage of Bluetooth IDs seen	BID $\% = [n(BID)/Total(BID)] * 100$

Table 3 Sample physical activity and sleep data

Date	Steps	Calories	Distance	Active minutes	Floors climbed	Hours slept	Times awake
8/25/2016	10976	2681	5.3	30	13	7.22	4
8/26/2016	4412	2245	2	10	13	6.85	7
9/5/2016	2185	2614	1	5	4	9.03	14

Table 4 Sample heart rate data

Date	Calories out	Max	Min	Minutes	Name	Resting heart rate
8/26/2016	250	132	94	46	Fat burn	70
8/28/2016	153	220	160	17	Peak	68

3.2 Physical Activity and Sleep Data

This data is collected from wireless activity tracker as shown in Table 3.

3.3 Heart Rate Data

The sample heart rate data collected from wireless activity tracker is shown in Table 4.

Table 5 Comparison of models

	Naïve Bayes		Decision tree		Random	
	Low	High	Low	High	Low	High
Sensitivity	0.94	0.2	0.84	0.28	0.41	0.24
Specificity	0.3	0.95	0.38	0.90	0.58	0.77
Precision	0.51	0.55	0.52	0.47	0.43	0.22
Accuracy	0.72		0.62		0.55	
Pearson cor.	0.52		0.43		0.24	

4 Classification Approach

In this work, machine learning approach is used to detect human stress and two classification models are used: (1) Naïve Bayes and (2) Decision Trees. To find which of the features contribute more toward increasing the accuracy of the classifier, we have used forward feature selection method, which add the predictors one by one to build the model, the predictor improves the classifier's accuracy is at each step and the highest accuracy is retained. The accuracy is compared with a baseline random classifier which randomly predicts the class based on prior probabilities. The Pearson correlation was weak in the random classifier, whereas it was strong for other models. With respect to the random classifier, the Naïve Bayes classifier had an increase of 17% in accuracy and 7% for decision tree. Our result shows correlation between high stress and higher smartphone usage of various users. DetectStress shows 72% accuracy of perceived stress recognition using both smartphone and wireless physical activity tracker data. But these preliminary results can be further improved with an increased number of participants and data (Table 5).

5 Conclusion and Future Work

Identifying stress is a challenging task, and it makes a difference in every individual's life. The proposed model is a simple model built by taking into consideration four major dimensions of well-being. The heart rate variability (HRV), sleeping pattern, social behavior, and physical activity of each person are combined together to give an inference of human stress level. This comes under reality mining research as it detects stress based on multiple dimensions of human well-being. This work opens up the possibility to use a wireless activity tracker to implement in a stress detection system rather than tracking physical activities only. This model is still in improvisation and in future, more dimensions of well-being will be considered. Our study will further expand the data collection period and include even more number of users in future. Also, our focus will be on analysis of specific stress situations from the smartphone data which will provide more insights into the users' behavior.

References

1. Otenyo, Jane Kate. Sleeping habits and sleep deprivation among college students. The University of Arizona. 2015.
2. T G Pickering. Mental stress as a causal factor in the development of hypertension and cardiovascular disease. Current Hypertension Reports. 2001 June, 3(3), pp. 249–254.
3. S Cohen, D Janicki Deverts, W J Doyle, G E Miller, E Frank, B S Rabin and R B Turner. Chronic stress, glucocorticoid receptor resistance, inflammation, and disease risk. Proceedings of the National Academy of Sciences, USA. 2012 April, 109(16), pp. 5995–5999.
4. J Herbert. Fortnightly review. Stress, the brain, and mental illness. BMJ: British Medical Journal. 1997 August, 315(7107), pp. 530–535.
5. N D Lane, E Miluzzo, H Lu, D Peebles, T Choudhury, A T Campbell. A survey of mobile phone sensing. IEEE Communications Magazine, 2010, 48(9), pp. 140–150.
6. W Dong, B Lepri, A Pentland. Modeling the co-evolution of behaviors and social relationships using mobile phone data. Proceedings of the 10th International Conference on Mobile and Ubiquitous Multimedia. ACM. 2011, pp. 134–143.
7. R LiKamWa, Y Liu, N D Lane, L Zhong. Moodscope (2013): Building a mood sensor from smartphone usage patterns. In Proceeding of the 11th annual international conference on Mobile systems, applications, and services, MobiSys '13, NY, USA, ACM, 2013, pp. 389–402.
8. Isha Ghosh, Vivek K Singh. Predicting Privacy Attitudes using Phone Metadata. International Conference on Social Computing, Behavioral-Cultural Modeling & Prediction (SBP-2016), Washington DC. 2016 July, pp. 1–13.
9. A Madan, M Cebri_an, S T Moturu, K Farrahi, A Pentland. Sensing the health state of a community. IEEE Pervasive Computing. 2012, 11(4), pp. 36–45.
10. Padmaja B, Rama Prasad V V, Sunitha K V N. Treenet analysis of human stress behavior using socio-mobile data. Springer Journal of Big data. 2016 November, 3(24), pp. 1–15.
11. V K Singh, L Freeman, B Lepri, A S Pentland. Classifying Spending Behavior using Socio-Mobile Data. HUMAN. 2013, 2(2), pp 1–13.
12. V K Singh, Burcin Bozkaya, Alex Pentland. Money Walks: Implicit Mobility Behavior and Financial Well-Being. PLoS Journal. 2015 August, 10(8).
13. Sai T Moturu, Inas Khayal, Nadav Aharony, Wei Pan, Alex (Sandy) Pentland. Using Social Sensing to Understand the Links between Sleep, Mood, and Sociability. IEEE International Conference on Social Computing, Cambridge, MA, 2011, pp. 208–214.
14. Muaremi A, Arnrich B, Trster G. Towards measuring stress with smart phones and wearable devices during workday and sleep. BioNanoSci. 2013 June, 3, pp. 172–183.
15. C Schubert, M Lambertz, RA Nelesen, W Bardwell, J B Choi, JE Dimsdale. Effects of stress on heart rate complexity - a comparison between short-term and chronic stress. Journal of Biological psychology, 2009 March, 80(3), pp. 325–332.
16. S Cohen, K R C, L U, Gordon. Measuring stress: A guide for health and social scientists. Oxford University Press, USA, 1997.
17. K Plarre, A Raij, S Hossain, A Ali, M Nakajima, M Al'absi, E Ertin, T Kamarck, S Kumar, M Scott, D Siewiorek, A Smailagic, L Wittmers. Continuous inference of psychological stress from sensory measurements collected in the natural environment. 10th international conference on information processing in sensor networks (IPSN). 2011 April, pp 97–108.
18. How to measure Stress in Humans. http://www.stresshumain.ca/documents/pdf/Mesures%20physiologiques/ CESH_howMesureStress-MB.pdf. Date accessed: 5/1/2017.
19. Cohen S, KamarckT, Mermelstein R. A global measure of perceived stress. Journal of Health and Social Behavior. 1983 December, 24(4), pp. 1027–1035.
20. Han K S, Kim L, Shim I. Stress and sleep disorder. Experimental neurobiology. 2012 December, 21(4), pp. 141–150.
21. Dinges D, Pack F, Williams K, Gillen KA, Powell JW, Ott GE, Aptowicz C, Pack AI. Cumulative sleepiness, mood disturbance, and psychomotor vigilance decrements during a week of sleep restricted to 4 – 5 hours per night. Sleep, 1997 April, 20(4), pp. 267–277.

22. Keith M Diaz, David J Krupka, Melinda J Chang, James Peacock, Yao Ma, Jeff Goldsmith, Joseph E Schwartz, Karina W Davidson. Fitbit®: An accurate and reliable device for wireless physical activity tracking. International journal of cardiology. 2015 April, 185, pp. 138–140.
23. A Angeli, M Minetto, A Dovio, P Paccoti. The overtraining syndrome in athletes: A stress-related disorder. Journal of Endocrinological Investigation. 2004, pp. 603–612.
24. Heart Rate Variability Research Review. http://www.8weeksout.com/2011/12/05/heart-rate-variability-research-review/. Date accesses: 2/2/2017.
25. G Alvarez, N Ayas. The impact of daily sleep duration on health: A review of the literature. Cardiovascular Nursing, 2004 March, pp. 56–59.
26. Nicholas D Lane, Mashfiqui Mohammod, Mu Lin, Xiaochao Yang, Hong Lu, Shahid Ali, Afsaneh Doryab, Ethan Berke, Tanzeem Choudhury, Andrew T Campbell. Bewell: A smartphone application to monitor, model and promote wellbeing. In 5th ICST/IEEE Conference on Pervasive Computing Technologies for Healthcare IEEE Press, 2011 May, pp. 23–26.
27. A Oulasvirta, T Rattenbury, L Ma, E Raita. Habits make smartphone use more pervasive. Personal and Ubiquitous Computing. 2012 January, 16(1), pp. 105–114.
28. Fox K R. The influence of physical activity on mental well-being. Public Health Nutritionl. 1999 September, 2(3A), pp. 411–418.
29. Paffenbarger RS Jr, Hyde RT, Wing AL, Hsieh CC. Physical activity, all-cause mortality, and longevity of college alumni. In New England journal of medicine, 1986 March, 314(10), pp. 605–613.

Impact of Term Weight Measures for Author Identification

M. Sreenivas, T. Raghunadha Reddy and B. Vishnu Vardhan

Abstract The rapidly growing data in the Web result in stolen, unidentified, and fraudulent data. Identification of such data is of a prime objective for forensic departments, researchers, and governments. In this context, authorship analysis is very useful to reveal the truth by analyzing the text. Authorship analysis is observing the properties of a text to predict authorship of a document. Stylometry is the root for authorship analysis, which is a linguistic research field that exploits the machine learning techniques as well as knowledge of statistics. Authorship Attribution is a type of authorship analysis technique, which is aimed at recognizing the author of an anonymous text within a closed set of authors or subjects. Most of the researchers in Authorship Attribution approaches proposed various set of stylistic features to differentiate the authors based on style of writing. It was observed from the literature the accuracy of author prediction was not satisfactory with stylistic features. In this paper, the experimentation carried out with various term weight measures identified in various text processing domains to predict the author of a new document. The results show that the term weight measures obtained good accuracies for author prediction when compared with most of the existing approaches.

Keywords Term weight measure · Author identification · Authorship attribution
BOW model

M. Sreenivas
Research Scholar of RUK, Sreenidhi Institute of Science and Technology, Hyderabad, India
e-mail: msreenivas@sreenidhi.edu.in

T. Raghunadha Reddy (✉)
Dept of IT, Vardhaman College of Engineering, Hyderabad, India
e-mail: raghu.sas@gmail.com

B. Vishnu Vardhan
Dept of CSE, JNTUH Jagtiyal, Karimnagar, India
e-mail: mailvishnu@jntuh.ac.in

© Springer Nature Singapore Pte Ltd. 2019
R. S. Bapi et al. (eds.), *First International Conference on Artificial Intelligence and Cognitive Computing* , Advances in Intelligent Systems and Computing 815,
https://doi.org/10.1007/978-981-13-1580-0_8

1   Introduction

The Internet is growing rapidly with textual information along with the cybercrimes also increased in the Internet. The people are sending harassing messages in social media, and the terrorist organizations send threatening mails without specifying their correct details. The researchers are attracted to know the author of these texts by analyzing the writing styles. Authorship Attribution is predicting the author of a text by processing the texts of several authors [1]. Various applications like security, research literature, and forensic analysis used Authorship Attribution techniques.

In general, researchers in Authorship Attribution proposed various types of features to discriminate the authors writing styles. The general approach followed in most of the existing approaches for Authorship Attribution was Bag of Words (BOW) approach for document representation. It was identified that the stylistic features were not sufficient to improve the accuracy of author prediction. In this work, different weight measures were analyzed from various domains to assign certain weights to the terms. In this paper, the experimentation carried out with reviews domain and obtained good accuracy for author prediction when term weight measures were used.

This paper is structured into five sections. The related work in Authorship Attribution is explained in Sect. 2. The dataset characteristics and evaluation measure used to evaluate the classifiers are explained in Sect. 3. The analysis of various term weight measures is explained in Sect. 4. The experimental results of term weight measures for author prediction using different classifiers are presented in Sect. 5. The conclusions and future scope are explained in Sect. 6.

2   Literature Survey

The style of writing is a primary indicator of an individual identity in predicting the author of a document in Authorship Attribution. In general, three steps are followed in Authorship Attribution approaches. First, the most discriminative features were identified to differentiate the authors' writing styles. Second, the document representation models were identified to represent the document with these features. Finally, the suitable machine learning classification algorithms were detected to generate the classification model for predicting the most probable author of an anonymous document [2].

Tanguy et al. extracted [3] rich set of language-specific features like contracted forms, character trigrams, POS trigrams, lexical generosity and ambiguity, phrasal verbs, syntactic complexity, syntactic dependencies, lexical cohesion, lexical absolute frequency, morphological complexity, quotations, punctuation, first-/third-person proper and narrative. They used maximum entropy technique for author identification. They noted that the performance of set of rich linguistic features was better for author prediction when compared with word frequencies and trigrams of characters.

In [4], the researchers extracted POS bigrams and trigrams, character trigrams, percentage of direct speech from the documents and syntactic complexity and structure from the documents to represent the vectors of documents. They obtained overall accuracy of 77% in authorship identification by using Sequential Minimal Optimization (SMO) classifier. It was found that the author prediction accuracy was improved when the application-specific features were added to existing feature set.

Akiva used [5] SVM light machine learning method to produce classification model. They used binary Bag of Words representation to represent the document vector, which captures the absence or presence of common words in a document. It was identified that the author prediction accuracy was improved when the number of texts was increased in the training data.

Vilarino experimented [6] with three supervised learning methods such as Naïve Bayes, Rocchio, and Greedy. They considered the original words in a document as features. It was observed that the Rocchio method performs well compared to Naïve Bayes and greedy methods. Hugo Jair Escalante performed [7] experimentation with a document occurrence representation for author prediction. They observed that their representation outperforms when compared with Bag of Words approach and also observed that this document representation works well for small datasets. Mikros et al. extracted [8] character bigrams, character trigrams, word unigrams, word bigrams, and word trigrams features from e-mail corpus. They used logistic regression and one-class machine learning methods for author prediction.

3 Existing Approach in Authorship Attribution

3.1 BOW Model

The design of BOW model is depicted in Fig. 1. In this model, preprocessing techniques such as stopword removal and stemming were applied on the dataset to remove the terms which are weak in text discrimination. The features were extracted from the updated dataset. In this work, most frequent terms were considered as features. Treat these most frequent terms as Bag of Words. The documents were represented with this Bag of Words. The term frequency was considered to represent the weight of the terms in document vectors. In this work, different term weight measures were identified to assign weight to the terms. The machine learning classifiers generates classification model.

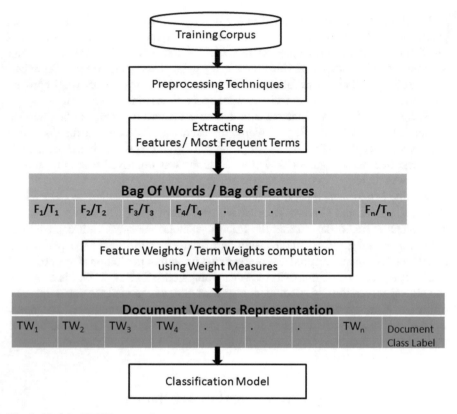

Fig. 1 Model of BOW approach

3.2 Dataset Characteristics

The dataset was collected from amazon.com, and it contains ten different authors' reviews on different products. The corpus is balanced in terms of number of documents in each author group, and each author group contains 400 reviews of each.

3.3 Evaluation Measures

Various measures are used such as precision, recall, F1 measure, and accuracy by the researchers in Authorship Attribution to test the accuracy of author prediction. In this work, accuracy measure is used to evaluate the efficiency of the author prediction. Accuracy measure was represented as

$$\text{Accuracy} = \frac{\text{Number of documents predicted their author correctly}}{\text{Total number of documents}}$$

4 Term Weight Measures

Traditional term weighting measures are term frequency (TF), binary, and term frequency–inverse document frequency (TF-IDF). Binary weight measure assigns 1 or 0 to the term based on the term presence or absence in a document. TF measure computes the frequency of a term in a document. TF may assign large weights to the common terms (a, an, the, of, etc.,) which are weak in text discrimination. To overcome this shortcoming, TF-IDF measure is proposed by the researchers to reduce the weight of common terms. In TF-IDF measure, the IDF allocates more weight to the terms that appeared in less number of documents. Although the TF-IDF was proved in information retrieval domain and several text mining tasks for quantifying the term weights, it is not most effective for author prediction because TF-IDF disregard the class label information of the training documents. Therefore, researchers are looking for alternative effective term weight measures in Authorship Attribution.

Based on the utilization of the class label information, the term weight measures were categorized into two types such as unsupervised and supervised term weight measures. An unsupervised term weight measure does not use information regarding class label. The supervised term weight measure use class label information.

4.1 Unsupervised Term Weight Measures

4.1.1 Nonuniform Distributed Term Weight (NDTW) Measure

The NDTW measure was proposed in information retrieval domain to assign suitable weight to the terms based on term distributions [9]. This measure assigns more weight to the terms which are distributed uniformly across the documents. Equation (1) shows the NDTW measure.

$$Wt_{ij} = w(t_i, p_j) = \log(\text{TOTF}_{ti}) - \sum_{k=1}^{m} \left(\frac{tf(t_i, d_k)}{\text{TOTF}_{ti}} \log\left[\frac{1 + tf(t_i, d_k)}{1 + \text{TOTF}_{ti}} \right] \right) \quad (1)$$

where $tf(t_i, d_k)$ is the count of term t_i in document d_k, TOTF_{ti} is the number of times the term t_i occurred in profile p_j.

4.1.2 Normalized Document Length Term Weight (NDLTW) Measure

An NDLTW measure was proposed in [10] to avoid the differentiation of small-sized and large-sized documents. This measure used a pivoted unique term normalization factor to balance the weight of the terms irrespective of the size of the document. Equation (2) represents the NDLTW measure.

$$W_{t_{ij}} = W(t_i, p_j) = \frac{(1 + \log(\mathrm{TF}_i))\big/(1 + \log(\mathrm{AVGTF}_i))}{\sum_{k=1}^{m}(1 - \mathrm{slope}) * \mathrm{AVGUT}_k + \mathrm{slope} * \mathrm{UT}_k} \tag{2}$$

where $W(t_i, p_j)$ is the term weight of t_i in p_j profile. UT_k is a frequency of unique terms in document d_k, TF_i (term frequency) is the count of term t_i in document d_k, slope value is 0.2, AVGUT_k is a ratio of frequency of unique terms in a document d_k to total number of terms in document d_k, AVGTF_i is a ratio of the frequency of term t_i to the total number of terms in document d_k.

4.2 Supervised Term Weight Measures

4.2.1 Relevance Frequency-based Term Weight (RFTW) Measure

RFTW measure was proposed in text classification domain to assign discriminative power to the terms [11]. This weight measure allots more weight to the terms which are having high frequency in positive category documents than negative category documents. The RFTW measure is represented in Eq. (3).

$$tf * rf = tf * \log\left(2 + \frac{a}{\max(1, c)}\right) \tag{3}$$

where a is the frequency of positive category of documents which contain the term t_i, c is the count of negative category documents which contain the term t_i.

4.2.2 Discriminative Feature Selection Term Weight (DFSTW) Measure

DFSTW measure allocates more weight to the terms that are having high average term frequency in class c_j and the terms with high occurrence rate in most of the documents of c_j [12]. The DFSTW measure is represented in Eq. (4).

$$W(t_i, c_j) = \frac{tf(t_i, c_j)/df(t_i, c_j)}{tf(t_i, \overline{c_j})/df(t_i, \overline{c_j})} \times \frac{a_{ij}}{(a_{ij} + b_{ij})}$$
$$\times \frac{a_{ij}}{(a_{ij} + c_{ij})} \times \left| \frac{a_{ij}}{(a_{ij} + b_{ij})} - \frac{c_{ij}}{(c_{ij} + d_{ij})} \right| \tag{4}$$

4.2.3 TF-Prob Measure

TF-Prob is a probability-based weight measure defined in [13], which combines A/B and A/C. The TF-Prob measure finds the weight of term t_k with respect to c_j as shown in Eq. (5).

$$w(t_k, c_j) = tf_k * \log\left(1 + \frac{A}{B}\frac{A}{C}\right) \tag{5}$$

4.2.4 ICF-based Term Weighting Schemes TF-IDF-ICSDF

Inverse class frequency (ICF) is similar to IDF in TF-IDF, which is defined as the ratio of the total classes to the number of classes which contains the term. TF-IDF-ICSDF measure was proposed in [14]. The TF-IDF-ICSDF weight is computed by Eq. (6).

$$w(t_k) = tf_k * \left(1 + \log\frac{N}{df_k}\right) * \left(1 + \log\frac{m}{\sum_{j=1}^{m}\frac{df_{kj}}{N_j}}\right) \tag{6}$$

where N_j and df_{kj} are the total numbers of documents in class c_j and the number of documents which contain the term t_k, respectively.

4.2.5 SUTW Measure

Supervised unique term weight (SUTW) measure [15] as in Eq. (7) combines inner-document distribution, inter-class distribution, and intra-class distribution information of terms to measure the weight of a term.

$$W_{t_{ij}} = W(t_i, p_j) = \sum_{k=1,d_k \in p_j}^{m} \left(\frac{tf(t_i, d_k)}{tf(t_i, p_j)}\left[\frac{\log(d_{tk})}{0.8 * \mathrm{AVGUT}_k + 0.2 * \mathrm{UT}_k}\right]\right)$$
$$\times \frac{a_{ij}}{(a_{ij} + b_{ij})} \times \frac{c_{ij}}{(c_{ij} + d_{ij})} \tag{7}$$

where UT_k is the number of unique terms in a document and AVGUT_k is the ratio of number of unique terms in a document to total number of terms in a document.

5 Empirical Evaluations

5.1 Results of Term Weight Measures

In this work, 8000 most frequent terms were extracted from the corpus to represent the document vectors. The BOW model is used to represent the document vectors with these 8000 terms. Various term weight measures were used to assign the weight to the terms in document vectors. Table 1 represents the accuracies of author prediction when different term weight measures were used to define the weight of the

Table 1 Accuracies of author prediction for various term weight measures

Classifier/term weight measures	Naïve Bayes multinomial	Random forest
TF	60.09	64.21
TF-IDF	66.29	69.45
NDTW	71.48	73.01
NDLTW	74.31	77.13
RFTW	87.39	88.92
TF-Prob	88.16	87.83
DFSTW	89.67	90.02
TF-IDF-ICSDF	90.71	91.11
SUTW	92.23	93.82

term. In Table 1, the SUTW measure obtained highest accuracy for author identification when compared with all other term weight measures. It was identified that the supervised term weight measures achieved best accuracies for author prediction when contrasted with accuracies of unsupervised term weight measures. It was also noted that the random forest classifier achieved good accuracies for most of the term weight measures when compared with Naïve Bayes multinomial classifier.

6 Conclusions and Future Scope

In this paper, various term weight measures were used to assign weight to the terms in document representation. The supervised term weight measures' performance is good when compared with unsupervised term weight measures. The SUTW measure obtained good accuracy of 95.82% for author prediction when compared with other term weight measures. In our future work, it was planned to propose a term weight measure to improve the accuracy of author prediction and also considered that proposal of a new document representation model to increase the prediction accuracy of authors.

References

1. Efstathios Stamatatos. "A survey of modern authorship attribution methods", Journal of the American Society for Information Science and Technology, 03/2009.
2. Juola, P.: Authorship attribution. Found. Trends Inf. Retr. 1 (2006) 233–334.
3. Ludovic Tanguy, Franck Sajous, Basilio Calderone, and Nabil Hathout. Authorship attribution: using rich linguistic features when training data is scarce, CLEF 2012 Evaluation Labs and Workshop, 17–20 September, Rome, Italy, September 2012.

4. Stefan Ruseti and Traian Rebedea. Authorship Identification Using a Reduced Set of Linguistic Features—Notebook for PAN at CLEF 2012. CLEF 2012 Evaluation Labs and Workshop, 17–20 September, Rome, Italy, September 2012.
 5. Navot Akiva. Authorship and Plagiarism Detection Using Binary BOW Features, CLEF 2012 Evaluation Labs and Workshop, 17–20 September, Rome, Italy, September 2012. ISBN 978-88-904810-3-1. ISSN 2038-4963.
 6. Darnes Vilariño, Esteban Castillo, David Pinto, Saul León, and Mireya Tovar. Baseline Approaches for the Authorship Identification Task, Notebook for PAN at CLEF 2011. CLEF 2011 Evaluation Labs and Workshop.
 7. Hugo Jair Escalante. EPSMS and the Document Occurrence Representation for Authorship Identification, Notebook for PAN at CLEF 2011. CLEF 2011 Evaluation Labs and Workshop.
 8. George K. Mikros and Kostas Perifanos. Authorship identification in large email collections: Experiments using features that belong to different linguistic levels, Notebook for PAN at CLEF 2011. CLEF 2011 Evaluation Labs and Workshop.
 9. Dennis, S.F., "The Design and Testing of a Fully Automated Indexing-Searching System for Documents Consisting of Expository Text", in Informational Retrieval: A Critical review, g. Schecter, editor, Thompson Book Company, Washington D.C., 1967, pages 67–94.
10. Singhal, A., Buckley, C. and Mitra, M., "Pivoted document length normalization", in Proceedings of the 19th annual international ACM SIGIR conference on Research and development in information retrieval, ACM., (1996), 21–29.
11. Lan, M., Tan, C. L., Su, J., & Lu, Y. (2009). Supervised and traditional term weighting methods for automatic text categorization. IEEE Transactions on Pattern Analysis and Machine Intelligence, 31 (4), 721–735. http://doi.org/10.1109/TPAMI.2008.110.
12. Wei Zong, Feng Wu, Lap-Keung Chu, Domenic Sculli, "A discriminative and semantic feature selection method for text categorization", International Journal of production Economics, Elsevier, Jan 2015, pp. 215–222.
13. Liu, Y., Loh, H. T., & Sun, A. (2009). Imbalanced text classification: A term weighting approach. Expert Systems with Applications, 36 (1), 690–701. http://doi.org/10.1016/j.esw a.2007.10.042.
14. Ren, F., & Sohrab, M. G. (2013). Class-indexing-based term weighting for automatic text classification. Information Sciences, 236, 109–125. http://doi.org/10.1016/j.ins.2013.02.029.
15. Raghunadha Reddy T, Vishnu Vardhan B, Vijayapal Reddy P, "Profile specific Document Weighted approach using a New Term Weighting Measure for Author Profiling ", International Journal of Intelligent Engineering and Systems, 9 (4), pp. 136–146, Nov 2016.

Automating WEB Interface in Relation to User Behaviour

Sasi Bhanu Jammalamadaka, B. K. Kamesh Duvvuri,
K. R. Sastry Jammalamadaka and J. Himabindu Priyanka

Abstract Most of the businesses in the world are turning to be global through making available their business to the customers, suppliers, and other state holders through specially designed websites that implements both e-commerce and m-commerce. The content being hosted on to the web site sometimes is quiet complex as it involves text, audio, video, and 2D/3D. The e-commerce/m-commerce-related websites are designed considering different complex user interfaces. The WEB sites are designed using several URL redirections and also involve deep navigation into the WEB site. The users get frustrated when they need to spend quite a bit of times in navigating through the websites in search of the content that they are looking for. There is a need for developing an interface that can provide right into home page the most important links that are most frequently visited by the users. The links are to be fetched based on the user behaviour. Frequently visited WEB links which are selected based on the user profile who entered into WEB site when displayed right into the home page makes the web surfing for information most effective. The utilization of the WEB sites will increase quite drastically when information needed is made available with ease. In this paper, two methods have been presented that allows the user to navigate through the WEB site to the desired content pages with ease and using the elemental level navigable links that show the content that is most frequently needed by the users.

Keywords User behaviour · WEB navigation · Clicked counts · Snippets
Elementary level links

S. B. Jammalamadaka (✉) · B. K. K. Duvvuri · J. H. Priyanka
Department of CSE, SMEC, Dhulapally, Kompally, Secunderabad 500014, India
e-mail: sasibhanukamesh1@gmail.com

B. K. K. Duvvuri
e-mail: kameshdbk@gmail.com

K. R. S. Jammalamadaka
Department of ECSE, KLEF University, Veddeswaram, Guntur 522502, India
e-mail: drsastry@kluniversity.in

© Springer Nature Singapore Pte Ltd. 2019   91
R. S. Bapi et al. (eds.), *First International Conference on Artificial Intelligence and Cognitive Computing* , Advances in Intelligent Systems and Computing 815,
https://doi.org/10.1007/978-981-13-1580-0_9

1   Introduction

Users navigate from one URL to the other in order to get through the information required which is contained in various resource files situated in the network of networks generally termed as World Wide Web. The information as such is stored as hypertext. The user interface as such is programmed in terms of hypertext. The browser resident on the client processes the hypertext and outputs the user interface through which the users interacts and browses through the content stored on the WEB. A designer of a web site designs the navigation using which the users will be able to navigate across the WEB. The usability of the WEB site as such is dependent on the designed navigational scheme. The navigational scheme as such may include local, contextual, supplemental or global access to the resource file situated across the WEB. The primary navigation is hierarchical, the ability to move from a higher level URL to a lower level URL. The user can navigate within the same site using hierarchical navigation which of course is a serious restriction. Since there is a serious limitation using the hierarchal navigation, additional navigational schemes are required.

Local navigation is achieved through navigating through a specific section of WEB site. Global navigation on the other hand is undertaken as an outline navigation to move from one section of a web site to another section or from one web site to the other. The navigational schemes provide for framework using which the user can navigate effectively and easily across the WEB site. There is still heavy scope to improve the WEB navigation providing excellent acceptance of the users. Many issues that include standard icons, convection, irrelevant links, reveal structure, buried information, slow loading of pages, clustering of data within a single web page, poor readability makes the accessing of content through web sites complicated.

Every user wants to get into the location of the web site where the desired content is located. The design of the WEB sites is done in such a way that the user knows the current location within the WEB site through provision of many navigation facilities which include navigation bar or tools which provide clear indication about the content indication the knowledge and depth of the same.

Looking at the user interface displayed by the browsers, the user will clearly know the content using which they can navigate. They also know the kind of output that they can get. The users will be able to explore more through navigational bars and links displayed as a part of user interface. Navigational tools can be included into the user interface. The entire web site can be broken into logical groups using the navigational tools. The user can easily navigate to the logical group in which they are interested. The users are more bothered about the way the navigate across the web site in search of the content. The users are not bothered about the way a web site is designed. The users are satisfied if they can get the information that they need with least navigation to be carried. The users will form an idea about the web site through carrying navigation of the web site. The users will be surfing the web site for longer period of time if the navigation is simple and understandable. They keep returning to the same site again and again. Proper navigation scheme will lead to more usability of the WEB site. Good navigation designed into a web site will help search engines to locate the URL containing the snippet words inputted by the users for searching the needed content. A search engine keeps the web sites with excellent navigation very high on the search order. A web site is accessible, and usable only when proper navigation scheme is implemented. It becomes easy for the search engines to automate the accessibility of the content based on the navigation carried by the users. Users will navigate the web sites and search for the content that they are looking for only when effective navigation is implemented.

Users have to spend lots of time to navigate to an area where the desired content is placed. Users negating a web site are not uniquely identified. Users are identified with a user ID which is valid for a browsing session. The ID is no more valid once the session expires. When the same user wants to navigate yet another time a new user ID is assigned. No user can thus be identified uniquely using IDs assigned by the WEB server. But to make the web interface available to the user as per the navigation requirements of the end user, user behaviour must be captured and the behaviour must be selected based on the initial clicks made by the user. The frequent URLs navigated by the user are associated with the user behaviour, and the same are to be made available as menu items using which the users will be able to navigate to the locations where the content desired is located.

2   Related Work

By seeing the user's browsing, we can trace out the interests of the user. Each of these browsing is weak by itself, but each can contribute to a judgment about the document's interest. One of the strongest behaviours is for the user to save a reference to a document, explicitly indicating interest. Following a link can indicate one of several things. First, the decision to follow a link can indicate interest in the topic of the link. However, because the user does not know what is referenced by the link at the time the decision to follow it has been made, that indication of interest is tentative, at best. If the user returns immediately without having either saved the target document, or followed further links, an indication of disinterest can be assumed. Hence, the user considerable time that would be wasted exploring those "dead-end" links.

Lieberman [1] has introduced a model called Letizia that serves as a user interface that serves the purpose of WEB browsing. The agent tracks the user behaviour and anticipates the user interests through concurrent, autonomous exploration of links from the user current position. The agent finds the browsing strategy based on best-first search which is backed by some heuristics. The agent derives the user interest through inferences derived out of browsing behaviour of the user.

Search engines are generally used by the users for getting useful information from World Wide Web (WWW). A search engine produces different results every time and the same queries in terms of snippets are submitted by the users regardless of the user who submitted the search. The information requirements of each user are different. The search results produced by the search engines must be adaptable to a user as per their information needs. Sugiyama et al. [2] have proposed several approaches for adapting the search results according to the user needs without the requirement of any of the user's interaction. They have verified the effectiveness of the approaches by experimentation. They have proved that search engines that adapt to the user preferences that can be modelled into user behaviour based on modified collaborative filtering with detailed analysis of users browsing history.

Personalization of user's web search can be undertaken through users' previous interactions in search of desired content. It is rather unrealistic that the users will be able to specify the keywords or snippets that well precisely fetch the content of their choice. Teevan et al. [3] have presented techniques using which implicit information related to the users is acquired. The implicit information acquired is used to re-rank web search results within a relevant feedback framework. They have explored models that have been built from previous searches undertaken by the user, previously visited web pages, the documents and emails that either created or read. Rich representation of the user and the corpus are quite important for personalization. The representations can be approximated based on which client-side algorithms can be developed and coded through which personalization of the when search can be carried. They have shown that personalized client-side algorithms have significantly improved the web searching mechanism.

Interface designers design the web interfaces through which users can initiate various web-related content searching. The user interface design can develop in such a way those users requirements are straight away met by knowing the search behaviour of the user. A tool to simulate user behaviour would be quite helpful for properly designing the user interface. The tool will help developing semantics which will help in predicting the search behaviour even in the absence of text matches in the content scanned. Kaur and Hornof [4] have considered three semantic models which include LSA, WordNet, PMI-IR and made a comparison to determine their suitability to predict a link that the users will select given some information related snippets and a WEB page. They have presented that PMI-IR semantic model could predict the human performance quite effectively compared to other semantic models.

Web users surf the WEB sites quite frequently for accessing frequently required information. There is no specific information stored in the WEB logs with which a user is uniquely identified. Every user navigates the WEB site leading to certain patterns of surfing. The web users being novice can only be recognized through patterns of surfing the web. Many algorithms have been presented in the literature that considers clustering the users based on some criteria generally leading to finding favourite links. The algorithms do not use any navigational patterns, and the favourite links provided in the sidebars do not provide any useful information. User behaviour cannot be predicated using just the favourite links. Most of the algorithms take more time to form clusters and present them to the user the favourite links in a sidebar. Kamesh et al. [5] have presented a method that quickly recognizes frequent navigational patterns of the novice user derived from a WEB log and clicks rendered by the user for navigating through URL links. The user's interestingness is captured and used to find the interesting patterns. The customized URLs are presented as toolbar icons using which the user can directly surf the interesting areas. The method has been built into large web site, and it has been proved that the surfing has been made efficient when compared with other methods presented in the literature.

WEB sites that make available the URLs which are frequently and usually accessed by the users for immediate access without the need for frequent navigation backward and forward is most importantly required. The usage of such type of web sites will be phenomenally very high. WEB logs provide a good source to determine the pattern of navigation that different users adapt for surfing. Many techniques exist in the literature that uses the concept of clustering through usage of one of the available clustering algorithms. All of them suffer from huge amount of time that it requires to generate the clusters and also none have recommended a method of customizing the user interface that renders the frequently required URL links right in the home page of large WEB sites. Sastry et al. [6] have presented a method that helps quickly recognizing the frequently navigated URLs with the help of the first elementary level of click rendered by the user. The method is simple, fast and stands the proof of concept. The method has been applied on to large web site, and it has been proved that it is efficient when compared with other methods presented in the literature.

Search engines are frequently used for obtaining some information hosted on different WEB sites. The correctness of the information searched is very much dependent on the keying in of the most appropriate WEB snippets and the order in which the WEB snippets are presented. Search engines literally visit the metadata of all the WEB sites before the content that matches the keyed in WEB snippets is presented to the end user through a set of URLs. Some of the business specific WEB sites are huge in size and therefore needs a search mechanism to be built as a part of the WEB site. A different mechanism has to be built for searching and integrate the search engine within the application specific WEB site. If search engines are to be used for searching an application specific web site, huge overheads are caused due to generation of unrelated WEB content and it also leads consumption of huge time due to the reason that many unrelated WEB sites are to be visited. Many methods exist that surf based on the concepts of navigation patterns, user behaviour, etc., but have not quite exploited the searching on the categories of concepts that are hidden with the content that is hosted on the WEB site. Kamesh et al. [7] have presented a method that implements a search engine within an application specific web site and that exploits the category of concepts that are hidden within the content hosted on the WEB sites. The concept-based searching leads to customized searching as well. The method produces the search results which are quite relevant and specific to a kind of customer who is surfing the WEB site.

Application-based WEB sites are becoming bulky day by day. Provision of search mechanism within each and every WEB site has become a necessity. Most of the search mechanisms that are being used in these days are WEB-snippet based, which are generated based on the query initiated by a user. Search engines are being used for implementing searching within WEB sites which causes very high overheads due to generation of many unrelated WEB content and visiting of many web sites to collect content is time-consuming when searching has to be limited to a single WEB site. The searching must also be provided considering that many novice users will be looking for the content existing in the application specific WEB site. Sastry et al. [8] have presented a method that help provision of searching within an individual application-specific WEB sites considering that the users looking for searching are novice. The user behaviour has been considered based on which customized search results which are more relevant and directly required are produced, thus making the process of searching within an application specific WEB site much faster.

Most users want their search engine to incorporate three key features in query results. Relevant results (results they are actually interested in), uncluttered (easy to read interface), helpful options to broaden or tighten a search for accuracy. Sastry et al. [9] have addressed another aspect with new improvement measures for an enhanced experience to the end user. A trivial query such as travel arrangement has to be broken down into a number of codependent steps over a period of time based on prior search patterns of the same user, thus providing customized holistic view. For instance, a user may first search on possible destinations, timeline, events, etc. After deciding when and where to go, the user may then search for the most suitable arrangements for air tickets, rental cars, lodging, meals, etc. Each step results in one or more further queries, and each query results in one or more clicks on relevant pages. Current search engines cannot support this kind of hierarchical queries. They have presented a random walk propagation method that can construct user profiles based on the credentials obtained from their prior search history repositories. Combined with click points driven click graphs of user search behaviour the IR system can support complex queries for future requests at reduced navigations. Random walk propagation over the query fusion graph methods supports complex search quests in IR systems at reduced times. For developing an interactive IR system, they have proposed to use the search quests as auto-complete features in similar query propagations. Biasing the ranking of search results can also be provided using ranking algorithms (top-k algorithms). Supporting these methods yields dynamic and improved performance in IR systems, by providing enriched user querying experience. A practical implementation of the proposed system validated their claim.

Many browsers are available for WEB surfing and mining. The users have the option to use a browser of their choice for either surfing or mining. Every browser has a kind of popularity especially considering different aspects of browsing and mining. Shukla and Singhai [10] have presented a method that considers a using a browser for surfing to find the content of their choice. The user switches to a different browser if the browser selected at the earlier instance did not provide the content desired by the user. The behaviour of the user in selecting a browser is modelled through a Markov chain from which the transmission probabilities are calculated. The user quitting a browser has been considered as a parameter of variation in the popularity of the browser. A graphical study is formed to find the relationship between the user behaviour and the market popularity of the web browser. This presentation could only find the browser to be used for searching and surfing.

Users have preference for searching WEB for want of specific content of their choice. The knowledge of these preferences is essential for development of search engines. Agichtein et al. [11] have presented a real-world study related to modelling the behaviour of a user who is involved in web searching. The model is used to predict the web search preferences of the user. Several applications, which include ranking, web search personalization, click spam detection, are dependent on predicting the user behaviour. Agichtein et al. [11] have presented a model that considers query dependent deviations from the noisy user behaviour. They have presented that the behaviour of the user is modelled beyond clickthrough interpretation. The model that depends on clickthrough interpretation is found to be more robust with more prediction accuracy when compared to standard clickthrough methods for predicting the user behaviour.

Personalized search results can be achieved through user profiles and based on the description of the user interests. Proxy servers and desktop bots have been used to capture browsing histories, form which user behaviour is derived. These methods need the participation of the user installing to wither the proxy servers or bots. Speretta et al. [12] have explored less invasive method to capture user-related information. They have considered building the user profiles based on the activity carried on web server itself. The information captured is used to provide personalized user-based search results. They have developed a wrapper through which Google can examine different sources of information from which the user profiles in terms of queries and snippets can be derived from the search results. These user profiles were created by classifying the information into concepts from the Open Directory Project concept hierarchy and then used to re-rank the search results. User feedback was collected to compare Google's original rank with our new rank for the results examined by users. They found that queries were as effective as snippets when used to create user profiles.

3 Investigations and Findings

A simple method that keeps track of user behaviour is completely basing on keeping track of the way the user navigates the web sites through making clicks. The sequence in which the clicks are made more or less depicts the user behaviour. The frequently visited URL as hyperlinks is displayed on top of the content shown in the home page. The user clicks on the desired URL as one of the URL displayed in the home page is frequently visited URL by the user. As the user selects an URL, further refinement can be done and the revised frequently visited URLs can be refreshed. The navigable URLs are displayed for the user to select, and the URLs are refreshed every time the user navigates to a particular URL.

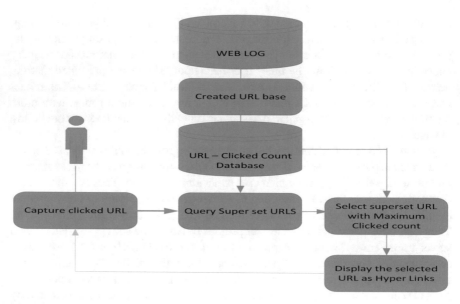

Fig. 1 Process flow for navigating through URL

3.1 Process for Developing Initial Database of Clicks Made by the User

Every WEB server maintains log which has the entries every time a section to the URL is made. The WEB log could be used as the source of information for getting the initial click count for each of the URL that is navigated. If a URL is selected, all the other URLs which are the superset of the selected URL can be found and out of them all the URLs that have clicked count more than the threshold values can be selected from the database and the same are displayed as hyperlinks for the user to navigate. By using this method, the user will be able to surf the WEB at a faster rate. Figure 1 shows the process flow for developing initial database of URLs that have been navigated prior implementing an automation algorithm that finds the superset of URLs that would be of immediate interest to the user.

WEB log has entries that have the data related to the navigation made by the user. Table 1 shows the details of data contained in the WEB log. The WEB log is processed to create database of URLs along with clicked count. The layout of the clicked count table is shown in Table 2. The process has been applied to an existing WEB log related to the existing WEB site, and the home page of which is shown in Fig. 2. A snapshot of the WEB log is shown in Fig. 3, and details of clicked counts computed for each URL are shown in Table 3.

The frequently visited URLs which are supersets of the current URLs are included into the dropbox dynamically and provided to the user as a part of main menu for the user to select to one to navigate to the desired URL. A sample home page that

Table 1 Details of WEB log entries

Serial number	Name of the field	Field type
1.	User ID	Text
2.	Date of entry	Text
3.	Time of entry	Text
4.	URL clicked	Text

Table 2 Details of clicked count database

Serial number	Name of the field	Field type
1.	URL	Text
2.	Clicked count	Number
3.	Date of counting	Text
4.	Time of counting	Text

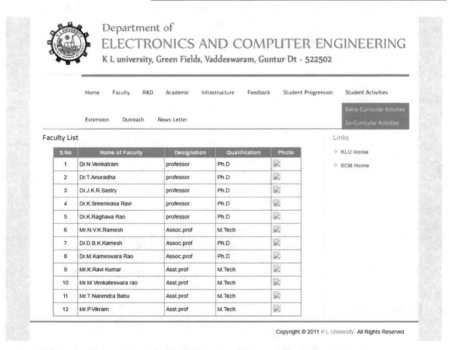

Fig. 2 A WEB page of an example WEB site

shows the frequently navigable URLs as a drop down box is shown in Fig. 4. Every time a user clicks on an URL, the clicked stored in the database is incremented. The clicked count table shows the latest navigational patter. The current experience of navigation is always used for determining the future navigational requirements.

	ActionID	ActionName	Url	IPAddress	TickTime
▶	2	Genr	http://localhost:.... :		2016-04-14 09:...
	3	Genr	http://localhost:.... :		2016-04-14 09:...
	4	Genr	http://localhost:.... :		2016-04-14 09:...
	5	Genr	http://localhost:.... :		2016-04-14 09:...
	6	Genr	http://localhost:.... :		2016-04-14 09:...
	7	Genr	http://localhost:.... :		2016-04-14 09:...
	8	Genr	http://localhost:.... :		2016-04-14 09:...
	9	Genr	http://localhost:.... :		2016-04-14 09:...
	10	Genr	http://localhost:.... :		2016-04-14 09:...
	11	Genr	http://localhost:.... :		2016-04-14 09:...
	12	Genr	http://localhost:.... :		2016-04-14 09:...
	13	Genr	http://localhost:.... :		2016-04-14 09:...
	14	Genr	http://localhost:.... :		2016-04-14 09:...
	15	Genr	http://localhost:.... :		2016-04-14 09:...

Fig. 3 A snapshot of WEB log

Table 3 Generated clicked count table

Serial number of the URL	URL	Clicked count
1.	my-site	25
2.	my-site/research	25
3.	my site/research/publication	20
4.	my site/research/sponsored-research	5
5.	my site/research/student-activities	25
6.	my site/research/student-progression	9

4 Conclusions

User behaviour can be captured in terms of clicked count on a URL. Frequently visited URLs by the users can be determined through the clicked count. The user can be made to move on to the lowest level of URL where the desired content is located with ease when the same are made available in the home page as drop-down list. The URLs that the user is expected to move are known by finding the superset of the current URL.

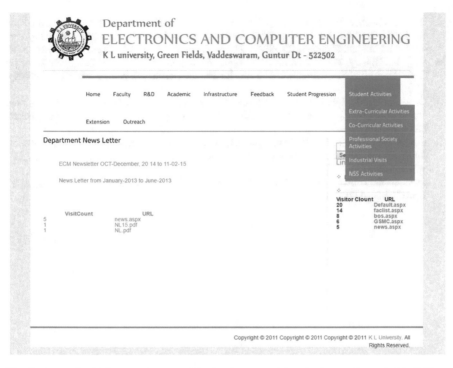

Fig. 4 A samples WEB page that shows the dropbox containing the frequently navigable URLs

References

1. Henry Lieberman," Letizia: An Agent That Assists Web Browsing" AAAI Technical Report FS-95-03. Compilation copyright © 1995.
2. Kazunari Sugiyama, Kenji Hatano, Masatoshi Yoshikawa, "Adaptive Web search based on User Profile Constructed without any effort from Users" WWW2004, May 17–22, New York, New York, USA, 2004.
3. Jaime Teevan, Susan T. Dumais, Eric Horvitz, "Personalizing Search via Automated Analysis of Interests and Activities" *SIGIR '05*, August 15–19, 2005, Salvador, Brazil. Copyright ACM 1-59593-034-5/05/0008…$5.00, 2005.
4. Ishwinde rkaur and Anthony J. Horn of, "A comparison of LSA, word net and PMI-IR for predicting user click behavior" April 2–7,2005.
5. D.B.K Kamesh, J.K.R. Sastry, M. Devi Kavya Priya, "Surfing Large Websites based on User Behavior" International Journal of Applied Engineering Research ISSN 0973-4562 Volume 9, Number 20, pp. 7717–7726, 2014.
6. Dr. JKR. Sastry, S. Pavani Snigdha, M. Devi Kavya Priya and Ms. J. Sasi Bhanu, "Customized WEB Searching within Application Specific WEB Sites" International Journal of Applied Engineering Research ISSN 0973-4562 Volume 9, Number 19, pp. 5313–5324, 2014.
7. DBK Kamesh, Dr. JKR Sastry, S. Pavani Snigdha, "Searching Large Application Specific WEB Sites through Concept based Categorization" International Journal of Applied Engineering Research ISSN 0973-4562 Volume 9, Number 21, pp. 10005–10021 © Research India Publications, 2014.

8. Dr. JKR. Sastry, S. Pavani Snigdha, M. Devi Kavya Priya and Ms. J. Sasi Bhanu, "Personalized and Customized WEB Searching within Application Specific WEB Sites" International Journal of Applied Engineering Research ISSN 0973-4562 Volume 9, Number 19 pp. 5313–5324, 2014.

9. Dr. J. K. R Sastry, M.V.B.T Santhi, S. Pavani Snigdha, A New Clustering Technique based on User Search Histories, International journal for development of computer science & technology, 1(3), 2013.

10. Diwakar Shukla, Rahul Singhai, Analysis of Users Web Browsing Behaviour Using Markov chain Model, Int. J. Advanced Networking and Applications Volume: 02, Issue: 05, Pages: 824–830 (2011).

11. Eugene Agichtein, Eric Brill, Susan Dumais, Robert Ragno, Learning User Interaction Models for Predicting Web Search Result Preferences, *SIGIR'06*, August 6–11, 2006, ACM 1-59593-369-7/06/0008.

12. MircoSperetta, Susan Gauch, Personalizing Search Based on User Search Histories, ACM 1-58113-000-0/00/0004.

Identification of Effective Parameters for Designing a Data Channel

S. Venkateswarlu, D. B. K. Kamesh, J. K. R. Sastry and Ch. Radhika Rani

Abstract Many models exist in the literature, and each model has used a set of parameters for the simulation. The models use either the conventional or modern approaches. Overall, there are 22 different parameters that should be considered when a model is to be designed. These 22 parameters make the system complex. There is a need to determine those few parameters that together give the combined effect of all the 22 parameters. There are in existence different design models that cater for designing the transmitter and receiver side of the systems. The designing of the models must be done independently considering the parameters that affect the communication on either side. Most suitable and comprehensive models are to be designed.

Keywords Parameters · Effective parameters · Channel models

1 Introduction

Few of the channel parameters that will affect the data that are being communicated to the receiver are considered here. Channel sub-carriers and time duration in between the transmission of bits are to be handled to mitigate the impairments such as the

S. Venkateswarlu · D. B. K. Kamesh (✉)
Sreyas Institute of Engineering and Technology, Hyderabad 500068, India
e-mail: kameshdbk@gmail.com

S. Venkateswarlu
e-mail: somu23pavan@gmail.com

S. Venkateswarlu · D. B. K. Kamesh
St. Martin's Engineering College, Dhulapally, Hyderabad 500014, India

J. K. R. Sastry
K.L. University, Vaddeswaram, Guntur District, India

Ch. R. Rani
Department of CSE, K.L. University, Vaddeswaram, Guntur District 522502, India

© Springer Nature Singapore Pte Ltd. 2019
R. S. Bapi et al. (eds.), *First International Conference on Artificial Intelligence and Cognitive Computing* , Advances in Intelligent Systems and Computing 815,
https://doi.org/10.1007/978-981-13-1580-0_10

absence of originality in the sub-carriers due to selectivity of time and intersymbol interference (ISI) which happens due to huge selectivity. The allocated power and signal constellation can be changed at the transmitter side for each of the sub-carrier so that throughput could be maximized.

The delay spread is a key parameter which is to be considered for focusing on the phase difference and path delay. If the source or the destination is moving from location to location, the Doppler shift will affect the signal phase and thus the data received. The transmission and reception when done simultaneously and several times, spectral efficiency can be further improved.

2 Channel Definition and an Overview of Channel Parameters

2.1 Key Features not Included in the Present Models

Many advances have been made relating to channel zones and propagation. Still there are many leftover issues that must be addressed, especially related to estimating the parameters that are to be considered in the model design [1–3]. Some models have already included into the model for estimating the keyhole impact, double dissipating, polarizations, dissemination of multiple paths, and time variances.

2.2 Single Versus Double Scattering

Geometric-based channel models consider scattering which is single bounce between the receiver and the transmitter [4]. Some scatterings [5] which are either diffraction bounced or multiple bounced are neglected. This implies that angular power that is common to transmitter and the receiver and coupling of direction of departure (DoDs) with DoDs which is immediate and cannot be separated.

Additional scattering that is double bounce is required for getting good system performance. Single bounce is inferior to multiple-bounce scattering, but the multiple-bounce scattering does not cause any separable DoDs and direction of arrival (DoAs) [6, 7]. Most of the models existing as date assume that equal wave propagation occurs between mobile scatterers and base station. It also implies that the angular power between transmitter and receiver is not separable.

2.3 Keyhole Effect

If the space between the base station (BS) and mobile station (MS) [4] scatterer radius is less than the space between the BS and MS, then it may lead to a small rank matrix and different statistics of amplitudes statistics for all BS and MS scatterers. This aspect is in agreement with the statistics of Rayleigh amplitude. The Rayleigh distribution due to many of the massive MS scatterers occurs many a time due to occurrence of these statistics.

Abstract assays appear that for rank-one channel matrices, the channel capacity is absolutely low commensurable to the capacity of single-antenna systems. Such channels have been known as **keyhole channels**. A hardly broader abstraction of rank-reduced channels, named "pinhole" channels.

Be that as it may, the ID of comparing certifiable proliferation situations and the estimation of rank-decreased channels have been observed to be greatly troublesome. A scientific model has been proposed by Gesbert et al. 2002 [8], that is appropriate for these sorts of channels and is basically an adjustment of the Kronecker model got by embedding a low-rank scatterer correlation matrix in the middle of the transmitter (Tx) and receiver (Rx) relationship matrices. On the other hand, physical- and geometry-based models up to now do not replicate keyhole (pinhole) impacts.

2.4 Diffuse Multipath Components

The radio channel can be shown by superimposing many of the propagation paths, especially considering the geometric-based channels [9]. The propagation paths are represented through the statistics which are spatial and derived from normal channel estimations. The exactness of the model can be controlled via number of propagation paths within the limits. The quality of the propagation paths should not be expanded beyond a limit as it results in over modelling and also effects the evaluation of the reliable measurements. To deal with such a situation, an extra component is added to the model that is related non-spatial requirements. These extra components are either thick or diffused multipath components (MPCs).

This kind of expansion could be relegated to various measurements that may be used to demonstrate the response due to channel impulse reaction that includes a set of strongly concentrated paths which are called specular MPCs. The impulse reaction also includes a larger number of parameters that influence the generation of larger number of MPCs which are diffusing.

2.5 Polarization

Dual-polarized arrays can be fabricated quite efficiently when compared to single-polarized arrays. Dual-polarized arrays offer twice the number of modes when compared to the number of modes supported by single-polarized arrays [10]. The issue of polarization has not been considered quite effective when it comes to channel modelling. Each propagation path has to be defined in agreement with polarization states which are orthogonal. Further depolarization takes place due to the existence of diffractions, reflections and scattering. Scattering as such becomes four polarization states from two admission states.

Oestges et al. (2004) have presented the model that is geometry-based model which caters for doubly polarized systems. The model presents each scatterer connected with a matrix coefficient to represent random passages. In the TGn model which is based on 802.11 standard, depolarization has been presented in a measurable manner and a random variable is used for representing the issue of cross-polarization displaying of the polarization is quite complicated as the coordinate the systems which should incorporate the coordinates of the environment, directions of the polarization states, and the coordinates of the receiver makes the model much more complex.

2.6 Time Variation

Channel time variation is because of movement of terminals or scatterers. In the setting, minimal experimental results have been gotten with regards to time variations, mostly on account of constraints in channel sounding hardware.

Analytical models tend to neglect these absolute concrete causes and capture time variations via statistical characterizations. Further measurement and experimental evidence is affirmed and is appropriate to see whether this is absolutely justified. Deterministic, ray tracing-based models or geometric-based channel models (GSCMs) can cover channel time variations absolutely by prescribing transmitter, receiver, or scatterer motion which can be characterized statistically. These will automatically reproduce realistic correlations for alternating channel snapshots.

3 Parametric Identification—Phase-A

The presentation of various types of models described in the previous section reveals the consideration of a set of model parameters for effective and efficient modelling of a channel. Table 1 shows the list of parameters and a mopping of the same to the models presented in the previous sections [11, 12].

Table 1 Parametric identification—phase-A

Parameter serial no	Name of the parameter	Ray tracing	GSCM	Saleh Valenzuela	Zwick	IID	Kronecker	Weichselberger
1	Single versus double scattering	Single bob disseminating accepted Numerous skipped scrambling dismissed	Single bob disseminating accepted Numerous skipped scrambling dismissed	Single bob disseminating accepted Numerous skipped scrambling dismissed	Single bob disseminating accepted Numerous skipped scrambling dismissed	This relates to the spatially white divert which happens just in rich disseminating situations described by MPCs consistently conveyed	Single bobbing dispersing Exceptional instance of Weichselberger model got with the rank-one coupling framework	Its analogue is based on the Eigen amount atomization of the Tx and Tx alternation matrices
2	Angular power spectrum (APS)	Joint Tx and Rx APS not separable	Joint Tx and Rx APS not separable	Joint Tx and Rx APS not separable	Joint Tx and Rx APS not separable	Characterize the channel matrix statistically in terms of the correlations between the matrix entries	Characterize the channel matrix statistically in terms of the correlations between the matrix entries	Popular correlation-based analytical channel model
3	Time variation	Incorporate channel time varieties unequivocally by endorsing TX, Rx movement which can be described factually	Incorporate channel time varieties unequivocally by endorsing TX, Rx movement which can be described factually	Channel time variety is because of development of terminals or scatterers	Channel time variety is because of development of terminals or scatterers	Actual physical and capture time variations via statistical characterizations are neglected	Actual physical and capture time variations via statistical characterizations are neglected	Actual physical and capture time variations via statistical characterizations are neglected

(continued)

Table 1 (continued)

Parameter serial no	Name of the parameter	Ray tracing	GSCM	Saleh Valenzuela	Zwick	IID	Kronecker	Weichselberger
4	Wave propagation	Accept that waves engender just as likely between any pair if BS and MS scatterers	Accept that waves engender just as likely between any pair if BS and MS scatterers	Accept that waves engender just as likely between any pair if BS and MS scatterers	Accept that waves engender just as likely between any pair if BS and MS scatterers	Portray the drive reaction of the channel between the individual transmit and got receiving wires in numerical/expository manner without expressly representing wave engendering	Portray the drive reaction of the channel between the individual transmit and got receiving wires in numerical/expository manner without expressly representing wave engendering	Portray the drive reaction of the channel between the individual transmit and got receiving wires in a numerical/expository manner without expressly representing wave engendering
5	Polarization	Single polarization used	Dual polarization used	Single polarization used	Single polarization used	Cross-polarization proportion is dealt with as an irregular variable	Cross-polarization proportion is dealt with as an irregular variable	Cross-polarization proportion is dealt with as an irregular variable
6	Diffuse multipath components	Channel actuation acknowledgment consists of several ell apply able paths	Channel actuation acknowledgment consists of several ell apply able paths	Diffuse MPC is incorporated to depict non-specular parts	Diffuse MPC is incorporated to depict non-specular parts	This compares to the spatially white divert which happens just in rich dissipating situations described by MPCs	It implements a distinct DoD–DoA range. It cannot replicate the coupling of a solitary DoD with a solitary DoA	It implements a distinct DoD–DoA range. It cannot replicate the coupling of a solitary DoD with a solitary DoA

(continued)

Table 1 (continued)

Parameter serial no	Name of the parameter	Ray tracing	GSCM	Saleh Valenzuela	Zwick	IID	Kronecker	Weichselberger
7	Multiple propagation paths	Increment in number of spread ways result in over displaying	Increase in number of propagation paths result in over modelling	Increment in number of spread ways result in over displaying	Increment in number of spread ways result in over displaying	Have not advised any ambit that represents the reproduction of diffusing MPCs reliably	Have not advised any ambit that represents the reproduction of diffusing MPCs reliably	Have not advised any ambits that represent the reproduction of diffusing MPCs reliably
8	Model characterization	Model is described by the physical engendering parameters in a deterministic way	Model is portrayed by drive reaction by the laws of wave proliferation	This model describes and determines physical parameters in a stochastic way	This model describes and determines physical parameters in a stochastic way	This model portrays the divert grid measurably regarding the connections between's the lattice passages	This model portrays the divert grid measurably regarding the connections between's the lattice passages	This model portrays the divert grid measurably regarding the connections between's the lattice passages

It can be seen from Table 1 that each model considers a parameter in different manner, and therefore, it becomes necessary to find the most effective representation of each parameter that must be included into the design of a channel model.

Apart from the parameters which are listed in Table 2, few more parameters are considered for designing the channel model. These additional parameters are not considered by any model. The additional parameters are very important for characterizing the channel. The details of these additional parameters are shown in Table 3.

Table 2 Small parameters' set that represents larger set of parameters

Parameter serial	Name of the parameter	Description of parameter	Represented parameter	Justification for representation
1	Type of scattering	Rayleigh scattering of signals is considered in the design. It also considers the diffusion that is caused by the signals travelling through the atmosphere	Scatterer locations	Multiple-bounced scattering is used. Hence, identification of scatterer locations is important
			Scatterer distributions	Motivation reaction of the channel is portrayed between the individual Tx and Rx receiving wire mathematically without the wave spread
2	Path delay	When a signal has been transmitted towards a receiving station, through the atmosphere, multiple signals reflected by several objects will be received by the receiving antenna. These signals do travel through different paths of various lengths. As a result, the receiver will received multiple signals with path differences, delays and with phase differences	Propagation delay	The receiver has to consider the propagation delay of signals received through various paths for calculating net received signal
			Propagation time	Propagation time has to be for knowing the path delay
			Time variation	Channel time variation has to be considered as there is a movement of terminals

(continued)

Table 2 (continued)

Parameter serial	Name of the parameter	Description of parameter	Represented parameter	Justification for representation
3	Wave propagation		Delay drifts	The signal will reach the receiver after getting multiple hops during propagation Hence, delay drift is considered
		While the signals are travelling through atmosphere, the noise, phase difference, and Doppler frequency shift will affect the propagation of waves. These parameters will affect the signal strength and the content of the data received through the signals	Polarization	Multiple polarizations is considered for effective reproduction of received signal
			Radio propagation	The application of wave propagation to radio communications is known as radio wave propagation
			Propagation paths	Multiple propagation paths are considered
4	Diffuse multipath components	Radio channel is a superposition of limited number of engendering ways	Antenna pattern	Antenna pattern is typical polar radiation plot
			Transmit antenna	Omnidirectional transmitting antenna is assumed
			Receive antenna	It is assumed that the receiving antenna will receive the signals from all the directions
5	Diffuse multipath components	Radio channel is a superposition of finite number of propagation paths	DoD	The DoD has to be considered for releasing the signals from the Tx

(continued)

Table 2 (continued)

Parameter serial	Name of the parameter	Description of parameter	Represented parameter	Justification for representation
			DoA	The DoA has to be considered for receiving the signals from all directions
6	Multiple propagation paths	Multiple propagation paths are considered	Multipath components	Diffuse MPC is included to describe non-specular components
7	Doppler shift	The dynamic behaviour of transmitter and receiver will cause the change of frequency of the signals being propagated. The change in frequency will cause loss of data or it affects the signals. Thus, the modulated signals while being demodulated at the receiver will get affected. The Doppler frequency shift needs to be considered	Overall attenuation	Unless we measure the overall attenuation, the SNR cannot be calculated
8	Effect of delay in path(s) on received signal strength	With the change in the length of the path, the path delay changes. Its effect on the received signal strength is observed	Power delay profile	Postponement force profile is tested, and every proliferation way is discretized for ascertaining got signal quality viably
			Laws of wave propagation	All the properties of waves are considered for effective representation of signals in the model
			Impulse response	Impulse response consists of several strong paths

Table 3 Small parameters' set that represents larger set of parameters

Parameter serial	Name of the parameter	Description of parameter
1	Band of operation	Most of the models proposed have fallen under narrowband or wideband spectrum models. It is supposed that there is a need for channel model that works for different frequencies ranging from narrowband to wideband
2	Security	The most important parameter of all is the security of the data. The data received at the receiver are 9 affected by several parameter either listed above or even more. All these will affect the content that is present in the data transmitted. Special care has to be taken for the content of the data. No data should be lost; even a bit is also important. Security parameter will thus be very important
3	Variable packet length	Size of the data packet is varied, and the effect of this variation on the received signal is observed

4 Conclusions

Many conventional models and advanced models exist in the literature for modelling the channels. Different types of parameters are used for undertaking the modelling of a channel. Each parameter is introduced into the model by using different methods of implementation. Using all the parameters and considering the effective method of each of the parameter lead to a complex method of undertaking the modelling of a channel. Fewer set of parameters and a specific method of implementation of the same have to be chosen and introduced into the model so that the model becomes simple.

References

1. S. Venkateswarlu and Dr. Sastry JKR, A Qualitative Analysis of Rayleigh Faded Frequency Selective Channel Simulator, IJAER Journal, Vol 9, No 22, PP. 10281–10286, 2014.
2. S. Venkateswarlu and Sastry JKR, Justification of Rayleigh faded channel for data transmission in wireless environment", IJETT journal, Vol. 14, No 4, pp. 106–109, Aug-14.
3. S. Venkateswarlu, Ch. Radhika Rani, Channel Modelling -Parameters and Conditions to be Considered, IJEMS Journal, Vol 7, Number 2 (20117), PP 319–325, ISSN 2249-3115.
4. F. Molisch, A. Kuchar, J. Laurila, K. Hugl, and R. Schmalenberger, "Geometry- based Directional Model for Mobile Radio Channels Principles and Implementation," European Trans. Telecomm., vol. 14, pp. 351–359, 2003.
5. J. Laurila, A. F. Molisch, and E. Bonek, "Influence of the Scatter Distribution on Power Delay Profiles and Azimuthal Power Spectra of Mobile Radio Channels," in Proc. ISSSTA'98, Sept. 1998, pp. 267–271.
6. Chong, C. Tan, D. Laurenson, M. Beach, and A. Nix, "A New Statistical Wideband Spatiotemporal Channel Model for 5-GHz Band WLAN Systems," IEEE J. Sel. Areas Comm., vol. 21, no. 2, pp. 139–150, Feb. 2003.
7. Q. Spencer, B. Jeffs, M. Jensen, and A. Swindlehurst, "Modeling the Statistical Time and Angle of Arrival Characteristics of an Indoor Multipath Channel," IEEE J. Sel. Areas Comm., vol. 18, no. 3, pp. 347–360, Mar. 2000.
8. T. Zwick, C. Fischer, and W. Wiesbeck, "A Stochastic Multipath Channel Model Including Path Directions for Indoor Environments," IEEE J. Sel. Areas Comm., vol. 20, no. 6, pp. 1178–1192, Aug. 2002.
9. Saleh and R. Valenzuela, "A Statistical Model for Indoor Multipath Propagation," IEEE J. Sel. Areas Comm., vol. 5, no. 2, pp. 128–137, Feb. 1987.
10. P. Soma, D. Baum, V. Erceg, R. Krishnamoorthy, and A. Paulraj, "Analysis and Modeling of Multiple-Input Multiple-Output (MIMO) Radio Channel Based on Outdoor Measurements Conducted at 2.5 GHz for Fixed BWA Applications," in Proc. IEEE Intern. Conf. on Comm., ICC 2002, vol. 1, Apr./May 2002, pp. 272–276.
11. S. Venkateswarlu and Dr. Sastry JKR, Design of Transmitter for Rayleigh Faded Channel for Data Transmission, IJAER Journal, Vol 9, No 22, pp. 12377–12386, 2014.
12. S. Venkateswarlu, Dr. Sastry JKR, Development of Data Recovery System for the Data Transmitted over Rayleigh Faded Channel, IJAER Journal, Vol. 10, No.1, pp. 1077–1082, 2015.

Cognitive-Based Adaptive Path Planning for Mobile Robot in Dynamic Environment

Dadi Ramesh, Syed Nawaz Pasha and Mohammad Sallauddin

Abstract Artificial intelligence plays a major role in robotics. The algorithms proposed so far are utilized to recognize static hindrances. The issue of way arranging of
robots from source to goal includes distinguishing the static and dynamic snags in the
detecting territory of the robot and decides a crash-free way. Most of the algorithms
are targeted on the trail-finding procedure in a very illustrious atmosphere and leave
higher-level functions like obstacle detection, but the cognitive-based adaptive path
planning (CBAPPA) is a reconciling and psychological feature-based mostly thinking system that it identifies the dynamic obstacles in an unknown surrounding. In this
paper, we tend to propose an associate approach to seek out optimum path in dynamic
surroundings. During this approach, the mechanism that has sensors processes the
data received through sensors and decides the path supported the data processed.

Keywords Cognitive science · Adaptive path · Dynamic environment

1 Introduction

One of the problems in path coming up with is to seek out the best path from supply to destination in static and dynamic atmosphere. So the pathfinding strategies
are playing a prime position in a dynamic environment [13, 15, 19]. Path planning
algorithms have the responsibility of finding optimal path from one coordinate to
another coordinate in any environment by taking a starting point and a destination
point. Then, they count one-by-one series of paths and constitute source to destination once upon a time arriving to the optimal path. For this, an art an adjunct of

D. Ramesh (✉) · S. N. Pasha · M. Sallauddin
Computer Science & Engineering, S R Engineering College, Warangal, India
e-mail: ramesh_d@srecwarangal.ac.in

S. N. Pasha
e-mail: snp786@gmail.com

M. Sallauddin
e-mail: sallauddin.md@gmail.com

© Springer Nature Singapore Pte Ltd. 2019
R. S. Bapi et al. (eds.), *First International Conference on Artificial Intelligence
and Cognitive Computing* , Advances in Intelligent Systems and Computing 815,
https://doi.org/10.1007/978-981-13-1580-0_11

117

sensors are unavailable into the route to run the robot. And these sensors dig into the past the flea in ear from the environment and pick up the coordinates to which the robot has to move in dynamic environment when we give start and goal positions. So cognitive-based path planning algorithm finds a path when using cognitive science and adaptive-based methodologies. On this clear the adaptive fashion of the Cro-Magnon man is implemented as people amount to be asked premiere where one is heading when they have protect to complementary information. Unlike analytical systems, human [12, 18, 20] will method the data in no time and effortlessly in their daily routine.

2 Related Work

There is a unit deviation of algorithms to tackle this issue of finding the best path from source to destination. The path planning has to be done keeping in mind the final destination. The outlook planning is light as a feather if there are no obstacles, notwithstanding if there is any complication in the orientation from source to destination before we prefer to notice for arbitrary paths. A home of that a way planning algorithms have been implemented.

2.1 Cognitive-Based Adaptive Path Planning

The approach of the CBAPPA [2, 14] is primarily based on cognitive-based steps. Every step is enforced by variety of heuristic algorithms [7]. Functionally, CBAPPA is countermined into two distinct stages:

In the first stage, it finds a straight line from source to destination; this straight line will become the optimal path at any cost for any algorithm. If the destination position is unknown, then robot tries to find the direction of the destination from source point of view. This straight line will become the guide to the robot in the second stage.

In the second stage, it starts moving toward the destination by setting intermediate goals by overcoming the obstacle and by using straight line as a guide [6, 8, 16]. The intermediate goal area is about sensory area of robot. While moving in a sensory area, it avoids the unnecessary obstacles which are far away from straight line. The robot will consider only the obstacle which really interrupts robot motion in sensory area. When an obstacle is intercepted within sensory range, then it forms a half circle around the obstacle and follow the half circle until it reaches the other side of the obstacle. So the half circle both the ends are on the straight line. Like that, it sets intermediate goals and avoids the obstacle until it reaches the final destination.

The CBAPPA finds the path same like how a human being finds a path in an unknown environment from source to destination. When the results are compared with A* algorithm [11] results, the CBAPPA algorithm finds an optimal path by

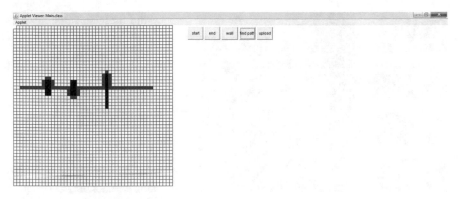

Fig. 1 Result CBAPPA for static obstacle

improving the efficiency and it reduces the searching area to find final path from source to destination. The efficiency has been improved from 17.18 to 58.33%, and the searching area is also reduced from 33.35% in total area to 17.5%.

The heuristic algorithmic programs [7] planned AN initial path with A* algorithm [3, 11] by victimization the previous map information, rapt the automaton on the e trail till it reached the goal. D* [4, 17] centered the price updates to reduce state expansions and additional reduced process prices. The algorithmic program used a heuristic operate the same as A* to each propagate value will increase and focus value reductions (Fig. 1).

A biasing operate was accustomed complete automaton motion between pre-planning operations. Supported a qualitative estimate, 40% of the entire space was searched by the algorithmic program to search out a path. The cognitive feature primarily based on accommodative path designing algorithmic program reduced the looking out space and noticed optimum path compared to different path designing algorithms.

3 Proposed System

We are proposing CBAPPA [2, 14] for avoiding dynamic obstacles. It is based on cognitive [1, 5] and adaptive methodologies. It is on determined behaviors of the biological units and paid attention to the behavior of ignoring the inapplicable info from surroundings and making an attempt to achieve the target quickly.

The system works on three steps to search out associate in nursing best path from supply to destination by avoiding excess obstacles. Here, it starts with primary path at the beginning, then finds refined path up to the sensory range, and generates the final path by considering all sensory information. In this robot processing, army plays sharp and flat role. It receives the sensory taste from generally told the sensors and stores the flea in ear in the fare format. From that, it divides the obstacle into two

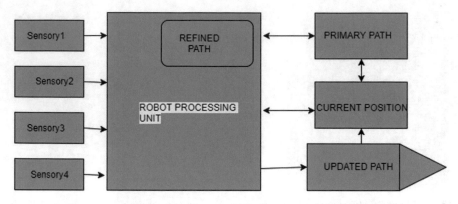

Fig. 2 System overview

categories, like static and dynamic. If there is a static obstacle on the primary path, robot avoids it by using CBAPPA and updates the current position. And if there is dynamic obstacle then follows cognitive-based adaptive path planning algorithm for dynamic obstacle (CBAPPAD) to avoid and update the current position.

3.1 Working of Cognitive-Based Adaptive Path Planning Algorithm for Dynamic Obstacles (CBAPPAD)

1. First, find the primary path from source to destination by setting intermediate goals.
2. Take the information from the sensors and find the obstacle positions.
3. Then, differentiate the obstacle as static and dynamic. If it is static obstacle and not on the primary then ignore that obstacle.
4. For the static obstacle and which is on the primary path apply CBAPPA.
5. Else consider the obstacle as dynamic.
6. Then, form a circle around the obstacle and label it with four points which is equidistance from the obstacle as P1, P2, P3, P4 (P1 should always be toward primary path) as shown in Fig. 2.
7. Calculate the distance from P1 to point X (which is on primary path and perpendicular to P1).
8. Then, the paths are refined using the following equations

D1 = distance from P1 to X.
D2 = distance from X to ROBOT's current position.
V1 is velocity of obstacle.
V2 is velocity of robot.
If D1/V1 > D2/V2
Then, final path is the primary path

Fig. 3 Work space

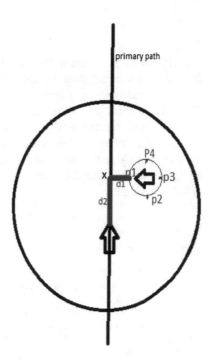

Else **D1/V1 < D2/V2**
Then, the final path is P2 → P3 → P4 → primary path.

9. Update the current position.

And if current position == goal position (Fig. 3).
Then stop.
Else go to step 2.

3.2 Searching Area and Efficiency

$$\text{Searched Area} = \frac{\text{total area searched to find final path}}{\text{total Area}}$$

$$\text{effiency} = \frac{\text{area used for final path}}{\text{searched Area}} * 100$$

4   Conclusion

CBAPPA finds path similar to how a biological organ behaves when finding its path from source to destination by taking decisions dynamically. It ignores inapplicable data and quickly method optimum and economical methods. In this paper, result of CBAPPA on static obstacles was compared with some existing algorithms and found that the searching area was decreased and the efficiency was righteous in finding an unassailable path from man to destination. Then, we applied the CBAPPA on zealous obstacle and strive to enliven the nonchalance by disturbing to charge the complaint along the head of the line path, and from this point, forward searching orientation will be decreased.

References

1. B. Krose, "Environment Representations for Cognitive Robot Companions", ERCM News, (53), 2003, pp. 29–30.
2. Cognitive Based Adaptive Path Planning Algorithm for Autonomous Robotic Vehicles, Adam A. Razavian, Junping Sun, ©2005 IEEE.
3. A. Stenz and M. Hebert, "Navigation System for Goal Acquisition In Unknown Environments," Proceedings of IROS 95, USA, 1995, pp. 425–453.
4. E. W. Dijkstra, A note on two problems in connexion with graphs, Numerische Mathematik. 1 (1959), S. 269–271.
5. A. Stevens and P. Coupe, "Distortions in Judged Spatial Relations," Cognitive Psychology, 10(4), 1978, pp. 526–550.
6. T. McNamara, "Mental Representations of Spatial Relations", Cognitive Psychology, 18(1), 1986, pp. 87–121.
7. Hart P. E., Nilsson N. J., Raphael B., A Formal Basis for the Heuristic Determination of Minimum Cost Paths, IEEE Transactions on Systems Science and Cybernetics SSC4 (2): pp. 100–107, 1968.
8. J. Hopfield and D. Tank, "Neural Computation of Decisions in Optimization Problems", Biological Cybernetics, 52, 1985, pp. 141–152.
9. O. Miglino, F. Menczer, and P. Bovet, "A Neuro-Ethological Approach for the TSP: Changing Metaphors in Connectionist Models", Journal of Biological Systems, 2(3), 1994, pp. 357–366.
10. D. Durand, "Rationale, Goal, and Scope" Journal of Neural Engineering. Retrieved January 20, 2005, from: http://www.iop.org/EJ/journal/-page=about/1741-2552/1.
11. Planar Map Pathfinding Based On The A* Algorithm Li Mingcui.
12. J. C. Latombe and J. Barraquand, "Robot Motion Planning: A Distributed Representation Approach.," International Journal of Robotics Research, 10(6), 1991, pp. 628–649.
13. A. Razavian, "The Numerical Object Rings Path Planning Algorithm (NORPPA)," Proceedings of 35th IEEE Conference on Decision and Control, 1996, pp. 4406–4411.
14. COGNITIVE THINKING FOR MOBILE ROBOT IN PATH PLANNING, D. Ramesh, International Journal of Latest Trends in Engineering and Technology 2016.
15. V. Akman, Unobstructed Shortest Path in Polyhedral Environments. New York: Springer-Verlag, 1987.
16. J. Nilsson, Principles of Artificial Intelligence, Morgan Kaufmann, 1994.
17. A. Stentz, "The Focussed D* Algorithm for Real-Time Replanning", Proceedings of International Joint Conference on Artificial Intelligence, 1995, pp. 1213–1221.
18. Autonomous Local Path-Planning for a Mobile Robot Using a Genetic algorithm Kamran H. Sedighi, Kaveh Ashenayi, Theodore W. Manikas, Roger L. Wainwright, Heng-Ming Tai.

19. Mobile Robots Path Planning Based on Dynamic Movement Primitives Library, Zhuang Mei, Yang Chen, 36th Chinese Control Conference.
20. Biologically Inspired Visual Odometry Based on the Computational Model of Grid Cells for Mobile Robots Huimin Lu 2016 IEEE.

Prediction for Indian Road Network Images Dataset Using Feature Extraction Method

Suwarna Gothane, M. V. Sarode and V. M. Thakre

Abstract Indian roads have various distress issues, enormous stress, and immense need of rejuvenation to handle the augmented need of the Indian economy, vast traffic, and heavy vehicle speed. To overcome the problem of road network, we developed a method to track roadworthiness. Identification of severity level and accuracy is achieved using MATLAB. This paper presents SVM evaluation on real road network images. Using GLCM feature extraction and support vector machine classifier, we achieved 81.3% accuracy results.

Keywords Road distress · Potholes · Patches · Road · Cracks
Classification · Accuracy

1 Introduction

The road network is a system of interconnected paved carriageways planned to communicate, conveyance and freight. It is the most essential transport infrastructure in urban areas and connects all other areas. The road network facilitates social and economic development of the country. By connecting geographic locations, road

---

Paper Work: Section 1 consists of introduction, Sect. 2 consists of literature survey, Sect. 3 includes architecture of proposed system, Sect. 4 includes implementation, and Sect. 5 includes conclusion and future work.

S. Gothane (✉)
CMR Technical Campus, Hyderabad, India
e-mail: gothane.suvarna@gmail.com

M. V. Sarode
Government Polytechnic Yavatmal, Maharashtra State Board of Technical
Education, Mumbai, India
e-mail: mvsarode2013@gmail.com

V. M. Thakre
Sant Gadge Baba Amravati University, Amravati, India
e-mail: vilthakare@yahoo.co.in

© Springer Nature Singapore Pte Ltd. 2019
R. S. Bapi et al. (eds.), *First International Conference on Artificial Intelligence and Cognitive Computing* , Advances in Intelligent Systems and Computing 815,
https://doi.org/10.1007/978-981-13-1580-0_12

networks smooth out progress of people mobility, goods, and services. It creates welfare and enables fairly high-speed individual transportation for the masses [1]. The capacity of a road is measured by vehicles traveled per hour and also considering width, number of lanes, and speed limit. Larger traffic on road leads to congestion. Safety of road users can be focused by prevention of accidents and following speed protocol, wearing seatbelt, etc. Proper planning is critical in ensuring road safety. On state and national highways, where speed limit is more than 50–60 km per hour, layouts of physically separated directions for separate lanes to motorways and larger urban roads aids in traffic safety. Indian Government requires focusing on social amalgamation and safety of the country through proper road network [2]. Around 65% of goods transport and 85% of passenger traffic travel through Indian road [3]. To handle the augmented necessities of the Indian economy modernisation needs creates enormous pressure. Expansion, i.e., widening of existing roads and mainte-nance, becomes increasingly important to overcome nasty traffic situation. Conges-tion on road reduces the fuel efficiency of the vehicles resulting in intense pollution as traffic runs inefficiently at low speeds [4]. Pollutants due to poor road condition cause health problems, affects climate, and causes environmental damage. DHL [5], a global logistics company, proved that the average time for truck shipments from New Delhi to Bangalore takes twice, in contrast to China, and one-third to European Union countries [6]. Road speed in India and lanes were enhanced in kilometers. The KPMG report noted that India's road network logistics and transportation issue obstruct GDP growth by 1–2%. Gothane [7] proved one of important measures for road accident is bad road condition using info gain attribute evaluator function. So deprived rural roads and passage congestion in the cities is India's recent biggest challenge.

2 Literature Survey

In the existing work, different techniques are introduced for detecting road dis-tress. Kim and Ryu [8] suggested data determined decision method on images taken from optical device, for pothole classification. Varadharajan et al. [9] worked with arrangement of data driven and computer vision with oversegmentation algorithm and concentrated environmental issues of data. Huidrom et al. [10] proposed heuris-tically derived decision logic for potholes, cracks and patches detection. Chambon et al. [11] focused on multi-scale extraction and Markovian segmentation and noticed cracks using morphological tool. Using MATLAB, Koch et al. [12] worked on geo-metric properties and identified pothole shape. Su et al. [13] proved inhomogeneous objects segmented effectively with efficient output. Liu et al. [14] identified crack by segment extending on complex pavement images with Visual Studio C++ 6.0. Sun et al. [15] worked with image smoothing and applied threshold segmentation concept for cracks identification. Mahmood et al. [16], used fuzzy logic to identify cracks, patches, bleed with if-then rules returned good accuracy and suggested maintenance treatment. Suman et al. [17] worked with low cost to achieve parameters of interest of distress. Sandra et al. [18] used fuzzy logic on input with expertise opinion to identify parameters of interest. Fwa and Shanmugam [19] examined fuzzy mathematics with analysis and uncertainty in pavement condition with severity, repair needs. Oliveira and Correia [20] focused on neural network, detected and characterized cracks with unsupervised concept on Portuguese road. Saar and Talvik [21] developed neural networks approach to identify and classify defects. Xu et al. [22] developed artificial neural network for pavement crack detection by processing images. Bhoraskar et al. [23] worked on sensors monitored road, traffic, breaking events using smart phones, accelerometer, GPS, and magnetometer. Mohan [24] investigated system sensed by piggybacking on smart phones using sensing component accelerometer, microphone, GSM radio, GPS sensors and detected potholes, bumps, braking, honking. Salman et al. [25] suggested a multidirectional crack detection by Gabor function with 95% precision. Gothane et al. [26] achieved highest accuracy of classification 98.84% by support vector machine classifier compared to other classification approach. Junga et al. [27] proposed crack identification and localization using image processing on normal camera images with edge identified using a percolation process, with efficient speed. Zhang et al. [28] worked on intensity feature, background information with neural network, classified 500 images of size 3264×2448 collected by mobile phone and obtained superior crack detection performance. In MATLAB, Wang et al. [29] proved on 120 asphalt pavement images using wavelet energy field concept, morphological processing, geometric criterions, and Markov random field and obtained accuracy of 86.7% for pothole identification.

3 Architecture of Proposed System

In the proposed work, we are using input images of Indian road network of Maharashtra state national highway. We have applied image preprocessing operation resizing function to have all images of same size and applied RGB separation, histogram equalization, and erosion operation. Gray-level co-occurrence matrix (GLCM) supports feature extraction. Support vector machine performed classification of image as pothole, cracks, and patches and identified severity of distress based on threshold values and performed result evaluation with classification accuracy. Architecture of proposed system is shown in Fig. 1. Steps with expressions are explained in detail:

Step1: Input Image:
        We used real-time images of photograph taken from surface condition of Nagpur–Pulgaon–Karanja–Aurangabad–Ghoti–Mumbai MSH-12 from Public Work Department. After every 200 m image has considered in worst road condition.
Step2: Preprocessing:
        Image resizing: Here, we resize image to have all images of the same size using the following equation.

Fig. 1 Architecture of road distress detection

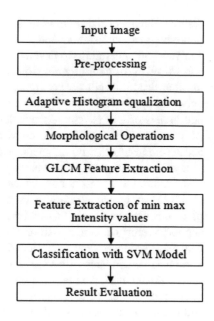

$$\vartheta I_x = \|\max\|\{(f_x, f_y), [w\ H]\}$$

Here I_x is input matrix, ϑ is type of matrix, f_x, f_y is collection of data, W is image width, and H is height.

RGB separation:

$$\begin{vmatrix} X \\ Y \\ Z \end{vmatrix} = \begin{vmatrix} X_r & X_g & X_b \\ Y_r & Y_g & Y_b \\ Z_r & Z_g & Z_b \end{vmatrix} * \begin{vmatrix} R \\ G \\ B \end{vmatrix} \tag{1}$$

where X, Y, Z are the image tri-stimulus values. RGB values are obtained from the transfer functions, and the 3×3 matrix is the calculated tri-stimulus values for total three channels; i.e., X_r, Y_r, Z_r are the considered tri-stimulus values with red channel as maximum information.

Step 3: Adaptive Histogram Equalization:

Color image enhancement is supported by adaptive histogram equalization. Intensity levels are continuous quantities normalized to the range [0, 1]; let $p_r(w)$ denote the probability density function. Following transformation on the input levels obtains processed intensity levels, s,

$$s = T(r) \int_0^1 p_r(w)\mathrm{d}w \quad p_s(s) = \begin{cases} 1 \text{ for } 0 \le s \le 1 \\ 0 \text{ otherwise} \end{cases} \tag{2}$$

where w is a dummy variable of integration and PDF of output is uniform.

Step4: Morphological Operations:

Morphological operation erosion is performed to remove the unwanted information by checking the structure of elements within the matrix. Grayscale binary images can be written as:

$$\varepsilon_B(X) = \{x \backslash B_x \subseteq X\}. \tag{3}$$

Can be rewritten into the intersections of the translated sets X_{-b}:

$$\varepsilon_B(x) = \cap_{b \varepsilon B} X_{-b}$$

which can be extended to include grayscale images with the following equation:

$$\varepsilon B(f) =_{\wedge b \varepsilon B} f - b \tag{4}$$

We can define the erosion of image

$$[E_B(f)](x) = \min_{b \in B} f(x + b) \tag{5}$$

X is calculated random information, and ε_B is data matrix, b denotes thresholding value of structure of elements, B denotes empty matrix [].

Step 5: GLCM (Gray-Level Co-occurrence Matrix Feature) Extraction:

GLCM $\{P(d, \theta)(i, j)\}$ represents the probability of occurrence of a pair of gray levels (i, j) separated by a distance d at angle θ. The commonly used unit pixel distances and the angles are $0°$, $45°$, $90°$, and $135°$.

$$p_x(i) = \sum_{j=1}^{N_g} p(i, j); \quad p_y(j) = \sum_{i=1}^{N_g} p(i, j)$$

$$p_{x+y}(k) = \sum_{i=1}^{N_g} \sum_{j=1}^{N_g} p(i, j), \quad k = 2, 3 \ldots 2N_g; \, i + j = k$$

$$p_{x-y}(k) = \sum_{i=1}^{N_g} \sum_{j=1}^{N_g} p(i, j), \quad k = 0, 1 \ldots N_g - 1; \, |i - j| = k \tag{6}$$

$p(i, j)$ gray-level co-occurrence matrix.

Step 6: Feature Extraction of Min—Max Intensity Values:

The features such as contrast, inverse difference moment, correlation, variance, cluster shade, cluster prominence, and homogeneity are calculated using the equations given below.

Contrast:

$$s_c = \sum_i \sum_j (i - j)^2 p(i, j) \tag{7}$$

The contrast is low if the gray levels of the pixels are similar. Contrast identifies an object distinguishable from other objects, background and varies in color and brightness.

Inverse different moment:

$$s_I = \sum_i \sum_j \frac{1}{1 + (i - j)^2} p(i, j) \tag{8}$$

Smoothness of the image is determined by IDM and is expected to be high if the gray levels of the pixel are similar.

Correlation:

$$S_o = \frac{\sum_i \sum_j (ij) p(i, j) - \mu_x \mu_y}{\sigma_x \sigma_y} \tag{9}$$

The gray-level linear dependence between the pixels at the specified positions relative to each other is determined by.

Variance:

$$S_v = \sum_{i,j-1}^{N} (i, j)^2 p(i, j) \tag{10}$$

The variance is calculated by this function.

Cluster shade:

$$S_{cs} = \sum_{i,j=1}^{N} \left(i - M_x + j - M_y\right)^3 p(i, j) \tag{11}$$

The measure of the skewness of the matrix identifies cluster shade. High value indicates the image is not symmetric.

Cluster prominence:

$$S_{cp} = \sum_{i,j=1}^{N} \left(i - M_x + j - M_y\right)^4 p(i, j) \tag{12}$$

where

$$M_x = \sum_{i,j=1}^{N} i p(i, j); \, M_y = \sum_{i,j=1}^{N} j p(i, j)$$

Low cluster prominence indicates peak in the co-occurrence matrix around the mean values and judges variation in gray scales.
Homogeneity:

$$S_H = \sum_{i=0}^{N-1} \sum_{i=0}^{N-1} \left(\frac{P_{ij}}{(1 + |i - j|)} \right) \tag{13}$$

Elements based on closeness to the GLCM diagonal and range $= [0\ 1]$ is termed as homogeneity. Homogeneity 1 indicates diagonal GLCM. Number of features used is 16. A low value of contrast is obtained with constant gray levels, opposite indicates high values.

$$\textbf{Mutual information} = - \sum_{i,j=1}^{N} P(i, j) \log \left(\frac{P(i, j)}{P_i(i) P_j(j)} \right)$$

$$\text{where } P_i(i) = \sum_{i=1}^{N} P(i, j) \text{ and } P_j(j) = \sum_{j=1}^{N} P(i, j). \tag{14}$$

Mutual information reduces uncertainty.

Step 7: Support Vector Machines for Binary Classification:
SVM finds out an optimal hyperplane and separates the data into two classes by boundaries between +ve and −ve classes.
The linear SVM: $f(x) = x'\beta + b$, where x is an observation of row X.

- The vector β determine coefficients for orthogonal vector of hyperplane. For separable data, the optimal margin length is $2/\leftarrow\beta\leftarrow0$.
- b is the bias term.

The root of $f(x)$ for particular coefficients is hyperplane. For a particular hyperplane, $f(z)$ is the distance from point z to the hyperplane.

$$\widehat{f}(x) = \sum_{j=1}^{n} \widehat{a}_j y_j x' x_j + \widehat{b}. \tag{15}$$

b is the estimate of the bias and αj is the jth estimate of the vector $\alpha, j = 1, \ldots, n$. It classifies a new observation, z using sign $f(z)$. Nonlinear SVM works in a transformed predictor space to find an optimal, separating hyperplane. The dual formalization for nonlinear SVM i

$$0.5 \sum_{j=1}^{n} \sum_{k=1}^{n} \alpha_j \alpha_k y_j y_k G(x_j, x_k) - \sum_{j=1}^{n} \alpha_j \tag{16}$$

0.5 is constant value applying SVM, $\alpha, j = 1, \ldots, n$. Calculation intensity values G is gain of the matrix.

$$\widehat{f}(x) = \sum_{j=1}^{n} \widehat{a}_j y_j G(x, x_j) + \widehat{b}. \tag{17}$$

with respect to $\alpha_1, \ldots, \alpha_n$, subject to sum $\alpha jyj = 0$, $0 \le \alpha j \le C$ for all $= 1, \ldots, n$, and the KKT complementarily conditions.

$G(x_k, x_j)$ are elements of the Gram matrix

$$\max_{\alpha} W(\alpha) = \max_{\alpha} -\frac{1}{2} \sum_{i=1}^{l} \sum_{j=1}^{l} \alpha_i \alpha_j y_i y_j (\mathbf{x}_i \cdot \mathbf{x}_j) + \sum_{i=1}^{l} \alpha_i$$

$$\max_{\alpha} W(\alpha) = \max_{\alpha} \left\{ \min_{w,b} L(w, b, \alpha) \right\} \tag{18}$$

$$\overline{\alpha} = \arg\min_{\alpha} \frac{1}{2} \sum_{i=1}^{l} \sum_{j=1}^{l} \alpha_i \alpha_j y_i y_j (\mathbf{x}_i \cdot \mathbf{x}_j) - \sum_{i=1}^{l} \alpha_i$$

4 Implementation of Proposed System

In MATLAB, we obtained the result of 385 images of real road network of MSH 12 after following steps of architecture is given in Table 1 as follows:

We classified with support vector machine classifier based on threshold and identified roads have potholes, patches, or cracks distress with the following condition.

```
if result_features<1.33
helpdlg('Select another Road Image');
elseif result_features>1.36 & result_features<1.48
helpdlg('Road have Cracks');
elseif result_features>1.50 & result_features<1.58
helpdlg('Road have Patches');
elseif result_features>1.59 & result_features<1.65
helpdlg('Road Have Potholes');
elseif result_features>1.75
helpdlg('Normal Road');
```

Severity can be obtained by considering average of every 15 images. Considering average of 15 images called as road 1 and continues till road n. Average of 15 images is calculated in Table 2.

Severity calculation of road values is possible with the following evaluation criteria. If obtained result features < 22.5, we have assigned road with top priority of

Table 1 SVM algorithm result using MATLAB tool

Imge	Value	Im	Value	Im	Value	Im	Value	Im	Value	Im	Value	Im	Value
1	1.573654715	56	1.942163169	111	1.543564443	166	1.528761721	221	1.431583267	276	1.673654715	331	1.528761721
2	1.272970858	57	1.79434669	112	1.498027887	167	1.41381685	222	1.395537806	277	1.357312811	332	1.637786964
3	1.513587736	58	1.968461009	113	1.498027887	168	1.585873293	223	1.543564443	278	1.528761721	333	1.637786964
4	1.599314237	59	1.827374478	114	1.802918261	169	1.86410219	224	1.717741809	279	1.528761721	334	1.513587736
5	1.200622324	60	1.835116495	115	1.528761721	170	1.337290067	225	1.835116495	280	1.249882293	335	1.661976064
6	1.337290067	61	1.896038233	116	1.376715003	171	1.625262405	226	1.543564443	281	1.543564443	336	1.465684865
7	1.625262405	62	1.842657877	117	1.558009263	172	1.513587736	227	1.625262405	282	1.295181743	337	1.907525069
8	1.870864469	63	1.842657877	118	1.802918261	173	1.999295444	228	1.988699408	283	1.316598681	338	1.748198189
9	1.661976064	64	1.757878463	119	1.776555785	174	1.757878463	229	1.988699408	284	1.482066148	339	1.249882293
10	1.802918261	65	1.811278124	120	1.558009263	175	1.513587736	230	1.69621226	285	1.572108447	340	1.923416339
11	1.877437311	66	1.802918261	121	1.395537806	176	1.717741809	231	1.748198189	286	1.785560292	341	1.728134238
12	1.802918261	67	1.757878463	122	1.585873293	177	1.738284866	232	1.513587736	287	1.465684865	342	1.974489403
13	1.316598681	68	1.827374478	123	1.431583267	178	1.625262405	233	1.572108447	288	1.200622324	343	1.200622324
14	1.558009263	69	1.842657877	124	1.785560292	179	1.376715003	234	1.41381685	289	1.661976064	344	1.513587736
15	1.850001078	70	1.97155667	125	1.337290067	180	1.685064776	235	1.482066148	290	1.685064776	345	1.785560292
16	1.337290067	71	1.965201699	126	1.272970858	181	1.272970858	236	1.685064776	291	1.357312811	346	1.707102827
17	1.933134317	72	1.937734294	127	1.14609425	182	1.295181743	237	1.661976064	292	1.41381685	347	1.41381685
18	1.558009263	73	1.965201699	128	1.116115075	183	1.979868757	238	1.493027887	293	1.431583267	348	1.558009263
19	1.625262405	74	1.717741809	129	1.316598681	184	1.357312811	239	1.337290067	294	1.376715003	349	1.69621226
20	1.482066148	75	1.918295834	130	1.41381685	185	1.498027887	240	1.528761721	295	1.543564443	350	1.717741809
21	1.225811174	76	1.982316608	131	1.41381685	186	1.14609425	241	1.431583267	296	1.376715003	351	1.748198189
22	1.637786964	77	1.728134238	132	1.357312811	187	1.802918261	242	1.465684865	297	1.357312811	352	1.482066148
23	1.431583267	78	1.982316608	133	1.272970858	188	1.650022422	243	1.41381685	298	1.999295444	353	1.465684865

(continued)

Table 1 (continued)

Imge	Value	Im	Value	Im	Value	Im	Value	Im	Value	Im	Value	Im	Value
24	1.968461009	79	1.992158662	134	1.337290067	189	1.543564443	244	1.599314237	299	1.585873293	354	1.785560292
25	1.901871924	80	1.599314237	135	1.295181743	190	1.395537806	245	1.41381685	300	1.558009263	355	1.661976064
26	1.357312811	81	1.448864489	136	1.625262405	191	1.599314237	246	1.395537806	301	1.225811174	356	1.738284866
27	1.337290067	82	1.912999214	137	1.316598681	192	1.637786964	247	1.543564443	302	1.69621226	357	1.896038233
28	1.249882293	83	1.842657877	138	1.465684865	193	1.41381685	248	1.174134873	303	1.748198189	358	1.977259887
29	1.482066148	84	1.572108447	139	1.225811174	194	1.685064776	249	1.685064776	304	1.376715003	359	1.890022377
30	1.083542888	85	1.819429253	140	1.465684865	195	1.513587736	250	1.316598681	305	1.558009263	360	1.572108447
31	1.225811174	86	1.431583267	141	1.316598681	196	1.337290067	251	1.395537806	306	1.543564443	361	1.673654715
32	1.272970858	87	1.513587736	142	1.612440949	197	1.498027887	252	1.357312811	307	1.200622324	362	1.685064776
33	1.465684865	88	1.376715003	143	1.337290067	198	1.249882293	253	1.14609425	308	1.431583267	363	1.912999214
34	1.395537806	89	1.661976064	144	1.625262405	199	1.482066148	254	1.661976064	309	1.482066148	364	1.637786964
35	1.528761721	90	1.842657877	145	1.431583267	200	1.376715003	255	1.498027887	310	1.685064776	365	1.543564443
36	1.395537806	91	1.637786964	146	1.543564443	201	1.673654715	256	1.225811174	311	1.748198189	366	1.625262405
37	1.200622324	92	1.612440949	147	1.558009263	202	1.41381685	257	1.558009263	312	1.14609425	367	1.918295834
38	1.69621226	93	1.572108447	148	1.599314237	203	1.482066148	258	1.357312811	313	1.728134238	368	1.572108447
39	1.482066148	94	1.448864489	149	1.748198189	204	1.513587736	259	1.572108447	314	1.543564443	369	1.249882293
40	1.316598681	95	1.572108447	150	1.612440949	205	1.585873293	260	1.558009263	315	1.498027887	370	1.543564443
41	1.585873293	96	1.174134873	151	1.650022422	206	1.802918261	261	1.738284866	316	1.543564443	371	1.738284866
42	1.942163169	97	1.528761721	152	1.14609425	207	1.625262405	262	1.870864469	317	1.431583267	372	1.992158662
43	1.448864489	98	1.827374478	153	1.661976064	208	1.585873293	263	1.707102827	318	1.599314237	373	1.912999214

(continued)

Table 1 (continued)

Imge	Value	Im	Value	Im	Value	Im	Value	Im	Value	Im	Value	Im	Value
44	1.498027887	99	1.572108447	154	1.513587736	209	1.41381685	264	1.225811174	319	1.249882293	374	1.819429253
45	1.337290067	100	1.295181743	155	1.543564443	210	1.528761721	265	1.498027887	320	1.673654715	375	1.954434003
46	1.738284866	101	1.174134873	156	1.585873293	211	1.498027887	266	1.79434669	321	1.685064776	376	1.988699408
47	1.599314237	102	1.685064776	157	1.337290067	212	1.543564443	267	1.465684865	322	1.990508349	377	1.937734294
48	1.448864489	103	1.395537806	158	1.612440949	213	1.465684865	268	1.599314237	323	1.337290067	378	1.835116495
49	1.465684865	104	1.757878463	159	1.599314237	214	1.857148437	269	1.431583267	324	1.295181743	379	1.767329719
50	1.395537806	105	1.748198189	160	1.513587736	215	1.482066148	270	1.448864489	325	1.543564443	380	1.802918261
51	1.513587736	106	1.625262405	161	1.558009263	216	1.572108447	271	1.316598681	326	1.585873293	381	1.907525069
52	1.850001078	107	1.465684865	162	1.650022422	217	1.498027887	272	1.637786964	327	1.448864489	382	1.585873293
53	1.827374478	108	1.707102827	163	1.448864489	218	1.767329719	273	1.174134873	328	1.376715003	383	1.928362072
54	1.850001078	109	1.295181743	164	1.431583267	218	1.912999214	274	1.528761721	329	1.776555785	384	1.748198189
55	1.890022377	110	1.376715003	165	1.625262405	220	1.612440949	275	1.650022422	330	1.528761721	385	1.827374478

Table 2 Average of every 15
images

Average of 15 images	Result
Average of 1–15	20.43977
Average of 16–30	23.37783
Average of 31–45	22.87557
Average of 46–60	27.85707
Average of 61–75	25.70682
Average of 76–90	23.00168
Average of 91–115	23.41345
Average of 116–130	20.47801
Average of 131–145	22.48374
Average of 146–160	22.87749
Average of 161–175	24.28252
Average of 176–190	22.79107
Average of 191–215	22.56961
Average of 216–230	24.13294
Average of 231–245	24.28334
Average of 246–260	21.49807
Average of 261–275	23.05114
Average of 276–290	21.8552
Average of 291–315	22.79911
Average of 316–330	22.61187
Average of 331–345	23.066378
Average of 346–360	24.4770002
Average of 361–375	25.77949
Average of 376–385	16.34043

repair. If result_features > 22.6 and result_features < 24, we assigned second priority repair. And if result_features > 24, we have assigned third priority of repair.

Accuracy, correction of result evaluation is given by formula: Accuracy = TP + TN/0.5,

where TP is called as true positive, and TN is called as true negative.

So we achieve 81.3% accuracy with the following calculation:

Accuracy total % = 31309/385 = 81.32207%.

5 Conclusion and Future Work

In this paper, we analyzed road condition using support vector machine classifier. Classification of images has been done as nature of pothole, cracks, and patches. By keeping threshold levels, we have identified road maintenance falls in high, medium, or low priority of repair. With our proposed algorithm, we have achieved accuracy of distress as 81.3%. Further, we can increase the classification accuracy by modern algorithm considering lighting and environmental conditions.

References

1. Rodrique, J.P. and T. Notteboom (2013): The Geography of Transport Systems. 3rd Edition.
2. "Report of Working Group on Road Transport" (PDF), Ministry of Road Transport, Government of India, 2011.
3. "An Overview: Road Network of India". Ministry of Road Transport, Government of India. 2010.
4. John Pucher; Nisha Korattyswaropam; Neha Mittal; Neenu Ittyerah, "Urban transport crisis in India" (PDF). Archived from the *original* (PDF) on 14 March 2007.
5. "Logistics in India". DHL. 2008. Archived from the original on 24 July 2012.
6. "Logistics in India, part 1 (A 3 part series)" (PDF). KPMG. 2010.
7. SuwarnaGothane,M.V. Sarode, "Analyzing Factors, Construction of Dataset, estimating importance of factor and generation of association rules for Indian road Accident", IEEE 6th IACC, 2728th Feb. 2016, Proc. IEEE Xplore, 18 August 2016.
8. Taehyeong Kim, Seung-Ki Ryu: System and Method for Detecting Potholes based on Video Data, Proc. Emerging Trends in Computing and Information Sciences, vol. 5 pp. 703–709, Sep. 2014.
9. S. Varadharajan, S. Jose, K. Sharma, L. Wander, C. Mertz: Vision for road inspection, Proc. IEEE Int'l Springs, Colorado, USA, pp. 115–122, Mar. 2014.
10. Lokeshwor Huidrom, Lalit Kumar Sud: Method for automated assessment of potholes cracks and patches from road surface video clips, Proc. Elsevier, 2nd Conf. Transportation Research Group of India, vol. 104, pp. 312–321, 2013.
11. Sylvie Chambon and Jean Marc Moliard: Automatic Road Pavement Assessment with Image Processing Review and Comparison, Proc. Int'l Journal of Geophysics, Hindawi Publishing Corporation, vol. 2011, pp. 1–20, June 2011.
12. Christian Koch, Ioannis Brilakis: Pothole detection in asphalt pavement images, Proc. Elsevier Ltd, Advanced Engineering Informatics, vol. 25, pp. 507–515, Aug. 2011.
13. Yijie Su, MeiQuing Wang: Improved C-V segmentation model based on local information for pavement distress images, Proc. 3rd International Congress, IEEE Conf. Image and Signal Processing, Yantai vol. 3, pp. 1415–1418 Oct. 2010.
14. Fanfan Liu, Guoai Xu, Yixian Yang, Xinxin Niu: Novel Approach to Pavement Cracking Automatic Detection Based on Segment Extending, Proc. IEEE Int'l Symp. Knowledge Acquisition and Modeling, Wuhan, IEEE, pp. 610–614, Dec. 2008.
15. Zhaoyun Sun, Wei Li, Aimin Sha: Automatic pavement cracks detection system based on Visual Studio C++ 6.0, Proc. Int'l Conf. Natural Computation, Yantai, Shandong, vol. 4, pp. 2016–2019, Aug 2010.
16. M. Mahmood, M. Rahman1, L. Nolle: A Fuzzy Logic Approach for Pavement Section Classification, Proc. Int'l Journal of Pavement Research and Technology, vol. 6, pp. 620–626, Sep. 2013.
17. S. K. Suman, S. Sinha: Pavement Maintenance Treatment Selection Using Fuzzy Logic Inference System, Proc. Int'l Journal of Engineering and Innovative Technology, vol. 2, pp. 172–175, Dec. 2012.
18. K. Sandra, V. R Vinayaka Rao, K. S. Raju, A.K Sarkar: Prioritization of Pavement Stretches using Fuzzy MCDM Approach fuzzy logic, Proc. Springer Soft Computing using Industrial Applications, vol. 39, pp. 265–276, 2007.

19. T. F. Fwa and R. Shanmugam: Fuzzy logic Technique for Pavement condition rating and maintenance, Proc. 4th Int'l Conf. on Managing Pavements, Center for Transportation Research, National University of Singapore, pp. 465–476, 1998.
20. Oliveira H. Correia P.L: Automatic Road Crack Detection and Characterization, Proc. IEEE Transactions on Intelligent Transportation Systems, vol. 14, pp. 155–168 Mar. 2013.
21. Saar T, Talvik O.: Automatic Asphalt pavement crack detection and classification using Neural Networks, Proc. IEEE Electronics Conference, 12th Biennial Baltic, Tallinn, pp. 345–348, Oct. 2010.
22. Guoai Xu, Jianli Ma, Fanfan Liu, Xinxin Niu: Automatic Recognition of Pavement Surface Crack Based on BP Neural Network, Proc. IEEE Int'l Conf. Computer and Electrical Engineering, Phuket, pp. 19–22, Dec. 2008.
23. Ravi Bhoraskar, Nagamanoj Vankadhara, Bhaskaran Raman, Purushottam Kulkarni: Traffic and Road Condition Estimation using Smartphone Sensors, Proc. IEEE COMSNETS, pp. 1–6, 2012.
24. Prashanth Mohan, Venkata N. Padmanabhan, Ramachandran Ramjee: Rich Monitoring of Road and Traffic Conditions using Mobile Smartphones, Proc. 6th ACM Conf. on Embedded network sensor systems, pp. 357–358, Nov. 2008.
25. Salman M, Mathavan S, Kamal K, Rahman M: Pavement crack detection using the Gabor filter, Proc. IEEE 16th Int'l Conf Intelligent Transportation Systems, vol. 93, pp. 2039–2044, Oct. 20.
26. Suwarna Gothane, M.V. Sarode, K. Sruja Raju, "Design Construction & Analysis of model Dataset for Indian Road Network and Performing Classification to Estimate Accuracy of Different Classifier with its Comparison Summary Evaluation", International Conference on Swarm, Evolutionary, pp 50–59, Proc. Springer Link Part of the Lecture Notes in Computer Science book series (LNCS), volume 9873, 01 December 2016.
27. HweeKwon Junga, ChangWon Leea, Gyuhae Parka, "Fast and non invasive surface cracks detection of press panel using image processing", Proc. Engineering 188 72–79, Science Direct, 6th Asia Pacific Workshop on Structural Health Monitoring, 2017.
28. L. Zhang, F. Yang, Y.D. Zhang, Y.J. Zhu, "Road crack detection using deep convolutional neural network", Int. Conf. on Image Processing (ICIP), IEEE, pp. 3708–3712, 2016.
29. Penghui Wang, Yongbiao Hu, Yong Dai, and Mingrui Tian, "Asphalt Pavement Pothole Detection and Segmentation Based on Wavelet Energy Field", National Engineering Laboratory for Highway Maintenance Equipment, 28 Feb 2017.

Design and Simulation of Capacitive MEMS Accelerometer

Yugandhar Garapati, G. Venkateswara Rao and K. Srinivasa Rao

Abstract In this paper, we have designed capacitive MEMS accelerometer. The sense element is a comb-type, inter-digitated one with a set of movable fingers suspended between another set of fixed fingers. The whole structure is made of doped polysilicon that enables most of the realization of the sense elements and structure dimension with conventional semiconductor processes used in a CMOS fabrication line. Of particular interest is the extremely low capacitance in the femtofarad range that the sense element provides as a pickup parameter. Improvement of this pickup parameter with the variation of various parameters of the structure and sense elements is to be explored. The dimensions and characteristics of the sense elements and structure are simulated and discussed.

Keywords Accelerometers · Capacitance · Displacement · MEMS
Polysilicon

1 Introduction

The application domain of these sensors such as aerospace and automotive industries demand their realization in miniature, compact form without compromising on aspects such as performance and cost [1–4]. It is in this context that one of the major

Y. Garapati (✉)
Department of Computer Science & Engineering, GITAM, Hyderabad, Andhra Pradesh, India
e-mail: yugandhar.garapati@gmail.com

G. Venkateswara Rao
Department of Information Technology, GITAM, Visakhapatnam, Andhra Pradesh, India
e-mail: vrgurrala@yahoo.com

K. Srinivasa Rao
Electrical Communication Department, KL Deemed to be University, Green Fields, Guntur, Andhra Pradesh, India
e-mail: srinivasakarumuri@gmail.com

© Springer Nature Singapore Pte Ltd. 2019
R. S. Bapi et al. (eds.), *First International Conference on Artificial Intelligence and Cognitive Computing* , Advances in Intelligent Systems and Computing 815,
https://doi.org/10.1007/978-981-13-1580-0_13

Fig. 1 MEMS comb
accelerometer

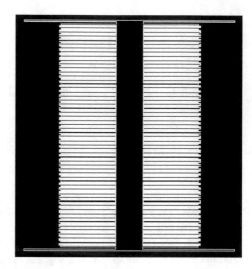

benefits of realization through MEMS is the amenability of integration of the sense element with the accompanying electronics gathers further importance.

In this paper, we have designed capacitive accelerometers by using COMSOL Multiphysics tool. They provide lower power, compact, and robust sensing. The sense element is a comb-type, inter-digitated one with a set of movable fingers suspended between another set of fixed fingers [5–9]. Improvement of this pickup parameter with the variation of various parameters of the structure and sense elements is to be explored. The dimensions and characteristics of the sense elements and structure are simulated and discussed.

2   Designing of Capacitive Accelerometer

The structure design of our polysilicon surface micro-machined MEMS comb accelerometer is shown in Fig. 1.

The dielectric material used is air. This structure consists of a set of movable fingers inter-digitated between another set of fixed fingers in a comb-like structure realized in doped polysilicon [10]. The fingers are suspended above the surface of a silicon substrate and are anchored only at the extreme ends through contact pads. The pickup parameter is the capacitance between the two sets of fingers, which varies according to the inertial acceleration to be applied in y direction. Figure 2 shows a single unit of the whole structure. Such units constitute the whole structure. The various design dimensions of the whole structure are listed below:

Centre beam (seismic mass or proof mass): $600 \, \mu m \times 75 \, \mu m$
Tether: $575 \, \mu m \times 3 \, \mu m$
Fingers (movable and fixed): $155 \, \mu m \times 2 \, \mu m$

Fig. 2 Single unit of
structure

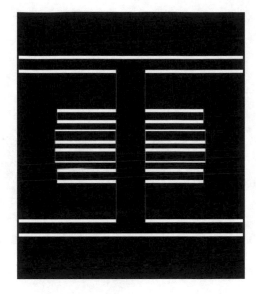

Fingers cells: 150 μm overlap and 2 μm spacing
Thickness of whole structure: 1 μm and 2 μm

2.1 Modeling and Simulation

For modeling and simulation [11], COMSOL Multiphysics has been used to compute
the static capacitance of the structure, without application of any acceleration.

2.2 Capacitance Calculation

In this paper, we have used electrostatics module of COMSOL Multiphysics simulation package to apply a voltage of 1 V to the proof mass with movable fingers and
0 V to all the fixed fingers for a 2D model with applied thickness of 1 μm [12–16].
The computations of static capacitance for both the small unit of the structure and the
whole structure have been done without application of any external perturbations.
For the small unit of the structure with 1 μm thickness, the capacitance value is 2.85
fF. As 48 such units are repeated for the whole structure and all the capacitances
are connected in parallel theoretically, the total capacitance can be calculated as
$48 \times 2.85 = 136.8$ fF [17–20]. The capacitance value computed by COMSOL Multiphysics simulation tool for the whole structure is 135.11 fF which is well within the
acceptable range. Similarly, the capacitance value is 5.70 fF for the small unit of the

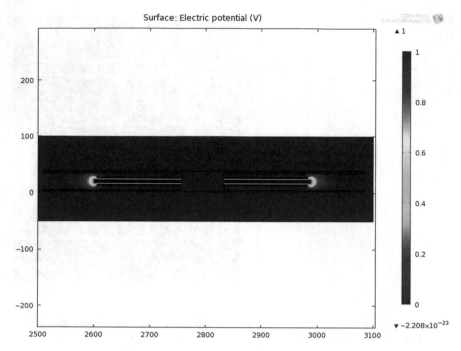

Fig. 3 2D model of small unit of the structure: electric potential

structure with 2 μm thickness and 270.33 fF for the whole structure. The simulation results are cited below:

For 2D model of small unit of the structure: electric potential (See Figs. 3 and 4). For 2D model of small unit of the structure with 1 μm thickness: capacitance value = 2.85 fF (See Figs. 5 and 6). For 2D model of whole structure with 1 μm thickness: electric potential (See Figs. 7 and 8). For 2D model of whole structure: capacitance value with 1 μm thickness = 135.11 fF (See Figs. 9 and 10).

3   Results and Discussion

The simulations were carried out using static conditions as well as dynamic conditions, i.e., by changing the thickness of the structures, at different displacements given in different directions and different acceleration inputs known as g levels, and the capacitance values are calculated and plotted.

As presented in figures, the capacitance value linearly increases with increase in thickness of the structure for different displacement values in y direction for small unit

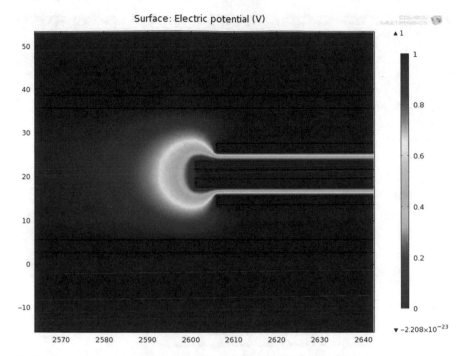

Fig. 4 2D model of small unit of the structure: electric potential

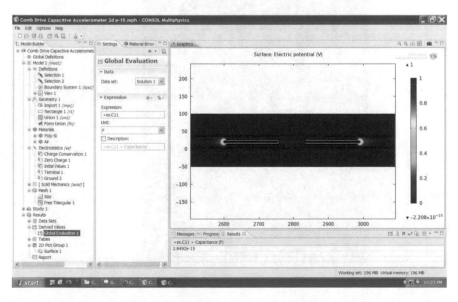

Fig. 5 2D model of small unit of the structure with 1 μm thickness : capacitance value

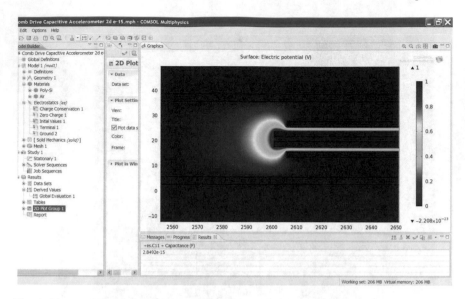

Fig. 6 2D model of small unit of the structure with 1 μm thickness : capacitance value

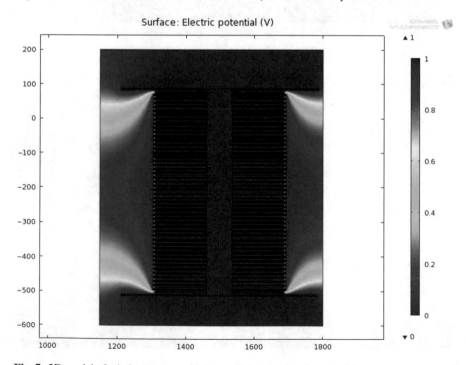

Fig. 7 2D model of whole structure with 1 μm thickness: electric potential

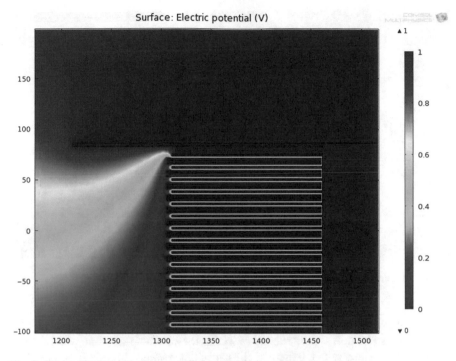

Fig. 8 2D model of whole structure with 1 μm thickness: electric potential

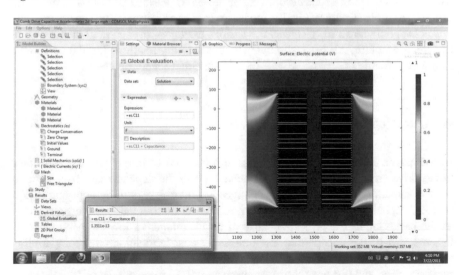

Fig. 9 2D model of whole structure: capacitance value with 1 μm thickness

of the structure as well as for the whole structure which is very much in accordance with the theory. Capacitance value increases with increase in displacement values for

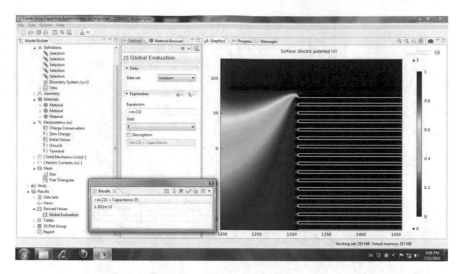

Fig. 10 2D model of whole structure: capacitance value with 1 μm thickness

different thickness of the structure, both in the positive and negative segments of y direction for small unit of the structure as well as for the whole structure which is very much in accordance with the theory. The capacitance value abruptly increases with time with increase in acceleration, i.e., 'g' levels for various thickness of the small unit of the whole structure. Finally, the increase in capacitance value can be observed with change increase in 'g' levels for various thickness of the small single unit of the small structure. All these simulated results are very much within the acceptable range of available theory of MEMS-based capacitive accelerometers.

4 Conclusion

In this paper, 2D device simulations have been used to investigate MEMS-based comb drive capacitive accelerometer behavior and the dependence of its static as well as dynamic characteristics on changes in dimensions and by applying disturbances. Capacitive sensing and its optimization of sensing value has been selected as the backbone of the project because of its advantages of low-temperature coefficients, low-power dissipation, low noise, low-cost fabrication, and compatibility with VLSI technology scaling. For these reasons, capacitive sensing has received the most attention and has been the most commercialized in recent years.

The optimization of the static characteristics of the structure by the variation of thickness was carried out. The resulting optimized device showed a very good response in terms of static capacitance as well as dynamic capacitance upon application of external disturbances. The optimization of the MEMS-based accelerometer resulted in improved value for static as well as dynamic capacitance. The change in the performance parameter value obtained under static conditions as well as dynamic conditions using simulation tool COMSOL Multiphysics 4.2a has been found to be within the acceptable range as per theoretical calculations. Also, the proposed device has been given disturbances in y direction which is an improvement over earlier designs which were subjected to disturbances in z direction. This change in direction of perturbations has led the solution of the problem of Stiction.

Thus, MEMS-based comb drive capacitive accelerometer with optimized device design by modifying the structure and improved sensing capacitance value has been proposed for high-performance, lightweight, and low-power applications. The proposed device can be considered as a future generation accelerometer with improved value of performance parameter which in turn is the pickup signal parameter for the signal conditioning circuitry and associated readout electronics.

References

1. Chang Liu "Foundations of MEMS" Special International Edition, Pearson Prentice Hall, USA, 2006.
2. Elena Gaura and Robert Newman "Smart MEMS and Sensor Systems" Imperial College Press, London, 2006.
3. Julian W. Gardner, Vijay K. Varadan and Osama o. Awadelkarim, "Microsensors, MEMS and Smart devices" 2nd Edition, John Wiley and Sons Ltd., England, 2002.
4. Larry K. Baxter, "Capacitive Sensors-Design and Applications", IEEE Press, New York, 1997.
5. Dr. Klaus C. Schadow, Professor Mehran Mehregany, Dr. Clyde Warsop, Mr. Paul Smith, "MEMS Aerospace Applications", North Atlantic Treaty Organisation, USA, 2004.
6. Mohamed Gad-el-Hak, "MEMS Introduction and Fundamentals" 2nd Edition, CRC Press, New York, 2006.
7. Marie-Ange Naida Eyoum, "Modularly Integrated MEMS Technology", University of California, Berkeley, 2006.
8. "An Introduction to MEMS (Micro-electromechanical Systems)" PRIME Faraday Partnership Wolfson School of Mechanical and Manufacturing Engineering Loughborough University, Loughborough, Leics LE11 3TU, London, 2002.
9. Matej Andrejaši˘, "MEMS Accelerometers", University of Ljubljana Faculty for mathematics and physics Department of physics, 2008.
10. John Doscher "Accelerometer Designs and Applications", Analog devices.
11. T.K. Sethuramalingam, Dr. A. Vimalajuliet, "Design of MEMS based capacitive accelerometer", International Conference on Mechanical and Electrical Technology (ICMET 2010), 2010.
12. Seiji Ayogi, Yu Chong Tai, "Development of surface micromachinable capacitive accelerometer using fringe electrical fields" Transducers '03, Boston, 2003.
13. Kanchan Sharma, Isaac G. Macwan, Linfeng Zhang, Lawrence Hmurcik, Xingguo Xiong, "Design Optimization of MEMS Comb Accelerometer", Department of Electrical and Computer Engineering, University of Bridgeport, Bridgeport, CT 06604.
14. Tolga Kaya, Behrouz Shiari, Kevin Petsch, David Yates, "Design of a MEMS Capacitive Combdrive Accelerometer", 1Central Michigan University, School of Engineering and Technology, Mount Pleasant, MI, US University of Michigan, Department of Electrical Engineering and Computer Science, Ann Arbor, MI, US.
15. Gang Zhang, "Design and simulation of a CMOS-MEMS accelerometer", Tsinghua University, 1994.

16. Ndu Osonwanne and Jason V. Clark, "MEMS Comb Drive Gap Reduction Beyond Minimum Feature Size: A Computational Study", Proceedings of the COMSOL Conference, Boston, 2010.
17. Jiangfeng Wu, "Sensing and Control Electronics for Low-Mass Low-Capacitance MEMS Accelerom".

Bioinformatics and Image Processing—Detection of Plant Diseases

N. Hanuman Reddy, E. Ravi Kumar, M. Vinay Reddy,
K. L. Raghavender Reddy and G. Susmitha Valli

Abstract This paper gives an idea of how a combination of image processing along with bioinformatics detects deadly diseases in plants and agricultural crops. These kinds of diseases are not recognizable by bare human eyesight. First occurrence of these diseases is microscopic in nature. If plants are affected with such kind of diseases, there is deterioration in the quality of production of the plants. We need to correctly identify the symptoms, treat the diseases, and improve the production quality. Computers can help to make correct decision as well as can support industrialization of the detection work. We present in this paper a technique for image segmentation using HSI algorithm to classify various categories of diseases. This technique can also classify different types of plant diseases as well. GA has always proven itself to be very useful in image segmentation.

Keywords Image processing · Bioinformatics · Image segmentation · Plant diseases · Classification

N. Hanuman Reddy · M. Vinay Reddy · K. L. Raghavender Reddy
Department of CSE, Vardhaman College of Engineering,
Hyderabad, Telangana, India
e-mail: hanuman@vardhaman.org

M. Vinay Reddy
e-mail: vinaykumarreddy@vardhaman.org

K. L. Raghavender Reddy
e-mail: raghu@vardhaman.org

E. Ravi Kumar (✉)
Department of IT, Vardhaman College of Engineering,
Hyderabad, Telangana, India
e-mail: ravikumar.e@gmail.com

G. Susmitha Valli
Department of Computer Science and Engineering, MLR Institute of Technology,
Hyderabad, Telangana, India
e-mail: susmita@mlrinstitutions.ac.in

© Springer Nature Singapore Pte Ltd. 2019
R. S. Bapi et al. (eds.), *First International Conference on Artificial Intelligence and Cognitive Computing* , Advances in Intelligent Systems and Computing 815,
https://doi.org/10.1007/978-981-13-1580-0_14

1   Introduction

We depend on agriculture not only for inside growth but also as one of the major exportable products. Indian economy to a large extent depends on agricultural growth and productivity. So it is the duty of the researchers and scientists to see the quality of agricultural production. There were techniques before to detect plant diseases and also to cure them. As science is progressing, the techniques are also modernized to a greater extent. These new techniques can speed up the process of disease detection and their rectification.

We need to detect plant diseases in their very initial stage. There are some automatic disease detection techniques as well. But modern tools like image processing and artificial intelligence can be applied with higher rate of accuracy [1]. These tools can be trusted for accurate detection. There are some diseases that are to be detected at an early stage, for example, a disease called as leaf disease that happens with pine trees. If these diseases are not adhered to an initial stage, they would affect the entire growth of agriculture.

Previously, researchers would detect the plant diseases with naked eyes. This kind of detection has two disadvantages. First, this method would require a large team of experts to forecast the results, and continuous monitoring is required. Another major disadvantage is that it is costly. Many experts give inaccurate predictions in this kind of method, and their forecasting is not to be relied upon. So we gradually shifted to automated techniques where we fully trusted only on the symptoms of the diseases.

As things are getting complicated, it was becoming difficult for the experts to rely only on the symptomatic treatment. We had to use other advanced techniques like image processing. Now experts are intelligent computer systems that can not only detect but also give a cure for the disease [2]. There was also grid computing method, but that was basically more costly. The intelligent computer system improves the rate of accuracy, with reduced time and increased throughput.

2   Related Work

Hue, saturation, and intensity color spaces can manipulate colors more efficiently. They can adjust the human interpretation of colors. The intensity component of HSI deals with brightness of the concerned image. This intensity component to a greater extent depends on the red, green, and blue component of the image. This method is also useful for implementing classical image processing functions such as histogram equalization and convolution. [3, 4]. This method provides with dynamic range of saturation, so it is preferred to shift colors or to compromise color. By the word saturation, we mean how much of white component is there in the pixel.

The hue component is defined at an angular range of $(0, 2\pi)$. The red is considered at an angle of 0, blue at $4\pi/3$, and green at an angle of $2\pi/3$. The HSI color space depends on the device, which means the color space is dependent on the color of the monitor screen [5]. The hue component helps to differentiate one color family from others. The saturation component helps to distinguish a strong color from a weak color [6]. The intensity component of the HSI model helps to distinguish a light color from a dark color.

Bioinformatics is a stream of science that merges information technology, biology, and computer science altogether. It is an advancing field that applies the science of computer application to recognize objects, manipulate, and analyze these objects. It also involves the concepts of biology to detect and cure diseases. In order to do this, it applies the emerging fields of computer applications like image processing, soft computing, and artificial intelligence [7, 8]. Bioinformatics along with computer science is also applied in the field of medical image processing. Bioinformatics uses complicated mathematical concepts to solve DNA and amino acid problems.

Bioinformatics is vital for storing, manipulating, and managing large biological datasets. Bioinformatics aims at discovering new inventions in the field of medical sciences. In the past few years, large advancements in the field of genomic and molecular biology have led to the advancement of bioinformatics. Bioinformatics aims at storing large dataset related to amino acid and DNA. The field of bioinformatics is thus responsible for analyzing and interpreting different data including amino acids and nucleotide and generates protein sequence and structure [9, 10].

Any abnormality that affects the growth of the plant is called a plant disease. Plant disease is carried by two types of agents called biotic and abiotic. Biotic means diseases caused by living organisms like bacteria, fungi, and nematodes. Abiotic means disease caused by non-living factors like toxicity, temperature, moisture. For example, the diseases in tomato are classified as Septoria leaf spot, late blight, and early blight. It is also seen that potato field is also infected by late blight pathogens in which the symptoms are generally shown after a week. The incubation of the pathogens takes a week time. These late blight pathogens nurture themselves on the host tissue. If the host dies, these pathogens also wither off. In case of seed tubers, these pathogens are difficult to see and they spread during cutting of the seeds. These pathogens grow in the home garden, and one of the most important sources of these pathogens is also home garden [10, 11].

In some part of the country, these pathogens can also grow independent of the host and are called oospores. This kind of structure can survive in the soil for a long period of time. They get their nutrition from the soil by affecting other plantations.

3 Proposed System

The proposed system first captures the image with the help of digital camera. The acquired image is preprocessed with the help of image enhancement techniques, and then clipping of unwanted area is done after which smoothing algorithm is applied to get a better result. The pixels which are green in color are maxed, and they considered being healthy. Other pixels whose green intensity value is more than a assumed threshold value are considered to be unhealthy. Then we apply HSI

Fig. 1 Potato leaf affected with early blight and symptoms of bacterial wilt

algorithm to divide the image into clusters of healthy and unhealthy pixels. The pixels that fall in the healthy region are unmasked.

The HSI algorithm concentrates on H (hue) and I (intensity) components of the algorithm. Sometimes it happens that, due to obstacle of luminosity, area to be considered for segmentation is not uniformly illuminated. So the hue value that is purely based on reflective property of the object is unaltered. If the image is a gray image, then we must stick to the "I" component of the image. We use fuzzy membership function to combine the "H" and "I" components of the image. We can use color occurrence method for feature extraction.

This color occurrence method considers color and texture to find a unique feature to decide on the parameters of health of the pixels of image (Fig. 1).

4 Result Analysis

We have captured the image with the help of digital camera. Then we clip the unwanted area of the image with the help of image processing techniques. We are considering threshold values of the pixels. The pixels whose green intensity values are more than the threshold values are considered to be unhealthy.

Our proposed method is better than normal eye detection, where the results are generally erroneous. As image processing tools are used, the results are accurate, and the approach is more scientific. The HSI algorithm along with fuzzy logic gives a better result as compared to other algorithm techniques.

5 Conclusions

We deal with a proposed system that helps us to recognize plant diseases. This proposed system makes us to recognize the plant diseases. It also helps to find out their coupled information like diseases name, their cures, and symptoms. Though color image segmentation is a critical task, our proposed system does it with much dexterity and skill. Segmentation is a challenging task in our proposed system as it is done on color images, where the color images vary from pixel to pixel. It is observed that the intensity images are clearer than the hue images.

We first acquire the image, and then as a part of preprocessing, we use some enhancement algorithms to enhance the area under consideration. Then the unwanted area is clipped after which we use image segmentation to divide the pixels into clusters of different intensities.

The image density around the bone areas in intensity image is close to each other than in the background areas. Using image segmentation, we also remove the noisy part of the image. This procedure of detecting the plant disease is robust and can be done with speed and accuracy. We also use color occurrence method for feature extraction and also to decide upon the health of the image.

As it is seen that image segmentation considering only Ne color component might not be successful, our algorithm is robust in the sense that it considers both the aspects of the image.

References

1. Monika Jhuria, Ashwani Kumar, Rushikesh Borse, Image Processing For Smart Farming: Detection of Disease and Fruit Grading, IEEE Proceedings of the 2013 IEEE Second International Conference on Image Information Processing, 2013, p. 521–526.
2. Shiv Ram Dubey, Anand Singh Jalal, Detection and Classification of Apple Fruit Diseases using Complete Local Binary Patterns IEEE, Third international conference on Computer and communication Technology, 2012, p. 247–251.
3. Ilaria Pertot, Tsvi Kuflik, Igor Gordon, Stanley Freeman, Yigal Elad, Identificator: A web-based tool for visual plant disease identification, a proof of concept with a case study on strawberry, Computers and Electronics in Agriculture, Elsevier, 2012, Vol.88, p. 144–154.
4. Xiaoou Tang, Fang Wen, IntentSearch: Capturing User Intention for One-Click Internet Image Search, IEEE transactions on pattern analysis and machine intelligence, 2012, vol.34, p. 1342–1353.
5. R.Gonzalez, R. Woods, Digital Image Processing, 3rd ed., Prentice- Hall, 2007. 7. CROP-SAP (Horticulture) team of 'E' pest surveillance: 2013: Pests of Fruits (Banana, Mango and Pomegranate) 'E' Pest Surveillance and Pest Management Advisory (ed. D.B. Ahuja), jointly published by National Centre for Integrated Pest Management, New Delhi and State Department of Horticulture, Commissionerate of Agriculture, Pune, MS. pp 67.
6. Parag Shinde, Amrita Manjrekar, Efficient Classification of Images using Histogram based Average Distance Computation Algorithm Extended with Duplicate Image Detection Elsevier, proc. Of Int. Conf. On advances in Computer Sciences, AETACS, 2013.
7. Patil, J. K., & Kumar, R. (2011). "Advances in image processing for detection of plant diseases". Journal of Advanced Bioinformatics Applications and research, 2(2), 135–141.
8. Wang, H., Li, G., Ma, Z., & Li, X. (2012, May). "Image recognition of plant diseases based on principal component analysis and neural networks". In Natural Computation (ICNC), 2012 IEEE Eighth International Conference on 29–31 May 2012, Chongqing pp. 246-251.
9. Wang,Z., Sun, X., Ma, Y., Zhang, H., Ma, Y., Xie, W., & Zhang, Y. (2014, July). "Plant recognition based on intersecting cortical model". In Neural Networks(IJCNN), 2014 IEEE. International Joint Conference 6–11 July 2014, Beijing pp. 975–980.

10. Novak, P., & Sindelar, R. (2013, November). "Ontology-based industrial plant description sup-porting simulation model design and maintenance". In Industrial Electronics Society, IECON 2013-39th Annual Conference of the IEEE 10–13 Nov. 2013, Vienna pp. 6866–6871.
11. Husin, Z. B., Shakaff, A. Y. B. M., Aziz, A. H. B. A., & Farook, R. B. S. M. (2012, February). "Feasibility study on plant chili disease detection using image processing techniques". In Intel-ligent Systems, Modelling and Simulation (ISMS), 2012 IEEE Third International Conference on 8–10 Feb. 2012, Kota Kinabalu pp. 291–296.

Hybrid Approach for Pixel-Wise Semantic Segmentation Using SegNet and SqueezeNet for Embedded Platforms

V. Mohanraj, Ramachandra Guda and J. V Kameshwar Rao

Abstract Recent research on convolutional neural network (CNN) has seen advances in accuracy improvements of object detection, recognition, tracking, and segmentation applications. Many CNN architectures have been developed in the last five years to achieve good amount of accuracy. To deploy trained CNN models on FPGA, Raspberry Pi and other hardware devices with reduced model size is a key area for research. In this paper, a hybrid approach for pixel level semantic segmentation using SqueezeNet and SegNet is proposed. The hybrid architecture enhances the accuracy to 76%.

Keywords Convolutional neural network · Segmentation · Encoder · Decoder
SqueezeNet · SegNet

1 Introduction

Image segmentation is the process of assigning a label to every pixel in an image such that pixels with the same label share certain characteristics. The applications of image segmentation are object detection, medical imaging, machine vision, and traffic control systems. Image segmentation can be done in multiple methods, namely thresholding, clustering, compression, histogram, and convolution neural network. Recently, CNN-based methods achieve better result in object detection, recognition, and image segmentation.

There are few areas where the segmentation plays a prominent role like industrial inspection, tracking of objects in sequence images, and some of the classifications

V. Mohanraj (✉) · R. Guda · J. V Kameshwar Rao
HCL Technologies Limited, Noida, India
e-mail: mohanraj4072@gmail.com

R. Guda
e-mail: grcksrgm@gmail.com

J. V Kameshwar Rao
e-mail: kameshjvkr@gmail.com

© Springer Nature Singapore Pte Ltd. 2019   155
R. S. Bapi et al. (eds.), *First International Conference on Artificial Intelligence
and Cognitive Computing* , Advances in Intelligent Systems and Computing 815,
https://doi.org/10.1007/978-981-13-1580-0_15

Fig. 1 Image segmentation
during autonomous car
driving [10]

on the satellite image. Due to the recent advancement of segmentation, it is also used in a medical imaging field to detect and measure the bone and tissues. Segmentation for advanced driver-assistance system holds prominence.

Existing semantic segmentation using CNN architectures is SegNet [1], ApesNet [2], SqueezeNet [3], FCN [4] and CNN architectures of visual recognition are VGG [5], AlexNet [6], GoogleNet [7], ResNet [8], ENet [9] (Fig. 1).

From the literature review [11], many CNN architecture has been developed for image recognition and segmentation. There is a need for better accuracy and reduced number of parameters for optimal performance on deep learning models of embedded platforms. To overcome these issues, this paper proposes a hybrid approach for pixel-wise segmentation on FPGA devices.

2 Hybrid Approach for Semantic Segmentation

In this paper, a hybrid approach for pixel-level semantic segmentation of a road scene understanding is proposed using SqueezeNet and SegNet-Basic. The architecture of hybrid semantic segmentation is shown in Fig. 2. Hybrid approach contains encoder, decoder, and softmax classifier layer stacked on the network sequentially. In the encoder network, SqueezeNet architecture is used by removing last convolution, averaging pooling, and softmax classifier from the network.

2.1 Encoder Network

The encoder network consists of 8 fire layer, 1 convolution and 3 max pooling layer stacked on the encoder network sequentially. The fire layer consists of a squeeze and expand convolution layer. A squeeze convolution layer (which has only 1*1 filters) feeding into an expand layer has a mix of 1*1 and 3*3 convolution filters. We expose three tunable dimensions (hyperparameters) in a fire module: s_{1*1}, e_{1*1}, and e_{3*3}. In

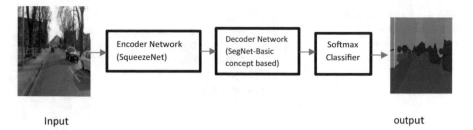

Input   output

Fig. 2 Hybrid semantic segmentation

Fig. 3 Fire layer

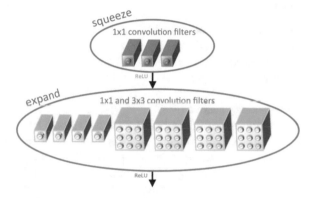

a fire module, s_{1*1} is the number of filters in the squeeze layer, e_{1*1} is the number of 1*1 filters in the expand layer, and e_{3*3} is the number of 3*3 filters in the expand layer. After each convolution layer, ReLU activation function is used. Maxpooling layer is used after the fire layer to reduce the spatial dimension of the image. In this paper, filter size 2*2 with a stride of 2 is used for maxpooling layer. Figure 3 shows squeeze and expand convolution in the fire layer [3].

Convolution (Conv) Layer
Convolution layers apply multiple filters to the raw input image to extract the high-level features. The convolution layer's parameters consist of a set of learnable filters. The size of every filter is smaller than the input width and height of volume. Slide each filter across the width and height of the input volume and compute the dot products between entries of the filter and the input at any position. It will produce a two-dimensional activation map that gives the responses of that filter at every spatial position. The network will learn the filters that activate when they see some type of visual feature.

In the first convolution layer, it detects the edges from raw pixel data. In the second convolution layers, it detects shapes or blobs from the previous layer output. In the subsequent convolution layer, it detects the high level from the last layer. Convolution layer accepts a volume size of $W1*H1*D1$ and requires four hyperparameters such as number of filters K, filter size F, stride S, and amount of zero padding P. The output volume size of the single convolution layer is $W2 * H2 * D2$.

Fig. 4 Maxpooling with 2*2 filter and stride of 2

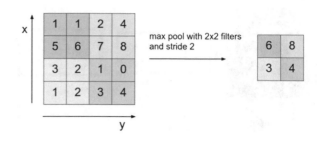

where

$$W2 = ((W1 - F + 2P)/S) + 1$$
$$H2 = ((H1 - F + 2P)/S) + 1$$
$$P = (F - 1)/2$$
$$D2 = K$$

Max-pooling Layer

The Max-pooling layer is inserted in between the convolution layers of encoder network. The Max-pooling layer is used to reduce the spatial size of image. The Max-pooling layer operates independently on every depth slice of the input, and it resizes the image using MAX operation. The pooling layer accepts input volume size of $W1*H1*D1$, and it requires two hyperparameters such as filter size F and stride S. It produces the output volume size of $W2*H2*D2$. Figure 4 shows the maxpooling operation for 4*4 matrix with 2*2 filter and stride of 2.

where

$$W2 = (W1 - F)/S + 1$$
$$H2 = (H1 - F)/S + 1$$
$$D2 = D1$$

2.2 Decoder Network

In decoder network, SegNet-Basic architecture concept is used to form decoder; the network consists of 8 fire layers, 3 upsampling, 2 convolution layers stacked on the network. Upsampling layer is used in the decoder network to get back the original dimension of the input image to perform the pixel-wise semantic segmentation.

UpSampling

Upsampling layer is used in between convolution layer of decoder network. During max pool operation, the positions of maximum value are stored in the max-pool indices. The decoder network upsamples its input feature map based on the max-pooling indices of the corresponding encoder feature map. Figure 5 shows the upsampling operation for the subfeature map.

Fig. 5 Upsampling for
subfeature map

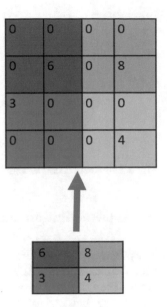

2.3 Softmax Classifier

Softmax classifier is used to get the probability of each pixel for classification. For each pixel K (number of classes), probability values are calculated. Among K probability values, the maximum value is the predicted class. Probability value for each pixel is calculated by softmax classifier as shown in below equation.

$$y_k = \frac{\exp(a_k)}{\sum_i \exp(a_i)}$$

$$a_k = X_i W_i + b$$

where X represents input vector, W—represents weight vector and b—bias value.

Figure 6 shows the detailed architecture of encoder and decoder network layer sequence for hybrid segmentation. Totally 16 fire layer, 3 convolution, 3 maxpooling, and 3 upsampling layers are used in the encoder and decoder network. In Fig. 6, SqueezeNet block represents encoder network of hybrid semantic segmentation for the proposed method.

Table 1 shows the layer-wise input and output dimension, filter size, and stride for the encoder and decoder network. Encoder network represented in the blue color and red color represents the decoder network of hybrid segmentation.

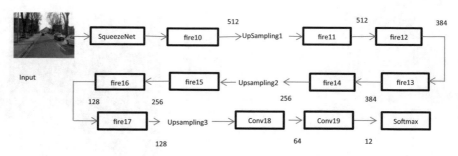

Fig. 6 Encoder decoder network for hybrid semantic segmentation

3   Implementation and Results

The proposed method is implemented in cloud-based GPU environment (Tesla K80). CamVid dataset is used to measure performance of the proposed segmentation technique. The CamVid is a road scene segmentation dataset consisting of 367 images for training and 233 images for testing RGB images (day and dusk scenes) with 11 semantic classes of manually labelled objects, e.g., car, tree, road, fence, pole, building [ref]. The original image resolution is 720×960, while we downsampled all images to 360×480 [1]. Figure 7 shows the training and validation loss/accuracy for CamVid dataset with 100 epochs. The proposed method uses OpenCV and Keras with backend Tensor flow based python libraries are used for implementation.

Figure 8 shows the test image result of pixel semantic segmentation for the proposed method. The first column shows original images of CamVid dataset; the second column shows pixel-wise semantic segmentation for a hybrid approach.

Fig. 7 Training and validation loss/accuracy

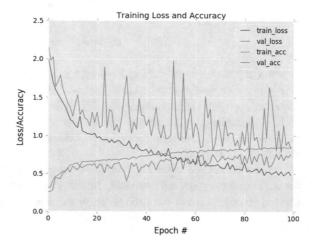

Table 1 Layer output size, filter size, stride, and number of filters for CNN architecture

Layer name/type	Output size	Filter size/stride	Squeeze s_{1*1}	Expand e_{1*1}	Expand e_{3*3}
Input image	360*480*3				
Conv1	360*480*64	3*3/1 (64)			
Maxpooling1	180*240*64	2*2/2			
Fire2	180*240*128		16	64	64
Fire3	180*240*128		16	64	64
Fire4	180*240*256		32	128	128
Maxpooling2	90*120*256	2*2/2			
Fire5	90*120*256		32	128	128
Fire6	90*120*384		48	192	192
Fire7	90*120*384		48	192	192
Fire8	90*120*512		64	256	256
Maxpooling3	45*60*512				
Fire9	45*60*512		64	256	256
Fire10	45*60*512		64	256	256
Upsampling1	90*120*512				
Fire11	90*120*512		64	256	256
Fire12	90*120*384		48	192	192
Fire13	90*120*384		48	192	192
Fire14	90*120*256		32	128	128
Upsampling2	180*240*256				
Fire15	180*240*256		32	128	128
Fire16	180*240*128		16	64	64
Fire17	180*240*128		16	64	64
Upsampling3	360*480*128				
Conv18	360*480*64	3*3/1 (64)			
Conv19	360*480*12	1*1/1 (12)			

Table 2 Performance analysis of proposed hybrid semantic segmentation

Model	Accuracy (%)
SegNet- Basic [1]	62.9
Hybrid segmentation (proposed method)	76

From the experimental analysis, the proposed method performs better in terms of accuracy, compared to existing semantic segmentation architecture such as SegNet-Basic. Table 2 shows the performance analysis of proposed semantic segmentation.

Fig. 8 Test images result in semantic segmentation of the trained CNN model

4 Conclusion

In this paper, pixel-wise semantic segmentation for road scene understanding is proposed using a convolutional encoder–decoder network. The proposed method is validated on CamVid dataset, and 76% accuracy is obtained on the test data. The proposed hybrid segmentation method achieved better accuracy compared to the existing segmentation techniques for embedded platforms. In future, new CNN architectures will be designed to enhance accuracies further.

References

1. Vijay Badrinarayanan, Alex Kendall and Roberto Cipolla "SegNet: A Deep Convolutional Encoder-Decoder Architecture for Image Segmentation." PAMI, 2017.
2. Chunpeng Wu, Hsin-Pai Cheng, Sicheng Li, Hai (Helen) Li, Yiran Chen, "ApesNet: a pixel-wise efficient segmentation network for embedded devices", IET Cyber-Physical Systems: Theory & Applications, ISSN 2398-3396, ISSN 2398-3396, 2016.

3. Forrest N. Iandola, Song Han, Matthew W. Moskewicz, Khalid Ashraf, "Squeezenet: alexnet-level accuracy with 50x fewer parameters and < 0.5mb model size", arXiv:1602.07360, 2016.
4. Long, J., Shelhamer, E., Darrell, T.: 'Fully convolutional networks for semantic segmentation'. Int. Conf. on Computer Vision and Pattern Recognition (CVPR), 2015.
5. K. Simonyan and A. Zisserman, "Very deep convolutional networks for large-scale image recognition," CoRR, vol. abs/1409.1556, 2014.
6. Alex Krizhevsky, Ilya Sutskever, Geoffrey E. Hinton, "ImageNet Classification with Deep Convolutional Neural Networks", NIPS'12 Proceedings of the 25th International Conference on Neural Information Processing Systems - Volume 1, Pages 1097–1105.
7. C. Szegedy, W. Liu, Y. Jia, P. Sermanet, S. Reed, D. Anguelov, D. Erhan, V. Vanhoucke, and A. Rabinovich, "Going Deeper with Convolutions", IEEE Conference on Computer Vision and Pattern Recognition (CVPR), 2015.
8. K. He, X. Zhang, S. Ren, J. Sun, "Deep Residual Learning for Image Recognition", IEEE Conference on Computer Vision and Pattern Recognition (CVPR), 2016.
9. Aabhish, sangpilkim, euge@purdue.edu "ENet: A Deep Neural Network Architecture for Real-Time Semantic Segmentation", arXiv:1606.02147v1.
10. https://wiki.tum.de/display/lfdv/Image+Semantic+Segmentation.
11. A. Garcia-Garcia, S. Orts-Escolano, S.O. Oprea, V. Villena-Martinez, and J. Garcia-Rodriguez, "A Review on Deep Learning Techniques Applied to Semantic Segmentation", arXiv:1704.06857v1, 2017.

Software Modernization Through Model Transformations

Prabhakar Kandukuri

Abstract The legacy software contains experienced and validated knowledge that is very useful. But, these legacy systems may be problematic in few aspects like compatibility, security, enhancement, complexity. Hence, software modernization is required to meet the current business needs. The existing approaches like re-factoring, re-architecture, re-engineering, white-box modernization, black-box modernization are not sufficient for handling above-mentioned problem in a cost-efficient way. The proposed approach is based on model transformations where legacy system automatically passes through different models to reach the executable code in modern technologies. In this approach, only database of legacy software is required to modernize. This approach is able to exploit the problem faced by the legacy systems as there is no availability of proper document to it. It provides a discriminative way in providing proper design document for legacy systems through model transformations.

Keywords Legacy database · Legacy modernization · Model transformations
Re-engineering · Re-factoring

1 Introduction

Presently, the legacy software is referred to as outdated software because usage of other software is increased instead of using available rationalized versions in the existing system. Generally, most of the data is stored in mainframes and minicomputers which became legacy. And software used in the legacy system may not associate with upcoming technologies that are applicable to the context, thus which creates a big confusion as there is no proper design documentation.

The legacy system is really been a major problem especially with business and insurance companies. Through the proposed technique, it results in the creation of

P. Kandukuri (✉)
Department of C. S. E., Vardhaman College of Engineering, Hyderabad, India
e-mail: kp.cs@rediffmail.com; prabhakarcs@gmail.com

© Springer Nature Singapore Pte Ltd. 2019   165
R. S. Bapi et al. (eds.), *First International Conference on Artificial Intelligence
and Cognitive Computing* , Advances in Intelligent Systems and Computing 815,
https://doi.org/10.1007/978-981-13-1580-0_16

the document, gives a report on each state, and also generates code for any legacy system. This approach definitely results in handling the legacy system and finally obtaining the exact code for the system in modern technologies. Using this approach, legacy modernization is experimentally done for "*Millionaire Recipes*" system. The following are the discussions which are made for the legacy system.

Transparent processes and tools are used to provide easy maintenance by this technique and it is observed that almost up to 40% of reduction occurs in the maintenance cost, so maintenance productivity can improvise.

According to this research, the rest of the paper is ordered as, Sect. 2 gives a clear introduction about the aim of the research-related issues and the overview of the basic strategies, activities, and techniques for modernization of legacy systems. Section 3 focuses on representation and the detail explanation regarding the conversion from the table to the entity–relationship model and further conversion from the entity–relationship model to the class diagram. Section 4 shows the model transformation using the legacy modernization for a system. In Sect. 5, results obtained for the work are discussed. Section 6 is about the conclusions and the future enhancement for the work is presented.

2 Related Work

Advanced software systems need to come into existence in order to face the evolution of new technologies and the currently changing business requirements. According to Malinova, nowadays the software systems are becoming legacy when they begin to resist change themselves [1]. Generally, the legacy systems are providing epochal business value that is the reason why they must be modernized and replaced [2]. Weiderman divided the software evolution as software maintenance, modernization, and replacement, respectively [3, 4]. He proposed a "Big-Bang" approach. In this regard, Lehman's first law [5] states that software has to be adopted otherwise it would not give satisfactory results. In such case, modernization helps to keep the things up-to-date and in use.

Sneed explained about wrapping and its levels of the legacy system [6]. He considered the process level, the simplest form of encapsulation [7]. Wrapping of the legacy system is a three-step process; they are wrapper construction, adoption of the target program, and finally the interaction between wrapper and target program.

Peetambera explained various legacy modernization techniques and its pros and cons [8–10]. Chikofsky E. explained re-engineering of legacy systems as reconstitution of legacy system by examination and alteration [11]. According to Demeyer, re-engineering includes three phases, and they are forward engineering, reverse engineering, and re-engineering [12]. According to Fowler M., re-factoring is a process of modifying the existing software in order to improve the performance without changing its behavior [13].

Comella-Dorda defined wrapping as a technique to carry out the re-engineering [14]. Further Mens, discussed different levels of legacy modernization like User Interface (UI), functional and data [15]. Venema T. proposed an automated migration approach without human intelligence in transformations [16]. Commercial-off-the-shelf (COTS) construction can be used in package implementation. But reusability of the legacy system is not possible with COTS construction [17].

The wrapping process mainly concentrates on interfaces and hides the complexities; thus, it provides flexibility [18]. Dijk presented a modernization process to create a Web interface of a legacy system [19]. The simplest way to modernize the legacy software is wrapping the existing system through SOA wrappers [20].

3 An Approach to Model Transformations

A model transformation is an automated way of modifying and creating models. The aim of using a model transformation technique is to reduce effort and errors by automating the building and modification of models where possible. Legacy systems are those which do not have a proper design document. Even the code exists for the system, it may not in an understandable format and there is a death of technology skills in it. The primary issue for legacy software is that it has only the database table with no proper design document. So, the database table of the legacy system is considered and it is converted to an entity–relationship diagram (E-R diagram). An E-R diagram shows the relationship between the entities present in the system. This diagram gives the information of the system along with the entities and the attributes. In this work, "DeZign for databases" open-source tool is used for conversion of database to E-R diagram.

Once the E-R diagram is obtained, it has to be transformed into class; it can be done through the open-source tool called "visual paradigm" as per the given procedure in Fig. 1. After generation of class diagram from E-R diagram, the class would not show the methods in its third compartment. Hence, the methods are provided using use cases. The use case diagrams are usually referred to as behavioral diagrams which shows the set of actions which are going to be performed by the system. Hence, each

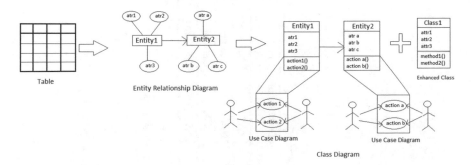

Fig. 1 An approach to model enhancement

Table 1 Event details

Id	Event name	Event information	Location	Organization name	Phone number	Fee	Date	Time	Picture
1	ZAX	Tech. event	BZA	KLCE	98765432	70	7-1-17	8:45	Poster
2	NAS	Coding	HYD	VMC	54321678	45	9-1-11	9:00	Picture
3	ENIX	Cultural	NRT	NEC	67890123	55	7-1-17	1:00	Imag1
4	AZIO	Marketing	VNK	NIEC	12345678	74	9-1-11	9:13	Imag2

Table 2 Event registries

Id	First name	Held date	Student name	College	Year	Branch	Phone No.	Mail Id
Null	AAAAA	07-12-2017	Prabhas	JNTU	2012	cse	7777777777	kp@z.com
Null	BBBBB	09-10-2011	Pramu	VCE	2017	IT	9999999999	pm@z.com
Null	CCCCC	07-12-2017	Koti	CMR	2012	cse	8888888888	bm@z.com
Null	DDDDD	09-10-2011	Ramesh	RVR	2017	IT	5555555555	rm@g.com

Table 3 Login data	User name	Password
	user	data

and every class of legacy system will get a use case diagram in place of methods. After obtaining class diagram, the enhanced classes can be added to perfect the system.

According to this research, an "Event Registration" Web service is considered; the database name is "student_info" which consists of three tables named as Event details, Event Registers, and Login Data. The tables for the Event Registration system are shown in Tables 1, 2, and 3, respectively (Fig. 2).

The conversion of E-R diagram to class diagram is shown in Fig. 3, which is drawn with the help of visual paradigm open-source tool.

After obtaining the class diagram with class name and attributes, the third compartment of the class has to be filled. There, the model integration comes into the picture. The model integration refers to the inclusion of a model (diagram) into another model. In this context, use case diagram is integrated into class diagram for the purpose of filling the third compartment of a class in order to provide methods.

The use case diagram is integrated into "Event" class for providing two methods, namely EventDisplay() and PublishEvents(). The use case diagram is obtained by the actions performed by the "EventRegistries" class and actions are turned as method, namely RegistrationForEvents(). Similarly, for the "LoginData" class the use cases act as methods in the class, namely Login() and Logout(). Now, the enhancement refers to increase in the modules that are adding extra features. For the given system, an additional class "MailingForRegistered" is been added to the class diagram, which is shown in Fig. 4.

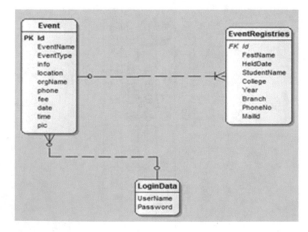

Fig. 2 Automatic transformation of database to E-R diagram through DeZign tool

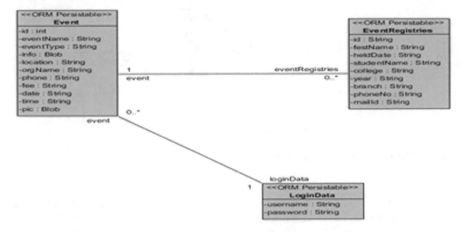

Fig. 3 Transformation of E-R diagram to class diagram through visual paradigm

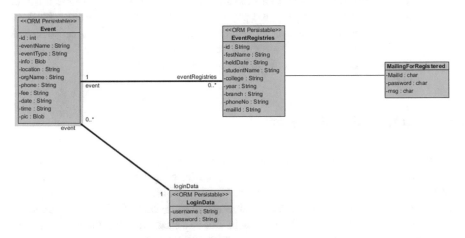

Fig. 4 Model enhancement through visual paradigm

Figure 4 shows the class diagram for the existing three classes (legacy system) and added additional features through visual paradigm by using model transformation technique.

4 An Approach to Legacy Software Modernization

The legacy modernization includes both existing functionalities and enhanced functionalities. In the modernization process, the existing functionalities are taken from the running system and included in the class diagram which is generated from the

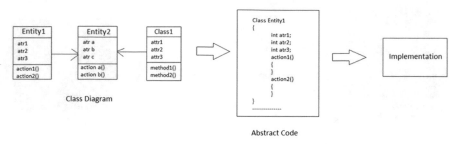

Fig. 5 Legacy modernization approach

E-R diagram of existing database. In this process, the new functionalities are added which is taken from the requirements.

After this integration process, the newly designed system has to be verified and then given to automatic code generation tool. Now, the tool will provide abstract code, and for this experiment, "visual paradigm" tool is used. This process is given in Fig. 5. With minimum modifications, the system will be modernized. Classes without methods can be derived through Extensible Markup Language (XML) tools. These classes are used for verification before modernization.

5 Results and Discussions

All were developed for a legacy system "Millionaire Recipes" and the code is been obtained by the approach model transformation, where the database present for legacy system is considered and further converted to the E-R diagram by using DeZign for databases tools and the obtained E-R diagram is transformed to the class diagram through the visual paradigm. Here, the behavior and state are being identified and the code is generated from the class diagram. This implementation is done for a legacy system and outputs were obtained experimentally successful.

According to the proposed approach, the Java code is obtained for the given system. Here, four Java files are generated because there are four classes in the class diagram for the system "Millionaire Recipes." Now, the obtained Java code generated is to be compared with the Java code generated by another process. This comparison is done to verify the code generated through our approach satisfies the code which is obtained through the XML code by using converters. In this process generally, the database table of a legacy system is considered by looking at the table, and it is easy to write the XML code.

Here, model transformations' approach results in reducing the user time spent on modernization of the legacy systems. It also reduces the human labor because the tools and techniques used in this approach can be used easily without any effect. The tools undergo automated migration. It integrates best tools for model transformations.

6 Conclusions and Future Enhancements

This paper addressing the problems faced by the legacy system and provided a holistic approach to solving the critical problem by providing design document for the legacy system. This work presented an easy way to add features like classes, packages to the legacy system, and finally generated code even though there is no code and design document present earlier for the legacy system. The performance of the approach is experimentally verified for a particular legacy system Millionaire Recipes. This approach is cost-effective because it takes less time, cost, and effort. The tools used are free wares and fully automated. Hence, it reduces the time to modernize the software system.

Further, this approach helps to generate the test cases easily for legacy software modernization and used in regression testers. Hence, it reduces the regression testing cost considerably [21]. It helps to improve the code readability when we integrate activity and sequence diagrams to the given approach in Fig. 1.

References

1. AnnaMalinova, "Approaches and Techniques for legacy software modernization" partially supported by the RS-2009-M13 project of the scientific Fund of the University of Plovdiv "PaisiiHilendarski, Bulgaria., 2011.
2. Malinova, A., V. Yordanov, J. van Dijk," Leveraging existing plasma simulation codes, International Book Series "Information Science &Computing", Number 5, pp. 136–142, Suppl. to the Int. J. "Information Technologies &Knowledge", Vol. 2/2008.
3. Weiderman, N., J. Bergey, D. Smith, Tilley, Scott R., "Approaches to Legacy System Evolution (CMU/SEI-97-TR-014), Pittsburgh, Pa.: Software Engineering Institute", Carnegie Mellon University, 1997.
4. Weiderman, N., L. Northrop, D. Smith, S. Tilley, K. Wallnau, "Implications of Distributed Object Technology for Reengineering (CMU/SEI-97-TR-005 ADA326945)". Pittsburgh, Pa.: Software Engineering Institute, Carnegie Mellon University, 1997.
5. Lehman, M. M., Ramil, J. F., Wernick, P. D., Perry, D. E., Turski, W. M, "Metrics and Laws of Software Evolution - The Nineties View", Proc. of the 4th Int. Symp. on Software Metrics, METRICS'97, IEEE Computer Society, Washington, 1997, p. 20.
6. Sneed, H., "Encapsulating Legacy Software for Reuse in Client/Server Systems", Proc. of WCRE-96, IEEE Press, Monterey, 1996.
7. Sneed, H., "Encapsulation of legacy software: A technique for reusing legacy software components", Ann. Softw. Eng, 9, 1–4 Jan. 2000, pp. 293–313.
8. PeetamberaandkarthikVenkatachalam: Legacy Modernization, 2007. www.infoysis.com.
9. Malinova, A. A., S. G. Gocheva-Ilieva, I. P., Iliev, "Wrapping legacy codes for Numerical simulation applications", Proc. of the III International Bulgarian-Turkish Conf. Computer science, Istanbul, Turkey, October 12–15, 2006, PartII, pp. 202–207, 2007.
10. Seacord, R., D. Plakosh, G. Lewis, "Modernizing Legacy Systems: Software Technologies, Engineering Processes, and Business Practices", Addison-Wesley, 2003.
11. Chikofsky, E., J. Cross II, "Reverse engineering and design recovery: A taxonomy, Software Reengineering", IEEE Computer Society Press, 1992, p. 54–58.
12. Demeyer, S., S. Ducasse, O. Nierstrasz, "Object-Oriented Reengineering Patterns, Square Bracket Associates", Switzerland, 2009.

13. Fowler, M., K. Beck, J. Brant, W. Opdyke, D. Roberts, "Refactoring: Improving the Design of Existing Code", AddisonWesley, 1999.
14. Comella-Dorda, S., K. Wallnau, R. Seacord, J. Robert, "A Survey of Black-Box Modernization Approaches for Information Systems", Proc. of the Int. Conf. on Software Maintenance, 2000, ICSM. IEEE Computer Society, Washington, p. 173.
15. Mens, T., S. Demeyer, "Software Evolution", Springer, 2008.
16. Venema, T., "The Oracle IT Modernization Series: The Types of Modernization, An Oracle" White per, http://www.oracle.com/technologies/modernization/docs/typesof moderniza-tion.pdf, 2008.c.
17. Venkatraghavan, N., "Legacy Modernization: Modernize and Scale", Infosys White Paper, http://www.infosys.com/microsoft/resourcecenter/Documents/legacy-modern.pdf, 2008.
18. Almonaies, A., J. Cordy, T. Dean, "Legacy System Evolution Towards Service-Oriented Archi-tecture", Proc. SOAME 2010, Int. Workshop on SOA Migration and Evolution, Madrid, Spain, March 2010, pp. 53–62.
19. Dijk, J. van, A. Malinova, V. Yordanov, Mullen J.J.A.M. van der, "NewInterfaces for the PlasimoFramework", 6th Int. Conf. on Atomic and Molecular Data and Their Applications, Beijing, China, 27–31 Oct. 2008, AIP Conf. Proc., Vol. 1125, 2009, pp. 176–187.
20. "Agile legacy life cycle", Illustrations: Alfredo Carlo, Capgemini Paper, www.capgemini.com.
21. Prabhakar K., "Cost Effective Model Based Regression Testing", "IAENG- World Congress on Engineering", Vol. I-WCE 2017, pp 241–246, July 5–7, 2017, London, U.K., 2017.

Semi-automatic Annotation of Images Using Eye Gaze Data (SAIGA)

Balavenkat Gottimukkala, M. P. Praveen, P. Lalita Amruta and J. Amudha

Abstract Eye gaze tracking is based on pupil movement and is an effective medium for human–computer interaction. This field is utilized in several ways and is gaining popularity due to its increased ease of use and improved accuracy. The main objective of this paper is to present a framework that would assuage the burden of image annotations and make it more interactive. Images are annotated by physically describing their metadata. The current system gives the user 100% freedom to label the images at his/her discretion but is very tedious and time consuming. Here, we propose semi-automatic annotation of images using eye gaze data (SAIGA), an approach that would assist in using the eye gaze data to annotate images. SAIGA—the proposed framework shows how time and physical efforts spent on manual annotation can be bettered by a large value.

Keywords SAIGA · Eye gaze tracking · Image annotation · Image ranking algorithm · Human–computer interaction

1 Introduction

In today's world, using a search engine for image retrieval has become a very common phenomenon and it happens on a massive scale and range. Search for images ranges from an object to a location to a person or practically anything imaginable. Every

B. Gottimukkala (✉) · M. P. Praveen · P. Lalita Amruta · J. Amudha
Department of Computer Science & Engineering, Amrita School of Engineering, Amrita Vishwa Vidyapeetham, Bengaluru, India
e-mail: balavenkatg.96@gmail.com

M. P. Praveen
e-mail: praveenmeenakshi56@gmail.com

P. Lalita Amruta
e-mail: pl.amruta@gmail.com

J. Amudha
e-mail: j_amudha@blr.amrita.edu

© Springer Nature Singapore Pte Ltd. 2019
R. S. Bapi et al. (eds.), *First International Conference on Artificial Intelligence and Cognitive Computing*, Advances in Intelligent Systems and Computing 815, https://doi.org/10.1007/978-981-13-1580-0_17

search engine or image retrieval system uses different algorithms to extract the images from its database. This extraction or retrieval is based on the indices or annotations of the images.

Image annotation is generally done by manually adding metadata to an image to describe it or automatically using image processing techniques. The amount of time spend on manual annotation is quite tedious; however, this is a mandatory one for many applications which require images to be labelled for training. To decrease the time consumption and to reduce the manual efforts behind annotation, we propose the use of eye gaze tracking into labelling of these images. Eye gaze tracking is a technique where a person's pupil movement is recorded by a hardware and integrated with a software which points to where the person looked at on the computer screen, how long he looked, etc. This data when processed help in learning about the interesting regions of an image. **SAIGA** is an implementation of eye gaze tracking to annotate images with an added image ranking method. To put it simply, it allows the user to annotate images just by looking at them.

There are few previous works which give a head start into the concepts used in **SAIGA** and other possible areas in which eye tracking could be used. Those have been briefly explained in Sect. 2. The system architecture of SAIGA has been detailed in Sect. 3. The framework-specific implementation has been briefly discussed in Sect. 4. The results and their analysis have been briefed in Sect. 5. The framework presents a very beneficial approach and has a lot of scope in numerous applications, some of which have been looked into Sect. 6.

2 Literature Survey

In the past few years, research on eye tracking has taken a huge rise and the latest advancements in technology have allowed manufacturing commercially viable and accurately measuring hardware devices. Eye trackers have been used in research on visual systems, marketing, psychology and also as an input device for human–computer interaction. The most widely used current designs of eye trackers are video-based.

Some of the important measurable of eye gaze data are fixations and saccades. In [1], the authors propose a taxonomy of fixation identification algorithms that classify algorithms in terms of how they utilize spatial and temporal information in eye-tracking protocols. Apart from reading the movement of the eyes and measuring the point of gaze of a user, another readable area is the blink of an eye [2].

There are numerous applications in the field of eye tracking, and it also has a few which are quite uncommon. We have come a long distance from fingerprint as a biometric to the use of retinal and heart rate scanners as biometrics. The author in [3] suggests that the gaze movement, velocity of eye movement, etc., could also be used as a biometric. Another interesting area is the application of eye-tracking technology in the study of autism [4]. For many years, eye-tracking has been used to investigate gaze behaviour in normal population. Recent studies have extended its use to individuals with disorders on the autism spectrum. Such studies typically focus on the processing of socially salient stimuli. The authors in [5] have extended this application into an android-based mobile eye gaze point estimation system for studying the visual perception in children with autism.

The focus of this paper lies in the application of eye tracking in the annotation of images. There have been several techniques used for the categorization and labelling of images [6]. With the currently available high-processing systems, automatic annotation of images is on the rise. The authors in [7] have briefly reviewed some of the popular automatic image annotation techniques. There are several tools available online for annotations as well. LabelMe is one such Web-based tool [8]. It also provides datasets which can be used for experimentation purposes. In this paper, we will look at a semi-automatic approach towards the categorization of images.

3 System Architecture

In this section, the contrast between the existing image annotation architecture and the proposed framework's architecture is depicted pictorially and the functioning of the system is discussed in detail. Fig. 1 represents the current existing system of annotating images where a lot of time and effort is spent. Fig. 2 gives a structural flow of SAIGA's architecture.

The circled region in Fig. 1 marks the manual annotation step in an elaborate architecture. As shown in Fig. 2, SAIGA provides a sophisticated architecture that helps in reducing the strenuous efforts behind manual annotation.

SAIGA functions in a simple fashion. An image dataset is given to the user and he's asked to view them based on a particular keyword to collect eye gaze data. This data are analysed to extract parameters which help in ranking the images based on prominence and also helps in categorizing them to label them at once.

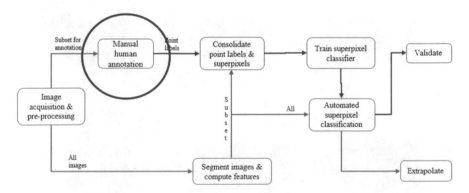

Fig. 1 Manual image annotation architecture

The first step in SAIGA's architecture is dataset representation. The proposed framework has been modelled to arrange the given images in the form of 3×3 grids. Each cell in the grid is called a region and is occupied by an image. Once this assorting is complete, the user gets to view the grids in a sequential fashion. The next step in SAIGA's architecture is data collection. As mentioned earlier, the hardware and software alliance of the eye tracking system collects a large amount of raw data in different criteria about the gaze of the person sitting in front of the screen. This raw data are passed to the analysis stage that identifies and extracts seven interesting parameters. Some of the major and frequently used terms in eye tracking are fixations and saccades. These are the primitive features that help in deriving the other interesting parameters or factors. A **fixation** is focusing the gaze at a particular region for a certain amount of time, while **saccade** is the large movement of gaze from one point to another. The other derived factors have been defined below.

Fixation Count (FC): Number of fixation points in the dedicated region of an image in the grid, where fixation points are the areas where users looked at for a time period.

Let $F: (x, y)$ represents each fixation point in a region r, FC of each region r is given by $FC_r = \sum F$ for each point $F: (x, y)$ in the coordinate space of each region.

Total Fixation Duration (TFD): The sum of the duration of each fixation point in the dedicated region of an image in the grid.

Let T be the time duration of each fixation point, then

TFD $= \sum T$ of fixation points F in region R.

Maximum Fixation Duration (MFD): The highest duration among the fixation points in the dedicated region of an image in the grid.

MFD $=$ max (List (T)), for each region R.

First Fixation Duration (FFD): The time from the beginning of the image's display to the first time a fixation occurs in the region of an image in the grid.

Let T_0 be the timestamp at beginning of viewing a grid G, and T_f be the timestamp of first fixation point F_1 in a region R, then

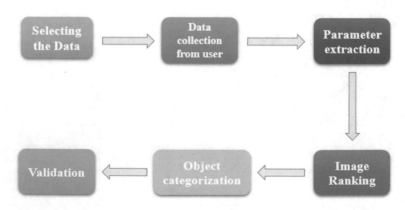

Fig. 2 Proposed SAIGA architecture

Let G_i be an image grid in the set of input images, where $i \in [1, n]$, $i \in Z$ where n is the number of input image grids.

Let M_{ij} be an image in region 'j' in image grid 'i' where $j \in [1, 9]$, $j \in Z$

D_{ijx} be the rank of image M_{ij} for each factor's coefficient F_x where $x \in [1, 7]$, $x \in Z$.

Rank measure of each image $M_{ij} = \frac{1}{N} \sum_{i=1}^{n} \sum_{j=1}^{9} \sum_{x=1}^{7} Fx * Dijx$

Final Rank of each image R_{ij} = Numerical index { Sort (M_{ij}) }

Fig. 3 Image ranking algorithm

$FFD = T_f - T_0$, for that particular region.

Visit Count (VC): The number of independent times the test subject visits the region of an image in the grid.

Let Ta be the timestamp value when the user's gaze enters a region R and Tb be the timestamp value when the user exits the region then,

$VC = \sum (Tb - Ta)$, where the summation is for each visit.

Maximum Visit Duration (MVD): The highest duration among all the visits in the dedicated region of an image in the grid.

$$MVD = \max(\text{List}(Tb - Ta)).$$

Mean Visit Duration (MnVD): The mean of each visit's duration in the dedicated region of an image in the grid.

Let N be the number of visits in region R, then

$$MnVD = \frac{1}{N} \sum (Tb - Ta).$$

The proposed framework uses a ranking system to identify those images which are more visually appealing or interesting in that particular category. The above-mentioned parameters are estimated for each image from the gaze information, and a rank is assigned as per the algorithm in Fig. 3.

For an image j, if the value of parameter i is $p(i)$ and the weight of respective parameter is $w(i)$, then rank of j is given by

$$R(j) = \sum_{k=1}^{7} p(i) * w(i) \tag{1}$$

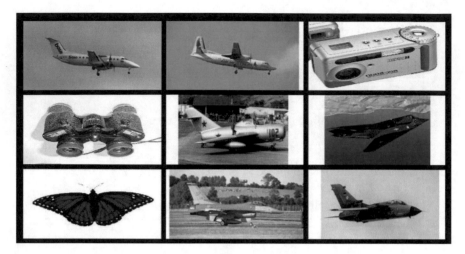

Fig. 4 Dataset image grid

where $j = 1$ to N (size of the image set) and $\sum w(i) = 1.0$.

The seven relevant parameters $p(i)$ are calculated for each grid and each image in the grid in the image stimulus. Those values are sorted accordingly for each grid and each image in the grid. The overall rank R_{ij} is identified by taking the weighted average of all the ranks $R(j)$ of an image for each parameter i mentioned earlier. The ranked images are then categorized based on a threshold value. The threshold is set at 75% of the average of 'Mean Visit Duration' parameter.

4 Implementation of SAIGA

This section discusses the implementation part of the proposed framework. EyeTribe eye tracker has been used for the experiments. EyeTribe device comes with dedicated software that assists in collecting the eye gaze data. The framework has been build using Python programming language.

The images used for the experiments have been extracted from the 101Object-Classifier dataset [9]. It is a standard image dataset used for annotation experiments. It has about 50 images on average for each category of images. The resolution of each image is roughly 300×200, and in every image, the object in question covers most of the image space.

These images are used in the experimental image stimulus in the form of grids as mentioned in the previous section. The appearance of an image grid in the experiments has been shown in Fig. 4.

For each experiment, the user or viewer is asked to look at the image stimulus for a dedicated time to collect the eye gaze data. Prior to this task, the EyeTribe device

Rating	Message	Description
★ ★ ★ ★ ★	Perfect	This is optimal calibration result. No recalibration needed
★ ★ ★ ★	Good	This calibration result is well suited for eye tracking
★ ★ ★	Moderate	This calibration result is acceptable but recalibration is advised
★ ★	Poor	This result is not optimal. Recalibration is needed to obtain better results.
★	Re-calibrate	This result is no good for eye tracking.
	Uncalibrated	Calibration is a must

Fig. 5 Calibration ratings

is calibrated to adjust for the user's eyes. Better the calibration, greater the accuracy of pinpointing the gaze position. Details on the calibration results are provided in the following section.

The user was made to sit in front of the computer along with the EyeTribe device connected to it. The UI of the device was launched, and the user was asked to undergo the calibration session to calibrate the EyeTribe to the user's eyes. This was repeated until a 5-star-rated calibration was achieved. A 5-star-rated calibration provides better and accurate raw data. The error in identifying the position of gaze is less than 0.5 cm. The following figure provides some details on different calibration ratings. Different ratings and their corresponding details are provided in Fig. 5.

The user is provided with a keyword and is asked to look at the images which have content corresponding to that keyword. Once the user completes looking at all the images in the dataset, the collected eye gaze data are passed on to an analysis programme. This programme extracts all the desired parameters and categorizes the images into 'Keyword present' and 'Keyword absent' directories. The analysis programme also contains the ranking algorithm detailed in the previous section. It gives the ranking measure of all the images which provides insight into the prominence of those particular images within the given dataset.

5 Results and Analysis

This section provides the results and their analysis for the experiments conducted towards the implementation of SAIGA. For this purpose, a dataset of about 150 images containing different objects was taken from 101ObjectClassifier dataset. This was used to create a set of 3×3 grids. Experiments were conducted for varying of dataset sizes. A 45-image dataset, 81-image dataset and 135-image dataset were

Fig. 6 Green circles on the images represent fixation points on the image grid

Table 1 Parameters extracted from an image grid

SI.	FC	FFT	TFD	MFD	VC	MVD	MnVD
1	1	13,556	66	66	1	66	66
2	0	0	0	0	0	0	0
3	0	0	0	0	0	0	0
4	9	1498	1334	300	6	367	222
5	2	399	600	533	1	600	600
6	0	0	0	0	0	0	0
7	6	6328	733	267	4	267	183
8	7	3530	1732	933	5	933	346
9	13	2664	1767	400	7	500	252

considered to perform the experiments. Experiments were conducted with over 32 users.

Due to the constraint of setting up the hardware device and calibrating it to every particular user and the time taken for experimentation, most of the experiments were restricted to a 45-image dataset. The experiment has been conducted on various other larger dataset (81 and 135 images). For the dataset which is given as input to the user, he was asked to look at images corresponding to the keyword 'Aeroplane'. Fig. 6 shows fixation map which indicates the region where the user has concentrated his gaze.

Table 1 gives the values for the seven different parameters for a single 3 × 3 image grid. Each row represents the data for a single image in the grid.

Applying the ranking algorithm, the rank obtained for the images in Fig. 6 corresponding to their regions are shown in Fig. 7 (Table 2).

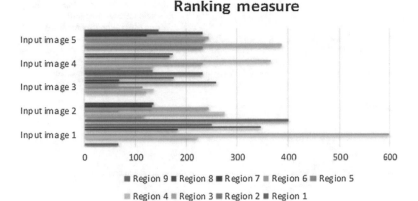

Fig. 7 Ranking of images in the grids

Table 2 Ranks of images in their particular regions

Region 7 → Rank 1	Region 9 → Rank 2	Region 8 → Rank 3
Region 4 → Rank 4	Region 5 → Rank 5	

The seven parameters have been given different priority levels and were given weights as follows:

Parameters	Weights
(1) Fixation count	0.15
(2) Total fixation duration	0.20
(3) Maximum fixation duration	0.05
(4) Visit count	0.15
(5) First fixation duration	0.10
(6) Maximum visit duration	0.25
(7) Mean visit duration	0.10

The regions 4,5,7,8 and 9 are ranked 4th, 5th, 1st, 3rd and 2nd respectively and are assigned with the target name 'Aeroplane' as the class label. Regions 1, 2, 3 and 6 have been declared to not have the target object in their images. The cumulative rank obtained by the ranking algorithm is shown in the Fig. 7. These rankings are then sorted in ascending order to know the prioritizations of the user.

Once the categorization was complete, the outputs were verified against the ground truth.

Figure 8 gives insight into the performance and accuracy levels for different calibration ratings. It can be observed that when the tracker is properly calibrated (5-star rated), then the accuracy of the output was seen to be greater than 80%. At

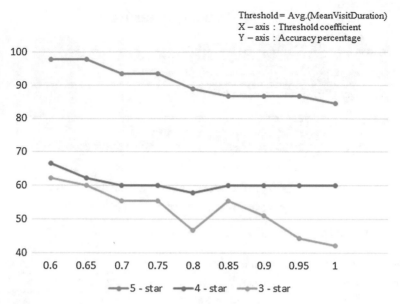

Fig. 8 Comparison of accuracy levels of the framework for various calibration ratings

the same time, a 4-star-rated calibration yielded results with accuracy in the range 60–70%. A 3-star-rated calibration provided poorer outputs with accuracy dipping to as low as 42%.

6 Conclusion and Further Improvements

SAIGA is beneficial in reducing the burden of manual annotation, and the framework makes the task at hand more interactive. With advancement in eye tracking devices, i.e., if they can be incorporated into mobile devices as well, **SAIGA** can be modified so as to personalize photographs and to tag people in them in a much easier fashion. Web applications can also be implanted with this kind of framework for numerous purposes. A better and more sophisticated hardware device with improved accuracy in pinpointing the gaze position would improve the performance of the proposed framework further.

References

1. Salvucci, Dario D., Goldberg, Joseph H. Identifying fixations and saccades in eye-tracking protocols in Proceedings of the 2000 symposium on Eye tracking research & applications (2000) https://doi.org/10.1145/355017.355028, ACM Digital Library, ISBN: 1-58113-280-8
2. J. Amudha, Roja Reddy S., Supraja Reddy Y., Blink analysis using Eye gaze tracker in Intelligent Systems Technologies and Applications 2016. ISTA 2016, vol 530, Springer, Cham (2016)
3. Bednarik Roman, Kinnunen Tomi, Mihaila Andrei, Fränti Pasi, Eye-Movements as a Biometric in Image Analysis. Springer Berlin Heidelberg (2005) ISBN: 978-3-540-26320-3, 978-3-540-31566-7
4. Boraston, Z. and Blakemore, S.-J., The application of eye-tracking technology in the study of autism in The Journal of Physiology, 581: 893–898 (2007) https://doi.org/10.1113/jphysiol.2007.133587
5. J. Amudha, Hitha Nandakumar, S. Madhura, M. Parinitha Reddy, Nagabhairava Kavitha, An Android-Based Mobile Eye Gaze Point Estimation System for Studying the Visual Perception in Children with Autism, in Smart Innovation, Systems and Technologies, New Delhi, vol 32. Springer (2014) ISBN: 978-81-322-2208-8
6. Reena Pagare and Anita Shinde, Image annotation techniques in A Study on Image Annotation Techniques, Maharashtra, vol 37 (2002)
7. Dengsheng Zhang, Md. MonirulIslam, Guojun Lu, A review on automatic image annotation techniques in Pattern Recognition. Volume 45, Issue 1, January 2012, Pages 346–362 (2011)
8. Russell BC, Torralba A, Murphy KP, Freeman WT, LabelMe: a database and web-based tool for image annotation in International Journal of Computer Vision, volume 77, issue 1–3, pages 157–173 (2008)
9. L. Fei-Fei, R. Fergus and P. Perona. Learning generative visual models from few training examples: an incremental Bayesian approach tested on 101 object categories. IEEE. CVPR 2004, Workshop on Generative-Model Based Vision (2004)

Optimizing Regression Test Suite Reduction

U. Sivaji, A. Shraban, V. Varalaxmi, M. Ashok and L. Laxmi

Abstract Software reliability is one of the essential features in the system quality. The purpose of the regression testing is to establish quality of the modified software before the software deployment. The main issue in this regression model is how to select a subset of test cases from the large test suite in the modified software version. Conventional regression subset selection models have addressed this issue on a small set of test suites or limited test cases. In this paper, we have studied the background work of different hybrid regression models and its limitations on the standard test suite. For different software applications, the proposed methodology is applied to find various combinations of faults randomly injected into various combinations of code lines as specified. We use selenium builder to test the software applications and run it in junit eclipse IDE to find faults. The tested output is interpreted with java randomly induced faults, and the system of methodology is evaluated in terms of APFD and APFD box plot.

Keywords Static suites · Graph models · Regression models · API programming · Test cases

U. Sivaji (✉) · A. Shraban · V. Varalaxmi · M. Ashok
IT Department, St. Martins Engineering College, Secunderabad 500014, Telangana, India
e-mail: sivaji.u117@gmail.com

A. Shraban
e-mail: shraban.tcs@gmail.com

V. Varalaxmi
e-mail: varalakshmi.vallabhuneni@gmail.com

M. Ashok
e-mail: ashokmandula01@gmail.com

L. Laxmi
Department of Computer Science and Engineering, MLR Institute of Technology, Hyderabad, Telangana, India
e-mail: laxmi.slv@gmail.com

© Springer Nature Singapore Pte Ltd. 2019
R. S. Bapi et al. (eds.), *First International Conference on Artificial Intelligence and Cognitive Computing* , Advances in Intelligent Systems and Computing 815,
https://doi.org/10.1007/978-981-13-1580-0_18

1 Introduction

As the software development grows, regression testing models are applied to different modified versions of the software to establish the changes that did not introduce unexpected terminations or errors [1]. If a defect occurs in the software development stage, then its corresponding revised version was developed for regression testing. It is difficult to perform regression testing to the entire test suite to find the buggy portions or test cases [2]. Generally, software engineers thus select a subset of all test cases for regression testing. Regression test selection is too hard to perform due to the existence of a large number of test cases [3]. Usually, the regression test manual process thus necessarily involves a certain amount of guesswork.

The increase of the software code increases new defects or procedures. As a result, whenever a software modification exists, we not only need to add new test cases to the test suite, but also add new defects simultaneously [4]. Also, we need to analyze the structure and dependencies of the function which is affected, to prove that no dependency defects are introduced. Therefore, regression test retests the entire system which is modified or changed (e.g., software updates, enhancement in new features).

Generally, there are common regression strategies used on the entire software system. The first is a complete regression model, which executes all the existing test cases in the test suite. The second one is partial regression, which observed the dependency of software modifications in the system and retests the affected portions.

For this reason, various models for cost reduction of regression methods have been used, including test suite minimization, prioritization, and selection techniques [5]. But the regression suite selection has limitations. For example, effective fault tolerance of the test suites can decrease by minimization.

2 Background and Related Work

There are different ways of doing test case optimization approaches. Some of these techniques such as coverage, similarity, search-based did not do well as the case of fault detection and did not produce satisfactory results [6]. Test case prioritization and reduction were the techniques used in regression testing to fulfill the faster sequence of executing test cases and at the same time attainment of optimality [7]. Multiobjective test case procedures play major role in order to tackle the situation. To improve the fault detection and cost-effectiveness, different criteria with combinations can be used and it is comparable to the approaches earlier [8]. To be more effective on fault detection, we experimented on studies with popular programs to the problem.

2.1 Regression Testing Models

Regression testing automation is implemented to improve the traditional regression model in two dimensions: One generates a set of test inputs that are targeted at the modified code, and another explicitly extracts the old and changed versions of the code. The final result is the set of changes between the previous and the new code versions. Unexpected modifications in the behavior and code information could help the regression faults.

2.1.1 Retest All

This approach involves all the test cases are running and are previously defined for fixing up bugs. However, this process uses retesting of entire application from beginning to end without loss of any generality. When compared to other techniques, this approach requires more time. This is the first and common technique foremost for kind of general regression.

2.1.2 Regression Test Selection

Retest all model is higher factor for time and cost, so regression test selection is used in place of retest all model. In this approach, only selected test cases are executed for results evaluation. This model minimizes the time and cost for test suite evaluation. Regression test selection divides the test suite into three parts of test cases.

Regression model for Web application:

The regression model in a Web application is summarized by the following steps:

Model the Web application and the modified Web application by using the event graph model.
Identify the changes in the graph nodes by comparing the nodes from both the computed graph.
Identify the newly affected nodes in the graph.
Select test cases that pass through the modified nodes and the newly affected nodes.

2.1.3 Graph-Based Regression Model

Let G and G' be the original application graph and a changed version graph. Compare both the graphs to reveal the node changes and edges which represent the dependencies. Select the graph G and perform the operations as:

For each node in the original graph G, find the corresponding node in the modified graph G'.

Fig. 1 Optimized regression
model

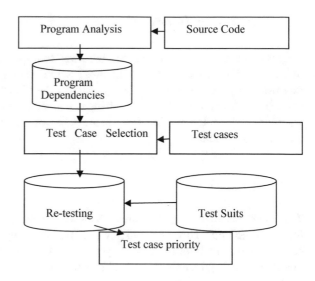

If the node is not found in the modified graph, this indicates that the node has been removed or null in the modified graph. Mark this node as modified one.

If the node exists in the graph G, then for each neighbor node in the G, find its corresponding neighbor in the modified graph.

If the neighbor is not available in the modified graph, this means that the relation has been deleted from the modified graph.

2.1.4   Optimized Regression Model Using Prioritization

See Fig. 1.

2.2   Procedure

In proposed regression technique methodology, we have included test case prioritization, suite reduction, and regression selection methodologies based on criteria such as test cost, coverage of code, and code length. In prioritization, test cases are ordered upon a certain criterion and test cases with high priority are one first executed to achieve high performance. In test case reduction stage, the test cases which possess redundant test results over time are removed from test suite to produce reduced set of test case in regression selection. As a result test cases which are smaller selected from the larger size of test suite. In our proposed work, the code coverage not only includes the statement coverage it also merges with event coverage and branch coverage to effectively calculate wider range of coverage. Test cases are ranked by their

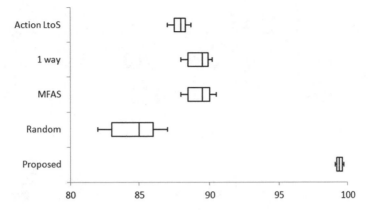

Fig. 2 Box plot APFD of Calc proposed system

coverage range of statements, branches, and events. The one with higher coverage value will be ranked high, and less coverage value will be ranked last.

Choice selection also supports the reduced test cases. The test cases that are essential to cover entire statements are only added in the test suite. For example, in the ranked list of five test cases only first three is enough to cover all the statements, events, and branches of the code. Then, those are only added in the test suite and other redundant test cases are removed. By that, the test suite size is reduced by removing unnecessary test cases. Finally, the obtained test suite from test suite is tested for its fault coverage. It should meet the condition that all the faults should be identified since all the statements, branches, and events in code are covered by the test suite.

2.3 Result

See Figs. 2 and 3.

For different software applications, the proposed methodology is applied to find various combinations of faults randomly injected into various combinations of code lines. We use selenium builder to test the software applications and run it in junit eclipse IDE to find faults.

3 Conclusion

In our proposed method, identifying test cases is ordered using multiobjective technique which is efficient and effective in terms of early discovering faults. The proposed method has been validated on several applications and run through platform which makes removal of redundant test cases. This approach is able to identify test

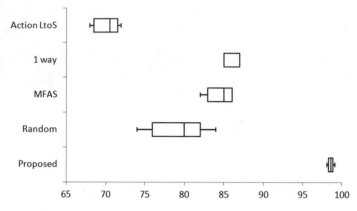

Fig. 3 Box plot APFD of paint proposed system

cases which cover all combinational faults that improve fault detection rate early regression prioritization technique. In this approach, the test cases will be prioritized and selected accurately effective within lower the execution time and higher the average percentage of faults detected (APFD).

References

1. Sapna, P. G and Arunkumar Balakrishnan, "An Approach for Generating Minimal Test Cases for Regression Testing", Procedia computer science, Vol.47, pp. 188–196, 2015.
2. Mirarab, Siavash, Soroush Akhlaghi and Ladan Tahvildari, "Size-constrained regression test case selection using multicriteria optimization", IEEE transactions on Software Engineering, Vol.38, No.4, pp. 936–956, 2012.
3. Hyunsook, Siavash Mirarab, Ladan Tahvildari and Gregg Rothermel, "The effects of time constraints on test case prioritization: A series of controlled experiments", IEEE Transactions on Software Engineering, Vol.36, No.5, pp. 593–617, 2010.
4. Zhai, Ke, Bo Jiang and W. K. Chan, "Prioritizing test cases for regression testing of location-based services: Metrics, techniques, and case study", IEEE Transactions on Services Computing, Vol.7, No.1, pp. 54–67, 2014.
5. Sampath, Sreedevi, Renee Bryce and Atif M. Memon, "A uniform representation of hybrid criteria for regression testing", IEEE transactions on software engineering, Vol.39, No.10, pp. 1326–1344, 2013.
6. Schwartz, Amanda and Hyunsook Do, "Cost-effective regression testing through Adaptive Test Prioritization strategies", Journal of Systems and Software, Vol.115, pp. 61–81, 2016.
7. Huang, Yu-Chi, Kuan-Li Peng and Chin-Yu Huang, "A history-based cost-cognizant test case prioritization technique in regression testing", Journal of Systems and Software, Vol.85, No.3, pp. 626–637, 2012.
8. Hettiarachchi, Charitha, Hyunsook Do and Byoungju Choi, "Risk-based test case prioritization using a fuzzy expert system", Information and Software Technology, Vol.69, pp. 1–15, 2016.

Application Monitoring—Active Learning Approach

Saurabh Mishra

Abstract In today's digital world, competition is growing among retail and e-commerce companies. Availability of Web application and high user satisfaction on application is the KPI for any successful business. For business-critical applications, especially for revenue-generating applications, transaction errors or unexpected downtime leads to revenue and productivity loss which leads to loss of customer base. Monitoring application health with hybrid IT architectures and complex structures is a big challenge. With the help of application monitoring, companies can overcome the complexities and improve the user experience. In this paper, we are presenting an analytical solution to monitor application health and predict the required action need to be taken automatically to maintain high availability and smooth performance of application. In this approach, applied active learning strategies are used to categorize the application health for required action to be taken. Active learning is a semi-supervised machine learning technique. The key concept behind this technique is, if an algorithm is allowed to interactively query the data from which it learns, it will perform better with less training.

Keywords Active learning · Uncertainty sampling · Query by committee
Random forest (RF) · Support vector machine (SVM)

1 Introduction

In today's digital world, competition is growing among retail and e-commerce companies. High performance and availability of business-critical applications, especially revenue-generating applications, are required. To maintain high availability, IT companies are using various tools to capture application logs and monitoring the application server health like CPU utilization, memory utilization, requests per second, interface transmission/reception like Ethernet. Due to the widespread access of the

S. Mishra (✉)
HCL Technologies, Hyderabad, India
e-mail: saurabh-mishra@hcl.com; saurabhmishra346@gmail.com

© Springer Nature Singapore Pte Ltd. 2019
R. S. Bapi et al. (eds.), *First International Conference on Artificial Intelligence and Cognitive Computing* , Advances in Intelligent Systems and Computing 815,
https://doi.org/10.1007/978-981-13-1580-0_19

Internet and increase in online presence of businesses, logs getting collected to monitor the application's health are increasing rapidly. Going through each log/record and marking, the health status requires a lot of manual effort, cost and resources. It requires deep expertise and time to verify each record. It is almost impossible to monitor the application health and decide the suitable action required manually in real time.

The objective of the study is to provide an analytical approach to monitor the application health and predict the action need to be taken automatically to maintain the high availability of application. In this approach, applied active learning strategies [1] are used to categorize the application health for required action to be taken.

For the prediction of application health, random forest is used as a feature selection technique [2] to choose the key variables. Classification model, SVM, is built using the key variables, and the accuracy is improved with the help of uncertainty sampling. To predict the required action need to be taken, build random forest [3] and support vector machine classifiers along with query by committee framework to improve model accuracy.

Apply uncertainty sampling to query the least certain data points to subject matter expert (SME) for labelling. To minimize the manual efforts and maximize the model accuracy, add these labelled uncertain data points and retrain the model.

2 Our Approach

2.1 Data

Data comprise of application's health monitoring variables related to four different applications across two different data centres. A total number of nine features related to application health like number of sub-systems, error reported, diagnostic error status are collected along with timestamp.

2.2 Feature Selection

All features may not be relevant for model building, and the presence of irrelevant features might reduce the model prediction power which ultimately leads to lower accuracies. Apply feature selection technique 'Random Forest' on labelled data to select the subset of features which are important.

2.3 Active Learning

Active learning is a semi-supervised machine learning technique. The key concept behind this technique is, if an algorithm is allowed to interactively query the data from which it learns, it will perform better with less training. Active learning is used when labelled data is very less as compared to unlabelled data. In this case, lot of unlabelled application logs and events are available as the data is generated very frequently. In active learning, we get the labels from an Oracle. Oracle can be experiment/human/crowd. In our case, we have used SMEs to label the application health based on given feature.

The main task in active learning is which data point to sample or query to Oracle. In our approach, we have used uncertainty sampling and query by committee technique to come up with sample data.

In uncertainty sampling, algorithm queries the data point for which it is least certain and selects the least certain unlabelled sample.

$$\vec{x}^* = \arg \min_{\vec{x}_i \in \{X\}} (p(\hat{y}_k|\vec{x}_i) - p(\hat{y}_j|\vec{x}_i)) \quad \text{for } j \neq k \tag{1}$$

We can ask Oracle for the corresponding label and retrain classifier with additional sample data.

Query by committee (QBC) [4] involves a committee of multiple classifiers, all trained on same label data. To measure the disagreement among classifier, we can use two approaches: First approach is Vote Entropy, i.e. number of votes that the labels receive.

$$x^*_{VE} = \operatorname*{argmax}_x - \sum_i \frac{V(y_i)}{C} \log \frac{V(y_i)}{C} \tag{2}$$

Second approach is Kullback–Leibler (KL) divergence [4].

$$x^*_{KL} = \arg\max_x \frac{1}{C} \sum_{c=1}^C D(P_{\theta^{(c)}}||P_C)$$

$$D(P_{\theta^{(c)}}||P_C) = \sum_i P_{\theta^{(c)}}(y_i|x) \log \frac{P_{\theta^{(c)}}(y_i|x)}{P_C(y_i|x)} \tag{3}$$

2.4 Building Classifier

2.4.1 Predicting the Health of Application

Build classification model, support vector machine (SVM), to predict the application health, and use uncertainty sampling to improve the model accuracy. Target label has two levels: Good and Bad. SVM classifier based on the initially available labelled data set with radial kernel and optimum parameters (gamma, cost and degree) is used to predict the health of an application with an accuracy of 92.1%. After retraining the model with additional data set received from uncertainty sampling and labelled by SME, accuracy has been improved to 97.43%.

Accuracy	0.9743
95% CI	(0.9661, 0.9809)
P-value [Acc>NIR]	<2e−16
Sensitivity	0.9752
Specificity	0.9721

2.4.2   Predicting the Action

For predicting the actions need to be taken based on the predicted application health, build random forest (RF) and support vector machine (SVM) classifiers and use query by committee strategy to improve the model accuracy.

For predicting actions, we have built random forest (RF) and support vector machine (SVM) classifiers on initially available labelled data and achieved 90.2% accuracy. After retraining the model by label data points obtained from query by committee framework, 97.14% accuracy has been achieved. Below is the accuracy matrix after adding sample data obtained from SMEs and retaining the model with additional data.

Accuracy	0.9714
95% CI	(0.9607, 0.9799)
P-value [Acc>NIR]	<2.2e−16
Sensitivity (ignore class)	0.9805
Sensitivity (investigate)	0.8736
Sensitivity (restart)	0.9909
Specificity (ignore class)	0.9842
Specificity (investigate)	0.9954
Specificity (restart)	0.9673

2.5 *Performance Monitoring*

Once the model has been deployed, we need to monitor the performance of the model continuously. Since we are using active learning and algorithm queries for the label from Oracle, we need to retrain our model with additional data to achieve better accuracy. In case of new patterns which are impacting the health of an application, we need to refit the model by considering the latest data.

3 Conclusion

In this study, we have presented an approach to monitor the health of an application and the required action need to be taken automatically using advanced analytical techniques. Applying active learning strategies to query the real-time data to Oracle will make the algorithm more robust and intelligent. Labelling of data manually to build supervised ML model is time-consuming and prone to manual errors, and it also requires a lot of expertise, time and cost. By applying this methodology, application health can be monitored and automatic actions can be triggered for smooth functioning of applications.

Acknowledgements The author sincerely thanks Mr. Suneev M and Mr. Kamesh JV for their continuous support and would also like to thank Mr. Srinivasarao V and his colleagues who encouraged him during this journey.

References

1. Donmez P., Carbonell J.G., Bennett P.N. (2007) Dual Strategy Active Learning. In: Kok J.N., Koronacki J., Mantaras R.L.., Matwin S., Mladenič D., Skowron A. (eds) Machine Learning: ECML 2007. ECML 2007.
2. Guyon, I., Elisseeff, A, (2003): "An Introduction to Variable and Feature Selection", Journal of Machine Learning Research.
3. Breiman, Leo, (2001): "Random Forests", Machine Learning 45(1):5–32.
4. Burr Settles – "Active Learning Literature Survey" (2010).
5. Pang-Ning Tan, Steinbach, M. and Kumar, V. (2014): "Introduction to Data Mining", 2nd Edition, Pearson Education.
6. Cohn, Ghahramani and Jordan, "Active Learning with Statistical Models" (1995).

An Automated Computer Vision System for Extraction of Retail Food Product Metadata

Venugopal Gundimeda, Ratan S. Murali, Rajkumar Joseph
and N. T. Naresh Babu

Abstract With the rapid growth in retail e-commerce industry, most of the traditional in-store retailers are focusing more on online and mobile channels. To stay competitive, retailers need quality metadata and powerful search platforms that entice customers make effective buy decisions. Many retailers have incomplete and inaccurate product information on their Web sites, and they use multiple manual-intensive methods for acquiring product information from suppliers and third-party sources. There is no one proven channel through which retailers can achieve high-quality metadata. Our study proposes an automation method to improve the extraction of unstructured product metadata from food product label images using computer vision (CV), machine learning (ML), optical character recognition (OCR), and natural language processing (NLP). We propose an automatic image quality classification system to identify images that give a high degree of metadata extraction accuracy, and we propose a technique to improve the quality of images using traditional computer vision algorithms to improve text detection and OCR- and NLP-based metadata extraction accuracy. Our results show 95% accuracy for attribute extraction from high-quality product images with machine-printed characters having contrasting backgrounds.

Keywords Automation quality assessment · Optical character recognition
Trademark detection · Machine learning · Natural language processing · Nutrition
facts extraction · Automatic image quality classification

V. Gundimeda (✉)
GTO-CDS-LAB, Cognizant Technology Solutions, 7th Floor, Building 20, Raheja Mindspace,
Hi-Tech City Road, Hyderabad 500081, India
e-mail: venugopal.gundimeda@cognizant.com

R. S. Murali · R. Joseph · N. T. Naresh Babu
GTO-CDS-LAB, Cognizant Technology Solutions, SEZ Ave, Elcot Sez, Sholinganallur, Chennai
600119, Tamil Nadu, India
e-mail: Ratan.SMurali@cognizant.com

R. Joseph
e-mail: rajkumar.joseph@cognizant.com

N. T. Naresh Babu
e-mail: nareshbabu.nt@cognizant.com

© Springer Nature Singapore Pte Ltd. 2019   199
R. S. Bapi et al. (eds.), *First International Conference on Artificial Intelligence
and Cognitive Computing* , Advances in Intelligent Systems and Computing 815,
https://doi.org/10.1007/978-981-13-1580-0_20

1   Introduction

Online retail industry is growing, and traditional brick-and-mortar retailers are expanding their online presence by making a transition from physical to virtual retail. According to Salsify [1] and Forbes [2], retail product discovery and recommendations are keys for consumers. About 90% or more consumers need clean product images with matching and accurate granular product metadata before they buy products online [3–5]. If we take the example of food products, consumers would need food product images, product title, weight/volume, nutrition facts, ingredients information, and special product warning messages explaining about allergies or precautions for making a purchasing decision.

To enhance online discovery of products, retailers do a lot of manual product metadata entry/validation activities to improve metadata quality on their e-commerce Web sites. In order to improve product metadata quality, retailers employ the following methods [1, 2]:

(a) Electronic Data Interchange (EDI): Rely on Suppliers to provide product images and metadata through various methods (e.g., Electronic, Printed or digital catalog) and various formats (e.g., Text, Excel, PDF and XML). Retailers often get inconsistent information/metadata and content from various suppliers for the same products. Few suppliers do not share any images, and sometimes, they supply inappropriate product images. Currently, for retailers, there is no simple way to validate metadata and images. They check and validate product information and images manually before storing the content in their respective systems.
(b) Purchase product information from third-party providers and online Universal Product Code (UPC) database access. They use product UPC as input and look at online UPC databases to validate available product metadata. Metadata available in online UPC databases is not always accurate, and the data differs from one UPC database to another.
(c) Use machine learning-based attribute prediction.

The key technical challenge associated with the content extraction is handling images with diverse product backgrounds, natural scenes, typographic and diverse fonts, cursive/handwritten text, various lighting conditions, low image quality, and camera artifacts. OCR is typically used for the extraction of text from images. It works well for scanned documents that have monochrome (single-color) backgrounds. However, OCR fails to extract text from the complex product label images.

The proposed computer vision system extracts food product attributes, brand, and food certification logos automatically from retail food product label images using CV, ML, OCR, and NLP techniques. We used image preprocessing, text detection, segmentation, and image enhancement to improve OCR-based text extraction accuracy. NLP-based techniques are used for automatic text correction and attribute extraction (unstructured text data to structured data). Currently, we have not focused on cursive and typographic font-based text extraction. The intent is to provide an alternative computer vision approach for retailers to alleviate the product metadata quality problem and reduce manual overhead.

2 Related Work

Researchers have done significant research on OCR, unstructured and structured document processing, logo detection, and document/image quality. Singh et al. [6], Smith et al. [7], and Singh et al. [8] used OCR extraction for document processing, and this approach will work for scanned documents. OCR recognition is most sensitive to geometric deformations, segmentation, layout analysis, text detection, and machine-printed characters of different font types and character sizes. To overcome some of these challenges, Bieniecki et al. [9] proposed preprocessing techniques which should be applied before doing OCR. To improve OCR accuracy, text detection [10, 11], paragraph/line/word/character segmentation, feature extraction [12, 13], classification and post-processing steps were proposed. Different segmentation techniques used are thresholding, color segmentation, edge-based segmentation, and region based segmentation. In our proposed system, we combined both color-based (MSER) [14, 15] and edge-based (Niblack) [16] segmentations to improve text detection accuracy. Gupta et al. proposed automated image quality test on printed documents [17].

Modern OCR systems are using deep learning techniques to bring higher degrees of accuracy. However, deep learning ecosystems require higher computational power and they are usually expensive.

In e-commerce industry, extracting product attributes from product images with high accuracy is difficult. Traditional methods such as dictionary-based approaches, bag of words, and named-entity recognition (NER) are used to extract attributes from textual information [3, 4, 18–22]. Recently, many methods use word embedding as additional information for extracting features for the identifications of brand detection, product title detection, and classification from product images. However, the methods cannot solve the problem of text product descriptions from product images.

Content-based image retrieval (CBIR) [23, 24] is used to retrieve similar images from the existing image database. CBIR-based logo detection has been implemented in [25, 26]. Local and global feature extraction is used to get prominent features to classify an image that is proposed in [12, 13]. Dalal et al. [27, 28] divided the input image into a number of overlapping sub-blocks. Histogram of gradients (HOG) directions is calculated in each sub-block. The key advantage of using HOG features is that they are color-invariant. The Euclidean distance is calculated from input image and bag-of-words model [29]. Most of the authors approached 'A rule-based approach' to extract nutrition facts form the images [30, 31]. Walter et al. [32] used multilayer perceptron (MLP) neural network for classifications. Backpropagation algorithm is used to train the MLP model.

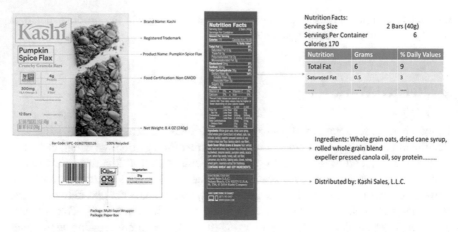

Fig. 1 Sequence of retail product label images

Fig. 2 Process flow diagram

3   The Proposed System

Many attributes have been discussed by many researchers. The proposed system relies on a sequence of retail product label images with a contrasting background with respect to the product color under a directed light ensuring that there are no shadows cast by the product as input as shown in Fig. 1.

The proposed computer vision system process flow diagram is shown in Fig. 2, which involves the following subsystems: (1) background removal, (2) automation quality classification, and (3) attribute extraction.

3.1   Background Removal

The background removal subsystem removes background color information from different product label images. This is done to improve character recognition accuracy as well as improve product label image acceptance. Product label images are referred to as documents. These images with different gradients, solid colors, and complex natural scenes undergo image preprocessing techniques such as background removal to identify and extract ROI (as shown in Fig. 3).

The illumination correction has been calculated by using morphological operation on grayscale images; before that, median filter is applied to remove noise from the

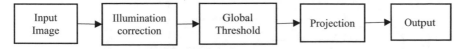

Fig. 3 Background removal subsystem

input raw image. The median filter is used to remove impulse (salt-and-pepper) noise from the image. In this method, pixel value is replaced by median value in neighborhood pixels 'w'.

$$MF(m, n) = median\{(I(i, j), (i, j) \in w\} \tag{1}$$

$$IC = MF - 0.25*Opening(MF, SE_1) - 0.25*Opening(MF, SE_2) - 0.5*I_G \tag{2}$$

where 'MF'—median filtered image, 'w'—neighborhood, centered around location m-rows,n-columns, IC—illumination corrected image, 'SE_1'—structural element 1, 'SE_2'—structural element 2, 'I_G'—Gaussian filtered image.

Otsu's threshold techniques method is a thresholding method which converts a given gray scale image to a binary image. The resultant image contains foreground and background regions. Using projection techniques, the ROI is cropped from input image. Histogram, vertical projection profile (VPP), and horizontal projection profile (HPP) are used as the features for the feature vector. The features are derived from the Otsu output image. VPP and HPP are calculated using Eq. (3). The number of histogram bins is 32, and they are normalized with respect to the total number of pixels. Similarly, horizontal and vertical projection bins are 32 and they are normalized with respect to height/width and n-bit gray scale, and it is derived from Eqs. (4) and (5). Therefore, the total feature vector size is 96 (32 × 3).

$$VPP = \sum_{l \le x \le m} f(x, y); HPP = \sum_{l \le y \le n} f(x, y) \tag{3}$$

$$NVPP = VPP/(Height * 2^{Nbitgrayscale}) \tag{4}$$

$$NHPP = HPP/(Width * 2^{Nbitgrayscale}) \tag{5}$$

where $f(x, y)$—input image, m and height—number of rows in the image, n and width—number of columns in the image, NVPP—normalized vertical projection profile, NHPP—normalized horizontal projection profile.

3.2 Automatic Image Quality Classification (AIQC)

The product label images have been put through automatic image quality classification subsystem to filter and classify images that would make sure that the product attributes are extracted reliably as shown in Fig. 4. The subsystem puts input documents into three buckets: *Accept, Need Manual Intervention, and Reject*. The

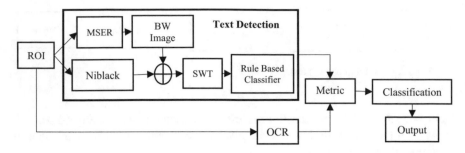

Fig. 4 Automatic image quality classification system

accepted document can be automatically processed using the proposed system without any manual intervention.

Multiple numbers of images have been captured for a single product. These images are nothing but front, back, top, bottom, left side view, right side view of retail product label images. Let us assume, the first image must front view and its most import one to identify the product name, remaining images are considered product label images as single image after applying mean of all images. The mean of all images is applied only on this automation quality classification block.

Traditional OCR system does not support cursive/typographic font text. To identify such text regions, a text detection module has been implemented. ROI image goes to text detection block as well as OCR block in parallel, and it is shown in Fig. 4. The input image has more color variations in nature. So, we are fusing texture and color features to get improved text detection accuracy. Text region detection has been done by combining MSER and Niblack algorithms, which results in low false text detection. Stroke width transform (SWT), the Euler number, and neighborhood-connected component methods are used to validate character/word regions, and the area of text detection regions is calculated. For the same image, the text is extracted using OCR engine and the area of extracted OCR character regions is calculated. Now, we have two areas of outputs: one from text detection block output and other one from OCR block output.

Figure 5 shows the outcome from OCR engine. Here, 'Region ID' describes image number. In this figure, possible text regions are tagged with <TxtRegion> tag, under that each word has tagged with <word> tag, in <word> tag, we are storing the following extracted metrics: word confidence, height, width, top and left along with extracted text.

The sample output is shown in Fig. 6, which consists of input front view image, text detection block output image, and OCR output image. The classification has been done by applying threshold on metrics, which results in automation quality test, and it is ready to proceed further. The classifier gives output as **ACCEPT** for good quality, **REJECT** for bad quality, and **NEED MANUAL INTERVENTION** for moderate quality.

Fig. 5 A piece of XML screenshot from OCR engine

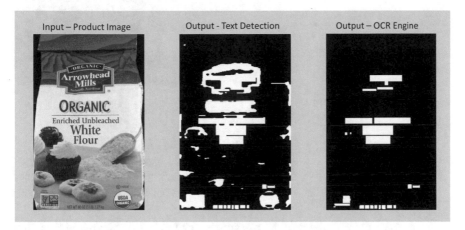

Fig. 6 Sample outcomes from text detection block and OCR engine block

These two outputs go to metric blocks. Metric block is key for the automation quality classification process. The proposed system uses the following automation quality classification metrics:

$$\text{Detection Percentage} = \frac{\text{Area of OCR character}}{\text{Area of text detection}} \tag{6}$$

$$\text{Mean character height: Based on the image resolution as low or high} \tag{7}$$

$$\text{Mean character confidence score} \tag{8}$$

Metrics/features are derived using the following approach. Initially, let us consider the front view image because it contains product name, brand name, and net weight. So, it is the most important image than the other view images. Split the image into nine segmented regions as shown in Fig. 7, and consider that the split has to be done with 40% of the input image with respect to height and width under overlapping conditions. The above three metrics are applied to each region, and the number of features is calculated as $9*3 = 27$. Later on, the remaining images are considered for the same product.

Fig. 7 Front view of nine
segmented regions

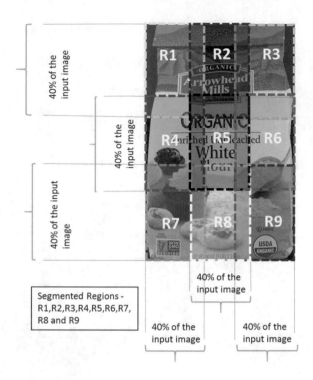

In this case, the referred product image has two more images: One is side view
image, and another one is bottom view image. These images are shown in Fig. 8a, b
and they are marked as region R10 and region R11 and so on respectively (here we
are not splitting these regions into 9 regions, because of uniformity (e.g., constant
font size and non-typographic character). Similarly, the three metrics are applied for
the below images and the mean of them is calculated. Now, the number of features
is calculated as mean $(2*3) = 3$. So, the total feature vector size is 30 (i.e., $27+3$).

3.3 Attribute Extraction

We are extracting attributes such as brand name, product title, net weight/volume and
nutrition facts, from input images with the help of various image processing techniques: artificial intelligence and NLP.

Fig. 8 a Side view image
(R10) and **b** bottom view
image (R11)

(a) Side view image (R10)   (b) Bottom view image (R11)

3.3.1 Brand Name Detection (Logo Detection) and Product Title Detection

Brand name and product title detection system is shown in Fig. 9. This system takes ROI as input and sends it to OCR engine to extract possible text, and it is stored in customized XML format. Customized XML file is fed to information extraction (IE) block to get brand name and product title. IE is the activity of obtaining information resources relevant to an information need from a collection of information resources. Information extraction is the science of searching for information in a document, also searching for metadata that describes data from databases of texts, images/videos. The results depend on the image quality and complexity.

If IE block is not able to identify brand and logo name, then the computer vision activity will come into picture. Trade Mark ™ and Registered ® detection is done using HOG cascade classifiers. HOG cascade requires both more positive images as well as negative images. Once we identify TM and R, we clearly know that brand/logo will be in a nearby segment and ROIs for brand/logo are derived. The proposed system used MSER-based segmentation to find possible text regions even when the input image is complex. The possible text regions are sent to OCR engine and customized XML is generated. Now, this XML will go to the second level of information extraction block to finalize brand and logo name. If we are still not able to identify brand/logo name, we used content-based image retrieval (CBIR). CBIR block diagram is shown in Fig. 10. Similarity measure has been done using the Euclidean distance (ED). If similarity measure is higher than threshold, then the top rank with the corresponding brand and product image is identified.

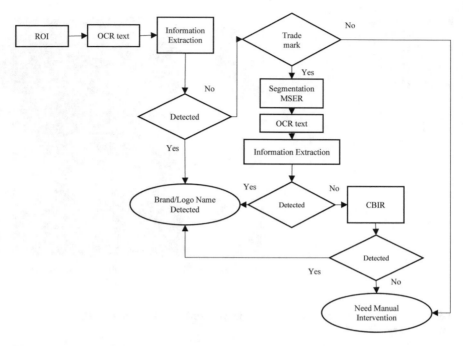

Fig. 9 Flowchart for brand/logo detection

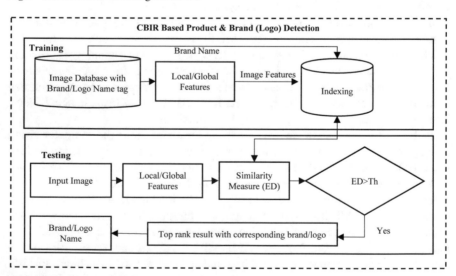

Fig. 10 Flow diagram for CBIR-based product and brand (logo) detection system

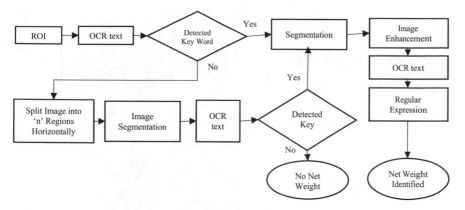

Fig. 11 Proposed net weight detection and extraction system

3.3.2 Net Weight Detection/Extraction

Net weight is one of the attributes in product metadata extraction. The proposed net weight detection and extraction system is shown in Fig. 11. Once we got the ROIs, OCR engine is used to extract possible text and it is stored in XML format. From the extracted text, we are searching the keyword 'net weight' and all possible combinations using regular expressions. If any one of the keywords is identified from the OCR output, then the corresponding coordinates are fetched from XML to find the exact segmented region of interest (net weight).

To improve accuracy, image enhancement technique is applied on net weight ROI image. Then, regular expression is formulated to get exact net weight volume/quantity. If keyword is not identified from the ROI, then we need to split image into three regions with respect to 40% of input image with overlapping conditions only on horizontal direction. We are concentrating only on three regions instead of nine regions, because net weight extraction is done by keyword search and regular expressions. If we segment image in vertical direction, we lose the information. The segmented region's split is shown in Fig. 12.

3.3.3 Nutrition Facts Detection and Extraction

The block diagram is shown in Fig. 13. The OCR is applied to input image, and the keyword 'nutrition facts' is searched for. Once we found keyword, the coordinates of it are taken to approximate nutrition facts bounding box. Coordinates and morphological operations are used to detect the horizontal and vertical lines. With horizontal line reference, each text subregion is cropped and the text is extracted using OCR. The extracted text is corrected with NLP-based dictionary/vocabulary management and text similarity. The below rule is used to extract the nutrition facts from the text:
Nutrition Fact Name *followed by* **Quantity** *followed by Percentage*

Fig. 12 Front view of three segmented regions for net weight

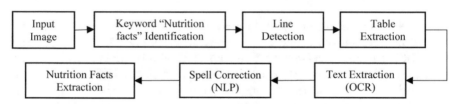

Fig. 13 Block diagram for nutrition facts detection

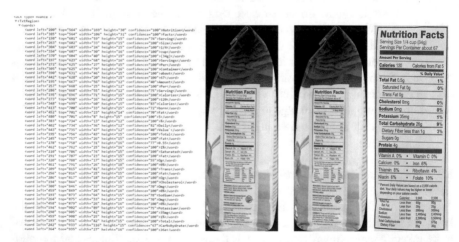

Fig. 14 Sample customized OCR output, nutrition facts bounding box, and line detection

The sample customized OCR output, nutrition facts bounding box, and line detection are shown in Fig. 14.

Table 1 Background removal process results

Product images with background removal	Number of products	Character extraction confidence score	Automatic image quality classification (accept)
Yes	352	71.61	83
No	352	70.16	57

3.4 Key Advantages

a. Product label images are a trusted source of metadata information. Naturally, the process would improve the quality of product metadata and data consistency.
b. Reduces the burden of validating product data provided by various vendors.
c. Provides additional information such as brand and certification logo information which is critical for consumer product discovery.

4 Results and Discussion

In order to examine the performance of the proposed solution, we evaluated an experiment, based on a real dataset with 352 food products of 53 brands which contain 955 images (include front, back, side view product images). The testing is done on real datasets. Background removal is used to improve OCR accuracy. We performed test on product images with/without background removal and evaluated character confidence score and automation quality. Background removal results for real dataset are given in Table 1. Background removal process has improved character-level extraction confidence by 2.06%, and the improved automation quality classification is 45.61%.

The automation quality classification filters out low-resolution, typo-graphic/cursive characters in document with low automation confidence. The mean character height of pixels in document is used to filter out low-resolution images, detection percentage is used to filter out typographic/cursive font characters, and higher mean character confidence score implies higher attribute accuracy which in turn results in higher automation quality classification. We used multilayer percep-tron, which is a type of neural network that is used to classify automation quality classifications. The parameter for classifications is given in Table 2. The automation quality classification training results and testing results are given in Tables 3 and 4, respectively. The training ROC curve for AIQC is shown in Fig. 15.

We used 95 and 868 images as positive and negative samples for training Trade Mark ™. We used 120 and 868 images as positive and negative samples for training Registered ®. Training parameters for Trade Mark ™ and Registered ® detection are given in Table 5. 83 real product images are taken for logo detection/extraction testing and accuracy is shown in Table 6.

Table 2 Parameter for MLP

Parameters	Number of perceptrons	Activation function	Training algorithm	Number of epochs	Learning rate
Value	64	tan h	Levenberg, Marquardt, BP	100	0.1

Table 3 AIQC results—training

	Target class			(%)
Output class	130	0	0	**100**
	0	130	10	**93**
	0	0	320	**100**
(%)	**100**	**100**	**97**	**98**

Table 4 AIQC results—testing

Number of product images	Accept	Need manual intervention	Reject
352	83	192	77

Fig. 15 Training ROC for AIQC

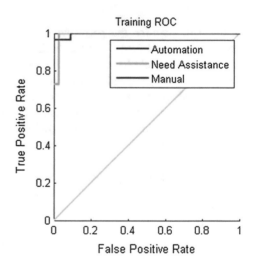

Table 5 Training parameters for Trade Mark ™ and Registered ®

Training parameter	
Number of cascades	20
False alarm rate	0.1
True positive rate	0.999

Table 6 Logo extraction results

Number of product images	Logo extraction using		Classification accuracy (%)
	Keyword	Image processing	
83	37	45	98.79

Table 7 Nutrition facts detection results

Types of backgrounds	Total images	Keyword errors	Boundary errors	Total errors	Accuracy (%)
White background	289	1	0	1	99.65
Dynamic background	63	2	1	3	95.24
Overall accuracy					**98.86**

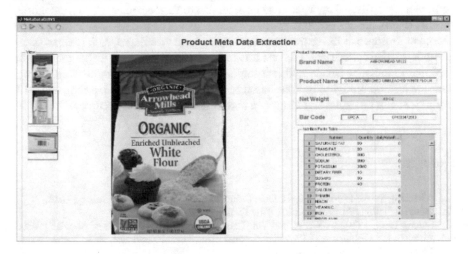

Fig. 16 Screenshot of consolidated result for a particular product

The proposed nutrition facts detection results are given in Table 7. Once nutrition facts region ROI is detected. Nutrition facts ROI is segmented. Attributes' segmentation-based text extraction gives better result compared to full-page OCR. The algorithm is tested with 83 real product images and two nutrition fact templates, and the attributes' accuracy of the algorithm is 98.10%.

The proposed system extracted the following attributes: brand name, product title, nutrition facts, net weight, and bar code, and it is shown in Fig. 16. The consolidated results for 83 real product images are given in Table 8.

Table 8 Consolidated results

Attributes	Number of products	Accuracy (%)
Product name	83	95.18
Net weight/volume	83	98.79
Logo extraction	38	98.79
Nutrition facts detection and extraction	83	98.10

5 Conclusion and Future Work

Computer vision and AI-based methods show clear automation potential to reduce data inconsistencies and improve metadata data quality, thereby improving retail industry's product data capture and metadata extraction processes. The proposed computer vision-based approach can be extended to various other product categories such as health, beauty, books, toys, and video games. The proposed approach can be extended to improve the extraction process for product images with diverse backgrounds, can be cylindrical, and can image labels. Deep learning techniques can be explored to enhance image and text region segmentation accuracy as well as support cursive and typographic font-based text extraction. Machine learning and NLP techniques can be explored to improve text attribute extraction accuracy.

References

1. "Cracking the Consumer Code: Product Content Drives Conversion" by Salsify, 2016.
2. "Five Predictions For Retail In 2017" by Forbes, Jan 4, 2017.
3. Lidong Bing, Tak-Lam Wong, and Wai Lam."Unsupervised extraction of popular product attributes from E-commerce Web sites by considering customer reviews". ACM Trans. Internet Technol. 16, 2, Article 12, (April 2016).
4. Petrovski, Petar, and Christian Bizer. "Extracting Attribute-Value Pairs from Product Specifications on the Web." Web Intelligence (WI'17), August 2017, Leipzig, Germany.
5. George, M., & Floerkemeier, C. (2014, September). Recognizing products: A per-exemplar multi-label image classification approach. In European Conference on Computer Vision (pp. 440–455). Springer, Cham.
6. Singh, Amarjot, Ketan Bacchuwar, and Akshay Bhasin. "A survey of OCR Applications." International Journal of Machine Learning and Computing 2.3 (2012): 314.
7. Smith, Ray. "An overview of the Tesseract OCR engine." Document Analysis and Recognition, 2007. ICDAR 2007. Ninth International Conference on. Vol. 2. IEEE, 2007.
8. Singh, Sukhpreet. "Optical character recognition techniques: a survey." Journal of emerging Trends in Computing and information Sciences (2013): 545–550.
9. Bieniecki, Wojciech, Szymon Grabowski, and Wojciech Rozenberg. "Image preprocessing for improving ocr accuracy." Perspective Technologies and Methods in MEMS Design, 2007. MEMSTECH 2007. International Conference on. IEEE, 2007.
10. Wang, Kai, Boris Babenko, and Serge Belongie. "End-to-end scene text recognition." Computer Vision" 2011 IEEE International Conference on. IEEE, 2011.

11. Huang, Xiaodong, and Huadong Ma. "Automatic detection and localization of natural scene text in video." Pattern Recognition (ICPR), 2010 20th International Conference on. IEEE, 2010.
12. Xi Wang, Zhenfeng Sun, Wenqiang Zhang, Yu Zhou, and Yu-Gang Jiang, "Matching User Photos to Online Products with Robust Deep Features". In Proceedings of the 2016 ACM on International Conference on Multimedia Retrieval (ICMR '16). ACM, New York, NY, USA, 7–14. 2016.
13. Hossain, M. Zahid, M. Ashraful Amin, and Hong Yan. "Rapid feature extraction for optical character recognition." arXiv preprint (2012).
14. Chen, Huizhong, et al. "Robust Text Detection in Natural Images with Edge-Enhanced Maximally Stable Extremal Regions." Image Processing (ICIP), 2011 18th IEEE International Conference on. IEEE, 2011.
15. Epshtein, Boris, Eyal Ofek, and Yonatan Wexler. "Detecting text in natural scenes with stroke width transform." Computer Vision and Pattern Recognition (CVPR), 2010 IEEE Conference on. IEEE, 2010.
16. Khurshid, Khurram, et al. "Comparison of Niblack inspired binarization methods for ancient documents." DRR 7247 (2009): 1–10.
17. Gupta, Anshul, et al. "Automatic Assessment of OCR Quality in Historical Documents." AAAI. 2015.
18. Zornitsa Kozareva. "Everyone Likes Shopping! Multi-class Product Categorization for e-Commerce". In The Annual Conference of the North Americal Chapter for the ACL. 1329–1333, 2015.
19. Yang, Shuo, and Jingzhi Guo. "A novel approach for cross-context document reasoning in e-commerce." Software Engineering and Service Science (ICSESS), 2015 6th IEEE International Conference on. IEEE, 2015.
20. Gabor Melli. 2014. Shallow Semantic Parsing of Product Offering Titles (for better automatic hyperlink insertion). In Proceedings of the 20th ACM SIGKDD international conference on Knowledge discovery and data mining. ACM, 1670– 1678.
21. Lienhart, Rainer, and Axel Wernicke. "Localizing and segmenting text in images and videos." IEEE Transactions on circuits and systems for video technology 12.4 (2002): 256–268.
22. Nadeau, David, and Satoshi Sekine. "A survey of named entity recognition and classification." Lingvisticae Investigationes 30.1 (2007): 3–26.
23. Sridhar R. Avula, Jinshan Tang and Scott T. Acton, "An object-based image retrieval system for digital libraries", Multimedia Systems, Springer Berlin, March 2006, vol 11 No. 3 pp: 260–270.
24. Andre Folkers and Hanan Samet, "Content based image retrieval using Fourier descriptors on a Logo database", IEEE Proceedings of 16th International Conference of Pattern Recognition, 2002, vol 3 pp. 521–524.
25. Romberg, Stefan, et al. "Scalable logo recognition in real-world images." Proceedings of the 1st ACM International Conference on Multimedia Retrieval. ACM, 2011.
26. Zhu, Guangyu, and David Doermann. "Automatic document logo detection." Document Analysis and Recognition, 2007. ICDAR 2007. Ninth International Conference on. Vol. 2. IEEE, 2007.
27. Dalal, Navneet, and Bill Triggs. "Histograms of oriented gradients for human detection." Computer Vision and Pattern Recognition, 2005. CVPR 2005. IEEE Computer Society Conference on. Vol. 1. IEEE, 2005.
28. Dalal, N., Triggs, B., & Schmid, C. (2006, May). Human detection using oriented histograms of flow and appearance. In European conference on computer vision (pp. 428–441). Springer, Berlin, Heidelberg.
29. Li, Teng, et al. "Contextual bag-of-words for visual categorization." IEEE Transactions on Circuits and Systems for Video Technology 21.4 (2011): 381–392.
30. Tanwir Zaman, "Vision Based Extraction of Nutrition Information from Skewed Nutrition Labels", Thesis submitted to Utah State University, 2016.
31. Pouladzadeh, Parisa, Shervin Shirmohammadi, and Rana Al-Maghrabi. "Measuring calorie and nutrition from food image." IEEE Transactions on Instrumentation and Measurement 63.8 (2014): 1947–1956.

32. Walter H. Delashmit and Michael T. Manry, "Recent Developments in Multilayer Perceptron Neural Networks", Proceedings of the 7th Annual Memphis Area Engineering and Science Conference, MAESC 2005, pp. 1–3.

Attacks on Two-Key Elliptic Curve Digital Signature Algorithm

N. Anil Kumar and M. Gopi Chand

Abstract In general, digital signature schemes have one parameter as secret and the remaining parameters as public. But in two-key signature scheme, we have two parameters as secret and the remaining as public. In this paper, we analyse the security of the two-key signature scheme and proposed an algorithm to attack two-key signature scheme and named it as Modified Shanks algorithm whose time complexity is $O(n)$.

Keywords Signature schemes · Elliptic curves · Shanks algorithm · ECDLP

1 Introduction

The elliptic curve digital signature algorithm (ECDSA) is a popular signing Algorithm. ECDSA is standardized elliptic curve-based signature scheme, appearing in ANSI X9.62 [1], ISO/IEC 19790:2012, etc. [4] has proposed a two-key signature scheme/algorithm. We will briefly discuss the scheme in Sect. 2. In this paper, we proposed the attack on the two-key signature scheme which will be described in Sect. 3.

2 Two-Key Elliptic Curve Digital Signature Scheme

Elliptic curve digital signature scheme/algorithm (ECDSA) is a signature where the signer uses his private key (one parameter) and hash of the message and generates

N. Anil Kumar (✉) · M. Gopi Chand
Department of IT, Vardhaman College of Engineering, Shamshabad, Kacharam,
Hyderabad 500018, Telangana, India
e-mail: anil230@gmail.com

M. Gopi Chand
e-mail: gopi_merugu@yahoo.com

© Springer Nature Singapore Pte Ltd. 2019   217
R. S. Bapi et al. (eds.), *First International Conference on Artificial Intelligence and Cognitive Computing* , Advances in Intelligent Systems and Computing 815,
https://doi.org/10.1007/978-981-13-1580-0_21

signature, which is a bit stream. Only the person who has the private key can only generate the valid signature. The person who wants to verify the signature uses the public parameters of the signer and hash of the message to determine whether the signature is valid or not. The security of the ECDSA is based on the assumption that the elliptic curve discrete log problem (ECDLP) is hard. ECDLP can be stated as finding d given P, R of the equation $dP = R$ where P and R are points on elliptic curve and d is a integer [2, 3].

Two-key elliptic curve digital signature algorithm (two-key ECDSA) is a signature scheme [4] where the signer uses his private keys, i.e. two keys (two parameters) and hash of the message, and generates the signature, which is a bit stream. Verification is same as ECDSA. The security is based on the hard problem. Let $dP + d'P' = S$ where P and P' are two base points on the elliptic curve. Given P, P' and S finding d and d' is hard problem.

We define the domain parameters as follows

Let the elliptic curve be $y^2 + x^3 + ax + b$. a, b are the coefficients that define the curve. The curve be defined on the group $\mathbb{F}_q$. The number of points on the elliptic curve is called the order of the curve. Let P and P' are two base point. Let the order of P be n and P' be n. Let i and d' be chosen at random be the private keys and $dP + d'P' = S$. So P, P' and S are the public parameters. H denotes a hash function. We take the hash of the message, if the bits in output of hash is greater than the number of bits of $n \times n'$ we truncate the output of hash so that it will be equal to the number of bits of $n \times n'$. We call the resulting as h.

2.1 Two-Key ECDSA Signature Generation

We use the private key and hash of the message to generate the signature for the given message. The signature generation procedure is as follows:

> **Input:** Domain parameters, private key, hash of message h .
> **Output:** Signature (x_1, s, s')
> **begin**
>
> 1   Select r_1 at random in the interval $[1, n-1]$ and r_2 at random in the interval $[1, n'-1]$;
> 2   Compute $r_1P + r_2P'$ and let it be the point (x_1, y_1) ;
> 3   **if** $(x_1, y_1) = \mathcal{O}$ **then**
> | goto step 1
>   **end**
> 4 $s = h r_1 + x_1 d$ and $s' = h r_2 + x_1 d'$;
> 5   **return** (x_1, s, s');
> **end**

Algorithm 1: Two key signature generation

2.2 Two-Key ECDSA Signature Verification

The signature can be generated by the person who has the private key but for veri-fication private key is not required. The message along with the signature is placed in the public domain. To verify that the message is accepted by the person who has signed, we have to use the signature verification algorithm. As the public parameters are made public, any person can verify that the signature is generated by the person who has the private key.

Input: Domain parameters, public key, hash of message h, signature
(x_1, s, s')
Output: validate or invalidate the signature
begin
  Compute $u = sP + s'P'$;
  Substitute x_1 in curve equation and generate two points W_1 and W_2 ;
  Then $W_1 = (x_1, y_1)$ and $W_2 = (x_1, y_1')$;
  Compute $u' = hW_1 + x_1S$;
  if $u = u'$ **then**
  | **return** *(The signature is valid)*
  else if $u' = hW_2 + x_1S$ **then**
  | OUTPUT (signature is valid)
  else
  | OUTPUT (signature is not valid)
  end
end

Algorithm 2: Two key Signature Verification

3 Attack on Two-Key Signature Scheme

In order to attack the cryptosystem, we have to find d, d' from (x_1, s, s'). The equa-tions which have the secret parameters are $s = hr_1 + x_1d$ and $s' = hr_2 + x_1d'$. The known variables are h, s, x_1 and s'. The unknown variables are r_1, r_2, d, d'. From two equations, we cannot solve for four unknown variables. This is an impossible case.

So in order to attack a public key cryptosystem, we have to attack the underlining hard problem. To attack the two-key ECDSA, we have to attack the underlining hard problem. The underlying hard problem is $dP + d'P' = S$.

We try to solve the underlying problem; we write the equation as

$$d \times P = S - d' \times P'$$

We calculate $S - i \times P'$ for each i in $[0, n_2]$ where n' is the order of P' and store the point and i in the List. The list looks like (point, number), i.e. $(S - 1 \times P', 1)$, $(S - 2 \times P', 2)$, $(S - 3 \times P', 3)$, ..., $(S - n \times P', n)$.

Then we compute $P, 2 \times P$ ... keep on checking the list for matching point. On finding a match, compute d and d'.

$S - y'P' = y'' \times P$ which implies

$$S = y'' \times P + y'P'.$$

So $d = y''$ and $d' = y'$.

3.1 Analysis

Let us analyse the time complexity. The time complexity of Algorithm 3 is $O(n) + O(n') = O(n + n') = O(2 * n) = O(n)$(let $n > n'$). As the time complexity is of the order of n, we can conclude that two-key signature is a safe algorithm to use.

Input: P, P', S, elliptic curve.
Output: d, d' satisfying the underlying hard problem.
begin
    let LIST be empty ;
    $S' = S$;
    for i *in* 1 *to* n_2 **do**
        $S' = S' - \times P'$;
        add to the LIST $\{(S', i)\}$;
    end
    The elements on the LIST look like [(x,y),i];
    sort the LIST based on (x,y);
    $T = \mathcal{O}$;
    for j *from* 1 *to* n' **do**
        $T = T + P$;
        search the LIST for T using binary search;
        if T *found in LIST* **then**
            $d = j$;
            $d' = i$;
            return d, d' ;
        end
    end
end

Algorithm 3: Algorithm to solve Two Key problem

4 Conclusion

In this paper, we have briefly explained about two-key ECDSA signature. We proposed an algorithm to attack two-key problem. We analysed its time complexity and found that two-key signature scheme is secure to use.

References

1. E. Barker, W. Barker, W. Burr, W. Polk, and M. Smid. Recommendation for Key Management –Part 1: General. NIST Special Publication, pages 800–857, 2005.
2. Joppe W Bos, J Alex Halderman, Nadia Heninger, Jonathan Moore, Michael Naehrig, and Eric Wustrow. Elliptic curve cryptography in practice. Microsoft Research. November, 2013.
3. Darrel Hankerson, Affred Menezes, and Scott Vanstone. Guide to Elliptic Curve Cryptography. Springer Verlag, 2004.
4. N.Anil Kumar and Chakravarthy Bhagvati. Two key signature scheme with application to digital certificates. In Recent Advances in Information Technology (RAIT), pages 19–22, ISM, Dhanbad, 2012.

Bioinformatics: An Application in Information Science

Parth Goel and Mamta Padole

Abstract Bioinformatics is an interdisciplinary subject of bonded relationship in between computer science, mathematics, and molecular biology. Biological information keeps growing tremendously. Molecular biologists are specialized in solving bioinformatics issues such as to store, analyze, and retrieve biological data by applying algorithm and techniques of computer science. This review is from the computer science perspective. The fundamental terminology of bioinformatics and its definition are essential to understand bioinformatics in depth. There are main three components of bioinformatics and data types. Data types are input format for tools or software. Real-life databases of bioinformatics are also discussed which are important for analyzing the algorithms. We then provide bioinformatics applications in various areas. As bioinformatics is a fusion from many disciplines, there are lots of research issues and challenges, but computational and biological research issues and challenges are quite significant. Nowadays, the tremendous amount of biological data are being generated, Due to them, bioinformatics has emerging future research trends in big data, machine learning, and deep learning which are presented at last.

Keywords Bioinformatics · Applications · Future trends in bioinformatics
Biological database · Research issues and challenges

P. Goel (✉)
Computer Engineering Department, Devang Patel Institute of Advance Technology and Research,
Charotar University of Science and Technology (CHARUSAT), Changa, Anand, Gujarat, India
e-mail: er.parthgoel@gmail.com

M. Padole
Department of Computer Science and Engineering, The M. S. University of Baroda, Vadodara,
Gujarat, India
e-mail: mpadole29@rediffmail.com

© Springer Nature Singapore Pte Ltd. 2019
R. S. Bapi et al. (eds.), *First International Conference on Artificial Intelligence and Cognitive Computing* , Advances in Intelligent Systems and Computing 815,
https://doi.org/10.1007/978-981-13-1580-0_22

1 Introduction

Bioinformatics is a multidisciplinary field in which biology is conceptualized by means of molecules and implementing informatics (computer science) methods to store, organize, understand, analyze, and visualize the details of these molecules significantly [1]. Computational management of biological data such as DNA, RNA, motifs, genes, genomes, proteins, and cells is explored using software or algorithms.

Bioinformatics can also be defined as an intersection of three fields which are molecular biology, computer science, and statistics. Mathematical representation of biological data can be organized, and feed these representations as an input to computer software for analysis.

Gregor Mendel had started the work in bioinformatics more than a century ago. He was an Austrian monk. He is also known as the "Father of Genetics." Mendel explained that the inheritance of traits was managed by factors inherited from one generation to next generation [2]. Paulien Higeweg and Ben Hesper invented the term bioinformatics in 1970 as "the study of informatic processes in biotic systems" [3].

This paper is outlined as follows: Section 2 presented basic terminologies in bioinformatics. Components and data types of bioinformatics are explained in Sects. 3 and 4, respectively. Section 5 describes NCBI [4]. Information of real-life databases is given in Sect. 6. Section 7 presents emerging fields and applications, respectively. Computational and biological research issues and challenges are discussed in Sect. 8. Finally, future research trends are described in Sect. 9.

2 Basic Terminologies in Bioinformatics

2.1 DNA

DNA is the abbreviation of deoxyribonucleic acid [5]. It is constructed from molecules. Bunch of atoms stuck together to form molecules like sugar and phosphate. These atoms combine and made a shape of long spiral ladder in double-helix structure, and it is also known as a blueprint of living things. It is made up of pairs of four different types of chemicals, and these are **A**—adenine, **C**—cytosine, **G**—guanine, and **T**—thymine which are known as nucleotides [5]. Each sugar molecule is joined with one of the nucleotides. To form DNA, nucleotides join together with each other in two fixed pair. Adenine binds with thymine (A–T or T–A), and cytosine binds with guanine (C–G or G–C). It may have trillions of pair or more than that. DNA lives in living cell as nucleons of it. Every DNA has specific information of living things. DNA makes RNA molecules; RNA generates amino acid. Amino acids combine together and they form proteins. Living cells are made up from proteins, and tissues are made up from living cells. All tissues create organs (e.g., heart, liver), and all organs make a body (body can be human body, any creature body, etc…).

2.2 Nucleotides

A nucleotide [5] is chemical molecule and is composed of three elements: chemical sugar, phosphate group and nitrogenous base. It is a building block for DNA. In double-helix form of DNA, bases are positioned in the middle and sugar and phosphate group arc on both sides. The bases join two strands using chemical bonds. Nucleotides come into five different chemicals. They are uracil base, adenine, cytosine, guanine, and thymine. DNA contains four bases: A, C, G, and T. RNA also contains four bases: A, C, G, and U. RNA does not contain thymine (T).

2.3 RNA

Partial copy of DNA which is made up of special chemicals inside the nucleus is known as ribonucleic acid (RNA) [6]. It is shorter than DNA, and it has only one side of DNA ladder (not in pair). It is also known as single strand of DNA. It comes out from the nucleus and interacts with ribosome. Ribosome is protein-building machine. They read every three consecutive RNA letters and make perfectly shaped protein chain. In each cell, three types of RNAs are present, namely messenger RNA, transfer RNA, and ribosomal RNA. These are also known as mRNA, tRNA, and rRNA, respectively.

2.4 Gene

Gene [7] is special starch/section of DNA. It is also sequences of A, C, G, and T and that code for specific traits. Gene has exact information that what it will do? Like eye gene, anger gene, hand size, single DNA contains thousands of genes. Humans have 20,000 genes together in body [8]. Some genes are small with only 300 pair of nucleotides, whereas some are large which contain more than millions of pairs. Different animals have different genes to work accordingly. But all are made up of four basic common nucleotides A, C, G, and T. Many creatures may have similarity in their genes, for example, human and chimpanzees [9].

2.5 Alleles

Alleles [10] are known as different forms of the same gene. For example, eye color gene has many variations like black eyes, blue eyes, and brown eyes. Gene specifies which trait, and allele specifies what form the gene takes.

2.6 Genomic Sequences

Genome sequencing [11] is finding the arrangement of nucleotides of DNA, RNA, and protein in a genome. Genome sequencing is done by advanced configuration machines, and these machines save them in a computer hard disk.

2.7 Genomics

Genomics [12] refers to the study of genes found out in human beings, creatures, plants, bacteria, viruses, and other living things. Particularly, it includes analysis of the genomes such as sequence analysis, structure analysis, expression analysis, and trait analysis.

2.8 Genetics

Genetics [12] is the study of inherited traits and variations from one generation to other. Every one inherits genes from their parents and passes the specific characteristics (traits) like eye color, skin color. Conversely, some genes may have certain diseases that may inherit from parents to their children. A scientist who studies genetics is called a geneticist.

2.9 Genome

A genome [13] consists of genetic information of organism. It has all necessary details to build and develop that organism. Entire genome of human has more than 3 billion DNA base pairs. Genome word is made from "gene" and "chromosome."

2.10   Chromosomes

Chromosome [14] is an entire chain of DNA along with a group of stabilizing proteins which carries genetic information. It is a thread-like structure. Twenty-three pairs of different chromosomes are available in human [14]. In this pair, one is come from mother and other from father. Male and female have same twenty-two pairs, and twenty-third pair is different in both which is sex chromosome. In total, 46 chromosomes are present in human body.

2.11   Protein

Protein [15] creates living cells. In our body, we have millions of different kind of proteins. When amino acid merges together, it must form a perfect shape protein in order to function properly. If it is wrongly shaped, they usually won't work. Amino acids live in cytoplasm of living cells. They help DNA to interact with cytoplasm to convert amino acids to protein. Protein sequences are built from 20 different amino acids. For example, pepsin protein is digestive protein which has particular shape which enables us to break up the food inside our stomach. So, food can be properly absorbed by body. Another example is keratin protein; they stuck together and make hard structure like nails, beaks.

2.12   Codon and Anti-codon

Codon [16] is 3-base sequence of nitrogenous bases in a row on mRNA. mRNA is single sequence of ploy nucleotides having **a**denine, **g**uanine, **c**ytosine, and **u**racil. Specific amino acids are coded for it. Information for proteins is also in it. There are in total 64 codons which have been discovered to be present on mRNA chain. A total of 61 codons are for specific amino acid, and the remaining three codons are known as stop codons indicating where the amino acid can be stopped. To start and stop protein synthesis is instructed by start and stop codons, and they act as punctuation marks in genetic code. The region between the start and stop codons (inclusive of them) is called open reading frame (ORF) or sometimes coding sequence (CDS). Usually, UAG is start codon and UAA, UGA, and UAG are stop codons. Anti-codon is 3-base sequence of nitrogenous bases in a row on tRNA. It recognizes the correct codon by its complementary sequence. For example, if codon: AUG, then anti-codon: UAC.

2.13 DNA Replication

Producing duplicate DNA from original DNA is known as DNA replication [17]. This process occurs in all living organisms. DNA has double-helix structure of two strands. These strands are detached during replication process. Each DNA sequence is one of the strands, and it works as a template for making its other strand. This process is also known as semiconservative replication.

2.14 Transcription

In transcription [18] process, particular part of DNA sequence is copied into RNA. It is the first process of gene expression.

2.15 Translation

In translation [18] process, mRNA is decoded to produce specific series of amino acids. It is the second process of gene expression.

2.16 Mutations in DNA

Any change in the sequence of nucleotide of DNA is known as mutations [19]. It occurs due to a change in environmental conditions like chemicals, ultraviolet rays, and random change in DNA replication. Even a minute change in the sequence of nitrogen base can change the amino acid leading to complete different protein. To put it differently, assume that sequence of nucleotide in DNA is a sentence and mutations are mistakes in spelling of words that comprise those sentences. Occasionally, mutations are insignificant similar to misspelled word in sentences, but they remain very clear whereas in some cases mutations have more powerful effects, such as a sentence whose meaning is totally changed.

Point Mutations. A point mutations is the change of only one nitrogen base in a DNA sequence, and it is least dangerous among all other types of DNA mutations. It normally happens when the process of DNA replication is occurred.

Frameshift Mutations. Frameshift mutations are more critical and lethal compared to point mutations. Despite the fact that just one nitrogen base is impacted similar to the point mutations, here the one particular base is either totally removed or an additional one is placed into the center of a DNA sequence. This particular change in sequence leads to shift in frame; hence, title is frameshift mutation. **Insertions**:

It is one type of frameshift mutation. Just like the name indicates, an insertion takes place when a particular nitrogen base will be unintentionally included in the middle of the sequence. **Deletions**: It is another type of frameshift mutation and happens when a nitrogen base is eliminated from the sequence.

3 Bioinformatics Comprises Three Components/Phases

3.1 Databases Creation/Generation

In the first phase, main objectives are to organize, store, and manage biological databases. These databases are freely available to know the existing information and submit new entries for researchers, for example, NCBI, GenBank [20], DDBJ [21].

3.2 Algorithms and Statistics Method Development

In the second phase involves software development process using statistics methods to mine the interesting knowledge from databases, for example, to know about unknown gene functionality comparing with known gene databases.

3.3 Data Analysis and Interpretation

The third phase involves visualization or interpretation of mined data from second phase given above in a biologically meaningful manner. It gives remarkable outcomes for future references.

4 Data Types in Bioinformatics

4.1 Primary Data

It is also known as raw data, for example, DNA or amino acid. These are building blocks of everything. Primary data are sequence data. Primary database can be found from GenBank, EMBL [22] (The European molecular Biology Laboratory), DDBJ (DNA Data Bank of Japan).

4.2   Secondary Data

It is made up from primary data. A bunch of amino acids merge together and build protein. Protein is stored as secondary data. It is also known as motif. Motifs are signatures to identify protein functionality.

4.3   Tertiary Data

These data are domains and folding units.

5   NCBI

National Center for Biotechnology Information (NCBI) is a research group which consists of mathematicians, computer scientists, biologists focused on molecular biology [23]. They work on biological problems which are solved by computational and mathematical methods. NCBI is responsible for updating GenBank, EMBL, DDBJ, and other databases. The NCBI allocates a unique id number to each species of organism. It also provides database in different format such as FASTA, BAM, SFF, and HDF5, etc.

6   Real-Life Database in Bioinformatics

See Table 1.

7   Applications of Bioinformatics

7.1   Sequence Analysis

Sequence analysis includes DNA sequencing—converting from original DNA sequence to computerized soft copy of DNA sequence; sequence similarity—finding same DNA sequence from organisms; sequence alignment—finding functional similarity of two DNAs; identifying repeat regions—finding subsequences of nucleotides which occurs frequently in given DNA; identifying coding regions—finding subsequence of DNA sequence which forms a gene. These all applications of sequence analysis can be done using computer tools.

Table 1 Types of databases and its web sources

Database	URL
Nucleotide sequences	
DDBJ EMBL GenBank	www.ddbj.nig.ac.jp www.ebi.ac.uk/embl www.ncbi.nlm.nih.gov/Genbank
Protein sequence (primary)	
SWISS-PROT [24] PIR-International [25]	www.expasy.ch/sprot/sprot-top.html www.mips.biochem.mpg.de/proj/protseqdb
Protein sequence (composite)	
OWL [26] NRDB [27]	www.bioinf.man.ac.uk/dbbrowser/OWL www.ncbi.nlm.nih.gov/entrez/query.fcgi?db=P rotein
Protein sequence (secondary)	
PROSITE [28] PRINTS [29] Pfam [30]	www.expasy.ch/prosite www.bioinf.man.ac.uk/dbbrowser/PRINTS/PR INTS.html www.sanger.ac.uk/Pfam/
Macromolecular structure	
Protein Data Bank (PDB) [31, 32] Nucleic acids database (NDB) [33] HIV protease database [34] ReLiBase [35] PDBsum [36] CATH [37] SCOP [38] FSSP [39] ASTRAL [40] HOMSTRAD [41]	ww.rcsb.org/pdb ndbserver.rutgers.edu/ www.ncifcrf.gov/CRYS/HIVdb/NEW_DATA BASE www2.ebi.ac.uk:8081/home.html www.biochem.ucl.ac.uk/bsm/pdbsum www.biochcm.ucl.ac.uk/bsm/cath scop.mrc-lmb.cam.ac.uk/scop www2.embl-ebi.ac.uk/dali/fssp http://astral.berkeley.edu/ http://mizuguchilab.org/homstrad/
Genome sequences	
Entrez genomes [42] GeneCensus [43] COGs [44]	www.ncbi.nlm.nih.gov/entrez/query.fcgi?db= Genomebioinfo.mbb.yale.edu/genome www.ncbi.nlm.nih.gov/COG
Integrated databases	
InterPro [45] Sequence retrieval system (SRS) [46] Entrez [47]	www.ebi.ac.uk/interpro www.expasy.ch/srs5 www.ncbi.nlm.nih.gov/Entrez

7.2 Prediction of Protein Structure

Complex protein structures such as secondary or tertiary structures are difficult to determine than primary structures of proteins; however, they can be identified using bioinformatics tools easily.

7.3 Genome Annotation

Genome annotation determines protein coding and the regulatory sequences in living organ. It has played very essential role in human genome project.

7.4 Health and Drug Discovery

Recently, drug discovery, diagnosis, and disease identification are done using bioinformatics tools. Initially, whole animal were used to test a drug. Nowadays, researchers made possible to work on specific genome using complete sequencing of genome.

7.5 Agriculture

Crop improvement. Special mutations can decrease crop diseases. It can survive in wide range of environmental conditions such as weather, sunlight, water.

Insect resistance. Ability of plants to resist insect bites can be increased by combining plants' genome molecules and resistance chemical molecules, and it will also improve nutritional quality of crop too.

7.6 Gene Therapy

Gene therapy is a technique that genetic disease is treated or prevented by inserting functioning genes into patient's cell which correct the effects of a disease-causing mutation.

7.7 Drug Development

Manual drug development process is costly and time-consuming. In bioinformatics, databases of drugs are available and software are useful for drug discovery, designing, and development processes which reduce cost and overall time.

7.8 Personalized Medicine

Personalized medicine can be created as per patient's genetic information. These information can help doctor to diagnose and treatment of disease. It is given for personalized cancer cure.

7.9 Microbial Genome Applications Such as Waste Cleanup

Deinococcus radiodurans bacteria have capability to repair damaged DNA and small fragments from chromosomes [48]. It happens because of additional copies of its genome. These are helpful for environmental cleanup.

7.10 Bioweapon Creation

Bioweapons are created in order to propagate disease among people, plants, and animals through a number of disease-causing agents like viruses, bacteria, fungi, or even toxins. Attackers may spread these agents through aerosols or food as well as water supplies.

8 Research Issues and Challenges

8.1 From the Perspective of Computer Science

Information Management of Biological Data. As the volume of data grows it is becoming critical to handle information of biological data. Researchers are facing challenges to store a huge biological data, to transfer data with rapid speed, to retrieve a piece of information from DNA and to visualize a whole DNA. DNA sequence can be represented in a binary format in a computer for all four base pairs. Combinations of two-bit binary format (0 or 1) are grouped together to make four different base pairs (00—AT, 01—CG, 10—GC, and 11—TA). Thus, four DNA base pair can be represented using 1 byte or 8 bits. The whole diploid human genome in bytes or bits can be calculated as following: 6×10^9 base pairs/diploid genome $\times$ 8 bits/4 base pairs $= 12 \times 10^9$ bits or (12×10^9 bits/8 $= 1.5 \times 10^9$ bytes) or 1.5 GB [49]. Similarly, we can calculate the size of the genetic data in human body. One of the estimation says that in the human body, 37.2 trillion cells are available [50]. Usually, in each cell, 1.5 GB data is stored. If we concentrate on just the cells that build human body and neglect the microbiome which are non-human cells that resides in human body, calculations of approx. amount of data kept in human body are as per given below: 1.5 GB $\times$ 37.2 trillion cells $= 55.8$ trillion GB or $55.8 \times 10^{12} \times 10^9$ bytes $= 55.8$ zettabytes (10^{21}).

Knowledge Discovery from Biological Data. The volume and diversity of biomedical data are increasing quickly, presenting numerous challenges and opportunities which range from data capture, data preprocessing, data mining, and data analysis. Data capture includes data collection and storage of data. Data preprocessing comprises data transformation, data normalization, data reduction, data cleanness, etc… It is required because data can be gathered from different sources in different formats. Data mining includes algorithms for finding interesting patterns from database. Data analysis includes visualization of mined data through reports and graphs. It is really difficult to analyze such a huge data using single machine. It requires supercomputer or parallel computation for executing data mining algorithms on a huge biological data.

Security and privacy of biological data. Biological data can be stored in computer hard drive. It is also important challenge to secure these data from threats, virus, and worms. While transferring a huge amount of biological data, encryptions and decryptions are the possible way to secure them. Usage of these techniques increase total processing time for data analysis because they are very time consuming process. Digital signature algorithm can be used for maintaining privacy, confidentiality, and authenticity of data.

8.2 From the Perspective of Biological

1. Comparisons of two genomes or DNAs or RNAs
2. To discover gene from DNA
3. Classification/Prediction of unknown molecular structures
4. Identify precisely, predictive model of transcription beginning and ending: to predict when and where transcription will take place in a genome
5. Predict precisely, splicing pattern RNA splicing/alternative splicing
6. Finding protein–protein, protein–DNA, and protein–RNA recognition codes very effectively
7. Finding the information from DNA, RNA, protein sequences, and structure
8. Generating pattern-matching techniques from DNA, RNA, and protein sequence analysis and structure analysis.

9 Future Research Trends

Biological data size is increasing significantly and collected from heterogeneous data source. These massive information can be stored and processed using big data tools such as Hadoop Distributed File System (HDFS) [51], spark [52], MongoDB [53]. To analyze them, machine learning techniques such as supervised learning like decision trees [54], naïve Bayes [55], support vector machines (SVMs) [56], neural networks [57], K-nearest neighbor (KNN) [58], and unsupervised learning like clustering [59] are the most commonly used methods in many applications. Due to advancement in computer resources, multicore programming is being used to solve big bioinformatics information, and graphical processing unit (GPU) is used for large matrix calculations which usually decrease execution time tremendously. Nowadays, biodata are also available in the form of images such as protein structure, gene expressions. Deep learning methods are proved in order to process images using convolutional neural network [60] and sequences using recurrent neural network [61].

10 Conclusion

Bioinformatics is a relatively young discipline, and advancement in this field has progressed very quickly in recent years. With the knowledge of biology, statistics, and computer programming language, bioinformatician can make tools and software packages which are used to verify hypotheses virtually before introducing expensive experiments. It is helpful to take more effective decision in less amount of time. In this survey paper, an introduction and overview are presented for the current state of the field from computer scientist's perspective. We discussed basic terms of bioinformatics to understand well about this field. We studied main components, data types and also discussed important database of different types of bioinformatics which are commonly used. We then looked emerging fields, several applications and described computational and biological research issues and challenges in bioinformatics. Finally, we examined various future trends of the field. Future of bioinformatics is impressive because of largely available databases which are either generic or specific, and they keep increasing which yields more accurate results and reliable interpretations.

References

1. Luscombe NM, Greenbaum D, Gerstein M, What is bioinformatics? An introduction and overview, NCBI, 83–99 (2001)
2. SABU M. THAMPI Introduction to Bioinformatics, CoRR (2009)
3. Hogeweg, Paulien, The Roots of Bioinformatics in Theoretical Biology, PLoS Computational Biology (2011)

4. National Center for Biotechnology Information, http://www.ncbi.nlm.nih.gov
5. Alberts, B., Johnson, A., Lewis, J., Raff, M., Roberts, K. and Walter, P. Molecular Biology of the Cell. 4th Edn, Annals of Botany, vol. 91.3 (2003)
6. Ribonucleic Acid, https://www.nature.com/scitable/definition/ribonucleic-acid-rna-45
7. Pearson H., Genetics: what is a gene?, Nature, 441, 398–401 (2006)
8. International Human Genome Sequencing Consortium, Finishing the euchromatic sequence of the human genome, Nature, 431, 931–45 (2004)
9. The Chimpanzee Sequencing and Analysis Consortium 2005, Initial Sequence of the Chimpanzee Genome and Comparison with the Human Genome, Nature, 37, 69–7 (2005)
10. Allele, https://www.nature.com/scitable/definition/allele-48
11. DNA Sequencing, https://www.genome.gov/10001177/dna-sequencing-fact-sheet
12. Griffiths AJF, Miller JH, Suzuki DT, An Introduction to Genetic Analysis-7th edition. W. H. Freeman, New York (2000)
13. Genome, https://www.ncbi.nlm.nih.gov/genome
14. Chromosomes, https://www.ncbi.nlm.nih.gov/pubmedhealth/PMHT0025047
15. Proteins, https://www.nature.com/subjects/proteins
16. J.Christopher Anderson, Thomas J Magliery, Peter G Schultz, Exploring the Limits of Codon and Anticodon Size, In Chemistry & Biology, Vol. 9, Issue 2, pp. 237–244, (2002)
17. Annunziato, A. T. Split decision: What happens to nucleosomes during DNA replication? Journal of Biological Chemistry, 280, pp. 12065–12068 (2005)
18. Ribosomes, Transcription, and Translation, https://www.nature.com/scitable/topicpage/ribosomes-transcription-and-translation-14120660
19. Genetic Mutation, https://www.nature.com/scitable/topicpage/genetic-mutation-1127
20. Benson DA, Karsch-Mizrachi I, Lipman DJ, Ostell J, Rapp BA, Wheeler DL. GenBank. Nucleic Acids Research, 28, (2000)
21. Okayama T, Tamura T, Gojobori T, Tateno Y, Ikeo K, Miyazaki S, Formal design and implementation of an improved DDBJ DNA database with a new schema and object-oriented library, Bioinformatics 14, (1998)
22. Baker W, van den Broek A, Camon E, Hingamp P, Sterk P, Stoesser G, The EMBL nucleotide sequence database. Nucleic Acids Research, 28, pp. 19–23 (2000)
23. The National Center for Biotechnology Information Programs and Activities, https://www.nlm.nih.gov/pubs/factsheets/ncbi.html
24. Bairoch A, Apweiler R. The SWISS-PROT protein sequence database and its supplement TrEMBL in 2000, Nucleic Acids Research, 28 (2000)
25. McGarvey PB, Huang H, Barker WC, Orcutt BC, Garavelli JS, Srinivasarao GY, et al. PIR: a new resource for bioinformatics. Bioinformatics, 16, pp. 290–291 (2000)
26. Bleasby AJ, Akrigg D, Attwood TK. OWL—a non-redundant composite protein sequence database. Nucleic Acids Research, 22, pp. 3574–3577 (1994)
27. Bleasby AJ, Wootton JC. Construction of validated, non-redundant composite protein sequence databases. Protein Eng, 3, pp. 153–159 (1990)
28. Hofmann K, Bucher P, Falquet L, Bairoch A. The PROSITE database, its status in 1999. Nucleic Acids Research, 27, pp. 215–219 (1999)
29. Attwood TK, Croning MD, Flower DR, Lewis AP, Mabey JE, Scordis P, PRINTS-S: the database formerly known as PRINTS. Nucleic Acids Research 2000, 28, pp. 225–227 (2000)
30. Bateman A, Birney E, Durbin R, Eddy SR, Howe KL, Sonnhammer EL. The Pfam protein families database. Nucleic Acids Research, 28, pp. 263–266 (2000)
31. Bernstein FC, Koetzle TF, Williams GJ, Meyer EF, Jr., Brice MD, Rodgers JR, The Protein Data Bank. A computer-based archival file for macromolecular structures. Eur J Biochem, 80, (1977)
32. Berman HM, Westbrook J, Feng Z, Gilliland G, Bhat TN, Weissig H, The Protein Data Bank. Nucleic Acids Research, 28, (2000)
33. Berman HM, Olson WK, Beveridge DL, Westbrook J, Gelbin A, Demeny T, The Nucleic Acid Database. A comprehensive relational database of threedimensional structures of nucleic acids. Biophys J, 63, pp. 751–759 (1992)

34. Vondrasek J, Wlodawer A. Database of HIV proteinase structures. TIBS, 22, (1997)
35. Hendlich M. Databases for protein-ligand complexes. Acta Cryst D 54, (1998)
36. Laskowski RA, Hutchinson EG, Michie AD, Wallace AC, Jones ML, Thornton JM. PDBsum: a Web-based database of summaries and analyses of all PDB structures. TIBS, 22, pp. 488–490 (1997)
37. Pearl FM, Lee D, Bray JE, Sillitoe I, Todd AE, Harrison AP, Assigning genomic sequences to CATH. Nucleic Acids Research, 28, pp. 277–282 (2000)
38. Lo Conte L, Ailey B, Hubbard TJ, Brenner SE, Murzin AG, Chothia C. SCOP: a structural classification of proteins database. Nucleic Acids Res, 28, pp. 257–259 (2000)
39. Holm L, Sander C. Touring protein fold space with Dali/FSSP. Nucleic Acids Research, 26, pp. 316–319 (1998)
40. Brenner SE, Koehl P, Levitt M. The ASTRAL compendium for protein structure and sequence analysis. Nucleic Acids Research, 28, pp. 254–256 (2000)
41. Mizuguchi K, Deane CM, Blundell TL, Overington JP. HOMSTRAD: a database of protein structure alignments for homologous families. Protein Science : A Publication of the Protein Society, 7, pp. 2469–2471 (1998)
42. Tatusova TA, Karsch-Mizrachi I, Ostell JA. Complete genomes in WWW Entrez: data representation and analysis. Bioinformatics, 15, (1999)
43. Lin J, Gerstein M. Whole-genome trees based on the occurrence of folds and orthologs: implications for comparing genomes on different levels. Genome Research, 10 (2000)
44. Tatusov RL, Koonin EV, Lipman DJ. A genomic perspective on protein families. Science, 278, (1997)
45. Attwood TK, Flower DR, Lewis AP, Mabey JE, Morgan SR, Scordis P, PRINTS prepares for the new millennium. Nucleic Acids Research, 27 pp. 220–225 (1999)
46. Etzold T, Ulyanov A, Argos P. SRS: information retrieval system for molecular biology data banks. Methods Enzymol, 266 (1996)
47. Schuler GD, Epstein JA, Ohkawa H, Kans JA. Entrez: molecular biology database and retrieval system. Methods Enzymol, 266, (1996)
48. Makarova, Kira S. Genome of the Extremely Radiation-Resistant Bacterium Deinococcus Radiodurans Viewed from the Perspective of Comparative Genomics. Microbiology and Molecular Biology Reviews, 65, pp. 44–79 (2001)
49. Samuel Levy, Granger Sutton, Pauline C Ng, Lars Feuk, Aaron L Halpern, Brian P Walenz, Nelson Axelrod, Jiaqi Huang, Ewen F Kirkness, Gennady Denisov, Yuan Lin, Jeffrey R MacDonald, Andy Wing Chun Pang, Mary Shago, Timothy B Stockwell, Alexia Tsiamouri, Vineet Bafna, Vikas Bansal, Saul A Kravitz, Dana A Busam, Karen Y Beeson, Tina C McIntosh, Karin A Remington, Josep F Abril, John Gill, Jon Borman, Yu-Hui Rogers, Marvin E Frazier, Stephen W Scherer, Robert L Strausberg, J. Craig Venter, "The Diploid Genome Sequence of an Individual Human", PLoS Biology, 5 (2007)
50. Eva Bianconi and Allison Piovesan and Federica Facchin and Alina Beraudi and Raffaella Casadei and Flavia Frabetti and Lorenza Vitale and Maria Chiara Pelleri and Simone Tassani and Francesco Piva and Soledad Perez-Amodio and Pierluigi Strippoli and Silvia Canaider, An estimation of the number of cells in the human body, Annals of Human Biology, Vol. 40, pp. 463–471 (2013)
51. K. Shvachko, H. Kuang, S. Radia and R. Chansler, The Hadoop Distributed File System, In: 26th IEEE Symposium on Mass Storage Systems and Technologies (MSST), Incline Village, NV, pp. 1–10 (2010)
52. Matei Zaharia, Mosharaf Chowdhury, Michael J. Franklin, Scott Shenker, and Ion Stoica., Spark: cluster computing with working sets. In: 2nd USENIX conference on Hot topics in cloud computing (HotCloud'10), USENIX Association, Berkeley, CA, USA, 10 (2010)
53. Naresh Kumar Gundla, Zhengxin Chen, Creating NoSQL Biological Databases with Ontologies for Query Relaxation, In: Computer Science, Vol. 91, pp. 460–469, (2016)
54. J. R. Quinlan. Induction of Decision Trees. Mach. Learn. 1, pp. 81–106 (1986)
55. Pedro Domingos and Michael Pazzani. 1997. On the Optimality of the Simple Bayesian Classifier under Zero-One Loss. Mach. Learn. 29, pp. 103–130 (1997)

56. Christopher J. C. Burges. 1998. A Tutorial on Support Vector Machines for Pattern Recognition. Data Min. Knowl. Discov. 2, pp. 121–167 (1998)
57. Hyunsoo Yoon, Cheong-Sool Park, Jun Seok Kim, Jun-Geol Baek, Algorithm learning based neural network integrating feature selection and classification, Expert Systems with Applications, Vol. 40, pp. 231–241 (2013)
58. T. Cover and P. Hart. 2006. Nearest neighbor pattern classification. IEEE Trans. Inf. Theor. 13, pp. 21–27 (2006)
59. John A. Hartigan. Clustering Algorithms, John Wiley & Sons, Inc., New York, NY, USA (1975)
60. Alex Krizhevsky, Ilya Sutskever, and Geoffrey E. Hinton. ImageNet classification with deep convolutional neural networks. In: 25th International Conference on Neural Information Processing Systems (NIPS'12), F. Pereira, C. J. C. Burges, L. Bottou, and K. Q. Weinberger (Eds.), Vol. 1. Curran Associates Inc., USA, pp. 1097–1105 (2012)
61. Zachary C. Lipton, John Berkowitz, Charles Elkan, A Critical Review of Recurrent Neural Networks for Sequence Learning, CoRR, (2015)

Leveraging Deep Learning for Anomaly Detection in Video Surveillance

K. Kavikuil and J. Amudha

Abstract Anomaly detection in video surveillance data is very challenging due to large environmental changes and human movement. Additionally, high dimensionality of video data and video feature representation adds to these challenges. Many machine learning algorithms failed to show accurate results and it is time consuming in many cases. The semi supervised nature of deep learning algorithms aids in learning representations from the video data instead of hand crafting the features for specific scenes. Deep learning is applied to handle complicated anomalies to improve the accuracy of anomaly detection due to its efficiency in feature learning. In this paper, we propose an efficient model to predict anomaly in video surveillance data and the model is optimized by tuning the hyperparameters.

Keywords Anomaly event classification · Deep neural networks · Video surveillance · Convolutional neural networks

1 Introduction

The use of surveillance cameras ensures a secure environment, and detecting anomaly in the prerecorded video data is an important aspect in this area. Anomalies in many scenarios are context specific or subjective, so modelling an anomaly is a challenging task. Detecting and localizing anomaly in video data is time consuming using hand crafted features. Instead of using highly complex features, recent approaches

K. Kavikuil (✉) · J. Amudha
Department of Computer Science & Engineering, Amrita School of Engineering,
Bengaluru, India
e-mail: kavikuild@gmail.com

J. Amudha
e-mail: j_amudha@blr.amrita.edu

K. Kavikuil · J. Amudha
Department of Computer Science & Engineering, Amrita Vishwa Vidyapeetham,
Bengaluru, India

© Springer Nature Singapore Pte Ltd. 2019   239
R. S. Bapi et al. (eds.), *First International Conference on Artificial Intelligence
and Cognitive Computing* , Advances in Intelligent Systems and Computing 815,
https://doi.org/10.1007/978-981-13-1580-0_23

propose an efficient way of extracting features from images and frames in a fully unsupervised manner. Learning good data representation is important for a machine learning algorithm. Deep learning marks a difference in feature learning in an unsupervised manner and outperforms traditional machine learning algorithms. Due to high dimensionality of video data, it is important to represent video data in an efficient way to analyze it. In this paper, we propose a novel approach which predicts appearance anomaly due to strange action or objects. The model proposed has been trained on normal as well as anomalous events. There are various soft computing techniques like supervised, unsupervised and apriori that can be employed to detect anomaly [1]. Each of these techniques has its own pros and cons. Anomaly detection method can be pixel or object based. Many artificial intelligent tasks like object detection, speech recognition, machine translation has been improved by the use of deep learning [2, 3]. The proposed model leverages the learning ability of deep architectures to classify anomalous frames.

2 Related Work

Typically for building an anomaly detection system, the model will be trained on normal events and anything that deviates normalcy will be predicted as anomaly. In this proposed approach, the model has been trained on both normal and anomalous events which makes the model more generalized. Convolutional neural network (CNN) is a type of artificial neural network that has shown remarkable performance in image classification [3]. CNN outperforms other models in image classification. CNN finds its application in various scenarios for feature extraction and classification. CNN in combination with other architectures can be used for predicting object detection, visual tracking [4]. Feature extraction from the high dimensional data plays an important role in anomaly detection of video data. Video is considered to be a sequence of frames, so a CNN based architecture performs feature extraction efficiently.

Deep CNN based approach was proposed in paper [5] to extract vision features. In paper [6] CNN classifier is used to predict emotions in faces. Many work had been proposed to detect regions specific to humans in surveillance video [7]. In paper [8] Itti Koch saliency model was proposed to detect unauthorized human activity. In paper [9] anomaly detection has been performed using deep cascaded CNN, to detect anomaly in crowded scenes and CNN was trained from scratch to extract the features. In paper [10] trained Fully connected CNN is fine tuned to an unsupervised CNN and a gaussian classifier is fit into classify anomalous frames. In this work the entire frame is passed to CNN and a classifier is fit to classify the frame and the model is optimized by hyperparameter optimizers. In paper [11] region proposal was marked to identify anomaly in specified regions. Anomaly detection in crowded scenes is extremely challenging and many approaches were proposed using spatial temporal patterns captures from the scenes [12]. A deep CNN model was proposed to detect anomaly in log data of distributed systems [13].

3 Methodology

The proposed model is based on the basic CNN architecture. CNN finds its use in large scale video classification due to its ability to learn powerful features from weakly labelled data and the connectivity in the architecture. The architecture proposed is able to learn spatial features from the video data. Videos vary in temporal content and each clip contains several contiguous frames in time and by extending the connectivity in the architecture, the model learns spatial features.

3.1 Anomaly Detection

Generally, anomaly detection models are trained on normal events during the training phase and trained model predicts abnormality if it does not fall under normal class of training samples. The proposed model learns both normal and abnormal events during training phase, so the model is robust in predicting various unusual scenarios. The spatial features learned makes the model more versatile in predicting anomaly at the frame level.

3.2 Deep Convolutional Neural Network

The model has stacked layers of convolutions, pooling layers and dense layers as shown in Fig. 1. A convolutional architecture during the training phase extracts features from small portion of the video frame and gradually moves across the larger portion of the video frame. In the later stage, we consolidate the features extracted from each convolution stage into a single feature map. These feature maps are aggregated by the pooling layers and passed on to the dense layers. Thus, the model learns the features automatically in a greedy manner. The deep architecture maintains a feature hierarchy through which relevant features are learned. To achieve this relevant feature extraction an appropriate kernel size and stride is chosen.

3.2.1 Kernel Size

Kernel size plays an important role in extracting features using conv layers. Choosing an appropriate kernel size helps in extracting relevant features. In this work 3×3 is the kernel's size chosen to convolve around 128×128 image. The number of filters are randomly fixed as 20 to train the model, so each layer generates 20 feature maps. The kernel size is fixed throughout in training this model. The model learns more discriminative features in the deeper layers to classify normal and abnormal frames efficiently. Stride size denotes the number of pixels the filters slide along the input

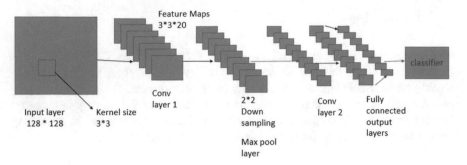

Fig. 1. High level design of the model

image to generate the feature maps. In this work, the stride size was fixed as 2, and the filter convolves around the input frame.

3.2.2 Convolution and Pooling Layers

The convolutional layers learn spatial features straightforwardly and represents 2-D spatial features efficiently. In this model, there are two convolutional layers which represents spatial features of the input frames. The feature maps from the convolutional layers has many trainable parameters and the pooling layers are responsible for reducing the trainable parameters by aggregating the feature maps. In the proposed model, max pooling is used and the aggregated features are passed to the next layer.

3.3 Model Implementation and Training

The proposed model has layers of convolutions followed by max pooling layers. RELU nonlinear activation function is used. All neurons from the previous layer are connected to each other in the fully connected layer and the SoftMax layer classifies the samples based on the probability distribution. The model was implemented using Keras on Tensorflow framework. The model's hyperparameters like kernel size, stride size, number of filters are fixed throughout this experiment wherein model's learning rate has been evaluated using adaptive learning rate schedulers like AdaDelta, Adam, rmsprop.

3.4 Parameter Optimization

To achieve the goal of anomaly classification task, we fine tune the parameters for better accuracy. The dataset is split into mini batches and the sensitivity of the hyperparameters are analyzed. Cross entropy is the loss function employed and the model's performance is checked with various optimizers. Weights are initialized randomly and the drop out regularization is done to improve the accuracy. The model is validated with random split and the accuracy corresponding to the changes in the hyperparameter settings are analyzed to build a robust model for anomaly classification task. The number of epochs are increased gradually and in each case model's processing time and the accuracy is examined. Learning rate is an important parameter in model's accuracy and determines the rate of updating the parameters. The optimizer AdaDelta scales the learning rate based on historical gradients. Adam (Adaptive moment estimation) optimizer is based on the estimation of first (mean) and second raw moments of historical gradients and schedules a learning rate for each parameter. RMSprop optimizer uses the magnitude of the gradients to normalize the gradients. The model's performance is inspected on these optimizers.

4 Experiments

We have conducted experiments to assess the performance of the model by various optimizers and hyper parameter settings. The kernel size of the model is fixed as 3×3 throughout the experiment. The stride size if fixed as 2, with which the kernel slides over the image to capture the spatial features. The model's trainable parameters are fine-tuned with adaptive learning rate schedulers and the model's sensitivity to these optimizers are checked. Drop out regularization has been adopted to avoid overfitting.

4.1 Dataset

The dataset chosen is Avenue dataset, it has 16 training and 20 testing video clips. The anomalies modelled are loitering and strange actions like throwing papers or bag.

The frames are separately extracted and annotated as normal and abnormal. The model is trained on the labelled dataset. The frames are of resolution 640×300. As part of video preprocessing, frames are extracted and are annotated. The frames are resized 128×128 and reshaped to suit the architecture. This entire preprocessing was done using OpenCV. The frames depicting the anomalous events are given in Fig. 2.

Fig. 2. Anomalous frames. **a** Strange action, **b** Strange action, **c** loitering

4.2 Visualizing the Hidden Layers

The hidden layers play a vital role in feature learning in a deep learning architecture. The deeper layers give more discriminative features from the frame. These learned features are used to classify normal and anamolous frames. The model's learned spatial features during its training is shown in Fig. 3.

5 Results and Discussion

The sensitivity of the optimizers and the performance of the model in all the cases are discussed in this section. The loss and accuracy of the model is also examined. These adaptive schedulers reduce the effort to set the learning rate manually. These optimizers improve the model's accuracy during training and validation by scheduling the learning rate appropriately.

5.1 Learning Rate

Learning rate is an important parameter for the model's accuracy, choosing a learning rate that is too small leads to slow convergence and that is too large can hinder convergence and the cost function may fluctuate. Model's performance is evaluated with different optimizers AdaDelta, Adam and RMSprop. The confusion matrix and the metrics are given in Table 1.

The plot of training and validation accuracy and loss is shown in Fig. 4. The graph depicts the gradual increase in accuracy. Testing has been performed by randomly split testing data from the dataset. As the plot of loss indicates as the number of iterations increases, the loss gradually reduces in training and validation. The plot

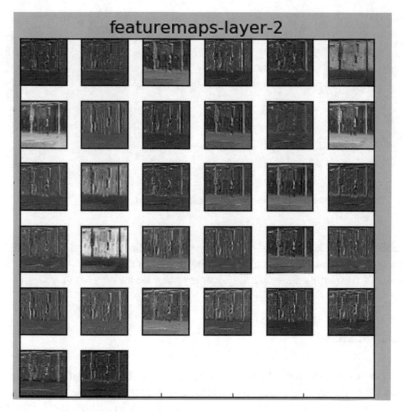

Fig. 3. Feature maps of hidden layer 2

Table 1 Precision, recall, f1-score of the model with different dropout probability set as 0.25

	Precision	Recall	F1 score	Support
Class 0 (Abnormal)	1.00	0.97	0.99	80
Class 1 (Normal)	0.97	1.00	0.99	66
Avg/Total	0.99	0.99	0.99	146

of accuracy shows that the accuracy of the model gradually increases and reaches to its best at the end of iterations.

5.2 Drop Out Regularization

Drop out probability determines the number of units to be dropped at each stage. The model was initially inspected with the drop out probability of 0.5. Later with altered

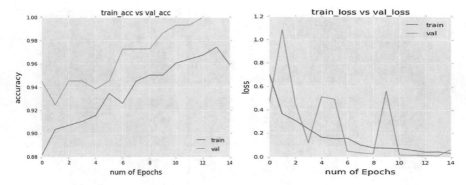

Fig. 4. Training and validation accuracy and loss for 15 epochs

probability of 0.25 same number of epochs were conducted. The plots show that the accuracy fluctuates and later converges. The change in the dropout probability alters the model's computing time and the accuracy steps down by a small margin.

6 Conclusion

The proposed model is a prototype tested on a particular dataset and it can be enhanced to detect frame level classification of anomalous events. The model is optimized by regularizing and tuning the hyperparameter optimizers. The experiment shows that the parameter changes effects very little in the model's performance. The resultant model outperforms other models in predicting anomalous events at the frame level. The proposed method is less sensitive to parameters and is easy to implement. Future work may include extending the model to predict wider dimensions of anomalous cases in temporal domain and integrating with hardware component to trigger alarms to indicate anomaly.

References

1. Dinesh Kumar Saini, Dikshika Ahir and Amit Ganatra.: Techniques and Challenges in Building Intelligent Systems: Anomaly Detection in Camera Surveillances'. Satapathy and S. Das (eds.), Springer International Publishing Switzerland 2016, Proceedings of First International Conference on Information and Communication Technology for Intelligent Systems: Volume 2, Smart Innovation, Systems and Technologies (2016).
2. A. Krizhevsky, I. Sutskever, G. E. Hinton.: ImageNet classification with deep convolutional neural network: Advances Neural Information Processing Systems (2012).
3. R. Ramachandran, Rajeev, D. C., Krishnan, S. G., and Subathra P.: Deep learning – An overview: International Journal of Applied Engineering Research, vol. 10, pp. 25433–25448, (2015).

4. Da Zhang, Hamid Maei, Xin Wang, and Yuan-Fang Wang: Deep Reinforcement Learning for Visual Object Tracking in Videos, Department of Computer Science, University of California at Santa Barbara, Samsung Research America. (2017).
5. K. Nithin. D and Dr. Bhagavathi Sivakumar P.: Learning of Generic Vision Features Using Deep CNN: In 2015 Fifth International Conference on Advances in Computing and Communications (ICACC), Kochi, (2015).
6. K. S. Sahla and Dr. Senthil Kumar T.: Classroom Teaching Assessment Based on Student Emotions: Intelligent Systems Technologies and Applications 2016. Springer International Publishing, Cham, pp. 475–486, (2016).
7. G. Sanjay, Amudha, J., and Jose, J. Tressa.: Moving Human Detection in Video Using Dynamic Visual Attention Model: Advances in Intelligent Systems and Computing, vol. 320, pp. 117–124, (2015).
8. D. Radha, Amudha, J., Ramyasree, P., Ravindran, R., and Shalini, S.: Detection of unauthorized human entity in surveillance video: International Journal of Engineering and Technology, vol. 5, (2013).
9. M. Sabokrou, M. Fayyaz, M. Fathy, R. Kettle, Deep-cascade: Cascading 3D Deep Neural Networks for Fast Anomaly Detection and Localization in crowded Scenes, IEEE Trans. Image Processing (2017) 1992–2004.
10. M. Sabokrou, M. Fathy, M. Hoseini, R. Klette.: Real-time anomaly detection and localization in crowded scenes. In: Computer Vision Pattern Recognition Workshops, pp. 56–62, (2015).
11. Shaoging Ren, Kaiming He, Ross Girshick, Jian Sun.: Faster R-CNN Towards Real Time Object Detection with Region Proposal Networks, IEEE Transactions Pattern Analysis Machine Intelligence, vol. 39, no., pp. 1137–1149, (2017).
12. Yandg Cong, Junsong Yuan, youdang Tang.: Video Anomaly Search in Crowded Scenes via Spatio-Temporal Motion Context. IEEE transactions on information Forensics and Security (2013), 8(10), 1590–1599, (2013).
13. Jiechao Cheng, Rui Ren, Lei Wang and Jian Feng Zhan.: Deep convolutional Neural Networks for Anomaly Event Classification on Distributed Systems. https://arxiv.org/abs/1710.09052 (2017).

Speaker Diarization System Using Hidden Markov Toolkit

K. Rajendra Prasad, C. Raghavendra and J. Tirupathi

Abstract Speaker diarization is ordinarily utilized as a part of the discourse acknowledgment application; it is portrayed in the writing. Then again, utilization of speaker homogeneous bunches, keeping in mind the end goal to adjust the acoustic model to speaker subordinate programmed discourse acknowledgment (ASR) framework that permits enhancing the acknowledgment execution. The i-vectors are likewise called as intermediate vectors. The progression of discourse bunching is as per the following: change over the discourse information in Mel-frequency cepstral coefficient (MFCC) form, model speaker voice by Gaussian mixture model (GMM), adapt the speaker's expression with normal speaker display, apply the VBEM-GMM, find contrast highlight of a few speakers information, and discover grouping comes about by utilizing traditional k-implies calculation. The Universal Background Model is a huge GMM, and this UBM presents the normal speaker display. The speaker's information which was at first changed over into crude shape Mel-frequency cepstral coefficient (MFCC) is utilized for speaking to the speaker voice characteristic. The progression of MFCC which is changing is finished by hidden Markov toolkit (HTK).

Keywords I-vectors · Mel-frequency cepstral coefficient · Gaussian mixture model · Hidden Markov toolkit

1 Introduction

Speaker diarization is normally used in the speech recognition application; it is described in the literature. On the other hand, use of speaker homogeneous clusters, in order to adapt the acoustic model to speaker dependent automatic speech recognition

K. Rajendra Prasad (✉) · C. Raghavendra · J. Tirupathi
CSE Department, Institute of Aeronautical Engineering, Hyderabad, India
e-mail: krprgm@gmail.com

C. Raghavendra
e-mail: crg.svch@gmail.com

© Springer Nature Singapore Pte Ltd. 2019   249
R. S. Bapi et al. (eds.), *First International Conference on Artificial Intelligence and Cognitive Computing* , Advances in Intelligent Systems and Computing 815,
https://doi.org/10.1007/978-981-13-1580-0_24

(ASR) system, allows improving the recognition performance. On the other hand, transcription system and rich speaker indexing use the speaker diarization output (probably) as one of the pieces of information, a lot of that has been extracted from the recording that enables the processing region further and other the automatic indexation.

Mainly, blind speaker diarization [1] problem that information is not known in advance about the number or identity of the person, which will focus on systems that propose a solution. While to punish the speaker segment allocated by any erroneous, it, in order to estimate the number of the loudspeakers correctly, the error measure is important for oriented system's rich transcription data. On the other hand, sufficient data, preferably, to have acoustic characteristics similar are grouped, therefore, that it has some speakers and more to accommodate the speaker model obtained exactly in the ASR system which becomes important. It can at the time of the high-level view, to distinguish between the systems of online and off-line. System before starting the processing thereof is treated with an off-line data can have access to all records. These are the most common in the reference [2]; they are the main focus of attention on this day. Online systems have access to the data recorded up to that point. They may allow you to allow a certain amount of data in order to the latency of the output becomes available for processing, but, in any case, there is no information on a complete record. Such a system begins with one speaker to increase the number of speakers repeatedly (anyone who starts talking at the start of the record), as they usually intervene. Typical systems are used for online processing, which exhibit the approximate results only.

In [3], they have been proposed in the clustering algorithm on the basis vector quantization (VQ) distortion measure and forest. Start the process with one speaker in the codebook, which adds a new speaker that exceeds the threshold in the codebook of the VQ distortion current in stages.

In [4], using a KL distance is calculated distance between two GMM models of two speech segments. Change point is detected to be assigned the audio is available, the data in accordance with the dynamic threshold, and exists in a speaker database, a new speaker is to either be created. Emphasis is placed on the high-speed classification of audio segments to speakers using decision tree structure for speaker model.

Some of the techniques presented [5] can be used online implementation also potentially possible; all systems that are presented below are based on the off-line processing. These systems are hierarchical clustering techniques, can be reached best diarization by repeated for several different potential clusters obtained by dividing or merging existing cluster, are classified into major groups of two while. On the other hand, clustering other techniques, get the diarization output without having to estimate the number of clusters of the first, to derive the cluster from large/small [6].

2 Speaker Diarization

Speaker diarization [7] is the process according to the identity of the speaker, to split the audio stream input to a homogeneous segment. We providing the identity of the speaker, when used with a speech recognition system, by configuring the audio streams on and can improve the readability of automatic speech transcription speaker. This is a combination of a speaker clustering and "spoke?" Speaker diarization is speaker segmentation that is used to answer questions. Initially, it is intended to find a change point of the speaker in the audio stream. The second, which aims to group together the speech segments [8], is based on speaker characteristics.

Standard meausres follows the National Institute of Standards and Technology, as has been revealed by the evaluation of specific, broadcast [7] collected every year, the number of voice mail and conference recording is increased, the speaker diarization is noted by voice community speech that has been, and meetings broadcast news.

With the help of the hidden Markov model [9], for each speaker, in order to model each of the speaker of the corresponding frame, one of the methods of the most popular is to use a Gaussian mixture model in the speaker diarization, it is to assign. There are two main types of the scenario of clustering. The first is due to the by far the most popular, it is called bottom-up. Begins with dividing the full audio content by successive clusters, each cluster, in order to reach the situation, corresponds to the speaker actually; the algorithm tries to merge the redundant cluster gradually.

3 Gaussian Mixture Model

In GMM, we model the speaker data [10] (feature vectors obtained from the above step) using a statistical variation of the features. Hence, it provides us a statistical representation of how speaker produces sounds. Gaussian mixture density is shown to provide a smooth approximation to the underlaying long-term sample distribution of observation obtain form utterances by a speaker. These are the important motivations for using GMM as modeling technique [11].

The following formulae are used to guarantee the above condition:

Mixture weights:

$$\overline{P_i} = \frac{1}{T} \sum_{t=1}^{T} p\left(i_t = i \mid \overrightarrow{x_t}, \lambda\right)$$

Means:

$$\overrightarrow{\mu_i} = \frac{\sum_{t=1}^{T} p(i_t = i \mid \vec{x}_t, \lambda)}{\sum_{t=1}^{T} p(i_t = i \mid \vec{x}_t, \lambda)}$$

Variance:

$$\vec{\sigma}_i^2 = \frac{\sum_{t=1}^{T} p(i_t = i | \vec{x_t}, \lambda) \vec{x_t^2}}{\sum_{t=1}^{T} p(i_t = i | \vec{x_t}, \lambda)} - \vec{\mu}_i^2$$

3.1 Normalized Mutual Information (NMI)

In probability theory and information theory, the mutual information (MI) [12] or (formerly) transformation of two random attributes is a measure of the attributes' mutual dependence. Not limited to real-valued random attributes like the correlation coefficient, MI is more general and determines how similar the joint distribution $p(X, Y)$ is in the products of factored marginal distribution $p(x)$ $p(y)$. MI is the expected value of the pointwise mutual information (PMI) [12]. The most common unit of measurement of mutual information is the bit.

$$I(X; Y) = \sum_{x,y} p(x, y) \ln \frac{p(x, y)}{p(x)p(y)} \tag{2}$$

4 Results

See Tables 1 and 2.

Tables 1 and 2 shows comparison between proposed methods for two-speaker data and three-speaker data respectively. It is observed that CVAT-based GMM clustering supports strongly than VAT-based GMM clustering.

Table 1 Results for two-speaker data clustering

Performance measure	VAT-based GMM clustering	CVAT-based GMM clustering
VAT-GoodNess	0.1588	0.5367
NMI	0.0018	0.6157

Table 2 Results for three-speaker data clustering

Performance measure	VAT-based GMM clustering	CVAT-based GMM clustering
VAT-GoodNess	0.2945	0.4586
NMI	0.0315	0.2909

5 Conclusion

The speaker diarization is first and foremost focused on speaker clustering. In speaker clustering, the study of front-end factor analysis is a very important task for achieving of well-organized speech clustering results by i-vectors. The i-vectors are also called as intermediate vectors. The steps of speech clustering are as follows: Convert the speech data in Mel-frequency cepstral coefficient (MFCC) form, model speaker voice by Gaussian mixture model (GMM), adapt the speaker's utterance with average speaker model, apply the VBEM-GMM, find difference feature of several speakers data, and find clustering results by using classical k-means algorithm. The Universal Background Model is a large GMM, and this UBM presents the average speaker model. Thus, the accurate speaker modeling is done by GMM-UBM instead of GMM. The speaker's data was initially converted into raw form Mel-frequency cepstral coefficient (MFCC) which is used for representing the speaker voice characteristic. The step of MFCC which is conversion is done by hidden Markov toolkit (HTK).

References

1. D. Blei and M. Jordan, "Variational inference for dirichlet process mix-tures," Bayesian Anal., vol. 1, no. 1, pp. 121–144, 2006.
2. S. Shum, N. Dehak, and J. Glass, "On the use of spectral and iterative methods for speaker diarization," in Proc. Interspeech, 2012.
3. S. Tranter and D. Reynolds, "An overview of automatic speaker diary-station systems," IEEE Trans. Audio, Speech, Lang. Process., vol. 14, no. 5, pp. 1557–1565, Sep. 2006.
4. S. Shum, "Unsupervised methods for speaker diarization," M.S. thesis, Mass. Inst. of Technol., Cambridge, MA, USA, Jun. 2011.
5. F. Valente, "Variational Bayesian methods for audio indexing," Ph.D. dissertation, Univ. De Nice-Sophia Antipolis—UFR Sciences, Nice, France, Sep. 2005.
6. T. Stafylakis, V. Katsouros, P. Kenny, and P. Dumouchel, "Mean shift algorithm for exponential families with applications to speaker clustering," in Proc. IEEE Odyssey, 2012.
7. J. Prazak and J. Silovsky, "Speaker diarization using PLDA-based speaker clustering," in Proc. IDAACS, 2011.
8. P. Kenny, D. Reynolds, and F. Castaldo, "Diarization of telephone conversations using factor analysis," IEEE J. Sel. Topics Signal Process, vol. 4, no. 6, pp. 1059–1070, Dec. 2010.
9. S. Shum, N. Dehak, E. Chuangsuwanich, D. Reynolds, and J. Glass, "Exploiting intra-conversation variability for speaker diarization," in Proc. Interspeech, 2011.
10. N. Dehak, P. Kenny, R. Dehak, P. Dumouchel, and P. Ouellet, "Front-end factor analysis for speaker verification," IEEE Trans. Audio, Speech, Lang. Process., vol. 19, no. 4, pp. Jul. 2010.
11. M. Rouvier and S. Meignier, "A global optimization framework for speaker diarization," in Proc. IEEE Odyssey, 2012.
12. H. Ning, M. Liu, H. Tang, and T. Huang, "A spectral clustering approach to speaker diarization," in Proc. ICSLP, 2006.

Survey: Enhanced Trust Management for Improving QoS in MANETs

Srinivasulu Sirisala and S. Ramakrishna

Abstract Mobile ad hoc network (MANET) consists of autonomous movable nodes lacking any centralized control, and nodes move arbitrarily within the self-configurable MANET environment. All the nodes must cooperate with each other for packet routing, and cooperating nodes must trust each other. We define trust as scale of faith on the nature of interacting nodes. In a network, the trust between nodes is very essential for interaction. And to estimate trustworthiness among interacting nodes of MANETs, we consider many parameters like resource constraints (e.g., computing power, energy, bandwidth) and instant elements (e.g., topology variations, continuous movement, crashs, transmission time, and channel condition). We consider trust in MANETs in many aspects like QoS, routing, and to mitigate the attacks. In this paper, we are presenting survey which comprises trust computational approaches for improved QoS in MANETs.

Keywords Mobile ad hoc network · Trust management · QoS · Routing

1 Introduction

A collection of wireless mobile nodes forms a central administration less network is called a mobile ad hoc network (MANET), where a node can communicate with other nodes which are in its access area. As of MANETs can be formed rapidly and easily, it became as an appropriate communication network for the applications like war zones, emergency, and rescue operations.

Ensuring the QoS in mobile ad hoc networks is a difficult task when compared with wired networks, the reason is node mobility, be short of administration and

S. Sirisala (✉) · S. Ramakrishna
Department of Computer Science, SVU College of CM & CS,
Tirupati 517501, Andhra Pradesh, India
e-mail: vasusirisala@gmail.com

S. Ramakrishna
e-mail: drsramakrishna@yahoo.com

© Springer Nature Singapore Pte Ltd. 2019
R. S. Bapi et al. (eds.), *First International Conference on Artificial Intelligence and Cognitive Computing* , Advances in Intelligent Systems and Computing 815,
https://doi.org/10.1007/978-981-13-1580-0_25

255

limited available resources. In QoS-aware routing protocols, the nodes along the path should have the less required energy, bandwidth, and stability to transfer the source data without loss to the destination node. By taking these constraints into the account, a node's QoS trust will be computed.

Trust is characterized as a level of conviction. The principle thought of trust is while setting up the correspondence how far a node can accept other node so that to build up the high reliable network. Trust is of two kinds; one is immediate trust and another is backhanded trust, i.e., the trust among adjacent devices is called as immediate trust. The devices gather the data from the proposal of neighbor nodes called as backhanded trust. Trust preserves by the parameters like honesty, capacity, and generosity. In light of various definitions, the way we look and count of trust will vary. The gauge of trust of a node is entangled for the reason of the mobility of nodes [1]. There are numerous trust administration methods [2] which are being used to supervise the trust in the system. Trust administration is conversely utilized as a part of reputation administration. Trust is a node conviction in view of the trust characteristics, and the notoriety is conclusion regarding a node. Trust administration embraces trust foundation that is gathering proper confirmations, trust production, trust circulation, trust sighting, and assessing trust facts. This paper overviews distinctive trust-based protocols, trust routing, and QoS in MANETs.

2 Survey on QoS and Routing Based on Trust Management

A. Fuzzy-Based Quality of Service Trust Model for Multicast Routing

The FQTM [3–5] is the advancement to the MAODV, which establishes path one node to another with high-trusted intermediate nodes. FQTM accomplishes it in three phases: (i) HELLO packets exchange with adjacent nodes to gather the information of node properties like node velocity, direction, free time slots, source data size, (ii) using node properties, FQTM computes the QoS metrics like link expiry time, available bandwidth, energy, and node reliability, (iii) computation of node trust value using fuzzy logic which takes QoS metrics as input values, (iv) formation of multicast tree which gives high-trusted path between any two nodes of the MANET (Fig. 1).

B. Recommendation-Based QoS Trust Aggregation and Routing [QTAR]

QTAR [6] is a framework for "Recommendation-Based QoS Trust Aggregation and Routing" in mobile ad hoc networks. QTAR performs the routing by computing trust which is happened in four phases like QoS trust computation, aggregation, propagation, and routing.

For trust computation, it is considering the QoS parameters like residual energy, available bandwidth, and node stability.

For trust aggregation, QTAR is making use of Dempster–Shafer Theory (DST) [7]. In MANETs, usually a node will have three possible trust evidences like trust $\{T\}$,

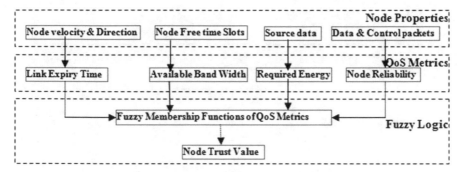

Fig. 1 Shows the FQTM architecture

Fig. 2 Aggregation of QoS
trust recommendations

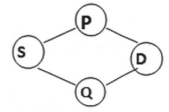

distrust $\{\overline{T}\}$, and uncertainty (trust/distrust) = $\{T, \overline{T}\}$. Using DST trust aggregation is computed by considering the trust evidence recommendations from neighbors on a node. For example, consider the following topology (Fig. 2).

The D node trust value is computed by aggregating the trust evidence recommendations of P and Q on D. This approach can be extended for aggregating n nodes' recommendations on node D.

Once like above trust has been calculated for every node in the network, the trust assessment of a node will be propagated to all other nodes of the network using HELLO packets, so that every node comes to identify the trust value of other node.

For the trust routing, the QTAR uses the AODV routing principles [8].

C. Trust-Based Job distribution with Multi-Objective Optimization in Service-Oriented MANETs [TMOS]

In the service-oriented MANETs, the main objectives are (1) improvise the mission trustworthiness which depends on job finishing proportion; (2) diminish the utilization fluctuation prompting high load adjust among all nodes, (3) reduce the holdup time to finish time-being tasks, thus improving the quality of service (QoS) and in service-oriented MANETs task assignment by satisfying these objectives is a challenging task. To accomplish this in [9] presented trust-dependent heuristic algorithm functions by announcing limited acquaintance of node condition to resolve node to job assignment problem with multi-objective streamlining prerequisites.

For this, the following architecture was considered.

The above architecture describes the system of nodes and jobs. Here it has been taken a service-arranged MANET which has to deal with animatedly inward jobs

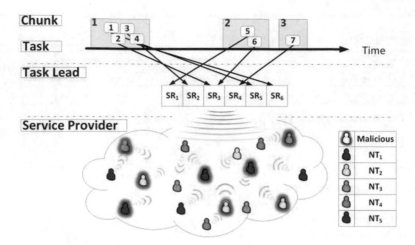

Fig. 3 System architecture for nodes and tasks. TL: Task Lead; CN: commander node

in an operation to accomplish various system objectives. A commander node (CN) administrates the mission group. Underneath the CN, several Task Leads (TLs) pilot the undertaking groups. The CN picks TLs in beginning of network foundation in the view of the trustworthiness of nodes known to CN earlier and the TLs (going about as SRs) every one selects trustworthy SPs instantly for completing the jobs allocated to them. Jobs that overlie in the same timeline are united as a chunk. In Fig. 3, jobs 1–4 are bent as a chunk 1 and jobs 5–6 are bent as chunk 2. There is a rush between the jobs in a chunk for the sake of services offered by SPs as each SP can partake in just a single occupation at any given moment. At final, it is hypothesized that the TLs picked by the CN are completely reliable.

In TMOS, the task assignment and execution will happen in four steps: (1) advertisement of task specification, (2) bidding a task, (3) member selection, and (4) task commitment by nodes

D. Green Trust Management to diminish Trust-bending Attacks on MANETs

In MANETs trust-bending attacks may endeavor to diminish nodes' belief on other nodes' honesty by creating unscrupulous suggestions or satisfying twofold face conducts as a consequence the energy consumption to build up the trust system may be increased so it decreases the network lifetime and cause the ecological contamination.

The GTMS [10] is vitality proficient trust framework which gives nodes a capacity to estimate other nodes reliability to diminish trust-bending assaults. To accomplish it, GTMS gets additional information from packets get from the promiscuous process, so that it is possible to recognize even minor level of mutilation assaults. The GTMS facilitates a node to instantly empower/incapacitate its promiscuous process as indicated by the system status, keeping in mind the end goal to diminish vitality utilization while adequate behavioral information is assembled. Besides, the rate of suggestion proliferation is adaptively attuned to decrease the transmitted overhead and accomplish advance vitality protection.

GTMS contained four primary phases: knowledge compilation, trust-level calculation, trust establishment, and self-adaptation.

The GTMS performs the knowledge collection using modified WatchDog mechanism [11]. Using it, a node gets the trust worthiness information of neighbor nodes by sending intended packets to its neighbors so that a node comes to know the neighbors' conduct (trust) toward itself and any of its neighbors. With this information, each node builds the local trust table (LTT). This table having one tuple per adjacent, i.e., quantity of the data has to be transmitted by that node and exactly passed through it and the local trust level projected for its adjacent.

GTMS using its self-adoption component can switch on/off working in a promiscuous mode to reduce energy consumption.

A node gets the indirect trust information by mutual exchange of recommendations for trust on neighbor. So that with these direct and indirect information, all the nodes in the MANET will build the global trust table (GTT).

Once GTT constructed, then trust level will be computed as **by** computing multiple trust values for a single node concerning its activities in the direction of its neighbors

In GTMS, trust establishment component is the responsible for make use of trust values of nodes is establishing the route between nodes. For this, it employs a maximum trust limit to spot whether a node is usual or mischievous. This maximum limit is mentioned as a policy in knowledge base.

Self-adoption component of GTMS will adjust the decision rules in the knowledge base dynamically by analyzing GTT, which is required to improve efficiency in establishing the route.

E. Bayesian Trust Model

In [12] trust model was formulated based on the SUNNY algorithm [13]. The communication between gadgets in the autonomous network like MANETs depends on potency of faith one is having on another. Such that one should not have malevolent nature and should execute requested service in acceptable way. Figure 4 shows the outline of the trust framework. The trust esteem on a node n signified by $P(n)$ is spoken to as probabilistic esteem extending from $0 \leq P(n) \leq 1$.

Choosing Trust method: The client is permitted to pick security level for the required administration. Here there are three types of security to be specific: (a) poor, (b) moderate, and (c) extreme. Based on the security type, the trust calculation plans can be picked which are direct, indirect or trust network. For poor security needed administrations, trust calculations are finished utilizing the direct approach in which trust esteem of a node will be consider as it is. To accept a node minimum esteem is 0.4 moderate security administrations calculated through indirect approach which gathers assessments from its adjacents and come to the conclusion on trust. Trust network is framed when security is essential for the administrations.

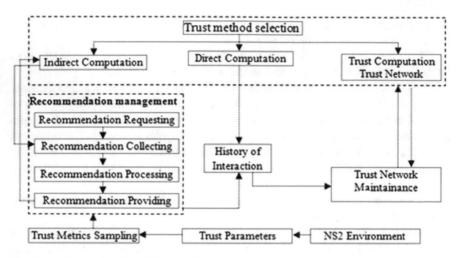

Fig. 4 Architectural outline of the trust framework

Recommendation administration:

Recommendation comprises of accompanying sub-stages: (a) testing the present network parameters, (b) filtering false suggestions acquired from the inspecting unit, (c) calculating the aggregate quantities for the estimations as expressed in the Bayesian approach [14], and (d) on getting the indirect suggestion request sends notice to the related indirect calculation unit.

Trust network preservation:

It deals with upholding of the trust network, network construction, and the trust recommendations' collection process.

Repository of communication:

The *Direct* and *Indirect* computation calculations are executed as given in [14]. Keeping in mind the end goal to keep up the data about the past connections, repository called the *Repository of communication* is kept up. Each node history values put away as *Htruster (trustee)*. Let assume A wants to send the data to then past interactions measure value is spoken to as HB(A) = {H1 … Hn}, where Hi records a connection from B to A. Every record contains a tuple {*ei, si, di*} where *ei* is the assessment of the communication, it equivalents to 1 if it is adequate interaction and 0 otherwise, *si* is the kind of communication between two points, and *di* is the time of in touch.

F. Trust elements in cyclic MANETs [TDCM]

In [15] it is noticed that in MANETs very little attention has paid toward nodes' movement and elements of trust. To address it, a changed AODV routing protocol is presented which includes the trust elements and overhead adjustment to handle the

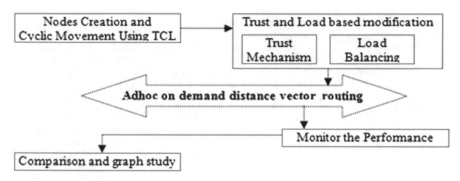

Fig. 5 TDCM architecture

traffic and guarantees safe and sound routing in cyclic mobile ad hoc network. The architecture is as follows (Fig. 5).

According to this work before exchanging, the packets between two nodes trust will be calculated as follows.

Every node in the MANET is given by the default trust esteem 0.5 and primarily a node is conscious of trust esteems of its just adjacent. The trust esteem of a node will be recalculated in light of the packet dropping by itself, if the dropping rate is high trust value of the node will be decremented or else rise the esteem of trust as indication of fair conduct.

Every node i in CMANET if node i need to speak with node j, total the estimations of trust to discover trust passageway Tp. In this way, calculate trust path for every pair of nodes. Then select the path with greatest Tp. This is how a path between source and destination with highest trust value will be selected.

Load balancing: When blockage was happened on the trusted path, load balancing tactic will be initiated. In accordance, traffic will be diverted through the other best trust path (Table 1).

3   Conclusion

In mobile ad hoc networks, trust and quality of service (QoS) are two major features to be addressed. In MANETs, a node has to participate and communicate with unknown neighbors to run the applications. Due to non-cooperation and malicious activities of nodes in the network, it is not a trivial task to run the applications with required level of QoS. Many authors in last decade tried to address the security and QoS issues independently, but lost their performance. Hence, the recent research focused on trust management techniques (soft security) to improve the QoS, in which every node maintained by trust esteem of all others in the network, so that it can participate with trustworthy nodes only to improve the quality of the applications. In this paper, we discussed different recent trust gathering and maintenance techniques in MANETs.

Table 1 A comparison of various trust finding schemes

S. No	Reference	Scheme	Metrics	Advantages	Disadvantages
1	Nageswararao et al. [3]	FQTM	Fuzzy logic to compute trust	Reduced path failures and improved throughput	Latency is high and accuracy to be improved
2	Nageswararao et al. [6]	QTAR	Dampster's theory for trust computation	Accurate trust computation	Better algorithm than the AODV can be used for routing
3	Yating Wang et al. [9]	TBMS	Local knowledge of the node status was used for task-to-node assignment	Node-to-task assignment was performed at run time	Member binding and node selection for task assignment can be done better
4	Zahra Hosseini et al. [10]	GTMT	Modified WatchDog mechanism, dynamic trust calculation	Reduced energy consumption and trust-distortion attacks and accurate estimation of nodes' trust-worthiness	Maintenance of global trust table in MANETs is difficult
5	Lydia et al. [12]	Bayesian trust model	Modified SUNNY algorithm	Ensures the trustworthi-ness of nodes when secure connection is required	This model might not give promised performance with increased network size
6	Vijaya Singh et al. [15]	TDCM	Trust elements and load balancing	Effective traffic handling	Few parameters are considered as trust dynamics

We did comparative study (in tabular form) of different trust methods by highlighting their merits and demerits. All the trust methods discussed in this paper are also applicable to the social networks, to curb the harmful activities of the unknown users.

References

1. V.L. Pavani and B. Sathyanarayana, "A Reliable Data Delivery Using Trust Management System Based on Node Behavior Predication in MANET", IEEE-*iCATccT*, May (2015).
2. M.S. Khan and M.I Khan. MATF: a multi-attribute trust framework for MANETs, EURASIP Journal on Wireless Communications and Networking, Vol.16, (2016), pp. 1–17.
3. Nageswara Rao Sirisala and C.Shoba Bindu, "Fuzzy Based Quality of Service Trust Model for Multicast Routing in Mobile Adhoc Networks", IJAER, Volume 10, Number 12 pp. 32175–32194, (2015).
4. Fei Hao; Geyong Min; Man Lin; Changqing Luo; Yang, L.T., "MobiFuzzyTrust: An Efficient Fuzzy Trust Inference Mechanism in Mobile Social Networks", IEEE Transactions on Parallel and Distributed Systems, vol.25, no.11, pp. 2944–2955, (2014).
5. Zhexiong Wei; Tang, H.; Yu, F.R.; Maoyu Wang; Mason, P., "Security Enhancements for Mobile Ad Hoc Networks With Trust Management Using Uncertain Reasoning," IEEE Transactions on Vehicular Technology, vol.63, no.9, pp. 4647–4658, (2014).
6. Nageswara Rao Sirisala and C.Shoba Bindu, "Recommendations Based QoS Trust Aggregation and Routing in Mobile Adhoc Networks". IJCNIS, Vol. 8, No. 3, December (2016).
7. Hui Xia and Jia Yu "Applying trust enhancements to reactive routing protocols in mobile ad hoc networks" journal of Wireless Networks, vol 22, pp 2239–2257, (2016).
8. Priya Sethuraman and N. Kannan, "Refined trust energy-adhoc on demand distance vector (ReTE-AODV) routing algorithm for secured routing in MANET" journal of wireless networks. Vol 22, pp 1–11, (2016).
9. Yating Wang, Ing-Ray Chen, "Trust-Based Task Assignment with Multi-Objective Optimization in Service-Oriented Ad Hoc Networks", IEEE Transactions on network and service management, Vol 14, pp 217–232, (2016).
10. Zahra Hosseini and Zeinab Movahedi, "A Green Trust Management Scheme to Mitigate Trust-distortion Attacks on MANETs", 2016 IEEE conferences on ubiquitous & computing, (2016).
11. B.-J. Chang and S.-L. Kuo, Z. Movahedi, Z. Hosseini, F. Bayan, and G. Pujolle, "Trust-distortion resistant trust management frameworks on mobile ad hoc networks: A survey," IEEE Communications Surveys Tutorials, 2016.
12. B. Lydia Elizabeth and R. Aaishwarya," Bayesian based Confidence Model for Trust Inference in MANETs", IEEE – ICRTIT, (2011).
13. U. Kuter, and J.Golbeck. SUNNY: A new algorithm for trust inference in social networks using probabilistic confidence models. In Proceedings of the National Conference on Artificial, (2007).
14. M. K. Denko, T. Sun, I. Woungang, "Trust management in ubiquitous computing: A Bayesian approach", Elsevier journal on Computer Communication, Vol 34, pp 398–406, (2010).
15. Vijaya Singh and Ms. Megha Jain, "Analysis of Trust Dynamics in Cyclic Mobile Ad Hoc Networks", IEEE – ABLAZE, (2015).

A Generic Survey on Medical Big Data Analysis Using Internet of Things

Sumanta Kuila, Namrata Dhanda, Subhankar Joardar, Sarmistha Neogy and Jayanta Kuila

Abstract In medical science, various medical parameters and post-operational data should be analyzed properly. Using Internet of things (IoT), the physicians can access the local and remote area patients. The goal of this work is that through Web- and Internet-based communication doctors can monitor and analyze patient's health data and parameters. Health data is a combination of different types of data, in different formats, thereby being referred to as big data. After patient interaction with the doctor, the medical record of the patient, which includes voluminous data regarding patient's case history, doctor's prescription, laboratory test report, diagnostic report, current treatment details, will be stored in electronic health records (EHRs). Other information, like pharmacy information, medical journals used to investigate and analyze the case, health insurance policies, may also be part of the record. This paper discusses the characteristics and challenges of medical big data. Medical data is vital since necessary and relevant information needs to be extracted for the well-being of the patient.

Keywords Internet of things · Big data · Electronic health records (EHRs) ECG · Hadoop distributed file system · Diabetic retinopathy (DR)

S. Kuila (✉) · S. Joardar
Haldia Institute of Technology, Haldia, India
e-mail: sumanta.kuila@gmail.com; sukuila@gmail.com

S. Joardar
e-mail: subhankarranchi@yahoo.co.in

N. Dhanda
Amity University, Noida, Uttar Pradesh, India
e-mail: ndhanda@lko.amity.edu

S. Neogy
Jadavpur University, Kolkata, India
e-mail: sarmisthaneogy@gmail.com

J. Kuila
Sunayan Advanced Eye Institute, Tamluk, India
e-mail: dr.jkuila@rediffmaill.com

© Springer Nature Singapore Pte Ltd. 2019 265
R. S. Bapi et al. (eds.), *First International Conference on Artificial Intelligence and Cognitive Computing* , Advances in Intelligent Systems and Computing 815,
https://doi.org/10.1007/978-981-13-1580-0_26

1 Introduction

In recent years, the Internet of things (IoT) and big data are emerging areas for research. These two are supplementary to each other. IoT enables a number of objects and devices having IP addresses linked or connected with each other. Considering the fact that trillions of devices are linked with each other and they are providing huge volume of data, so the efficiency of data collection mechanism becomes questionable [1]. Not only that, IoT uses real-time or close to real-time system for communicating information. Dealing with this massive amount of data is indeed very challenging. In the digitized world, vast amount of data is generated. Our goal is that we have to analyze and use the relevant data from the huge data set using big data concept [2]. The study reveals that 40% of IoT-related technology will be in healthcare domain by 2020 [3]. Also by 2025, huge IoT-related market will be generated in human health and fitness domain [4]. As per Fusion Analytics World, the current market of big data in medical sector (globally) is at $10.1 billion. By 2020, the estimated business will go up to $22.3 billion [5]. So the scope is huge. According to information analyzed by Medikoe, a 2012 report reveals that already health-related data has reached a volume of 400 PB (peta bytes) and in all possibility it will reach 25,000 + PB by the year 2020 [5]. This paper is organized as follows: Sect. 2 discusses IoT in health care, Sect. 3 describes big data in health care with case studies, and Sect. 4 concludes the paper.

2 Internet of Things in Health Care

Today, the trend in healthcare sector is that patients prefer routine medical checkup and other healthcare services from medical clinic at the home environment. Especially in the case of emergency when quick action is required, it is observed that treatment from home is more effective than shifting the patient to the hospital. Thus, expenditure on the treatment can be reduced. On the other side, hospital can reduce patient pressure by shifting the possible and relatively easy treatment to home environment [6].

2.1 Architecture

The architecture of IoT health care can be designed into two parts, network layers and technological frameworks.

2.1.1 Network Layers in Health Care

The IoT network, referred to as IoThNet in [7] which covers the healthcare domain, works as a backbone that facilitates to transmit and receive medical data and enables the healthcare-tailored communications for end users. IoT itself follows a logical framework to serve all medical-related issues in efficient and convenient way.

Topology: To create the seamless healthcare environment, IoT components refer to the collection of different elements that collect vital signs and sensor data such as body temperature, blood pressure (BP), oxygen saturation, and electrocardiograms (ECG) and form a typical IoT environment for health care. It transforms the heterogeneous computing and storage space capability of static and portable electronic devices such as laptops, smartphones, and health terminals into hybrid computing grids [8].

Architecture: The architecture describes the organization and the system as a whole, including the physical part of the system, with their practical organization, and operating principles and techniques. The key issues discussed in IoThNet are the interoperability of the IoT gateway and the WLAN, multimedia streamings, etc. It also describes the relevance in healthcare domain.

Platform: Platform includes networking platform and computing platform. This structure describes methodical hierarchical model of how different users or agents can access different databases in application layer with the help of support layer. The application layer supports different healthcare-related services such as healthcare means suppliers, application designers, platform providers, content providers. More specifically, this will support the repository where all health-related information is stored. The repository will support patients, doctors, and medical support staffs to act upon and analyze the big data collected by IoT devices [9].

2.1.2 Technological Framework

Technological framework provides mutually non-interoperable application-specific solutions for health monitoring devices. The important activity such as the process of accumulating data from sensors is taken care of by wireless sensor networks (WSNs). The framework also contains high-speed network connectivity and standard user interfaces with proper display facility [10]. As we have earlier discussed, different IoT devices like smartphone, laptop are required to receive medical data and deliver it to the cloud. In tier 2, sensor gateway is there to handle network and database-related issues. At tier 3, Web application which is IoT-based health monitoring system using big data can be implemented. In middle tier, the use of application server and repository server will take care of the database-related issues. Local sensor networks will support Wi-fi, BTLE, radio frequency identification (RFID), near-field communication (NFC). In addition, advanced machine-to-machine (M2M) communication allows direct communication of the remote healthcare sensors with the Internet. IoT implements specific protocols for inter-device communication which offers real-time access to device data and allows the remote user to access the device. This has a scope to develop a Web application that is scalable, globally accessible, and able to provide communication interface to external applications [11].

2.2 Applications

IoT health care has several business applications. Remote patient monitoring and ubiquitous health care are among them.

2.2.1 Remote Patient Monitoring

It is required that ready access to different types of health monitoring system is often required to avoid health risk. Without this facility, different medical conditions may go undetected. Small powerful wireless solutions based on IoT make it possible to provide health condition monitoring and possible treatment in remote locations also. India has quite a few remote (hilly, inaccessible) areas where it is quite difficult to access doctors in emergency [12].

Figure 1 shows that using a couple of implantable and other devices it is possible to create a setup to facilitate the healthcare services [12]. Patient data (regarding health condition and others) is received using pill camera. Using implantable transceiver, this data is transmitted to physician. IoT-based cloud computing is the motivator here [12].

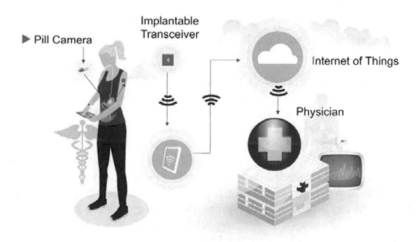

Fig. 1 Remote healthcare monitoring [12]

2.2.2 U-Healthcare System

Ubiquitous health care uses a large number of environment and patient's sensors, and it is used to monitor and improve patient's health condition. Tiny sensors accumulate and analyze clinical information to diagnose health problem [13]. With the help of laptops, PDAs, and smartphones, ubiquitous computing has become easy and efficient. Figure 2 describes that body area network (BAN) system in which sensors are connected in human body captures bio-signals from human body. The sensors sense different health parameters like pulse and breathing, body temperature, blood pressure, etc.

The work in [13] provides an intelligent medical service (IMS) that receives information from BAN and works as a hub between patients and hospitals. The hospital system maintains the data repository. Physicians in the hospital can take necessary preventive or corrective actions for the patient from that medical repository.

2.3 Security Issues

The personal and private medical data which is collected through tele-monitoring needs security and privacy. When "everything" is connected, a number of different

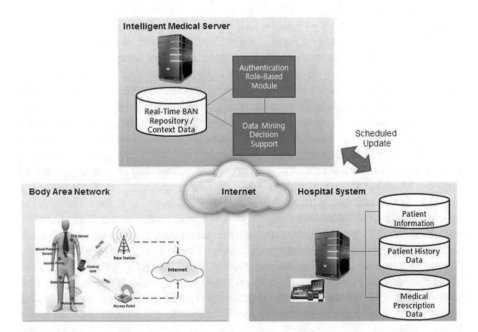

Fig. 2 Body area network [13]

security and privacy issues arise [14]. The standard security issues are mentioned below.

Resilience to Attacks—The system is to be designed such that it avoids single point failure and should regulate itself to node failures. Data authentication and access control should be properly dealt with. So retrieved address and object information should be authenticated [14].

Client Privacy—Necessary steps should be taken to confirm that information should be dealt with by authorized user only. Data security and privacy should be properly monitored by database administrator [15].

3 Big Data in Health Care

Different authors describe big data in various ways and from different perspectives. If we analyze some of them, it is like big data is high volume, high velocity, high variety, and so on. They are information assets that require new forms of processing to enable enhanced decision making, insight discovery, and process optimization [16]. Big data is a term which defines hi-tech, high speed, high volume, complex, and multivariate data to capture, store, distribute, manage, and analyze information. Big data is the amount of data beyond the ability of technology to store, manage, and process efficiently [16]. So big data is a huge data, and sometimes, it is difficult to maintain it in the repository properly. In the previous discussion, it is apparent that healthcare data is not only generated using IoT but also delivered and received. Many a times, the health data contains images, sounds apart from text. So, technically it becomes difficult to maintain in the database repository. So healthcare data, which is very important for detection of disease and which is the key for healthcare solution and medical treatment, should be maintained properly using big data technology. The ten properties (known as 10v) and characteristics will bring the advantages and challenges of big data. **Volume**: Bulk amount of real-time data is generated and stored, where records are stored in terabytes. **Velocity**: The speed of the data generation is considered here. As an example, the online shopping giant, Amazon, operates millions of requests everyday and maintains speed of data for processing. **Variety**: Structured and unstructured different data formats. **Value**: The specific information is used from data repository. It is mainly used in big data analytics. **Veracity**: Veracity provides accurate data for processing. Quality and understandability are checked here. **Variability**: Data variability deals with different data types. **Visualization**: Dozens of variables and parameters contained by visualization. Data visualization identifies the area that requires functionality, scalability, response time in the database. **Volatility**: Volatility is the time required to consider that which data is not useful, irrelevant, or historic. Volatility also calculates how long data needs to be maintained. **Vulnerability**: Vulnerability is the security concern for big data. **Validity**: Validity refers to the correctness of the big data. Sixty percent of data scientist's effort is required for cleansing data to be ready for analytics [17, 18]. Data processing challenge includes data collection, filtering, processing, resolving similarities, data analysis, and output description. These are the important characteristics of big data.

3.1 Capability Profile of Big Data in Medical Sector

Healthcare organizations need to access their capabilities provided by big data solutions that will prove their organizational efficiency and operational gain which increase their competitive benefits and produce business value and productivity [19]. Traceability is the capacity to trace output data from all IT components of the system. Medical output data such as clinical data, behavior and sentiment data, activity and cost data, pharmaceutical R&D data are commonly collected at real time, and the goal is to trace the relevant medical data when it is required. From several IoT-based applications, we have a huge volume of medical data with enormous flow. Here, we will discuss the main big data techniques and challenges in healthcare sector.

Data Acquisition: An important issue to consider here is that, while implementing big data, we have to define online filters in order to remove redundant data without loss of valuable information. The huge input stream produced by millions of sources from IoT at healthcare-related applications needs to be properly collected. Selection of data sources, collection of information from those sources, filtering and cleaning data are the activities related to data acquisition [20].

Data Cleaning: Data cleansing or data cleaning is the process of identifying and correcting (or removing) damage or erroneous data from a record set, table, or database. This refers to identifying unfinished, wrong, mistaken, or unrelated parts of the data. The dirty unwanted data may be replaced or modified or deleted. Detecting and repairing dirty data are one of the permanent challenges in data analysis. Failure to do so can result in inaccurate analytics and unreliable decisions. The idea is focused on qualitative data cleaning of big data which uses different rules, constraints, and patterns to detect errors. There has been growing interest in both academia and industry over the last few years in data cleaning. This problem includes new abstractions, interfaces, and new approaches for scalability [21].

Data integration, Aggregation, and Representation: The basic properties of DBMS are applicable here. The properties of data are that it is always very heterogeneous and most of them have different metadata. Data integration requires huge effort as it is quite difficult to integrate big data due to its robustness. New approaches can improve the automation of big data, and several such approaches, procedures, and tools are available. Different data aggregation tools and techniques and implementation are needed for different data analysis tasks [22].

Query Processing, Data Modeling, and Analysis: The Web-based big data in healthcare data analysis is an important topic in cloud computing that is specifically tied up with resource description framework (RDF). The storage structured is properly analyzed, and with it, the storage structure is able to deal with different query processing and data mining techniques. This is needed to increase knowledge discovery from within the data set. Here, RDF data model [24] represents different association and binary relations. Hadoop MapReduce-based data handling is treated as the Map and Reduce model for processing huge scale data. Also, most of the existing techniques do not perform like MapReduce, and it is frequently used for optimization [23].

3.2 Use of Big Data in Medical Science

Large-scale analysis for specific case requires the collection of heterogeneous data from various sources. Analysis of big data will enhance epidemic spread prediction, pharmaceutical development, personalized patient care, fraud detection, to name a few [10]. Some of the specific applications are as follows.

Genomics Analysis: Patients genomic data and clinical data are important because cancer treatment analysis can be done by combining these data. By using patient's genomic information, doctors can use specific genomic drug for the treatment of a disease. Big data analysis uncovers hidden patterns, unknown correlations, and other characteristics through large-scale data sets. Using big data, infrastructure for manipulating various genomic data and comprehensive electronic health records (EHRs) is challenging. But they also provide a possible chance to build up a well-organized and effective view to recognize clinically actionable genetic variants for individualized diagnosis and treatment [24]. Practical big data toolsets are also available to spot clinically actionable genetic variants.

Flu Outbreak Prediction and Control: Continuous analysis of public health data helps to detect and control possible disease outbreak. Analysing Web-based social media information prediction of flu outbreaks is possible [25]. This is based on consumer search, social content, and query activity. The collected data from the Web is analyzed, and necessary treatment may then be prescribed by the physicians.

Analysis of Clinical Output: In USA, Blue Cross and Blue Shield of North Carolina has provided some promising example—how clinical analysis of big data can be used to reduce people's health risk and update clinical output [10]. Some important activities in this analysis are genomics and sequence analysis, visualization, health text analytics, health ontology, long-term care, patient empowerment impacts, etc.

Medical Device Design and Manufacturing: Big data and computational methods can play an important role in design and manufacturing of medical devices. Input will come from various case studies, and set of anatomical configuration, operational methods, device architecture can be designed based on that. Growth in medical device design and development is data-driven and simulation-based. This is an emerging work in the areas of both medical science and big data. Big data technologies permit a wider set of device equipments, anatomical design, delivery methods, and tissue interactions to be examined [26].

Personalized Patient Care: Using big data in health care has changed the healthcare perspective from disease-oriented model to patient-centric model. An individual can now participate in his own well-being and care and obtain services focused on individual requirements and choice. The healthcare industry has generally been an incrementally advancing field. Here, data gathered from the electronic records is analyzed properly and treatment is recommended accordingly. Advanced tool in the line of recommendation systems has been developed to deliver personalized model of a patient's health profile [27]. However, irrespective of their early achievements, these tools face some significant problems. The volume of collected medical-related data is increasing rapidly, and the operational time for analysis will soon not be limited in a practical time frame anymore. So it is challenging to use big data in real medical applications. But personalized care is a very sought-after issue, and hence, much research is going on at present.

E-consultation and Tele-Diagnosis: The aggregated ECG- and ultrasound-based diagnostic images from medical clinics worldwide generate a huge amount of data (big data). This data is used to develop an e-consulting program that is being uploaded on Web sites to deliver suitable recommendation/treatment. Analysis of Internet household reporting discloses that there are huge differences in the economic levels of different countries. Thus, in 2015, Europe had coverage of 81.2%, against coverage of 10.7% for Africa in the domain of e-consultation in healthcare sector. Actually, a third (32%) of the population of developing countries uses the Internet, against 82% in developed countries [28]. As a result, problems of isolation and accessibility to health facilities and professionals are given priority in most developing countries. Technological innovations such as tele-diagnosis, tele-expertise, or tele-consultation that constitute effective solutions are not yet sufficiently implemented [28].

Pharmaceuticals and Medicine: Integration of data regarding clinical, patients, health care, safety, and public research will generate the knowledge for decision making and target selection through big data analytics for drug discovery. Usually, healthcare industry has lagged behind other industries in the use of big data. This is sometimes because of the fact that people are resistant to change. Healthcare providers generally make treatment decisions independently, using their own medical result. They would not like to rely on protocols based on big data. Other problems are more structural in nature. Many healthcare stakeholders do not want to invest in information technology as they fear underperformance. With their previous system's operation, they have limited capacity to standardize and accumulate data. The nature of the healthcare industry itself is also competitive. Though there are many vendors, there is no way to easily share information among providers or facilities. This is due to privacy concerns. Within a single hospital also it is difficult to share medical information within one group or department because organizations may not have proper procedures for data integration and communicating findings [28].

3.3 Case Studies: Big Data in Healthcare Services

The presence and impact of big data are already felt in real-life healthcare sector. Some of them are discussed here.

Diabetic Retinopathy: Around 422 million people around the world are living with diabetes and one-third of them are moving to diabetic retinopathy (DR), a common condition that can lead to permanent blindness, if neglected. Early detection and treatment dramatically reduce the risk at very low level. It is needed that people in low-income group and medically underserved areas receive this treatment.

New research from IBM uses IoT and big data technologies to solve this problem in medically backward regions. The research considers deep learning, conventional neural network, and visual analytics technologies. It analyzes 35,000 images accessed via EyePACs. The IBM technology learned to identify lesions and other marks of damages to retina's blood vessels, collectively assessing the presence and severity of treatment [29]. So using big data, the hospitals can analyze data and provide treatment for diabetic retinopathy. Reports mention that the method was able to classify DR severity with 86% accuracy, thereby suggesting doctors to analyze the big data components to avoid late treatment.

Glaucoma Treatment: Glaucoma is one of the leading causes of blindness. The purpose of the study [30] was to investigate the impact of an IoT-based glaucoma care support system, where the physician will analyze the cases, comparing with the huge data (glaucoma-related cases) [30]. Monitoring glaucoma patients represents a significant burden on clinical services, since the work involves huge volume of data. The study says that the most important test in glaucoma is visual field (VF) test, and in London only, three different glaucoma clinics performed around 250,000 VF tests. The doctors then analyzed this huge data to make the decision. An estimated, one million patients in England annually visit clinics for glaucoma treatment. Big data analysis in such situations can help physicians for a follow-up in glaucoma and also provide solution for the treatment [31].

Ophthalmic Research: Several ophthalmic researches are benefitted from the use of medical databases to derive knowledge about disease etiology, disease observation, health services deployment, and health outcomes. Also, the quantity of data required for research is increasing and the volume has reached such a level that we have to implement big data concepts here. The database and data repository system used in ophthalmic sector along with the variety of data reveals that the study of big data in eye research has huge scope in big data using IoT [32]. Presence of large volume and variety of data in ophthalmic research area always requires IoT-based big data support.

4 Conclusion

This paper presents a review about IoT and big data and its impact on medical sector. Here, we also study the outcome of IoT and big data and associated technologies and the real-time implementation of these in the healthcare domain. The business scope and its cost-effective implementation are also covered in our survey. Our goal is to study a digital health framework where voluminous data is analyzed by using IoT and big data technology. It includes patient's previous medical data, current treatment, laboratory results, diagnostic reports, doctor's prescription, health insurance related data. The aim is to generate an automated medical checkup system.

References

1. K.R. Kundhavai, S. Sridevi, "IoT and Big Data- The Current and Future Technologies: A Review", International Journal of Computer Science and Mobile Computing, Vol.5 Issue.1, January- 2016, pp. 10–14.
2. Madhura A. Chinchmalatpure, Dr. Mahendra P. Dhore, "Review of Big data challenges in healthcare application", IOSR Journal of Computer Engineering (IOSR-JCE), 2016, pp. 06–09.
3. Dimiter V. Dimitrov, MD, PhD, "Medical Internet of Things and Big Data in Healthcare", Healthcare Informatics Research, July 22, 2016, pp. 156–163. http://dx.doi.org/10.4258/hir.2 016.22.3.156.
4. McKinsey & Company, McKinsey Global Institute, "The Internet of Things: Mapping the value beyond the hype", June 2015, Executive summary. https://www.scribd.com/document/2 88859625/McKinsey-Unlocking-The-Potential-Of-The-IoT-pdf.
5. Wrik Sen, "What's The Future of Big Data In Healthcare Services", Oct 20,2016 CXO today.com http://www.cxotoday.com/story/how-big-data-will-determine-the-future-of-health care-service-delivery/.
6. Alok Kulkarni, Sampada Sathe "Healthcare applications of the Internet of Things: A Review", (IJCSIT) International Journal of Computer Science and Information Technologies, Vol. 5(5), 2014, pp. 6229–6232.
7. S. M. Riazul Islam, "The Internet of Things for Health Care: A Comprehensive Survey", IEEE Access, June 4, 2015, volume 3, pp. 678–708.
8. Kumar Keshamoni, Mayank Tripathi, "Deliberation and exertion of wireless body area networks for exclusive health care supervision over IoT", International Journal of Advanced Technology in Engineering and Science, Vol No 5, Issue No 3, March 2017, pp. 31–47.
9. Zhibo Pangab, Qiang Chenb, "Ecosystem Analysis in the Design of Open Platform based Mobile Healthcare Terminals towards Internet of Things" 15th International Conference on Advanced Communications Technology (ICACT), Jan 27–30, 2013, IEEE Xplore digital library.
10. Lidong Wang and Cheryl Ann Alexander, "Big Data in Medical Applications and Health Care", American Medica Journal, pp. 1–6, 1.8.2015.
11. C. Doukas, I. Maglogiannis, "Bringing IoT and Cloud Computing towards Pervasive Healthcare", Sixth International Conference on Innovative Mobile and Internet Services in Ubiquitous Computing, 4–6 July 2012. IEEE Xplore digital library.
12. David Niewolny, "How the Internet of Things Is Revolutionizing Healthcare", https://www.n xp.com/docs/en/white-paper/IOTREVHEALCARWP.pdf.
13. Yvette E. Gelogo, Ha Jin Hwang & Haeng-Kon Kim, "Internet of Things (IoT) Framework for u-healthcare system", International Journal of Smart Home Vol. 9, No. 11, (2015), pp. 323–330. http://dx.doi.org/10.14257/ijsh.2015.9.11.31.

14. Liane Margarida Rockenbach Tarouco, Leandro Márcio Bertholdo, "Internet of Things in Healthcare: Interoperatibility and Security Issues", International Workshop on Mobile Consumer Health Care Networks, Systems and Services, 9th December 2013, pp. 6121–6125.
15. Sachin Babar, Parikshit Mahalle, Antonietta Stango, "Proposed Security Model and Threat Taxonomy for the Internet of Things (IoT)", © Springer-Verlag Berlin Heidelberg, 2010, pp. 420–429.
16. Hakan Özköse, Emin Sertac Ari, Cevriye Gencer, "Yesterday, Today and Tomorrow of Big Data", Science Direct, Procedia - Social and Behavioral Sciences 195 (2015),World Conference on Technology, Innovation and Entrepreneurship. pp 1042–1050.
17. Rekha J.H, Parvathi R, "Survey on Software Project Risks and Big Data Analytics", *2nd* International Symposium on Big Data and Cloud Computing (ISBCC'15) Procedia Computer Science 50 (2015), pp. 295–300, www.sciencedirect.com.
18. https://tdwi.org/articles/2017/02/08/10-vs-of-big-data.aspx.
19. Peter Groves, Basel Kayyali, "The 'big data' revolution in healthcare" Mckinsey & Company Center for US Health System Reform Business Technology Office, January 2013.
20. Data acquisition, Axel Ngonga, Lead Data Acquisition, BIG Data PPF, http://big-project.eu.
21. Xu Chu, Ihab F. Ilyas, Sanjay Krishnan, Jiannan Wang, Data Cleaning: Overview and Emerging Challenges. https://www.ocf.berkeley.edu/~sanjayk/wp-content/uploads/2016/04/datacleaning-tutorial.pdf.
22. Elisa Bertino, "Big Data - Opportunities and Challenges", 37th Annual Computer Software and Applications Conference, IEEE., 2013, pp. 479–480.
23. Pravinsinh Mori, A.R. Kazi, Sandip chauhan, "A Survey on an Efficient Query Processing and Analysis on Big Data (RDF) Using Map Reduce", IJIRST –International Journal for Innovative Research in Science & Technology, Volume 1, Issue 7, pp 142–142, December 2014.
24. Karen Y. He, Dongliang Ge, and Max M. He, "Big Data Analytics for Genomic Medicine", International Journal of Molecular Science 2017 https://www.ncbi.nlm.nih.gov/pmc/articles/P MC5343946/.
25. David Lazer, Ryan Kennedy, Gary King, Alessandro Vespignani, Big data, "The Parable of Google Flu: Traps in Big Data Analysis", 14 March, 2014 VOL 343, pp. 1205–1205, www.sc iencemag.org.
26. Lidong Wang and Cheryl Ann Alexander, "Big Data in Design and Manufacturing Engineering", American Journal of Engineering and applied science, 2015, 05–05-2015.
27. Keith Feldman, Nitesh V. Chawla, "Scaling personalized health care with big data", 2nd International Conference on Big Data and Analytics in Healthcare, Singapore 2014. pp. 1–14.
28. Draft Report of the Big data and Health, International Bioethics committee United Nations Educational, Scientific and Cultural Organization, Paris, 19 April 2017, pp. 2–28.
29. Heather Mack, MobiHealthNews. The Dotson Report, IBM Researchers use Big Data to screen for Diabetic Retinopathy with 86% Accuracy, April 24, 2017, dostonreport.com.
30. Fred Fred Gebhart, "Big Data transforming glaucoma care, research". June 01, 2015 ophthalmology Times, Ophthalmologytimes.modernmedicine.com.
31. D Crabb, "Using big data to examine visual field follow up in Glaucoma", Acta Ophthalmologica, Volume 91, Issue 252, 6th August 2013, European Association for Vision and Eye Research Conference.
32. L Antony Clark, Jonathon Q. N, Nigel Morlet, James B. Semmens, "Major review Big data ophthalmic research", Article in Survey of Ophthalmology" February 2016. pp 444–465: https://www.researchgate.net/publication/292680199.

Mean Estimation Under Post-stratified Cluster Sampling Scheme

M. Raja Sekar and N. Sandhya

Abstract Post-stratification is used when the size of each stratum is known, but frame of each stratum is unknown. Assumed strata contain clusters of unequal size; a random sample of some clusters is drawn and post-stratified according to the existing stratum of the population. This paper considers the mean estimation problem under the post-stratified cluster sampling setup. A modified weight structure is proposed to combine different cluster means. Attempt is made to obtain the optimum variance and estimate of the variance. The efficiency comparison of estimator is numerically supported by database study. All clustering-related results were obtained from Web Accessible Genome cluster open dataset. Mean estimation under post-stratified cluster sampling scheme shows overall accuracy of 92%.

Keywords Clusters · Estimators · Sampling methods · Comparison and scheme

1 Introduction

In sample surveys, when stratified sampling could not apply for the estimation purpose because of unknown frames of each stratum, then the use of post-stratification is well recognized [1–4].

In stratified setup, stratum sizes are manageable but list of stratum units is often hard to get. Moreover, stratum frames may incomplete or overlapping or several population units may fall under multiple strata while classification. Under these circumstances, post-stratification is a useful sampling design. According to Shukatme et al., the post-stratification is as precise as the stratified sampling with proportional allocation subject to condition of a large sample. Jagar et al. advocated that the

M. Raja Sekar (✉) · N. Sandhya
Department of CSE, VNRVJIET, Hyderabad, India
e-mail: rajasekar_m@vnrvjiet.in

N. Sandhya
e-mail: sandhya_n@vnrvjiet.in

© Springer Nature Singapore Pte Ltd. 2019
R. S. Bapi et al. (eds.), *First International Conference on Artificial Intelligence and Cognitive Computing* , Advances in Intelligent Systems and Computing 815,
https://doi.org/10.1007/978-981-13-1580-0_27

post-stratification with respect to relevant criteria may improve estimation strategy subsequently over the sample mean or ratio estimator [5–8].

Singh and Singh [23] proposed a class of estimators in cluster sampling. Consider stratified setup of population and suppose every stratum contains clusters of unequal size; a random sample of clusters is post-stratified according to the structure of stratum in the population [9, 10]. This constitutes post-stratified cluster design useful and closer to the real-life survey situations. This paper considers the estimation problem under this design.

2 Notation

Let a finite population of N clusters each of unequal size, divided into K strata, ith stratum contains N_i clusters and random sample of n clusters is drawn from the population $(n < N)$ using SRSWOR [21]. The sample is post-stratified such that n_i clusters are from N_i. In what follows notations are as under:

Y variable under study.
Y_{ijl} lth value of jth cluster of ith stratum in population.
M_{ij} size of jth cluster of ith stratum.
y_{ij} mean of jth cluster of ith strata included in sample.
$\bar{y}_i$ mean in sample of ith stratum.
$\bar{\bar{Y}}_i$ mean of cluster means in population.
W_i $\frac{N_i}{N}$ population proportion of clusters in ith strata.
p_i $\frac{n_i}{n}$ sample proportion of clusters from ith strata.
$\bar{Y}_i$ population means of ith strata.
$\bar{Y}$ grand mean of population.
$\bar{\bar{Y}}_{N_i}$ mean of cluster means in population of ith stratum.

3 Proposed Estimator

Assume probability of n_i being zero, the usual post-stratified estimator for mean [11, 12]

$$\sum_{i=1}^{k} W_i \bar{Y}_i \tag{1}$$

$$\overline{y_{ps}} = \sum_{i=1}^{k} W_i \overline{y^*}_i \tag{2}$$

Agarwal and Panda [25] suggested to utilize the proportion

$$p_i = n'_i/n \quad \text{in addition to } W_i \tag{3}$$

In order to design a new weight structure for combining different strata means like [22, 23]

$$W_i^* = [\alpha p_i + (1 - \alpha)W_i], \quad \alpha \text{ being a constant} \tag{4}$$

we propose an estimator

$$\overline{y_{\text{PSC}}} = \sum_{i=1}^{k} W_i^* \overline{y_i^*} \tag{5}$$

where PSC stands for "post-stratified cluster design" [24, 25] and

$$\bar{y}_i = \sum_{j=1}^{n_i} M_{ij} \bar{y}_{ij}/n_i \bar{M}_i \tag{6}$$

y_i^* is an unbiased estimator of $\overline{Y_i}$ under the condition of given n_i.
$\overline{Y_{\text{PSC}}}$ is unbiased for $\sum_{i=1}^{k} W_i \bar{Y}_i$ with variance.

$$\text{Var}(\overline{y_{\text{PSC}}}) = \left(\frac{1}{n} - \frac{1}{N}\right) \sum_{j=1}^{n_i} M_{ij} \overline{y_{ij}}/n_i \overline{M_i} \tag{7}$$

$$E(W_i^*) = E\{E(W_i^*/n_i)\} \tag{8}$$

$$E[E\left\{\alpha\left(\frac{n_i}{n}\right) + (1 - \alpha)\frac{N_i}{N}/n_i\right\} \tag{9}$$

$$E(\overline{y_{\text{PSC}}}) = E[E\left\{\alpha\left(\frac{n_i}{n}\right) + (1 - \alpha)\frac{N_i}{N}/n_i\right\} \tag{10}$$

To evaluate variance expression, following standard results are used.

$$E\left(\frac{1}{n_i}\right) = \frac{1}{nW_i} + \frac{(N - n)(1 - W_i)}{(N - 1)n^2 W_i^2} \tag{11}$$

$$E\left(\frac{n_i}{n}\right)^2 = \frac{1}{nW_i} + \frac{(N - n)(1 - W_i)}{(N - 1)n^2 W_i^2} + w_i^2 \tag{12}$$

$$V\left(\frac{n_i}{n}\right) = \frac{(N - n)}{(N - 1)} \cdot \frac{W_i(1 - W_i)}{n} \tag{13}$$

$$\text{Cov}\left(\frac{n_i}{n}, \frac{n_j}{n}\right) = \frac{(N - n)}{(N - 1)} \cdot \frac{W_i W_j}{n} \tag{14}$$

$$S^2 = \sum_{i=1}^{k} \sum_{j=1}^{S1} \left[\frac{M_{ij}}{M_i} - \bar{Y}_i\right]^2 /N - 1 \tag{15}$$

$$S_{b_i}^2 = \frac{1}{N_i - 1} \sum_{j=1}^{N_i} \left(\frac{M_{ij}}{M_i} Y_{ij} - \bar{Y}_i \right)^2 \tag{16}$$

Now expanding the first term or RHS of expression 16, we have

$$E[V(\overline{y_{PSC}}/n_i)] = E\left[V\left(\sum_{i=1}^{k} W_i^* \bar{y}_i^* \right) / n_i \right] \tag{17}$$

$$E[V(\overline{y_{PSC}}/n_i)] = V\left[\sum_{i=1}^{k} W_i * \bar{y}_i \right] \tag{18}$$

Adding Eqs. (17) and (18), we get the final expression
The minimum variance of the estimator

$$V_{mn}(\overline{y_{PSC}}) = \left(\frac{1}{n} - \frac{1}{N} \right) V\left[\sum_{i=1}^{k} W_i * \bar{Y}_i \right] + \left\{ \sum_{i=1}^{l} n_i \right\}_j \tag{19}$$

$$T1 = \left(\frac{1}{n} - \frac{1}{N} \right) \left(S^2 - \sum_{i=1}^{k} s_i \right)^2 \tag{20}$$

The optimum value of α could be obtained by differentiating the variance expression with respect to α and evaluating it to zero. The equation would provide the value of $\alpha_{opt} = (1 + p)^{-1}$ whose substitution provides required results [13–16].

The term P is a ratio of two quantities, and therefor, in repeated surveys of relatively shorter duration over the same characteristics, it would be a stable quantity and could be guessed by an expert survey practitioner [17–20].

4 Estimate of Variance

Unbiased estimator of

$$V(\overline{y_{PSC}}) = [n(N - 1) - (1 - \alpha)^2 (N - n)^{-1}] \tag{21}$$

$$V(\overline{y_{PSC}}) = \frac{(1 - \alpha)^2 (N - n)}{n(N - 1)} + \left[n(N - 1) - (1 - \alpha)^2 (N - n)^{-1} \right] \tag{22}$$

$$V(y_{PSC}) = \frac{(1 - \alpha)^2 (N - n)}{n(N - 1)} \sum_{i=1}^{k} W_i (Y_i - \bar{Y})^2 + [n(N - 1) - (1 - \alpha)^2 (N - n)^{-1}] \tag{23}$$

$$V(\overline{y_{PSC}}) = \frac{(1 - \alpha)^2 (N - n)}{n(N - 1)} \sum_{i=1}^{k} W_i (Y_i - \bar{Y})^2 + S^2 \tag{24}$$

4.1 Robustness of Estimator

Consider a very small quantity ε and replace P by

$$P \pm \varepsilon \quad \text{then}$$

$$(\alpha_{\text{opt}})_\varepsilon = (1 + (P + \varepsilon))^{-1} \quad \text{and}$$

$$[V_{\min}(\overline{y_{\text{PSC}}})]_\varepsilon = \left(\frac{1}{n} - \frac{1}{N}\right) \sum_{i=1}^{k} W_i S_{bi}^2 + \frac{(N-n)}{(N-1)n^2} \tag{25}$$

$$\text{Percentage gain} = \frac{V(\overline{y_{PS}})}{V(y_{PS})} \times 100 \tag{26}$$

In the initial phase of analysis, the data set is divided into six partitions, and again, these partitions are subdivided into two parts. All these partitions are compared with four parameters. All these parameters are related to population under consideration. These population parameters are used to calculate percentage gain which plays an important role in the analysis.

In the second phase of the analysis, the data set is partitioned into six groups and these groups are compared with four parameters. Efficiency comparisons can be performed with the help of these four parameters.

Results in Table 3 explain the robustness, and the efficiency of post-stratification is utilized when the measure of every stratum is known yet casing of every stratum is obscure. Tables 4, 5, and 6 explain the second phase of the results. In the context of this paper, strata are formed with help of clusters of different sizes. Two parameters, mean estimation and post-stratified sample, form the basis for this study. Precision and recall are improved by a modified weight structure.

5 Conclusion and Future Work

Table 1 describes the results associated with percentage gain in which parameter yps is used. In initial phase, database is divided into six groups, and these groups are further divided into strata into different sizes. These strata are analyzed with different estimated parameters N, X, Y, and S. Table 2 describes the efficiency parameter evaluation with the help of v and v_{min} parameters in which database is divided into six

Table 1 Population parameters

Data set no.	Strata no.	N_i	X_i	Y_i	S_{bi}^2
I	I	105	10.465	10.465	11.65
	II	19	18.136	18.136	14.11
	Total	124	11.64	11.64	21.64
II	I	120	15.336	15.336	17.36
	II	32	19.087	19.087	18.07
	Total	152	16.1493	16.1493	17.13
III	I	105	12.465	12.465	13.45
	II	19	17.363	17.363	18.363
	Total	124	13.2155	13.2155	23.2155
IV	I	180	10.1772	10.1772	13.12
	II	40	23.87	23.87	25.87
	Total	220	12.09	12.09	22.09
V	I	70	9.098	9.098	19.98
	II	24	12.654	12.654	11.64
	Total	94	9.432	9.432	19.42
VI	I	105	7.983	7.983	8.9
	II	38	18.98	18.98	19.98
	Total	143	9.09	9.09	29.09

Table 2 Efficiency comparison

Data set no.	V	V_{min}	μ	Gain
I	294	244	5	20.3
II	412	360	6	14.7
III	294	244	5	20.1
IV	340	293	8	16
V	566	463	6	23
VI	560	495	7	13.3

clusters. Table 3 describes robustness and efficiency comparison of estimators with several parameters in which data set is divided into eight clusters. Table 4 discusses the calculation of population parameters which plays an important role in calculation of estimator phase 1. Table 5 explains the evaluation of population parameters which play an important role in calculation of estimators in phase 2. In these two phases, data set is classified into six clusters. Table 6 describes the efficiency comparison of estimated parameters in phase 3 in which dataset is classified into eight groups.

Table 3 Robustness and efficiency comparison

Data set no.	μ	Ω	Ø	ß	Δ
I	0	0.78	6	244	21
	−0.01	0.88	4.32	240	21
	−0.002	0.98	4.98	242	22
	0.001	0.68	5.11	245	24
	0.002	0.58	5.16	247	23
II	0	−0.006	6.02	234	14
	0.002	−0.007	5.98	359	16
	−0.01	−0.321	5.93	362	20
	−0.02	−0.114	6.04	370	18
	0.001	−0.209	6.10	422	19
III	0.00	−0.240	5.06	245	21
	−0.01	−0.110	5.78	320	22
	−0.02	−0.017	4.98	321	23
	0.01	−0.023	6.01	298	22
	0.02	−0.01	5.15	231	21
IV	0.000	−0.04	8	290	16
	−0.01	−0.005	7.32	206	16
	−0.02	−0.006	8.01	202	17
	−0.01	−0.02	7.62	282	18
	0.112	−0.04	6.32	206	17
	0.011	−0.07	7.23	278	16
V	0.011	0.58	4.98	275	23
	−0.102	0.44	4.42	234	25
	−0.102	0.84	4.87	228	20
	0.111	0.54	8.09	208	23
	0.012	0.321	9.16	254	24
VI	0.11	−0.123	6.06	231	25
	−0.11	−0.100	5.78	221	26
	−0.13	−0.013	4.98	209	24
	0.11	−0.021	6.01	241	23
	0	−0.012	5.15	231	27
VII	0	−0.116	7.13	321	28
	0.012	−0.117	5.98	332	26
	−0.11	−0.118	7.01	323	22
	−0.22	−0.115	7.04	401	24
	0.011	−0.004	7.10	434	23
VIII	0	−0.062	7.01	234	24
	0.112	−0.073	6.98	359	23
	−0.11	−0.182	6.93	362	21
	−0.12	−0.005	6.04	370	19
	0.011	−0.114	6.10	422	19

Table 4 Evaluation of population parameters in phase1

Data set no.	Strata no.	N_i	X_i	Y_i	S_{bi}^2
I	I	112	11.65	11.00	12.32
	II	20	19.16	20.36	15.21
	Total	132	30.81	31.36	27.53
II	I	123	16.336	15	18.36
	II	42	20.12	20.07	19.07
	Total	165	36.456	35.07	37.43
III	I	113	14	10	13.45
	II	17	16	18	18.363
	Total	130	30	28	23.2155
IV	I	190	12	11	14
	II	40	24	24	26
	Total	230	36	35	40
V	I	80	10	10	19.98
	II	30	12	12	11.64
	Total	110	32	22	19.42
VI	I	110	10	9	9
	II	40	20	15	20
	Total	150	30	24	29

Table 5 Efficiency comparison of phase2

Data set no.	V	V_{min}	μ	Gain
I	304	321	6	21.3
II	400	265	9	15.7
III	234	302	6	22.1
IV	440	283	9	17
V	436	363	5	27
VI	660	402	8	12.3

Figure 1 describes the results of population parameters with different values of intensity parameter. Figure 2 explains the efficiency comparison of contraction factor with number of individual population with different forms of accuracy. Finally, Fig. 3 describes the sample size and sampling errors which is used estimate robustness and

Table 6 Robustness and efficiency comparison in phase3

Data set no.	μ	Ω	$\emptyset$	ß	Δ
I	0	0.62	7	304	22
	−0.02	0.78	5.42	240	25
	−0.012	0.88	5.08	232	27
	0.012	0.78	6.01	145	23
	0.010	0.68	5.19	347	24
II	0	−0.106	7.01	204	25
	0.110	−0.107	6.88	229	18
	−0.021	−0.018	4.98	282	21
	−0.002	−0.015	5.29	120	17
	0.010	−0.024	7.10	202	18
III	0.002	−0.083	6.06	201	20
	−0.02	−0.321	6.78	231	23
	−0.03	−0.020	5.88	204	21
	0.10	−0.198	7.01	251	17
	0.23	−0.231	6.32	208	21
IV	0.109	−0.212	9	221	17
	−0.180	−0.920	6.98	237	18
	−0.023	−0.001	7.43	245	17
	−0.432	−0.120	8	232	18
	0.321	−0.003	8.73	207	16
	0109	−0.007	7.94	208	17
V	0	0.732	7	204	22
	−0.012	0.788	5.12	241	23
	−0.152	0.873	5.08	207	26
	0.032	0.731	6.23	249	25
	0.541	0.601	6.16	273	24
VI	0.431	−0.181	6.46	204	22
	−0.491	−0.10	6.38	205	29
	−0.321	−0.02	5.98	223	27
	0.301	−0.04	7.01	204	23
	0.109	−0.02	7.13	252	24
VII	0	−0.05	5.92	218	16
	0.023	−0.06	4.08	328	17
	−0.021	−0.07	6.93	278	21
	−0.921	−0.04	5.04	270	19
	0.121	−0.05	5.10	205	17
VIII	0	−0.013	4.02	204	16
	0.202	−0.016	5.00	299	18
	−0.101	−0.025	6.23	342	20
	−0.102	−0.024	5.14	360	22
	0.021	−0.029	5.10	392	21

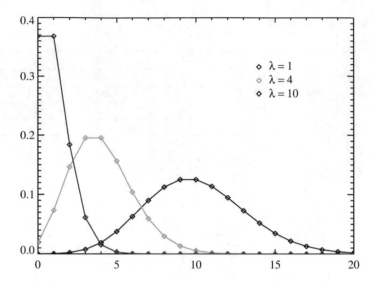

Fig. 1 Population parameters' graph

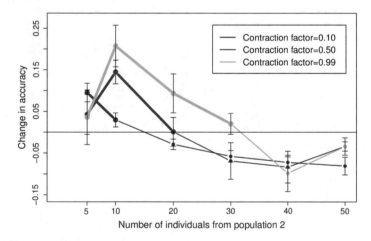

Fig. 2 Efficiency comparison graph

performance of the proposed estimator. Proposed estimator shows significant classification accuracy compared to classical estimators like shrunken estimator which gives 88%. Mean estimation under post-stratified cluster sampling scheme shows overall accuracy of 92%. These results can be compared with estimators related to probability distribution as a part of future work.

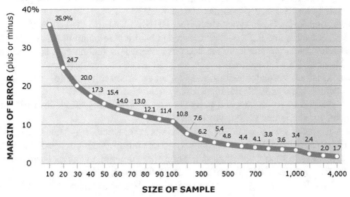

Fig. 3 Robustness and performance graph

References

1. Dr. M Raja Sekar *et al.*, " Tongue Image Analysis For Hepatitis Detection Using GA-SVM," *Indian Journal of computer science and Engineering, Vol 8 No 4,* pp. , August 2017.
2. Dr. M Raja Sekar *et al*, " Mammogram Images Detection Using Support Vector Machines," International Journal of Advanced Research in Computer Science ",Volume 8, No. 7 pp. 329–334, July – August 2017.
3. Dr. M Raja Sekar *et al.*, " Areas categorization by operating support Vector machines", ARPN Journal of Engineering and Applied Sciences", Vol. 12, No.15, pp. 4639–4647, Aug 2017.
4. Dr. M Raja Sekar, "Diseases Identification by GA-SVMs", International Journal of Innovative Research in Science, Engineering and Technology, Vol 6, Issue 8, pp. 15696–15704, August 2017.
5. Dr. M Raja Sekar., "Classification of Synthetic Aperture Radar Images using Fuzzy SVMs", International Journal for Research in Applied Science & Engineering Technology (IJRASET), Volume 5 Issue 8, pp. 289–296, Vol 45, August 2017.
6. Dr. M Raja Sekar, "Breast Cancer Detection using Fuzzy SVMs", International Journal for Research in Applied Science & Engineering Technology (IJRASET)", Volume 5 Issue 8, pp. 525–533, Aug,2017.
7. Dr. M Raja Sekar , "Software Metrics in Fuzzy Environment" , International Journal of Computer & Mathematical Sciences(IJCMS) , Volume 6, Issue 9, September 2017.
8. Dr. M Raja Sekar, "Interactive Fuzzy Mathematical Modeling in Design of Multi-Objective Cellular Manufacturing Systems", International Journal of Engineering Technology and Computer Research (IJETCR), Volume 5, Issue 5, pp 74–79, September-October: 2017.
9. Dr. M Raja Sekar, "Optimization of the Mixed Model Processor Scheduling", International Journal of Engineering Technology and Computer Research (IJETCR), Volume 5, Issue 5, pp 74–79, September-October: 2017.
10. Dr. M Raja Sekar, "Fuzzy Approach to Productivity Improvement", International Journal of Computer & Mathematical Sciences , Volume 6, Issue 9, pp 145–149, September 2017.
11. Dr. M Raja Sekar *et al.*, "An Effective Atlas-guided Brain image identification using X-rays", International Journal of Scientific & Engineering Research, Volume 7, Issue 12, pp 249–258, December-2016.
12. Dr. M Raja Sekar, *"Fractional* Programming with Joint Probability Constraints", International Journal of Innovations & Advancement in Computer Science, Volume 6, Issue 9, pp 338–342, September 2017.

13. Dr. M Raja Sekar, "Solving Mathematical Problems by Parallel Processors", "Current Trends in Technology and Science", Volume 6, Issue 4, PP 734–738.
14. Seth, G.S., and S.K. Ghosh. Proc. Math. Soc. BHU, Vol. 11, p 111–120,1995.
15. Seth, G.S., N.Matho and Sigh, S.K., Presented in National seminar on Advances in mathematical, statistical and computational methods in science and Technology, Nov. 29–30, Dept of Applied Maths., I.S.M. Dhanbad.
16. Fonseca, "Discrete wavelet transform and support vector machine applied to pathological voice signals identification in IEEE International Symphosium", p 367–374.
17. Dork, "Selection of scale invariant parts for object class recognition. In IEEE International Conference on Computer Vision, Vol 1, pp. 634–639.
18. P.Ganesh Kumar, " Design of Fuzzy Expert Systems for Microarray Data Classification using Novel Genetic Swam Algorithm", Expert Systems with Applications, pp 1811–1812, 2012.
19. T.S. Furey, et al., "Support Vector classification and validation of cancer tissue sampling using micro array expression data", Bioinformatics, pp 906–914, 2000.
20. Chun-Fu Lin, Wang. Sheng –De, "Fuzzy Support Vector Machines", IEEE Transaction on Neural Networks, pp. 13–22, 2002.
21. Shukhatme et.al., "Sampling Theory of Surveys with Applications", Iowa State University Press, Ames, Iowa, USA, PP 34–39, 1984.
22. Smith, T.M.F., "Post-stratificatio, ", The Statisticians, pp 31–39, 1991.
23. Singh, Rajesh and Singh, H.P., "A class of unbiased estimators in cluster sampling", Jour. Ind. Soc. Ag.Stat., pp 290–297, 1999.
24. Singh, D. and Choudhary, F.S., "Theory and anlysis of sample survey designs", Wiley Eastern Limited, New Delhi, 1986.
25. Agarwal, M.C. and Panda, K.B., "An efficient estimator in post-stratificaton", Journal of statistics, pp 45–48,2001.

Privacy-Preserving Naive Bayesian Classifier for Continuous Data and Discrete Data

K. Durga Prasad, K. Adi Narayana Reddy and D. Vasumathi

Abstract In data analysis, providing privacy to the customer data is an important issue. In this paper, a large number of customers are surveyed to learn the rules of classification on their data while preserving privacy of the customers. Randomization techniques were proposed to address this problem. These techniques provide more accuracy for less privacy of customers, conversely less accuracy for more privacy of the customers. In this paper, we propose a cryptographic approach with strong privacy and no loss of accuracy as a cost of privacy for continuous data. Our approach uses naive Bayes classification algorithm using frequency mining. The result shows the efficiency of our approach.

Keywords Privacy preserving · Cryptography · Naïve Bayes · Continuous data

1 Introduction

Every day, a lot of data is generating with fast growth on the Internet, and it is easy to collect the data from customers. To extract the useful knowledge from vast data, we use data mining. However, the privacy is the top concern in medical, financial and other data. Due to this, whenever the data is collected from the customers of large population, many customers may not give sensitive information or may give false

K. Durga Prasad (✉)
Research scholar, JNTUK, Kakinada, India
e-mail: dpcse007@gmail.com

K. Durga Prasad
BVRIT, Narsapur, India

K. Adi Narayana Reddy
ACE Engineering College, Hyderabad, India
e-mail: aadi.iitkgp@gmail.com

D. Vasumathi
CSE Department, JNTUHCEH, Hyderabad, India
e-mail: rochan44@gmail.com

© Springer Nature Singapore Pte Ltd. 2019
R. S. Bapi et al. (eds.), *First International Conference on Artificial Intelligence and Cognitive Computing* , Advances in Intelligent Systems and Computing 815,
https://doi.org/10.1007/978-981-13-1580-0_28

information. Applying data analysis or data mining algorithms on such data produces the low accuracy. Many customers are willing to provide the correct information when the sensitive data is protected from disclose. Consequently, the success of data collection and data mining depends on the measure of privacy protection. The customer sensitive data is protected by randomization, anonymization and cryptographic techniques. When the accuracy is low, the randomization and anonymization provide the privacy of the customer data. Cryptographic techniques provide privacy with high accuracy. The homomorphic encryption technique is used for privacy-preserving data analysis. The introduction of Fully Homomorphic Encryption (FHE) by Gentry [1] made the feasibility of privacy-preserving analysis on encrypted data without compromising the security of customer information.

The traditional cryptographic encryption techniques ensure the safety of data long term, but whenever an analysis of data is required the data must decrypted, and then the analysis is performed. Earlier Rivest et al. 1978 [2] shown that it is possible to apply some mathematical computations of an encrypted data by designing an encryption algorithm without decryption. However, Gentry [1] proposed that the FHE supports the mathematical computations theoretically. The mathematical operations such as multiplication and addition performed on the encrypted data (cipher text) yield the encrypted result which on decryption produces the same result as it is performed on the original data (plain text).

In this paper, a fully distributed data exchange is considered. The data of each customer is stored as row of data in the database. This is also called horizontally partitioned database. In this setting, the customer and miner communicates only once. It is efficient that the communication overhead is less compared to garbled circuits. Each customer sends the surveyed data, and the data miner applies some analysis and then learns the rules for classification. For classification, we use naive Bayes classification algorithm. This algorithm uses the frequency of attribute values and classifies the new instance with probability. To provide privacy to the customer data, we use cryptographic algorithms. Cryptographic algorithms are too expensive to use in data mining algorithms. However, homomorphic encryption is efficient in this scenario. We present the literature in Sect. 2, proposed the model in Sect. 3 and concludes the paper in Sect. 4.

2 Literature Work

Over the past several years, different techniques were proposed to preserve the privacy. In this section, we give the introduction of the privacy-preserving methods, namely randomization, anonymization, secure multiparty computation and homomorphic encryption techniques.

2.1 Randomization Technique

The randomization technique uses data perturbation. The data perturbation includes the techniques resampling [3], matrix multiplicative, k-anonymization [4], like additive, multiplicative [5], categorical data perturbation [6, 7], data swapping [8], micro-aggregation [9], data shuffling [10]. In these techniques, the original records of data are added with a noise data. These techniques are applied on the statistical databases. Agarwal and Agarwal [11], Agarwal and Srikanth [12], Evfimievski et al. [13], Evfimievski et al. [14], Du and Zhan [15] proposed data mining algorithms to protect the privacy using randomization on each customer data. These techniques provide privacy to the customer sensitive data, but the accuracy is low. Kargupta et al. [16] demonstrated that the random data distortion preserves little privacy and accuracy.

2.2 Secure Multiparty Computation

In distributed privacy-preserving data mining, two or more parties holding confidential information share to a trusted party. The trusted party computationally analyses and sends the result to the participants. There will not be any privacy to the shared information if the trusted party compromises. Secure multiparty computation provides the computational analysis between two or more parties. Yao [17] has introduced the secure multiparty computation concept. Goldreich [18] proved that secure multiparty computation provides a solution for any polynomial function. Lindell and Pinkas [19] introduced secure multiparty computation to data mining and build a decision tree by two parties, and both parties learning nothing from the data. Later on this technique has been applied to association rule mining, k-nearest neighbour and clustering.

Vaidya and Clifton [20, 21, 22] proposed to protect the privacy of customer data using association rule mining, k-means clustering and naive Bayes classifier on vertically partitioned data. Kantarciglu and Vaidya [23, 24] proposed an architecture to protect the privacy of client information. Wright and Yang [25] proposed privacy-preserving technique on distributed heterogeneous data using Bayesian network structure. All these techniques provide strong privacy, but the performance cost is high. These techniques are applied on small number of parties, say two parties. In these techniques, the communication between the parties is multiple numbers of rounds. The communication overhead is high.

2.3 Homomorphic Encryption Technique

Yang et al. [26] proposed classification algorithms to protect the privacy of customer data without loss of accuracy. This algorithm is developed based on additive homomorphic encryption of ElGamal public key cryptosystem with fully distributed communication. It is efficient than secure multiparty computation. It is secure against miner and n-2 corrupted customers and efficient for discrete attributes but does not consider the continuous attributes. Zhan [27] proposed privacy-preserving collaborative decision tree classification using homomorphic encryption on vertical data. Qi and Attalah [28] proposed efficient privacy-preserving k-nearest neighbour search using homomorphism. Chen and Zhong [29] proposed privacy-preserving backpropagation neural networks. Aslett et al. [30, 31] presented a new privacy-preserving method using the encrypted statistical machine learning and also presented the review on software tools for encrypted statistical machine learning. Vaidya et al. [22, 23] proposed privacy preserving of customer data based on vertical and horizontal data using naive Bayes classifier. In this method, the communication overhead is high. Kaleli and Polat [32] proposed recommendations on the distributed data using privacy-preserving naive Bayes classifier. Huai et al. [33] proposed privacy-preserving naive Bayes classifier with low communication overhead with high performance. This method is presented only for discrete data. We propose naive Bayes classifier to protect the privacy of customer data for both continuous and discrete attributes.

3 Proposed Privacy-Preserving Naive Bayes Classifier

In this setting, consider n customers $U_1, U_2, \ldots, U_n$ and each customer U_i has either a continuous or discrete value to be shared with the miner. The discrete value is represented by a Boolean value, and continuous values are converted into fixed precision values. These values are then converted into sequence of binary bits.

3.1 Naive Bayes Classifier

The naive Bayes classifier is a simple probabilistic learning algorithm based on Bayes theorem. It assumes the attributes are independent. We consider m attributes $\{A_1, A_2, \ldots, A_m\}$ and one class attribute V. Naive Bayes classifier assigns a class label to the new instance with attribute values $(a_1, a_2, \ldots, a_m)$ are given

$$V = \text{argmax } p(v^l) \prod_{i=1}^{m} p(a_i | v^l) \text{ for } v^l \in V \tag{1}$$

The term $p(a_i|v^l)$ is estimated as number of distinct terms in attribute value a_i in class label v^l if the attribute is discrete-valued attribute. When the attribute is continuous valued, then we use Gaussian/Normal distribution to estimate the $p(a_i|v^l)$

$$p(a_i|v^l) = \frac{1}{\sqrt{2\pi\sigma_l^2}} e^{-\frac{(a_i-\mu_l)^2}{2\sigma_l^2}} \qquad (2)$$

where μ_l and σ_l^2 are mean and variance associated with attribute A_i and class v^l. Naive Bayes classifier classifies the new instance but does not provide the privacy to the customer information. We propose privacy-preserving naive Bayes classifier for discrete and continuous attributes.

3.2 Privacy-Preserving Naive Bayes Classifier

Assume that each attribute is either continuous or discrete. Each discrete attribute A_i has r attribute values as $(a_{i1}, a_{i2}, \ldots, a_{ir})$, and the class attribute V has p classes as $(v^1, v^2, \ldots, v^p)$. The attributes with continuous values are converted into binary values. For all discrete attributes, the miner computes the sum $d = \sum_{i=1}^n d_i$ without revealing each d_i, where d_i is either 0 or 1. For all continuous attributes, the miner computes sum $sum = 2^{t-1}\sum_{i=1}^n d_{it-1} + 2^{t-2}\sum_{i=1}^n d_{it-2} + \ldots + 2^0\sum_{i=1}^n d_{i0}$ where t is number of bits used to represent continuous value. In this procedure, each customer sends one row of information and no customer communicates with other customer.

The privacy of data is protected by additive homomorphic encryption algorithm ElGamal public key algorithm. The privacy of ElGamal cryptosystem is based on the discrete logarithm problem. Let g be a primitive root of a large prime number P. Assume that each customer U_i has two key pairs x_i, $X_i = g^{x_i} \mod P$ and y_i, $Y_i = g^{y_i} \mod P$ where x_i and y_i are private keys and X_i and Y_i are public keys of customer U_i. Define two common keys X and Y as

$$X = \prod_{i=1}^n X_i \qquad (3)$$

$$Y = \prod_{i=1}^n Y_i \qquad (4)$$

Every customer uses the common key to send the data to the miner. Each customer U_i sends the Boolean value and the miner learns, $d = \sum_{i=1}^n d_i$. Each customer U_i sends the following pair to the miner

$$m_i = g^{d_i}.X^{y_i} \qquad (5)$$

$$h_i = Y^{x_i} \tag{6}$$

The miner learns d with the received pair of information as

$$r = \prod_{i=1}^{n} \frac{m_i}{h_i} \tag{7}$$

For $d = 1$ to n
if $g^d = r$ output d
The pair $\left(g^{d_i}.X^{y_i}, Y^{x_i}\right) = (m_i, h_i)$ is cipher text of the plain text d_i using ElGamal additive homomorphic encryption algorithm. Using the learned information, we rewrite the naive Bayes classifier for sensitive attributes S and non-sensitive attributes A-S as

$$v = \arg\max_{v^l \in V} \frac{\#v^l}{n} \prod_{i \in DA \& i \in S} \frac{\#\left(a_i, v^l\right)}{\#v^l} \prod_{i \in CA \& i \in S} p\left(a_i | v^l\right) \prod_{i \varepsilon A - S} p(a_i | v^l) \tag{8}$$

where DA—discrete attribute, CA—continuous attribute and $\#v^l$ and $\#\left(a_i, v^l\right)$ denotes the number of occurrences of class v^l and attribute a_i, class v^l, respectively. The term $p\left(a_i | v^l\right)$ is computed by using Gaussian distribution. The detailed privacy-preserving naive Bayes classier algorithm is presented as

For Discrete and Sensitive Attribute

- User U_i Private Keys:

$$x_{ij}^l, y_{ij}^l \text{ where } 1 \le i \le n, 1 \le l \le p$$

- User U_i Public Keys:

$$X_{ij}^l = g^{x_{ij}^l}, Y_{ij}^l = g^{y_i^{jl}} 1 \le i \le n, 1 \le l \le p$$

- Common Keys:

$$X_j^l = \prod_{i=1}^{n} X_{ij}^l, Y_j^l = \prod_{i=1}^{n} Y_{ij}^l, 1 \le l \le p, j \in S$$

- Representation of Discrete Attribute values

$$a_{i,j}^{k,l} = 1 \text{ if } \left(a_{i,j}^k, v^l\right) = \left(a_j^k, v^l\right)$$
$$= 0$$

- User U_i computes for $j \in S, 1 \le k \le r, 1 \le l \le p$

$$b_{i,j}^{k,l} = g^{a_{i,j}^{k,l}} \cdot X_j^{l^{y_i^l}}$$

$$h_{i,j}^{k,l} = Y_j^{l^{x_i^l}}$$

- User U_i sends $\left(b_{i,j}^{k,l}, h_{i,j}^{k,l} \right)$ for $j \in S$ to miner
- Miner learns sum as
  For $j \in S$ and $1 \le l \le p, 1 \le k \le r$

$$d_j^{k,l} = \prod_{i=1}^{n} \frac{b_{i,j}^{k,l}}{h_{i,j}^{k,l}}$$

For $\#\left(a_j^k, v^l \right) = 1$ to n

If $\left(g^{\#\left(a_j^k, v^l \right)} = d_j^{k,l} \right)$ return $\#\left(a_j^k, v^l \right)$

- For $l = 1$ to p count $\#v^l$

For Continuous and Sensitive Attribute

- Assume t bits are used to represent the value of a continuous attribute, and v bits are used to represent square of continuous attribute value.
- User U_i sends t bits (value of continuous attribute). For $k = 0$ to $t - 1$ and $1 \le l \le p, j \in S$

$$b_{i,j}^{k,l} = g^{a_{i,j}^{k,l}} \cdot X_j^{l^{y_i^l}}$$

$$h_{i,j}^{k,l} = Y_j^{l^{x_i^l}}$$

- Miner learns sum as
  For $j \in S$ and $1 \le l \le p, 0 \le k \le t - 1$

$$d_j^{k,l} = \prod_{i=1}^{n} \frac{b_{i,j}^{k,l}}{h_{i,j}^{k,l}}$$

- For $0 \le k \le t - 1$
  - For $\#\left(a_j^k, v^l \right) = 1$ to n
    If $\left(g^{\#\left(a_j^k, v^l \right)} = d_j^{k,l} \right)$ return $\#\left(a_j^k, v^l \right)$
- For $1 \le l \le p$

$$\mu_l = (2^{t-1}\#\left(a_j^{t-1}, v^l\right) + 2^{t-2}\#\left(a_j^{t-2}, v^l\right) + \cdots + 2^0\#\left(a_j^0, v^l\right))/\#v^l$$

- User U_i sends v bits (square of continuous attribute value). For $k = 0$ to $v - 1$ and $1 \leq l \leq p, j \in S$

$$c_{i,j}^{k,l} = g^{a_{i,j}^{k,l}} \cdot X_j^{l_i^l}$$

$$f_{i,j}^{k,l} = Y_j^{l_i^l}$$

- Miner learns sum as
  For $j \in S$ and $1 \leq l \leq p, 0 \leq k \leq v - 1$

$$d_j^{k,l} = \prod_{i=1}^{n} \frac{c_{i,j}^{k,l}}{f_{i,j}^{k,l}}$$

- For $0 \leq k \leq v - 1$

  - For $\#\left(a_j^k, v^l\right) = 1$ to n
    if $\left(g^{1 \leq l \leq p} = d_j^{k,l}\right)$ return $\#\left(a_j^k, v^l\right)$

- For $1 \leq l \leq p$

$$s\text{sum}_l = 2^{v-1}\#\left(a_j^{v-1}, v^l\right) + 2^{v-2}\#\left(a_j^{v-2}, v^l\right) + \cdots + 2^0\#\left(a_j^0, v^l\right)$$

$$\sigma_l^2 = \left(s\text{sum}_l - \left(2 * \#v^l - 1\right)\mu_l^2\right)/\#v^l$$

$$p\left(a_j|v^l\right) = \frac{1}{\sqrt{2\pi\sigma_l^2}} e^{-\frac{\left(a_j - \mu_l\right)^2}{2\sigma_l^2}}$$

3.3 Algorithm Analysis

The privacy-preserving naive Bayes classifier protects the privacy of honest customer against the miner and up to $n - 2$ corrupted customers. The corrupted customers and miner jointly learn only d. This d is same as the final result that the miner receivers from the n customers. The security the ElGamal public key cryptosystem depends on the random numbers used for encryption of plain text. Each plain text must be encrypted by a new random number. For this, we use a different key for every sensitive attribute.

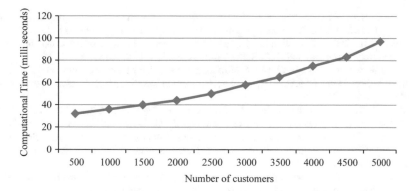

Fig. 1 Result analysis

3.4 Result Analysis

The privacy-preserving naive Bayes classifier is executed on i3 processor with 4 GB RAM in Java. We measured the computational time for customers from 500 to 5000 by varying the number of privacy-preserving discrete and continuous attributes. Figure 1 shows the relationship between number of customers and the time to compute the classification. As the number of customers increases and number of continuous attributes increases, the computational time also increasing.

4 Conclusion

The privacy-preserving naive Bayes for continuous and discrete attributes is developed using additive homomorphic encryption algorithm. It uses fully distributed communication for data sharing. The miner and customers communicate once as the customers send their data only once. The discrete data is encoded as 0 or 1, the continuous data is also encoded as corresponding 0 or 1 bits and each of these bits are exchanged with the miner. The miner only receives the total sum but cannot know the original values of the customers.

References

1. Gentry, C. A fully homomorphic encryption scheme. PhD thesis, Stanford University, 2009. crypto.stanford.edu/craig.
2. Rivest, Ronald. L, Leonard Adleman, and Michael L. Dertouzos. "On Data Banks and Privacy Homomorphisms", chapter On Data Banks and Privacy Homomorphisms, pages 169–180. Academic Press, 1978.

3. Liew C. K., Choi U. J. & Liew C. J. (1985). A Data Distortion by Probability Distribution. ACM Transactions on Database Systems (TODS), Vol 10, No 3, pp. 395–411.
4. Sweeney L.(2002). k-Anonymity: a Model for Protecting Privacy. International Journal on Uncertainty, Fuzziness and Knowledge-based Systems, Vol 10, No 5, pp. 557–570.
5. Kim J. J. & Winkler W. E. (2003). Multiplicative Noise for Masking Continuous Data. Technical Report Statistics #2003-01, Statistical Research Division, U.S. Bureau of the Census, Washington D.C.
6. Verykios V.S., Bertino E., Fovino I.N., Provenza L.P., Saygin, Y. & Theodoridis Y. (2004a). State-of-the-art in privacy preserving data mining, SIGMOD Record, Vol. 33, No. 1, pp. 50–57.
7. Verykios V. S., Elmagarmid A. K., Bertino E., Saygin Y. & Dasseni E. (2004b). Association Rule Hiding. IEEE Transactions on Knowledge and Data Engineering, Vol 16, Issue 4, pp. 434–447, ISSN 1041-4347.
8. Fienberg S. E. & McIntyre J.(2004). Data Swapping: Variations on a Theme by Dalenius and Reiss. Privacy in Statistical Databases (PSD), pp. 14–29, Barcelona, Spain.
9. Li X.B. and Sarkar S. (2006). A Tree-based Data Perturbation Approach for Privacy-Preserving Data Mining. IEEE Transactions on Knowledge and Data Engineering, Vol 18, No 9, pp. 1278–1283, ISSN 1041-4347.
10. Muralidhar K.& Sarathy R.(2006). Data shuffling a new masking approach for numerical data. Management Science, Vol 52, No 5, pp. 658–670.
11. Agarwal, D and Agarwal, C. On the design and quantification of privacy preserving data mining algorithms. In Proc. of the 20th ACM SIGMOD-SIGACT-SIGART Symposium on Principles of Database Systems, pages 247–255 ACM Press, 2001.
12. Agarwal, R and Srikant, R. Privacy preserving data mining. In Proc. of ACM SIGMOD Conference on Management of Data, pages 439–450 ACM Press, May 2000.
13. Evfimievski, A, Srikant, R, Agarwal, R and Gehrke, J. Privacy preserving mining of association rules. In Proc. of the Eighth ACM SIGKDD International Conference on Knowledge Discovery and Data Mining, pages 217–228. ACM Press, 2002.
14. Evfimievski, A, Gehrke, J and Srikant, R. Limiting privacy breaches in privacy preserving data mining. In Proc. of the 22nd ACM SIGMOD-SIGACT-SIGART symposium on Principles of database systems, pages 211–222. ACM Press, 2003.
15. Du, W and Zhan, Z. Using randomized response techniques for privacy-preserving data mining. In Proc. of the Ninth ACM SIGKDD International Conference on Knowledge Discovery and Data Mining, pages 505–510. ACM Press, 2003.
16. Kargupta, H, Datta, H, Wang, Q and Sivakumar, K. On the privacy preserving properties of random data perturbation techniques. In The Third IEEE International Conference on Data Mining, 2003.
17. Yao, A. Protocols for Secure Computation. FOCS 1982, 1982.
18. O. Goldreich (1998) "Secure multi-party computation", (working draft).
19. Lindell, Y and Pinkas, B. Privacy preserving data mining. J. Cryptology, 15(3):177–206, 2002.
20. Vaidya, J and Clifton, C. Privacy-preserving k-means clustering over vertically partitioned data. In Proc. of the Ninth ACM SIGKDD international conference on Knowledge discovery and data mining, pages 206–215. ACM Press, 2003.
21. Vaidya, J and Clifton, C. Privacy preserving association rule mining in vertically partitioned data. In Proc. of the Eighth ACM SIGKDD International Conference on Knowledge Discovery and Data Mining, pages 639–644. ACM Press, 2002.
22. Vaidya, J and Clifton, C. Privacy preserving naive Bayes classifier on vertically partitioned data. In 2004 SIAM International Conference on Data Mining, 2004.
23. Kantarcioglu, M and Vaidya, J. Privacy preserving naive Bayes classifier for horizontally partitioned data. In IEEE Workshop on Privacy Preserving Data Mining, 2003.
24. Kantarcioglu, M and Vaidya, J. Architecture for privacy-preserving mining of client information. In IEEE ICDM Workshop on Privacy, Security and Data Mining, pages 37–42, 2002.
25. Wright, R & Yang, Z. Privacy-preserving Bayesian network structure computation on distributed heterogeneous data. In Proceedings of the tenth ACM SIGKDD international conference on Knowledge discovery and data mining Pages 713–718.

26. Yang, Z, Wright, R. Privacy-Preserving Classification of Customer Data without Loss of Accuracy. In Proceedings of the 2005 SIAM International Conference on Data Mining.
27. Zhan, J. Using Homomorphic Encryption For Privacy-Preserving Collaborative Decision Tree Classification. IEEE Symposium on Computational Intelligence and Data Mining 2007.
28. Yinian Qi and Atallah, M. Efficient Privacy-Preserving k-Nearest Neighbour Search. IEEE ICDCS '08, 2008, pp. 311–3193.
29. Chen, Tingting and Zhong, Sheng. Privacy-preserving backpropagation neural network learning. Neural Networks, IEEE Transactions on, 20(10):1554–1564, 2009.
30. Aslett, Louis JM, Esperanc¸a, Pedro M, and Holmes, Chris C. Encrypted statistical machine learning: new privacy preserving methods. arXiv preprint arXiv:1508.06845, 2015a.
31. Aslett, Louis JM, Esperanc¸a, Pedro M, and Holmes, Chris C. A review of homomorphic encryption and software tools for encrypted statistical machine learning. arXiv preprint arXiv: 1508.06574, 2015b.
32. Kaleli, C and Polat, H. Privacy-Preserving Naïve Bayesian Classifier–Based Recommendations on Distributed Data. Computational Intelligent, Vol. 31 Nov 1 2015.
33. Huai Mengdi, Huang Liusheng, Yang Wei, Li Lu and Qi Mingyu. Privacy Preserving Naive Bayes Classification. In Proc. of International Conference Knowledge Science, Engineering and Management, Volume 9403 of Lecturer Notes in Computer Science, pages 627–638, 3rd November 2015.

Learning Style Recognition: A Neural Network Approach

Fareeha Rasheed and Abdul Wahid

Abstract In adaptive and intelligent e-learning systems, amidst other parameters that help to provide tailored instruction learning style plays a very pivotal role, and adaptive learning systems tend to revolve around this predominant parameter. And therefore deducing learning style is requisite that will help in the adaptation process. Learner behavior is observed and recorded, and certain parameters are used to deduce learning style. This paper presents an inference engine which is actually a neural network with backward propagation which uses certain observed parameters to infer the learning style. The engine so developed incorporates a popular and widely accepted learning style theory. The system will be a part of a larger adaptive learning and assessment system.

Keywords Learning style inference · Neural networks · E-learning Multiple intelligences

1 Introduction

Personalization in the e-learning systems has driven researchers to identify different parameters or learner characteristics that tutors look in a student in an off-line classroom and simulate it and model it in an online scenario. Learner characteristics can be defined as mental factors that may affect the learning process [1]. However, these characteristics are also dependent on the developmental stage the learner is in. For example, an adult learner is more resistant to change than a child who works with an open mind. The learner characteristics stated in educational psychology can

F. Rasheed (✉) · A. Wahid
Department of Computer Science and Information Technology,
Maulana Azad National Urdu University, Telecom Nagar, Gachibowli,
Hyderabad 500032, India
e-mail: fareeha_zarish22@yahoo.com

A. Wahid
e-mail: wahidabdul76@yahoo.com

© Springer Nature Singapore Pte Ltd. 2019
R. S. Bapi et al. (eds.), *First International Conference on Artificial Intelligence and Cognitive Computing* , Advances in Intelligent Systems and Computing 815,
https://doi.org/10.1007/978-981-13-1580-0_29

be divided into groups such as physical and physiological, cognitive, morals or values, emotional or personality, social-interactional, approach to learning [2]. System designers in an e-learning system are able to model only a few of these learner characteristics in an online scenario. Few of the learner characteristics that are being modeled online are motivation, efficacy, thinking style, emotions, and personality. [3, 4].

Learning style of the learner is the most important and interesting attribute in the learning scenario. Learning style is the preferred way of using one's ability to learn [5]. There are many different learning style models given by psychologists. Some of them are: David Kolb's model, Peter Honey and Alan Mumford's model, Learning modalities, Neil Fleming's VAK/VARK model, Anthony Gregorc's model, Cognitive approaches, NASSP model, Felder–Silverman learning style, and Gardner's theory of multiple intelligences. Some of these learning styles are explained below.

David Kolb's model is also called as the experiential learning, and Kolb's learning style defines different approaches to experiences, they are: *Concrete Experience* and *Abstract Conceptualization*, as well as two related approaches toward transforming experience: *Reflective Observation* and *Active Experimentation* [6].

Honey and Mumford's learning style was assembled above the Kolb's learning style theory. It consisted of four learning styles: *Activists, Theorists, Pragmatists, and Reflectors* [7].

Learning modalities refer to how learners use their senses to learn, the commonly considered modalities are four: visual (seeing), auditory (hearing), kinesthetic (moving), and tactile (touching). The auditory ones enjoy to read or listen, the visual ones often learn by seeing things, the kinesthetic learn by doing things rather than just observing and the tactile learn better when they touch things and experience its structure and feel [8].

Neil Fleming's VAK/VARK model is the most widely accepted and used model which defines how a learner likes to receive information. The different sensory modalities Fleming defines are Visual (see), Auditory (listen), Reading–Writing (text), Kinesthetic (doing). [9].

Felder–Silverman learning style was initially researched by Richard Felder and Linda Silverman to help professors in their teaching methodology to engineering students. But now it has become a widely accepted learning style and is increasingly being adopted in many online learning systems. The four dimensions in which this learning style model is defined are perception toward information, input preferred, processing information, and understanding information. Perception has two variants sensing (concrete) and intuitive (abstract/imaginary), inputs can be of two types visual or verbal, the information processing can be done in two ways, processing of information is usually done in two ways active (experimentation) and reflective (observing), and understanding information is again done in two ways sequential (follows a sequence) and global (does not work in any order) [10].

Gardner's theory of multiple intelligences is a cognitive approach toward learning. This theory has emerged from cognitive research and "documents the extent to which students possess different kinds of minds and therefore learn, remember, perform, and understand in different ways," [11]. Gardner exposes the challenges of educational systems that assume that every learner can learn and understand the same material, and no methodology is used to support the theory of differentiated instruction which emphasizes on developing course content or presentation that will be beneficial to every learner, the concept of tailored instruction. The learning styles associated with Gardner's theory of multiple intelligence are as follows:

Visual–Spatial: Learners with this intelligence are always able to create a picture in their mind and also look at the environment as jigsaw puzzle and try to learn things by fitting them in appropriate places [12].

Bodily–kinesthetic: Intelligent learners in this domain can skillfully control their body movements and are able to handle objects exceptionally. They learn by doing rather than just listening or observing [12].

Musical: Musically intelligent learners show sensitivity to rhythm and sound. They learner better with music in the background. Tools include musical instruments, CD-ROM, and multimedia [11].

Interpersonal: An interpersonally intelligent person is a people's person. These learners learn better through interaction with people and in groups. Tools for them may be forums, blogs, and the likes. [12].

Intrapersonal: These are solitary learners, they can learn better when they work alone, talking to themselves, they are independent learners. Tools for them are journals and diaries [11].

Verbal–Linguistic: These are word players, they often have the ability to think in words, they have high auditory skills, listen effectively and can make up poetry and stories. Tools usually used by them are word puzzles [12].

Mathematical–logical: Mathematically intelligent people are well-reasoned and calculated people, they think conceptually, abstractly and are able to see and explore patterns and relationships. They like to experiment and solve puzzles [12].

There have been various attempts in [13–17] to find the link between learning styles, and multiple intelligences and researchers have been successful in finding the link. The advantage of using the multiple intelligence theory is that it caters to various domains of students and does not heavily depend on the perceiving, input, and processing of information which forms the basis of the Felder–Silverman model [18]. Educationists over the years have proposed various activities that can be included in the curricula to comply with Gardner's theory of multiple intelligences. A few of them are domain-specific but others are generalized activities that can be incorporated in any domain. The book by Thomas Armstrong [19] is the most comprehensive book for instructional designers that describe every aspect of classroom teaching but with respect to the theory of multiple intelligences. Although many critics have tried to put this theory in bad shape, but empirical evidences were provided in [20].

Educational researchers in the E-learning domain have attempted to design and test systems based on the multiple intelligence theory and have reported the findings which prove that multiple intelligence theory is a good base for both adaptive

Table 1 Preferred learning activities and tools for MI theory

Intelligence	Essence	Online learning activities	Tools in online scenario that support the learning activities
Verbal–Linguistic	Learns easily through reading, writing, memorizing, and telling stories	Voice annotations, readable documents, encyclopedias, discussions and debates, story creation and telling, creative writing	Audio media, e-books and wiki projects, podcasts, text messaging course notes, voice over Internet, guided reading
Mathematical–Logical	Learns through reasoning, problem solving, and logic	Data collection and surveys, animations of experiments, math skill tutorials, critical thinking	Concept mapping, charts and graphics, spreadsheets, summarizing tools, vodcasts
Visual–Spatial	Visualizes and learns best through maps, charts, and puzzles	Timelines and maps, board games, clipart Online chess playing. PowerPoint presentations, arcade games	3D tools, drawing online, interactive news, albums and photos, virtual tours, video conferencing
Bodily–Kinesthetic	Learns best through doing or moving their body parts and role-playing	Online role plays and demos, motion-simulation and hand–eye coordination games, virtual reality systems	Flash animations, trials, online simulation tools and games, virtual labs
Interpersonal	Learns better in groups and can understand and group people effectively	Collaborative activities Group learning, brainstorming sessions, group calls	Blogs, bulletin boards, forums, social networks
Intrapersonal	A solitary learner and is usually a reflective person	Self-paced activities, e-portfolios, personal achievement tools	Online journaling, reflective activities, roleplay reflections

instruction and also adaptive assessment [21]. The following Table 1 lists activities in e-learning which are compliant with the MI theory and has been taken from [22–24].

2 Learning Style Inference

Traditional approaches both in the off-line and the online scenario use questionnaires that are to be filled by the learner and are then marked and checked using the key given to infer the learning style. However, a better technique that solves the issues of the above method is observing the learner in both the online and the off-line scenarios. In the online scenario, various systems have attempted to observe the learner activities in the systems and comparing them to a set of pre-decided parameters and using algorithms were able to infer the learning style of the learner. The different methods used in the inference of the learning style are decision trees, fuzzy logic systems, neural networks, and other intelligent techniques such as swarm intelligence and gene computation. For example, in [25] and [26], the authors have proposed feedforward multi-layer neural networks based upon the Felder–Silverman theory of learning styles. The classification is done according to the three dimensions processing, processing, and understanding. In [25], the authors have used 1 neuron per observed attribute in the system and 24 hidden units and 3 output neurons to map the three dimensions of the Felder–Silverman theory.

In this paper, we attempted to identify all such parameters used to observe the e-learner and can be used to infer the dominant intelligence of the learner that again corresponds to the learning style. For simplicity, only six of Gardner's divisions have been used, the omitted ones being the naturalistic and the musical intelligence. Learning management systems track and store all the paths the learner takes from the homepage and with tweaks and event observers can be initialized in the system to observe the learner more carefully and without any interference to the learner. Initially, the learner should be presented the questionnaire to fill it so the base style can be easily found and the hypothesis of whether the inferred learning style is same as the questionnaire deduced learning style or not.

Few parameters identified in this regard are given below (Table 2):

Neural networks are used to infer the learning style in this regard. The two different sets used in the learning style are the learning style and learning methodology preferred. These two sets are elucidated below:

$$\text{Learning style(LS)} = \left\{ \begin{array}{l} \text{Visual - Spatial(Vs)} \\ \text{Verbal - Linguistic(Vl)} \\ \text{Logical - Mathematical(L)} \\ \text{Bodily - kineasthetic(Bk)} \end{array} \right\},$$

and

$$\text{Preferred Learning mode(LM)} = \left\{ \begin{array}{l} \text{Interpersonal(Inte)} \\ \text{Intrapersonal(Intr)} \end{array} \right\}$$

The ordered pairs using these two sets will be of the form:

$$L = (\langle Vs, Vl, L, Bk \rangle, \langle Inte, Intr \rangle)$$

Neural networks are composed of basic units somewhat analogous to neurons in a human brain. These units are linked to each other by connections whose strength (weight) is modifiable as a result of a learning process or algorithm which is similar to

Table 2 Parameters identified in compliance with the Gardner's theory of multiple intelligences

Type of intelligence	Attributes that need to be extracted	Notation of the parameter
Verbal–Linguistic	No of audio files accessed	N_a
	No of wiki visits to the wiki links	N_w
	No. of podcasts accessed	N_p
	No. of e-books accessed	N_{eb}
	No. of documents created in course notes	N_{dc}
	ratio of learning goals completed using verbal LO's/total learning goals	R_{lgve}
Mathematical–Logical	No. of concept maps accessed	N_{cm}
	No. of charts and graphics files accessed	N_{cg}
	Ratio of learning goals completed using mathematical LO's/total learning goals	R_{lgml}
	No of problems solved correctly	N_{ps}
	No of failed attempts to solve problem	N_{fa}
	No of analogical questions attempted	N_{aq}
Visual–spatial	No. of photos accessed	N_{pa}
	No. of albums created	N_{ac}
	No. of doodle files created	N_{df}
	The ratio of learning goals completed using verbal LO's/total learning goals	R_{lgvi}
	No. of video conferences attended	N_{ca}
	No. of concept maps created	N_{cmc}

(continued)

Table 2 (continued)

Type of intelligence	Attributes that need to be extracted	Notation of the parameter
Bodily–Kinesthetic	No. of flash animation file accessed	N_{fa}
	No. of times simulators were used	N_{su}
	No. of times video podcasts accessed	N_{vp}
	the ratio of learning goals completed using bodily–kinesthetic/total learning goals	R_{lgbk}
	No. of application-level questions attempted	N_{apq}
	No. of game levels successfully completed	N_{gl}
Interpersonal–intrapersonal	Ratio of time spent in forums/total time spent in the system	R_{tf}
	No. of times social network sites were accessed	N_{sn}
	No. of group assignments completed	N_{ga}
	No. of questions answered in discussion	N_{qa}
	No. of conference calls made	N_{cc}
	No. of collaborative games played	N_{cg}
	No. of single player games played	N_{sg}
	No. of teaching assignments attempted	N_{ta}

the thickness of the myelin in neurons. Each of these units integrates independently (in parallel) the information provided by its synapses in order to evaluate its state of activation and similarly each of the neurons is fired when it reaches some threshold level. Analogous to the lighting of a bulb in your brain when you see a thing which you have already seen [27].

The entire learning style inference consists of two parts:

Part 1: infer the learning style modality, which is one of the following: visual–spatial, verbal–linguistic, mathematical–logical, bodily–kinesthetic.
Part 2: infer the learning mode preferred which is one of the two: intrapersonal or interpersonal

3   Modeling Learning Style with the Multi-layer Feedforward Neural Network

Part 1: *Inference of learning style modality*

Various attributes have been identified which can be used in the learning style inference and have been divided to form vectors for each learning style. These are all comprising of the number of hits to a particular type of learning object or the use of collaborative learning resources in a learning management system.

To complete the classification in the step 1, a multi-layer feedforward-backpropagation algorithm with the following structure has been proposed.

Input layer-hidden layer 1-leaky ReLU-hidden layer-leaky ReLU-output layer-Softmax.

Initially, the input neurons are proposed to be 4, 18 hidden units, and 4 output units, and when the network is implemented and trained using the education analytics dataset, the end network may contain less than the proposed units or extra neurons may be added to further improve the network.

The algorithm involves the following pre-processing steps:

1. Observe the different attributes in the system associated with the learning style.
2. The 38 attributes so identified are stored in the form of following vectors

$$v1 = \left[N_a, N_w, N_p, N_{eb}, N_{dc}, R_{lgve} \right]$$
$$v2 = \left[N_{cm}, N_{cg}, R_{lgml}, N_{ps}, N_{fa}, N_{aq} \right]$$
$$v3 = \left[N_{pa}, N_{ac}, N_{df}, R_{lgvi}, N_{ca}, N_{cm} \right]$$
$$v4 = \left[N_{fa}, N_{su}, N_{vp}, R_{lgbk}, N_{apq}, N_{gl} \right]$$

3. And these vectors are normalized in the range [0,1].
4. Assign the value 0 to the attributes which were not observed or with incomplete information.
5. The output vectors for each type of the verbal–linguistic is [0 0 0 1], for mathematical–logical it is [0 0 1 0], for visual–spatial it is [0 0 1 1], and for bodily–kinesthetic it is [0 1 0 0].

The feedforward-backpropagation algorithm for the proposed neural network is as follows:

Step 1:   Initialize initial weights randomly in the vector form to all the nodes in the network.

Step 2:   Compute the linear function by summation of initial weights and inputs.

Step 3:   The hidden layer 1 uses a leaky ReLU activation function of the following form

$$o_j = 1(x < 0)(\alpha x) + 1(x \geq 0)(x) \tag{1}$$

where o_j is the jth element of the output vector, x is the input to the neuron, and α is a small constant.

Step 4: The result of the hidden layer 1 is passed to another hidden layer with a leaky ReLU activation function.

Step 5: The result of the previous layer is passed to the output layer by applying the Softmax activation function which neatly classifies into different classes by calculating the probabilities of presence in each class.

$$\sigma(o)_j = \frac{e^{o_j}}{\sum_{k=1}^{K} e^{o_k}} \tag{2}$$

where o_j is the jth element of the output vector O and o_k is the kth element of the output vector O.

Step 6: De-normalize the data back to achieve the desired result.

The backpropagation of error is as follows:

Step 1: From output layer to hidden layer 2, calculate the error by using the Softmax's derivative as follows:

$$\text{if } i = j : \frac{\partial o_i}{\partial x_i} = \frac{\partial \frac{e^{x_i}}{\sum_K}}{\partial x_i} = \frac{e^{x_i} \sum_K - e^{x_i} e^{x_i}}{\sum_K^2} = \frac{e^{x_i}}{\sum_K} \frac{\sum_K - e^{x_i}}{\sum_K} = \frac{e^{x_i}}{\sum_K} = \left(1 - \frac{e^{x_i}}{\sum_K}\right) = O_i(1 - O_i)$$

$$\text{if } i \neq j : \qquad \frac{\partial o_i}{\partial o_j} = \frac{\partial \frac{e^{x_i}}{\sum_K}}{\partial x_j} = \frac{o \quad e^{x_i} e^{x_j}}{\sum_K^2} = -\frac{e^{x_i}}{\sum_K} \frac{e^{x_j}}{\sum_K} = O_i O_j \tag{3}$$

where x_i, x_j are ith and jth terms of the input vector, respectively, and o_i and o_j are ith and jth terms of the output vector, respectively

Step 2: On each node, update the weight matrix.

Step 3: From hidden layer 2 to hidden layer 1, calculate the leaky ReLU's derivative as follows.

$$f'^{(x)} = \begin{cases} 0.01, & x < 0 \\ 1, & x \geq 0 \end{cases} \tag{4}$$

Step 4: Update the weights in the hidden layer 2 nodes.

Step 5: From hidden layer 2 to hidden layer 1, calculate the leaky ReLU's derivative again using the Eq. (4).

Step 6: Restart the training process using the updated weights.

Part 2: inference of learning modality

The second part of learning style inference is the identification of mode of learning the learner prefers, whether he or she is a social person (interpersonal) or a solitary learner (intrapersonal).

The various attributes related to the class of interpersonal and intrapersonal intelligence which are R_{tf}, N_{sn}, N_{ga}, N_{qa}, N_{cc}, N_{cg}, N_{sg}, N_{ta} are directly sent as inputs $x0$, $x1$... to the neural network of the form.

Input layer-hidden layer 1-leaky ReLU-output layer-SoftMax.

The output of this neural network in [1 0] for interpersonal and [0 1] for intrapersonal.

The feedforward-backpropagation algorithm for the proposed neural network for inference of learning mode is:

Step 1:   Initialize initial weights randomly in the vector form to all the nodes in the network.

Step 2:   Compute the linear function by summation of initial weights and inputs.

Step 3:   The hidden layer 1 uses a leaky ReLU activation function described in Eq. (1).

Step 4:   The result of the previous layer is passed to the output layer by applying the SoftMax activation function which neatly classifies into different classes by calculating the probabilities of presence in each class described in Eq. (2).

The backpropagation of error is as follows:

Step 1:   From output layer to hidden layer 2, calculate the error by using the Softmax's derivative described by Eq. (3).

Step 2:   Update the weights on the connections.

Step 3:   From hidden layer to input layer, use the leaky ReLU's derivative described in Eq. (4) and update the weights on the connections.

Step 4:   Restart the training process with the updated weights.

4   Results and Analysis

For training the algorithm and testing it, we created a dataset, as a dataset with similar attributes was not available, we assigned random values to the variables and then calculated the total in each group of variables and assigned classes to it. We used MATLAB 2013a to code the neural network, trained it, and found out that even after 250 epochs the neural network does not converge. It is either the data set which is the problem or the design issues. As future work, our main aim will be to optimize the results which were achieved in this section. Further, the attributes being recorded can be reduced, and the networks can be trained with less number of inputs; however, the classes or outputs will remain the same.

5   Conclusion and Future Work

In this paper, we attempted to identify different attributes which can be used to infer the learning style and preferred learning mode of the learner in accordance with Gardner's theory of multiple intelligences. A total of 32 attributes were identified, and two different feedforward-backpropagation neural networks were proposed, one for the learning style and another for learning mode preferred. Future work would involve implementation and train the neural network by using the distance education dataset available in the Open Machine Learning repository and further a linear ensemble will be designed, implemented, trained, and tested to combine the outputs of the two different neural networks.

References

1. M. Nakayama and R. Santiago, "Learner characteristics and online learning.," in Encyclopedia of the Sciences of Learning, N. M. Seel, Ed., Springer US, 2012, pp. 1745–1747.
2. J. Snowman, "Educational Psychology: What Do We Teach, What Should We Teach?," Educational Psychology Review, vol. 9, no. 2, pp. 151–170, 1997.
3. D. H. Lim and H. Kim, "Motivation and learner characteristics affecting online learning and learning application," Journal of Educational Technology Systems, vol. 31, no. 4, pp. 423–439, 2003.
4. F. R. Prinsen, M. Volman and J. Terwel, "The influence of learner characteristics on degree and type of participation in a cscl environment," British Journal of Educational Technology, vol. 38, pp. 1037–1055, 2007.
5. R. J. Sternberg, "Allowing for thinking styles," Educational Leadership, vol. 52, no. 3, pp. 36–40, 1994.
6. A. Kolb and D. A. Kolb, "Kolb's Learning Styles," in Encyclopedia of the Sciences of Learning, Springer US, 2012, pp. 1698–1703.
7. P. Honey and A. Mumford, The learning styles helper's guide, Peter Honey Maidenhead, 2000.
8. M. J. Meyers, "The Significance of Learning Modalities, Modes of Instruction, and Verbal Feedback for Learning to Recognize Written Words," Learning Disability Quarterly, vol. 3, no. 3, pp. 62–69, 1980.
9. N. D. Fleming, Teaching and learning styles: VARK strategies, IGI global, 2001.
10. R. M. Felder and L. K. Silverman, "Learning and teaching styles in engineering education," Engineering education, vol. 78, no. 7, pp. 674–681, 1998.
11. H. E. Gardner, Intelligence reframed: Multiple intelligences for the 21st century, Hachette UK, 2000.
12. H. E. Gardner, Multiple intelligences: New horizons in theory and practice, Basic books, 2008.
13. D. W. Chan, "Learning and Teaching through the Multiple-Intelligences Perspective: Implications for Curriculum Reform in Hong Kong," Educational Research Journal, vol. 15, no. 2, 2000.
14. E. Giles, S. Pitre and S. Womack, "Multiple Intelligences and Learning Styles," Georgia, 2003.
15. H. F. Silver, R. W. Strong and M. J. Perini, So each may learn: Integrating learning styles and multiple intelligences., 1703 North Beauregard Street, Alexandria, VA 22311–1714: ERIC, 2000.
16. L. Campbell, B. Campbell and D. Dickinson, Teaching & Learning through Multiple Intelligences., Needham Heights, MA: Allyn and Bacon, Simon and Schuster Education Group, 2000.
17. S. J. J. Denig, "Multiple intelligences and learning styles: Two complementary dimensions," Teachers College Record, vol. 106, no. 1, pp. 96–111, 2004.
18. H. E. Gardner, Frames of Mind: The Theory of Multiple Intelligences, Basic Books, 2011.
19. T. Armstrong, Multiple intelligences in the classroom, Alexandria, USA: ASCD, 2009.
20. T. Armstrong, "MI Theory and Its Critics," in Multiple Intelligences in the Classroom, Alexandria, Virginia USA, ASCD, 2009, pp. 190–198.
21. C. Green and R. Tanner, "Multiple intelligences and online Teacher Education," ELT Journal, vol. 59, no. 4, pp. 312–321, 2005.

22. K. Zhang and C. Bonk, "Addressing diverse learner preferences and intelligences with emerging technologies: Matching models to online opportunities," Canadian Journal of Learning and Technology, vol. 34, no. 2, 2009.
23. K. B. Mankad, "The Role of Multiple Intelligence in E-Learning," International Journal for Scientific Research & Developmentl, vol. 3, no. 5, pp. 1076–1081, 2015.
24. S. Tangwannawit, N. Sureerattanan and M. Tiantong, "Multiple Intelligences Learning Activities Model in E-learning Environments," in Proceedings of the Fifth International Conference on eLearning for Knowledge-Based Society, Bangkok, 2008.
25. J. E. Villaverde, D. Godoy and A. Amandi, "Learning styles' recognition in e-learning environments with feed-forward neural networks," Journal of Computer Assisted Learning, vol. 22, no. 3, pp. 197–206, 2006.
26. S. V. Kolekar, S. J. Sanjeevi and D. S. Bormane, "Learning Style Recognition using Artificial Neural Network for Adaptive User interface in e-learning," in conference proceedings of International Conference on Computational Intelligence and Computing Research, Coimbatore, 2010.
27. H. Abdi, "A Neural Network Primer," Journal of Biological Systems, vol. 2, no. 3, pp. 247–281, 1994.

An Enhanced Efficiency of Key Management Method for Wireless Sensor Networks Using Mobile Agents

Ramu Kuchipudi, Ahmed Abdul Moiz Qyser, V. V. S. S. S. Balaram,
Sk. Khaja Shareef and N. Thulasi Chitra

Abstract Security in wireless sensor networks is very crucial especially whenever they deployed in military applications. Cryptography is used to protect sensitive information from disclosure. Key management is important component in cryptography. Cryptography is not useful if keys are disclosed to attackers. Designing an efficient key management for WSN is a difficult task because of scarcity of computing, communication, and memory resources. An efficient key distribution approach is proposed by using mobile agent paradigm rather than client–server model. The proposed approach will use good features of both symmetric and asymmetric cryptography. Mobile agents are used to generate public and private key pairs, update keys, and revocation of keys. The proposed scheme in the first level will use mobile agents for key distribution and coordination for asymmetric keys, and in the second level, sensor nodes can involve in constructing symmetric keys for secure communication through mutual authentication and encryption with those keys. The proposed

R. Kuchipudi (✉)
Department of CSE, Vasavi College of Engineering,
Shamshabad, Hyderabad, India
e-mail: kramupro@gmail.com

A. A. M. Qyser · V. V. S. S. S. Balaram
Department of CSE, Muffakham Jah College of Engineering and Technology,
Banjara Hills, Hyderabad, India
e-mail: aamoiz@gmail.com

V. V. S. S. S. Balaram
e-mail: vbalaram@sreenidhi.edu.in

A. A. M. Qyser · V. V. S. S. S. Balaram
Department of IT, Sreenidhi Institute of Science and Technology,
Yamnampet, Hyderabad, India

Sk. Khaja Shareef · N. Thulasi Chitra
Department of Computer Science and Engineering,
MLR Institute of Technology, Hyderabad, India
e-mail: khaja.sk08@gmail.com

N. Thulasi Chitra
e-mail: thulasichitra@gmail.com

© Springer Nature Singapore Pte Ltd. 2019
R. S. Bapi et al. (eds.), *First International Conference on Artificial Intelligence and Cognitive Computing* , Advances in Intelligent Systems and Computing 815,
https://doi.org/10.1007/978-981-13-1580-0_30

313

method is implemented using NS2 simulator, and results are compared with existing similar methods in terms of evaluation parameters like communication, computation overhead, memory overhead, and resiliency. The performance of proposed method is improved when it is compared with similar existing methods.

Keywords Wireless sensor network · Key distribution · Mobile agents

1 Introduction

An efficient key distribution mechanism in wireless sensor network (WSN) is required to provide secure communications. WSN is a collection of sensor nodes that are used to capture information from surroundings. Some of the applications of WSNs are healthcare monitoring, property surveillance, forest fire detection, study of wildlife habitat, and enemy intrusion detection, etc.

Different approaches are proposed in the existing literature like key predistribution schemes, self-enforcing schemes, and trusted server schemes to overcome resource scarcity problem. Keys are predistributed in sensor node's memory before their deployment, and secret keys are established between sensor nodes. Optimized asymmetric key cryptography is used in order to establish keys in self-enforcing schemes. Secret keys are generated and distributed among sensor nodes using server in trusted server schemes.

Mobile agents for key distribution in WSN are the research area little explored. Mobile agent-based key management scheme has been proposed for reducing communication overhead of Sahingoz [1].

Mobile agent is a piece of software which can migrate from one system to another and perform assigned task. In proposed method, mobile agent is used to distribute and update secret keys.

2 Related Work

The pair-wise and triple key establishment is used to have secure data aggregation in WSN. Then the scheme was applied to key management and secure forwarding in WSN. It was proved to be resilient against node capture attacks. The main application of the proposed scheme is that it can be applied to both static and dynamic networks with clustering. However, the security of this scheme is limited by the use of degree of polynomials.

The work in this paper is close to the work of Sahingoz [1] who proposed hybrid dynamic key management scheme for WSN where UAV is the mobile certification authority used to distribute public keys. In this paper, we proposed a scheme that makes use of mobile agent for public key distribution. The scheme reduces communication overhead, memory overhead, and computational overhead. Besides, it is resilient against node capture attacks.

BS is trusted node which stores details of sensor nodes [2–13]. Mobile agent is also assumed trusted component. The nodes of sensor network can be both static and dynamic. Signal range of all nodes does not exceed the threshold value. Other attacks like black hole, hello flood, and wormhole attack possible.

In the proposed system, the initial stage is to run System Initialization Agent (SIA) to generate and stores public and private key pairs in all sensor nodes. Then LCA is executed to store location information of sensor nodes. Later Key Distribution Agent (KDA), and Node Compromise Detection Agent (NCDA) algorithms are run to establish, update, and revoke keys, respectively.

3 Proposed Method

The mobile agent manager layer takes care of mobile agents and other activities such as communication, coordination, resource management, and data management with respect to the mobile agents. The MAM manages System Initialization Agent (SIA), Location Calculation Agent (LCA), Key Distribution Agent (KDA), and Node Compromise Detection Agent (NCDA) (Fig. 1).

System Initialization Agent (SIA) generates and stores public and private key pairs in all sensor nodes using RSA key generation algorithm. RSA key generation algorithm is executed for to produce public and private keys.

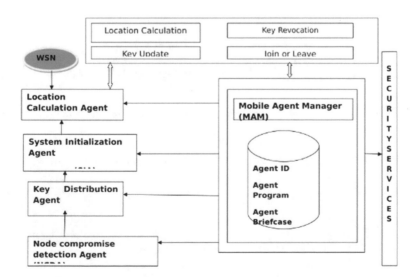

Fig. 1 Overview of the proposed methodology

LCA is responsible for to find positions of sensor nodes, cluster heads, and base station. LCA agent updates location information of sensor nodes whenever it visits them. The location information is used by the KDA which is responsible to provide public keys of neighbor nodes to sensor nodes.

Key Distribution Agent (KDA) is used to key establishment between nodes. It checks public key information on sensor nodes and compares it with its briefcase if they are not same public key information will be updated on sensor nodes. Then shared keys are established between a sensor node and neighbor nodes using those public keys. Sensor nodes will PRF to generate shared keys.

Key Distribution Agent (KDA) is used to update secret keys and public and private keys periodically, and it is also invoked by agent manager whenever nodes join or leave the sensor network. Node compromise Detection Agent (NCDA) is used to detect and isolate them from network. The NCDA calls KDA to update keys.

Cluster formation is based on the average energy of rest of the nodes compared with a candidate CH node besides the distance specified. An algorithm is proposed to achieve this. The algorithm is responsible to ensure that right SN is chosen as CH.

3.1 Location Calculation Agent (LCA) Algorithm

This is the mechanism used by LCA which is employed by MAM. Any pair nodes are decided as neighbor nodes based on the following steps. Any node within the transmission range of a node is called its neighbor.

Let (x_i, y_j) and (x_k, y_m) be the original positions of two sensor nodes A and B. Calculate distance between two sensor nodes

$$d = \sqrt{(x_k - x_i)^2 + (y_m - y_j)^2} \tag{1}$$

$$\text{Calculate } \frac{R - d}{V} \text{ Where V} = \text{Average speed of the node} \tag{2}$$

If $\left(\left(\frac{R-d}{V}\right) <= \text{Transmission\_Range}\right)$, then A and B are treated as neighbor sensor nodes.

3.2 Key Distribution Agent Algorithm

Select two large primes p and q such that $p \neq q$.

$n \leftarrow p \times q$

$\phi(n) \leftarrow (p-1) \times (q-1)$

Select e such that $1 < e < \phi(n)$ and e is coprime to $\phi(n)$

$d \leftarrow e^{-1} \bmod \phi(n)$ // d is inverse of e modulo $\phi(n)$

public_key $\leftarrow (e, n)$ //   To be announced publicly

private_key $\leftarrow d$ //   To be kept secret

return $SN_{PU} \leftarrow$ public_key and $SN_{PR} \leftarrow$ private_key

RSA_Encryption(M, e, n)

{

$C \leftarrow M^e \bmod n$

return C

}

RSA_Decryption(C, d, n)

{

$M \leftarrow C^d \bmod n$

return M

}

 KDA updates public keys of Neighbor nodes

Secret key K_{AB} generated by SN using PRF-MD5.

$SN \leftarrow N : E(N_{PU}, (K_{AB} \| RN))$

$D_{N_{PR}}(E(N_{PU}, (K_{AB} \| RN)))$

Data M is transmitted between SN and N encrypted using secret key K_{AB}

$SN \rightarrow N : E_{K_{AB}}(M)$

$N \rightarrow SN : D_{K_{AB}}(M)$

The key update is done based on the time elapsed based on a threshold. For the purpose of key update, there is no need to contact MAM. The reason behind this is that nodes have the public keys of neighbors.

For every 24 h KDA is initiated by Agent manager to update key information table of sensor nodes.

4 Results Analysis

The performance of proposed scheme is compared with existing schemes UAV-based scheme and HKM Scheme. When number of nodes is increased, the memory usage

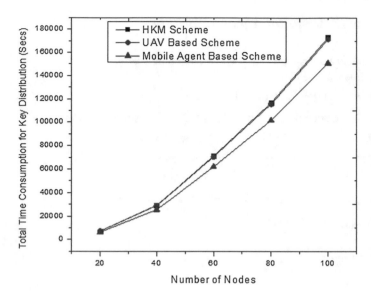

Fig. 2 Computation overhead

is increased as shown in Fig. 2 for the proposed scheme and UAV-based scheme [1] scheme and HKM scheme [13]. It is less in proposed scheme.

Similarly, when number of nodes is increased, computation overhead and communication overhead are also increased as shown in Figs. 3, 4 and 5 for the proposed scheme and UAV-based scheme [1] scheme and HKM scheme [13]. They are less in proposed scheme.

4.1 Computation Overhead

$$T_{KE} = N \times (K \times (1 \times T_{RSAE} + 1 \times T_{RSAD} + 1 \times T_{PRF}))$$

where

K	Number of neighbor nodes of a node
N	Number of sensor nodes
T_{RSAE}	Time consumption for RSA encryption
T_{RSAD}	Time consumption for RSA decryption

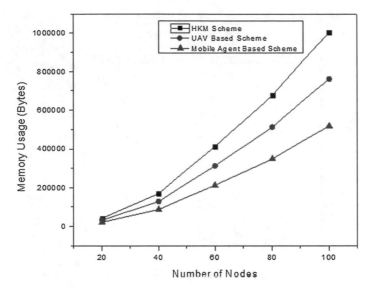

Fig. 3 Memory overhead

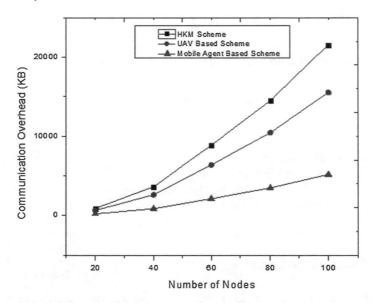

Fig. 4 Communication overhead

T_{PRF} Time consumption for pseudorandom function (PRF)
T_{KE} Total time consumption for key establishment.

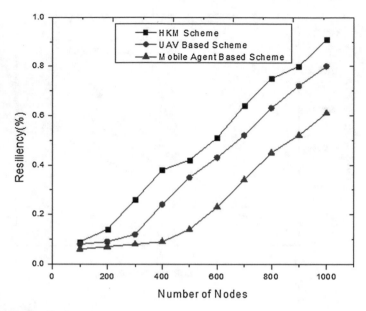

Fig. 5 Resiliency

4.2 Communication Overhead

$$C_{\text{overhead}} = N \times (K \times (P \times M_{\text{size}}))$$

where

N Number of sensor nodes
K Number of neighbor sensor nodes
P Number of messages exchanged between nodes
M_{size} Size of messages exchanged between nodes.

4.3 Memory Overhead

$$M_{total} = N \times (K \times (\text{size of } (K_{\text{PU}}) + \text{size of } (K_{SH}))) + size\,of\,(K_{\text{PR}})$$

where K Number of neighbor sensor nodes
K_{PU} Public key of sensor nodes
K_{SH} Shared key between any two sensor nodes
K_{PR} Private key of sensor nodes.

4.4 Resiliency

$$\text{Resiliency}(R) = \frac{\text{Number of Neighbor Nodes of Compromised Node}}{\text{Total Number of Nodes in Network}}$$

5 Conclusion

Mobile agent-based key distribution method scheme is proposed using mobile agents that are used to generate, update, and revoke keys. The dynamic key management is obtained by applying private and public key cryptography. In the first level, agent-based key distribution and coordination for asymmetric keys have been introduced while the second level is sensor nodes can involve in constructing symmetric keys for secure communication through mutual authentication and encryption with those keys. Agent-based public key dissemination and update of shared keys could reduce communication complexity, memory complexity besides improving resiliency when compared to the schemes in [1, 13].

References

1. Ozgur Koray Sahingoz. (2013). Large scale wireless sensor Networks with multi-level dynamic key management scheme. *Elsevier*, p. 20–30.
2. Eschenauer L, Gligor BD. A key-management scheme for distributed sensor networks. In: Proceedings of the 9th ACM conference on computer and communication security, Washington, DC, USA, 2002. p. 41–7.
3. Du W, Han YS, Chen S, Varshney PK. A key management scheme for wireless sensor networks using deployment knowledge. In: Proceedings of IEEE INFOCOM04. Hong Kong: IEEE Press; 2004. p. 586–97.
4. Liu D, Ning P. Establishing HKM keys in distributed sensor networks. In: Proceedings of 10th ACM conference on computer and communications security (CCS03). Washington, DC: ACM Press; 2003. p. 41–7.
5. Blom R. Theory and application of cryptographic techniques. In: Proceedings of the Eurocrypt 84 workshop on advances in cryptology. Berlin: Springer; 1985. p. 335–8.
6. Wallner D, Harder E, Agee R. Key management for multicast: issues and architectures, June 1999, RFC 2627.
7. F. Anjum, Location dependent key management using random key predistribution in sensor networks, in: Proceedings of WiSe'06.
8. S. Zhu, S. Setia, S. Jajodia, LEAP: efficient security mechanisms for large-scale distributed sensor networks, in: Proceedings of The 10th ACM Conference on Computer and Communications Security (CCS'03), Washington D.C., October, 2003.
9. Camtepe SA, Yener B. Combinatorial design of key distribution mechanisms for wireless sensor networks. IEEE/ACM Transactions on Networking (TON) 2007; 15(2):346–58.
10. Eltoweissy M, Heydari H, Morales L, Sudborough H. Combinatorial optimization of Group key management. Journal of Network and System Management 2004; 12(1).
11. Boneh D, Franklin M. Identity-based encryption from the weil pairing. In: Advance in cryptology-crypto, Lecture notes in computer science, vol. 2139, 2001. p. 213–29.

12. Seo, Seung-Hyun, Won, Jongho, Sultana, Salmin, and Bertino, Elisa, Effective Key Management in Dynamic Wireless Sensor Networks. IEEE Transactions on Information Forensics And Security, 10 (2) (February 2015), 371–383.
13. Pengcheng Zhao, Yong Xu, Min Nan, A Hybrid Key Management Scheme Based on Clustered Wireless Sensor Networks, Scientific Research (Aug 2012),197–201.

A Survey on Emotion's Recognition Using Internet of Things

K. P. L. Sai Supriya, R. Ravinder Reddy and Y. Rama Devi

Abstract Emotions play an important role in human life, because the emotions allow other people to understand the feelings. Emotions are obtained due to some physiological changes in human. When a person is in a situation where he is unable to speak, then their emotions can be used to understand the feelings. By using Internet of things, the emotions are going to be detected. In the first step, the sensors are placed on the human body. These sensors will capture the data, and real-time monitoring can be done. The data which is collected from the sensors is used for the emotion detection. Here the different works on IoT which is used for human emotions capturing and detection have been discussed. The analysis of these emotions will give significant results for various frame of mind of a person.

Keywords Internet of things (IoT) · Wireless sensor network · Linear discriminant analysis

1 Introduction

Nowadays, the Internet of things (IoT) became popular around worldwide and it is playing a major role in human life. IoT is nothing but interconnection of hardware and software devices like actuators, sensors, microcontrollers, Raspberry Pi. As human want to lead a luxurious life to maintain health and easy monitoring of the devices, they want to use the things which are portable in such a way that they can be carried out to any place and can be accessed from any place. Hence, IoT is seen everywhere and anywhere. Few applications of IoT are smart agriculture, smart city, smart homes,

K. P. L. Sai Supriya (✉) · R. Ravinder Reddy · Y. Rama Devi
Chaitanya Bharathi Institute of Technology, Hyderabad, India
e-mail: supu4895@gmail.com

R. Ravinder Reddy
e-mail: ravi.ramasani@gmail.com

Y. Rama Devi
e-mail: yrdcse.cbit@gmail.com

© Springer Nature Singapore Pte Ltd. 2019
R. S. Bapi et al. (eds.), *First International Conference on Artificial Intelligence and Cognitive Computing* , Advances in Intelligent Systems and Computing 815,
https://doi.org/10.1007/978-981-13-1580-0_31

health care, etc., [1]. Here the things or the devices are used in the smarter way. The healthcare applications of IoT play an important role, and it keeps patients' safe and healthy, and continuous monitoring of patient can be done when necessary.

The sensors play an important role because they are used to capture and store the data at every stage. The sensors are kept together in a network perfectly. After connection is implanted, the raw data is collected near the data center and real-time monitoring can be done. The data center can be accessed via Internet from anywhere and everywhere in order to retrieve, analyze, and process data [2].

The current research is focusing on the human health monitoring system in IoT to utilize the resources in the smarter way, and this may help to improve the human life in the less amount of time. Emotions occupy a major role in human life, and according to some theories, there are states of feelings that result in physical and psychological changes that influence the behavior [3].

In this paper, we reviewed the work on "physiological changes" of a human. A person if he is a dumb or a patient that they cannot express feelings toward the other person, then in this case detection of emotions plays an important. In this emotion detection system, consideration of the respiration, pulse rate, heart rate, and skin conductivity of the person is important. The real-time data monitoring is to be considered, the data is continuously monitored with the help of sensors, and the data from the sensor is sent to the device with the help of Internet. The sensors can be connected to the device with the help of Bluetooth (within range for limited bandwidth), WiFi with the standard IP. The main intent of this paper is to transistorize an overview on emotion's recognition system using IoT, wireless sensor networks, and big data tools [4].

2 Literature Review

2.1 Emotion Recognition System

Emotions can be recognized using facial expression, text, speech, body moments, and gestures. The most radical idea of emotion acceptance using facial expression is to subdivision of the facial images into various regions. In this method, the most commonly the facial nerve regions like chick, chin, eyebrows are considered. Here firstly the image of the person is captured, and few points are marked on the image at different regions. Next the image of the person is captured when they observe the changes in the face. The points are considered on the same positions as previous image and comparison are done between both the images. Here the points will change when we compare it with the normal image.

Using speech levels emotions were detected, here when the levels of sound increased or decreased based on the situation he is in. When the person is angry, he may raise his voice and speak loudly, and if at all he is sick at this time, he may talk slowly. The difference in the voice can be considered as a measure and can be used for emotion detection. Similarly, the emotions were detected from text [5]. Here the sentiment analysis was done on the text when the person text and post in the social media such as Twitter, Facebook, etc.

2.2 Emotion Recognition While Listening Music

The music is also used to detect the emotion of the person. The emotions which we observe in our daily life are the same emotions which are going to be observed when we listen to different kinds of music [6]. In this, they have gathered a group of people who were experienced in music. The selection of song was left to them, and each person had selected one song and the monitoring was done for few days. They observed that the blood pressure and respiration levels were increased when the person was listening to the beat song. This occurs because the person feels excited when he listens to beat song, and his body functionality differs from the normal state of his health condition. If a person listens to melodious song, then he feels pleasant and we observe no change in the blood pressure, respiration level remains constant, and skin conductance also remains same. If the person seems to be sad or if he weeps, at this scenario heart rate, respiration, and skin conductivity increased.

Here the wearable sensors were considered like electromyogram, skin conductivity, electrocardiogram, and respiration where continuous monitoring was done. The output is obtained from the sensor as a waveform. The raw data is obtained to perform data analysis, and the data cleaning is done because some unwanted data may be occurred when the person faces some environmental disturbances. The data cleaning is done using Fourier transformations. The peak values of the waveforms are considered, and classification was done. After classification, the emotions were detected.

2.3 Emotion Recognition Using Bio-Signals

In this, four biosensors were used, like electromyogram, skin conductance, electrocardiogram, and respiration, which were used to repossess the physiological signals with the help of Procomp2. Here they have considered two scenarios such as arousal and valence. With the help of the neural network classification, the emotions were being classified. The accuracy rates were around 96.6 and 89.9%, respectively [5]. Arousal means waking up from the sleep, and at this situation, the emotion is going to be checked, by considering the positive and negative valence and classified the emotions. The stance is describing the person when he is in the accepting stage. It says that the person is in a good situation. When we look at Fig. 1, the three lines which intercept at single point are seemed to be considered as the person is in the neutral state which represents that the person is calm. Here for the classification of Emotion Marquardt Backpropagation algorithm (MBP) was used [6].

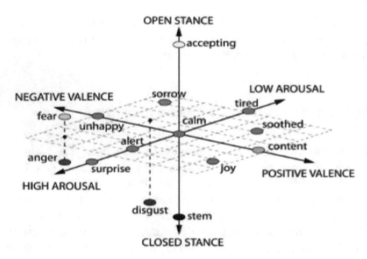

Fig. 1 Arousal, valence, and stance graph

2.4 Emotion Recognition While Driving

When the person is driving a vehicle, at this scenario the stress levels were detected. A real-time monitoring was done, and the stress levels were observed with the help of the physiological changes of the person [7]. The questionnaire analysis was used to detect the stress of the person when he is driving on the different roadways. When the person is driving his video is captured continuously with the help of the camera and later on the obtained data the analysis was done. The Beck Depression Inventory-II algorithm was used to detect the level of stress in the person at different situations.

As real-time monitoring is done, huge amount of data is obtained due to this and the time and space complexity may be more. So linear discriminant analysis (LDA) [8] can be used for real-time monitoring data and Pseudoinverse Linear Discriminant Analysis (PLDA) [9] can be applied to deal with small size problems. The Table 1 gives different algorithms which were used to detect and classify the emotion for different activities.

3   Proposed System

Internet of things is playing an important role in remote monitoring systems, and it is able to access some megabytes of the data, and real-time monitoring can be done. The sensors like respiration, skin conductivity, temperature, heart rate sensors are used for human health monitoring system. The above-mentioned sensors are placed at the specific places such as the respiration sensor [10] is placed at chest region, the heart rate sensor is placed at the wrist and this sensor is designed with the green light

Table 1 Algorithms for different activities

Algorithms	Activity	Emotions
Linear discriminant analysis + Sequential backpropagation [6]	Music	Excited, sad
Marquardt Backpropagation [6]	Bio-signals	Arousal, valency
Beck Depression Inventory-II [7]	Driving	Stress
K-nearest neighbor [12]	Facial	Angry, happy, sad
Backward feature selection technique [5]	Text, voice	Angry, happy

which acts as a radiator to checks the heartbeat of the person. The skin conductivity sensor can be placed either in the palm region or on the foot of the person, and the temperature sensor can be placed either in oral mouth or auxiliary (armpit).

The evolution of IoT system depends on protocol stacks [2]. The sensors are connected to the device wirelessly with the help of WiFi. The data from sensors to the device is sent with the help of Zigbee protocol. Zigbee is one of the low power consumptions, and it is now specially designed for IPV6 which is a standard protocol. The data is stored in the database of the system and retrieved whenever it is necessary to perform data analytics.

4 System Architecture

The sensors are connected at the specific places of the person, and the continuous monitoring is done. Here the data from the sensors is collected at the connected device. In the IoT technologies, all the sensors are together connected to the single device with the help of WiFi as shown in Fig. 2 [11]. Here the data is sent in the form of packets from the sensor to the device [4]. If it is for the longer distances, the relay nodes are to be added so that data can be transferred correctly. Here mostly the standard IPv6 protocol has to be used to avoid the data loss.

After collection of the sensor data perform data analytics and perform classification. Classification is to be done on the data set to detect the emotion of the person. Sometimes, the nonessential value may occur in the data set due to some environmental disturbances. So it would become difficult task to recognize the correct emotion of the person. Hence to avoid this perform data cleaning to eliminate missing or unwanted data. Fig. 3 shows how step-by-step processes are done for emotion classification.

To classify the emotion, linear discriminate analysis (LDA) can be used. The main objective behind the LDA is dimensionality reduction [8].

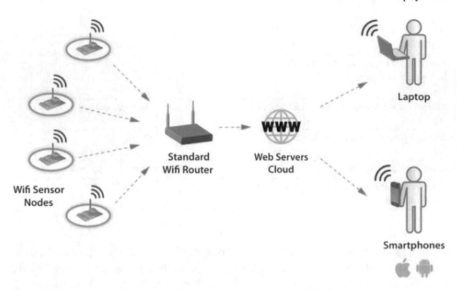

Fig. 2 System architecture

Fig. 3 Block diagram

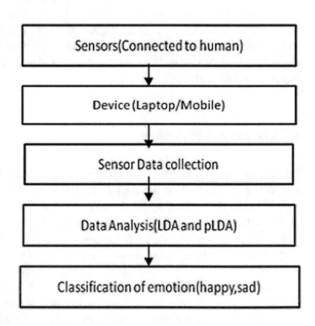

Here for a set of D-dimensional samples $\{d^1, d^2, d^3, \ldots d^n\}$. By protruding the samples on a line, $Y = W^T X$ can be obtained by the scalar values. Define separation between the projections, and we first need to find the projection vector and mean vector of each class in x and y feature space as

$$\mu i = \frac{1}{N} \sum_{x \in \omega} X$$

$$\mu i' \frac{1}{N} \sum_{y \in \omega} Y = \frac{1}{Ni} \sum_{x \in \omega} W^T X = W^T \mu i$$

In the LDA, the three scatter matrices are to be considered such as within-class M_w, between-class M_b, and mixture scatter matrices M_m.

$$M_b = \sum_{i=1}^{c} N_i (\mu_i - d')(\mu_i - d')^T = \emptyset_b \emptyset_b^\wedge T \tag{1}$$

$$M_w = \sum_{i=1}^{c} \sum_{j \in Ci} (d_j - \mu_i)(d_j - \mu_i)^T = \emptyset_w \emptyset_w^\wedge T \tag{2}$$

$$M_m = M_b + M_w = \emptyset_m \emptyset_m^\wedge T \tag{3}$$

N—number of all samples; N_i—number of samples in class; μ_i-mean of samples in the specific class; X'-mean of all samples.

To find an optimal transformation by Fisher, W satisfies

$$J(W) = \frac{W M_b W^T}{W M_w W^T} \tag{4}$$

According to Fisher's criterion, the separation between the classes has to be maximized and variance within the class is minimized. To find the maximum of $J(w)$, differentiate Eq. (4) and equate to zero.

$$\left[W M_w W^T \right] \frac{d}{dw} \left[W M_b W^T \right] - \left[W M_b W^T \right] \frac{d}{dw} \left[W M_w W^T \right] = 0 \tag{5}$$

Divide Eq. (5) with $W M_w W^T$, we get

$$M_b W - J M_w W = 0 \text{ i.e., } \frac{1}{M_w} S_b W - J W = 0 \tag{6}$$

For the generalized eigenvalue problem

$$W* = \arg \max \left\{ \frac{W S_b W^T}{W S_w W^T} \right\} \tag{7}$$

LDA classification is done based on the results, and the emotion of the person is predicted. But in LDA, whenever new features are added to the data it does not give the exact results, and this is considered as one of the disadvantages of the LDA. In the classification method, when we use LDA the scatter matrix within the class needs to be nonsingular and the main problem behind using the LDA is that it cannot deal with the small size samples (SSS). So, here Pseudoinverse Linear Discriminant Analysis (PLDA) [9] can be used to deal with the small size samples. In the PLDA, it finds the orientation matrix W with the help of computation on the eigenvalue decomposition (EVD). $M_w^+ M_b$ where M_w^+ is the pseudoinverse of M_w.

The W orientation is computed by finding the range space M_w. The range space within-class matrix is followed by the between-class scatter matrix M_b. Hence in the PLDA, the orientation W is worked out by finding the range space of total scatter matrix of M_T. Now the total scatter matrix is followed by the range space within-class scatter matrix of M_w. The advantage of PLDA method is eliminating the null spaces which help in improving the classification.

Let $X = \{d_1, d_2, \ldots d_n\}$ where $d_j \in R^d$, μ_j centroid of d_j, μ centroid of d. The between-class scatter matrix M_B is

$$M_B = \sum_{j=1}^{c} n_j (\mu_j - \mu)(\mu_j - \mu)^T \tag{8}$$

The within-class scatter matrix M_w is defined as

$$M_w = \sum_{j=1}^{c} \sum_{d \in d_j} (d - \mu_j)(d - \mu_j)^T \tag{9}$$

The total scatter matrix is M_T which is given as

$$M_T = \sum_{j=1}^{n} (d_j - \mu)(d_j - \mu)^T \tag{10}$$

The matrix M_T can also be formed as $M_T = XX^T$ which is a scatter matrix and where $X \in R^{d \times n}$

$$X = [(d_1 - \mu), (d_2 - \mu) \ldots (d_n - \mu)] \tag{11}$$

In a similar way, M_B can be formed as $M_B = YY^T$, where rectangular matrix $Y \in R^{d \times n}$ can be defined as

$$Y = [\sqrt{n_1}(\mu_1 - \mu), \sqrt{n_2}(\mu_2 - \mu) \ldots \sqrt{n_c}(\mu_c - \mu)] \tag{12}$$

Let the ranks of matrices M_T, M_B and M_w be a, b, and c, respectively. The W can be taken out by finding the range space of M_T followed by M_B, i.e. if eigenvalue decomposition of M_T is
$M_T = U_1^\wedge U_1^T$ where $U_1 \in R^{d \times n}$ which is corresponded to range space M_T, $^\wedge \in R^{d \times n}$ it is a diagonal matrix, then
$M_T' = U_1^T M_T U_1$ and $M_B' = U_1^T M_B U_1$ where the orientation matrix W can be found by finding EVD of $M_T^{'+} M_B^{'+}$

To find the range of M_T, we need to compute EVD of $XX^T A \in R^{d \times n}$; instead $M_T = XX^T \in R^{d \times n}$, this will reduce computational complexity.

$$X^T X = EDE^T \quad \text{or} \quad E_1 D_1 E_1^T \tag{13}$$

The orthonormal eigenvectors U_1 define M_T which is a range space and can be given as

$$U_1 = X E_1 D_1^{-1/2}$$

As we discard the null space M_T, we will not find any loss in the information. Use $U_1 \in R^{d \times n} t$ to transform the original dimensional space to a lower t-dimensional space. The matrices X and Y can be given in low-dimensional space.

$$\widehat{X} = D_1^{\frac{1}{2}} E_1^T \tag{14}$$

The matrix $\widehat{Y}$ can be obtained from $\widehat{X}$. With the transformed matrix $\widehat{X}$ as $[V_1, V_2, \ldots V_n]$

$$\widehat{Y} = \frac{1}{\sqrt{n_1}} \sum_{j=1}^{n_1} V_j, \frac{1}{\sqrt{n_2}} \sum_{j=n_1+1}^{n_1+n_2} V_j, \ldots, \frac{1}{\sqrt{n_c}} \sum_{j=n_1+n_2+\cdots+n_{c-1}+1}^{n} V_j \tag{15}$$

The orientation matrix W can be obtained as follows

$$W = X E_1 D_1^{\frac{-1}{2}} \widehat{W} \tag{16}$$

After getting the orientation matrices based on the results which are obtained, the emotions are going to be classified using LDA and pLDA.

5 Conclusion

The importance of IoT in emotion recognition in human life at different aspects is identified. The emotions in the different fields such as when the person is listening music, driving, speaking the emotions were detected using different classification methods by different machine learning techniques. The usage of different IoT devices for different emotion recognition is classified. Hence, it helps the user to know the status of the person, i.e., either he is in stress or relaxed, etc. This work can be further extended to the real-time health monitoring system in IoT.

References

1. S. M. Rizul Eslami, Daehanquack, MD.Humanu kabir.: The Internet of Things for Health Care: A Comprehensive Survey, IEEE Transaction, page: 678–708, June 1 2015.
2. C. Jr. Arcadius Tokognon, Bin Gao,: Structural Health Monitoring Framework Based on Internet of Things: A Survey, IEEE Internet of Things Journal, Volume: 4, No. 3, page: 619–635, June 2017.
3. Binbin Yong, Zijian Xu, Xin Wanga, Libin Cheng, Xue Li, Xiang Wu, Qingguo Zhou.: IoT-based intelligent fitness system, Elsevier Inc 2017.
4. Usman Raza, Parag Kulkarni, and Mahesh Sooriyabandara.: Low Power Wide Area Networks: An Overview, IEEE Wireless Communications Letters, page: 1–19,2017.
5. A. Haag, S. Goronzy, P. Schaich, and J. Williams,: Emotion Recognition Using Bio-Sensors: First Steps Towards an Automatic System, Proc. Ninth Int'l Conf. Reliable Software Technologies, page: 36–48, 2004.
6. Jonghwa Kim,: Emotion Recognition Based on Physiological Changes in Music Listening, IEEE Transitions on pattern analysis and machine intelligence, Volume: 30, No. 12, page: 2067–2083, December 2008.
7. Jennifer A. Healey and Rosalind W. Picard.: Detecting Stress During Real-World Driving Tasks Using Physiological Sensors, IEEE Transactions on Intelligent Transportation Systems, Volume: 6, Issue: 2, page: 156–166, June 2005.
8. Max Welling.: Fisher Linear Discriminant Analysis, www.ics.uci.edu. 2005.
9. Kuldipk. Paliwal and Alok Sharma.: Improved Pseudoinverse Linear Discriminant Analysis method for Dimensionality Reduction, International Journal of Pattern Recognition and Articial Intelligence Vol. 26, No. 1, 2012.
10. Philippe Arlotto, Michel Grimaldi, Roomila Naeck and Jean-Marc Ginoux.: An UltrasonicContactless Sensor for Breathing Monitoring, www.mdpi.com/journal/sensors, article on sensors ISSN 1424–8220.
11. Vlasios Tsiatsis, Catherine Mulligan, Stamatics Karnouskos, David Boyle.: Machint-to-Machine to the Internet of Things Introduction to a new age of Intelligence, Esilver Ltd, 2014.
12. Carlos Busso, Zhigang Deng, Serdar Yildirim, Murtaza Bulut, Chul Min Lee,Abe Kazemzadeh, Sungbok Lee, Ulrich Neumann, Shrikanth Narayanan.: Analysis of Emotion Recognition using Facial Expressions, Speech and Multimodal Information, 6th International Conference on Multimodal Interfaces, ICMI 2004.

KAZE Feature Based Passive Image Forgery Detection

D. Vaishnavi, G. N. Balaji and D. Mahalakshmi

Abstract Copy-move image forgery is a most common tampering artifact. It can be carried out by copy-pasting a region of the same image; thus, it has become a challenging one to find. So, this paper put forwarding a method to detect such forgery by extracting the KAZE features. RANSAC algorithm is functioned to get rid off the false matches such as outliers, and then the forged image is disclosed. The experiment is carried out using the publically available datasets, and their performances are quantitatively assessed using the true positive rate and false positive rate. A comparative analysis is also done with state-of-the-art methods, and it is certified that the proposed method produced good results than the other methods.

Keywords Copy-move forgery · KAZE · RANSAC · Clustering · TPR · FPR

1 Introduction

Image forgery has very much populated in our digital world. Since the images are also adequate in the fields such as an evidence in the court of law, news, and medical reports, image forgery has become a serious concern. This leads to establish the methods to discern such forgeries [1]. Image forgery is of three types: enhancement, splicing, and copy-move forgery. Among these, copy-move forgery is a demanding one to disclose, because it can be done by duplicating a part of an image and pasting

D. Vaishnavi (✉)
Department of CSE, Vardhaman College of Engineering, Hyderabad, Telangana, India
e-mail: vaishume11@gmail.com

G. N. Balaji
Department of IT, CVR College of Engineering, Hyderabad, Telangana, India
e-mail: balaji.gnb@gmail.com

D. Mahalakshmi
Department of IT, A.V.C. College of Engineering, Mayiladuthurai, Nagapattinam, Tamil Nadu, India
e-mail: dmahaa1985@gmail.com

© Springer Nature Singapore Pte Ltd. 2019   333
R. S. Bapi et al. (eds.), *First International Conference on Artificial Intelligence and Cognitive Computing* , Advances in Intelligent Systems and Computing 815,
https://doi.org/10.1007/978-981-13-1580-0_32

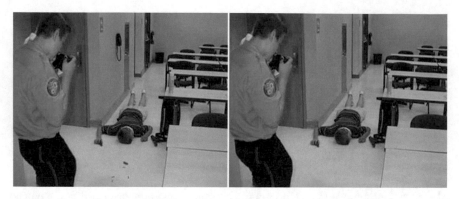

Fig. 1 Instance of copy-move forgery (original image on left and forged image on right)

it into some other portion of the same image. Detection of image forgery can fall into two: active approach and passive approach. The active approach requires a prior information to detect the forgery, but the passive approach does not [2–4]. Nowadays, passive approach is a most recent research area. Figure 1 shows an instance of digital image forgery, where one object is omitted in the forged image [5].

As the duplicate portion is made from the same image, it has analogous properties like illumination, texture, noise,. Thus, it makes us difficult to detect the forgery. There are numerous kinds of techniques in the literature, and it can be classified into block-based techniques and keypoint-based techniques.

In [6], scale-invariant feature transform (SIFT) is performed to obtain the feature descriptors and keypoints were matched by selecting the points which are far from the 11×11 pixel window to detect forgery. In [7], the authors have detected the forgery more effectively using the SIFT features by offering a g2NN test and the same features are used in the method [8] by improving the detection algorithm. The scheme in [9] extracted MPEG-7 visual invariant signatures. The method proposed in [10] utilized SURF algorithm to extract the features and best bin search algorithm to find the matching features. In [11], the keypoints of accelerated segment test (FAST) are identified and descriptors of binary robust independent elementary features are obtained to detect the forgery. The authors extracted the rotation-invariant DAISY descriptor to detect a cloned region [12]. The authors of [13] have improved the DAISY descriptors and extracted keypoints using the difference of Gaussian (DoG). The methods presented in [14, 15], extracted the SIFT features from approximate coefficients of discrete wavelet transform (DWT) and dyadic wavelet transform (DyWT), respectively. The method proposed in [16] extracted the contrast context histogram features and employed k-means clustering to detect the forgery. In [17], authors detected the forgery using the maximum stable extremal regions (MSER) features. This paper proposes a method to perceive image copy-move forgery using KAZE features.

2 Proposed Method

The proposed forgery detection method constitutes feature extraction, feature matching, and tamper detection steps. The detailed description of the same is illustrated in subsequent sections.

2.1 Feature Extraction

This work adopts KAZE algorithm to extract the feature points and to describe it. The normalized Hessian matrix is employed in nonlinear scale space to find local maxima points and elite its dominant orientation. The nonlinear scale space is formed with variable conductance diffusion and efficient additive operator splitting techniques [18]. Here, an evaluation of the luminance of an image with different scales is described as a divergence of flow function by nonlinear diffusion filter model [19]. The divergence is described by Eq. (1).

$$\frac{\partial L}{\partial t} = \text{div}(c(x, y, t) \cdot \nabla L) \tag{1}$$

where L denotes luminance of the image, ∇ represents gradient operators, div represents divergence, and $c(x, y, t)$ denotes the flow function and is given by Eq. (2).

$$c(x, y, t) = g(|\nabla L_\sigma(x, y, t)|) \tag{2}$$

where ∇L_σ refers gradient of L_σ and g is conductivity function,

$$g = \frac{1}{1 + \frac{|\nabla L_\sigma|^2}{k^2}}$$

where k refers control factor (diffusion level) and it is correlated with edge information. The greater value of k leads to a smaller amount of edge information. So, an empirical value for gradient histogram ∇L_σ is chosen as 70%.

To construct the nonlinear scale space, the octave and sub-levels are mapped to their respective scale σ by the given formulation.

$$\sigma_i(o, s) = \sigma_0 2^{0 + \frac{s}{S}}$$

where $i \in [0, \ldots, N]$, $o \in [0, \ldots, O - 1]$, $s \in [0, \ldots, S - 1]$ and N refers to a number of filtered images. Then, the feature points are extracted with Hessian matrix by Eq. (3).

$$L_{Hessian} = \sigma^2(L_{xx}L_{yy} - L_{xy}^2) \tag{3}$$

where L_{xx}, L_{yy}, L_{xy} are second-order derivatives and σ is an integer value of scale factor σ_i.

To obtain descriptors as rotation invariant, it is necessary to compute the dominant orientation for the feature points. So, a circular area covering the angle of $\pi/3$ is constructed around the keypoints, then derivative responses are characterized as points, and then dominant orientation is calculated by summing all responses contained by a sliding circle segment and selecting the longest vector as dominant orientation [4].

First-order derivatives Lx and Ly are computed for the scale factor σ_i of feature point over a rectangular window. Then, the window is isolated into 4×4 subregions. Each subregion's derivative responses are weighted with the Gaussian kernel and summed up as descriptors.

$$v_{\text{subregion}} = \left[\sum L_x, \sum L_y, \sum |L_x|, \sum |L_y| \right] \tag{4}$$

Subsequently, each sample in the grid is rotated by means of determined dominant orientation and each subregion is again weighted with Gaussian window to compute the derivative responses. Finally, $4 \times 4 \times 4 = 64$-sized normalized descriptor of features is attained.

2.2 Feature Matching

Feature matching can be done by determining the ratio between the nearest neighbors, and it is considered as initial matches whose ratio is less than that of the ratio threshold $T1$. Further, the final matches are stored if a distance between the neighbors is greater than the distance threshold $T2$. Then, the hierarchical clustering is applied to isolate the points of pasted and copied region using 'ward' linkage method.

2.3 Tamper Detection

False matches may lead to misinterpretation on the genuineness of an image, and it directs to inaccurate evaluation of transformation. Therefore, to avoid such false matches random sampling consensus (RANSAC) [20] has been employed. This algorithm first selects two or more matched points at random and computes the homography matrix. The rest of the points are transformed with respect to homography matrix and are analyzed by calculating the distance about corresponding matching points. Then, it is certified either outliers or inliers depend on the distance that lies over or below a threshold. It is continued for N number of times until the maximum possible number of inliers found. In our experiment, the value of N is fixed to 500 and distance threshold for 0.05.

3 Experimental Results

To quantify the performance of the proposed system, tampered images and original images are adopted from MICC-F220 and MICC-F2000 databases, where MICC-F220 dataset has 110 original and 110 tampered images and MICC-F2000 dataset has 700 tampered and 1300 original images of 2048 pixels × 1536 pixels. The size of forged parts of this dataset covers on average 1.12% of an image. This proposed system is evaluated on the basis of false positive rate (FPR) and true positive rate (TPR) and is given in Eqs. (5) and (6).

$$\text{TPR in } \% = \frac{\text{\# tampered images correctly detected}}{\text{\# tampered images}} \tag{5}$$

$$\text{FPR in } \% = \frac{\text{\# orignal images detected as tampered}}{\text{\# original images}} \tag{6}$$

The sample result of tamper detection is exhibited in Fig. 2, in which Fig. 2a gives an original image. Figure 2b shows tampered image which is forged by copying and pasting the photograph frame in the region. Figure 2c shows matched features that are detected by the proposed method, and Fig. 2d provides the final result of detected forged image.

To quantify the performance of the proposed system, the experiment is carried out on MICC-F220 database and it produced 89.09% of TPR and 9.09% of FPR. The results of the proposed system are analyzed with previous copy-move forgery detection methods [13–15, 21] and are given in Table 1. The proposed method attained 9.09% of FPR, which is a considerable result when comparing with the result of method DyWT + SIFT [14]. However, it secured a good result of 89.09% of TPR which is superior to other methods.

The experiment is also carried out on the MICC-F2000 dataset, and performance is evaluated using TPR and FPR measures. The proposed system achieved 6.92% of FPR and 92.85% of TPR. These results are analyzed with the existing systems, and its results are given in Table 2. It furnishes that the performance of the proposed method is well improved in terms of FPR when compared with all the methods, and it is improved to some extent in terms of TPR compared to the Areej et al. [17].

Table 1 Performance comparison with previous methods on MICC-F220 dataset

Methods	FPR %	TPR %
Proposed method	9.09	89.09
DOG + DAISY [13]	9.09	85.91
DyWT + SIFT [14]	10.00	80.00
DyWT [21]	4.00	74.00
DWT + SIFT [15]	2.00	66.00

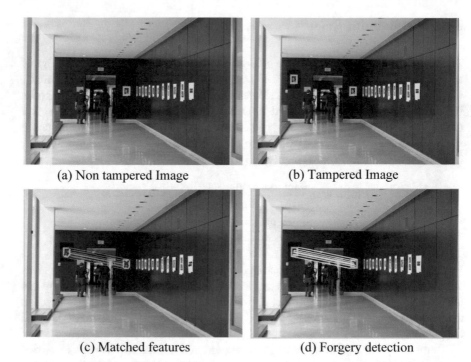

(a) Non tampered Image   (b) Tampered Image

(c) Matched features   (d) Forgery detection

Fig. 2 Result of proposed method on MICC-F220

Table 2 Performance comparison with previous methods on MICC-F2000 dataset

Methods	FPR %	TPR %
Proposed method	6.92	92.85
Areej et al. [17]	8.00	92.00
Amerini et al. [8]	9.15	94.86

4 Conclusion

A method to detect copy-move image forgery is proposed by adopting the KAZE features. And also, RANSAC and clustering algorithms are utilized to eradicate the outliers and identify the forged image. The experiment is carried out using the publically available datasets, namely MICC-F220 and MICC-F2000. Also, their performances are quantitatively assessed using the true positive rate and false positive rate. It produced a result of 89.09% TPR and 9.09% FPR on MICC-F220 dataset. Also, it attained the results 6.92 % and 92.85% of FPR and TPR respectively on MICC-F2000. However, the proposed method achieved a somewhat lower result in terms of FPR, it produced 0.85% superior results in terms of TPR than the other methods. This can be considered in future, and method may be developed to overcome this issue.

Acknowledgements The authors express their gratitude and credits for the use of the MICC-F220 and MICC-F2000 databases.

References

1. A. J. Fridrich, B. D. Soukal, and A. J. Luk, "Detection of copy-move forgery in digital images," in *in Proceedings of Digital Forensic Research Workshop*, 2003.
2. D. Vaishnavi and T. Subashini, "Image Tamper Detection Based on Edge Image and Chaotic Arnold Map," *Indian Journal of Science and Technology*, vol. 8, no. 6, pp. 548–555, 2015.
3. D. Vaishnavi and T. Subashini, "Fragile Watermarking Scheme Based on Wavelet Edge Features," *Journal of Electrical Engineering & Technology*, vol. 10, no. 5, pp. 2149–2154, 2015.
4. F. Yang, J. Li, W. Lu, and J. Weng, "Copy-move forgery detection based on hybrid features," *Engineering Applications of Artificial Intelligence*, vol. 59, pp. 73–83, 2017.
5. M. Puri and V. Chopra, "A survey: Copy-Move forgery detection methods," *International journal of computer systems*, vol. 3, 2016.
6. X. Pan and S. Lyu, "Region duplication detection using image feature matching," *Information Forensics and Security, IEEE Transactions on*, vol. 5, no. 4, pp. 857–867, 2010.
7. I. Amerini, L. Ballan, R. Caldelli, A. Del Bimbo, and G. Serra, "A sift-based forensic method for copy–move attack detection and transformation recovery," *Information Forensics and Security, IEEE Transactions on*, vol. 6, no. 3, pp. 1099–1110, 2011.
8. I. Amerini, L. Ballan, R. Caldelli, A. Del Bimbo, L. Del Tongo, and G. Serra, "Copy-move forgery detection and localization by means of robust clustering with J-linkage," *Signal Processing: Image Communication*, vol. 28, no. 6, pp. 659–669, 2013.
9. P. Kakar and N. Sudha, "Exposing postprocessed copy paste forgeries through transform invariant features," *Information Forensics and Security, IEEE Transactions on*, vol. 7, no. 3, pp. 1018–1028, 2012.
10. P. Mishra, N. Mishra, S. Sharma, and R. Patel, "Region Duplication Forgery Detection Technique Based on SURF and HAC," *The Scientific World Journal*, vol. 2013, 2013.
11. Y. Zhu, X. Shen, and H. Chen, "Copy-move forgery detection based on scaled ORB," *Multimedia Tools and Applications*, pp. 1–13, 2015.
12. J.-M. Guo, Y.-F. Liu, and Z.-J. Wu, "Duplication forgery detection using improved DAISY descriptor," *Expert Systems with Applications*, vol. 40, no. 2, pp. 707–714, 2013.
13. P. L. Jiming ZHENG, "Detection of Copy-move Forgery in Digital Image using DAISY Descriptor," *Journal of Computational Information Systems*, vol. 10, pp. 9369–9377, 2014.
14. V. Anand, M. F. Hashmi, and A. G. Keskar, "A copy move forgery detection to overcome sustained attacks using dyadic wavelet transform and sift methods," in *Intelligent Information and Database Systems*, Springer, 2014, pp. 530–542.
15. M. F. Hashmi, A. R. Hambarde, and A. G. Keskar, "Copy move forgery detection using DWT and SIFT features," in *Intelligent Systems Design and Applications (ISDA), 2013 13th International Conference on*, 2013, pp. 188–193.
16. D. Vaishnavi and T. Subashini, "A passive technique for image forgery detection using contrast context histogram features," *International Journal of Electronic Security and Digital Forensics*, vol. 7, no. 3, pp. 278–289, 2015.
17. A. S. Alfraih, J. A. Briffa, and S. Wesemeyer, "Cloning localization based on feature extraction and k-means clustering," in *Digital-Forensics and Watermarking*, Springer, 2014, pp. 410–419.
18. J. Weickert, B. T. H. Romeny, and M. A. Viergever, "Efficient and reliable schemes for nonlinear diffusion filtering," *IEEE transactions on image processing*, vol. 7, no. 3, pp. 398–410, 1998.

19. P. Perona and J. Malik, "Scale-space and edge detection using anisotropic diffusion," *IEEE Transactions on pattern analysis and machine intelligence*, vol. 12, no. 7, pp. 629–639, 1990.
20. M. A. Fischler and R. C. Bolles, "Random sample consensus: a paradigm for model fitting with applications to image analysis and automated cartography," *Communications of the ACM*, vol. 24, no. 6, pp. 381–395, 1981.
21. G. Muhammad, M. Hussain, and G. Bebis, "Passive copy move image forgery detection using undecimated dyadic wavelet transform," *Digital Investigation*, vol. 9, no. 1, pp. 49–57, 2012.

Efficiency-Based Analysis of Homomorphic Encryption Implications

Gouthami Velakanti and P. Niranjan

Abstract Homomorphic-based encryption incorporates authenticated encryption techniques which permits for computational principles generated on encrypted data with exclusive of necessitate execution of the decryption-based key. The traditional encryption schemes are used to provide the private environment for outsourcing the data storage to the third party end users, which will not work out for proving high-level security for avoiding data leakage-based uncertainties while the data is in different modes and also especially when the data without initially decrypting it. Huge data loss can occur with the latest TLS and SSH based attacks with lack of proper implementation of encryption methods with concern precautions. There are some other complex problems by using the traditional computational techniques such as privacy is concern when processing the confidential data through the third party providers; integrity of confidential data to facilitate the secure access and association from the illegal execution of entire data and secreted unofficial accessibility. Homomorphic-based hybrid encryption will overcome these complexities and perform the encryption-based derivations by exclusive of initial decrypting the data. The numerous dedicated environments such as searchable-based encryption, deterministic-based encryption, order-preserving-based encryption, partial and fully homomorphic-based encryptions consent to precise group of computations to be derived on encrypted data for implications in real life circumstances for real-life circumstances. This paper describes the hybrid-based encrypted approaches, methodology implications, applications and lagging issues and focuses on the efficiency analysis of homomorphism encryption implications along with how homomorphic encryption-based algorithms have composed for realistic complex problems and described the analysis of past designed homomorphic-based encryption approaches with their implications and advances.

G. Velakanti (✉) · P. Niranjan
Department of Computer Science and Engineering,
Kakatiya Institute of Technology and Sciences, Kakatiya University,
Warangal, Telangana, India
e-mail: gautami.velakanti@gmail.com

P. Niranjan
e-mail: npolala@yahoo.com

© Springer Nature Singapore Pte Ltd. 2019   341
R. S. Bapi et al. (eds.), *First International Conference on Artificial Intelligence and Cognitive Computing* , Advances in Intelligent Systems and Computing 815,
https://doi.org/10.1007/978-981-13-1580-0_33

Keywords Cloud computational-based encryption (CCbE) · Homomorphic-based encryption (HbE) · Fully homomorphic-based encryption (FHbE)
Searchable-based encryption (SbE) · Deterministic-based encryption (DbE)
Order-preserving-based encryption (OPbE) · Internet of things (IoT)
Multi-authority attribute-based encryption (MAAbE) · Simple encrypted
arithmetic library (SEAL)

1 Introduction

Homomorphic-based encryption (HbE) consents to build the precise procedure on confidential data holding by encrypted, it intended to facilitate the computations in the encrypted-based field, and it is an active important invention for the extensive and fast-rising variety of environments such as cloud computing (CC), cyber security (CS), distributed sensor (DS) fields, and Internet of things (IoT). While applications such as cloud computing require to have a practical solution, the encryption scheme must be secure [1]. The HbE incorporates authenticated encryption techniques which permit for computational principles generated on encrypted data with exclusive of necessitate execution of the decryption-based key. It best fits to protect the cloud based data; it derives the scheme which used to involve the software to active on data exclusive of decrypting it [2]. Homomorphic-based encryption is the best active technologies derived from move ahead in cryptographic-based research [3]. It incorporates authenticated encryption techniques which permit for computational principles generated on encrypted data with exclusive of necessitate execution of the decryption-based key. It best fit to protect the cloud based data; it derives the scheme which used to involve the software to active on data exclusive of decrypting it [4]. In current technology, the numerous approaches are proposed based on the partial and fully homomorphic based encryption environment, excluding to date the storage and time complexity of the related schemes has not allowed in their usage in the experimental practical motto [5]. In present scenario, secure cloud computing implementation is based on the fully homomorphic-based encryption (FHbE) protection which is the innovative idea. In this methodology, the end user client encrypt the data using client-based private key and that formed encrypted data is received by the concern server. Then, the server performs the required operations without decrypting that secure-data and sends result back to the end user client. Finally, the end user client decrypts that secure-data and gets the result. Security problem is overcome through this FHbE algorithm. Data privacy is supervised by implementing this algorithm [6]. In recent days, there is an interesting innovation thing upcoming in cryptography implications such as homomorphic encryption scheme appeared recently. It is a way of encoding p into $f(p)$ such that you can compute $f(p + q)$ easily knowing $f(p)$ and $f(q)$ even though you cannot easily restore p and q, and it will give the similar meaning for $f(p*q)$ [7]. Identifying the Complex Problems in cloud based computation of confidential data, the traditional encryption schemes are used to provide the private environment for outsourcing the data storage to the third party end users [4]; it will not work out for proving high-level security for avoiding data leakage-based uncertainties while the data are in different modes and also especially when the data without initially decrypting it. The huge data loss can be happen with the latest TLS and SSH based attacks lack of proper implementation of encryption methods with concern precautions. There are some complex problems by using the traditional computational techniques such as, privacy is concern when processing the confidential data through the third party providers; integrity of confidential data; for facilitate the secure access and association from the illegal execution of entire data and secreted unofficial accessibility.

(1) Privacy is concern when processing the confidential data through the third party provider
(2) Integrity of confidential data
(3) For facilitate the secure access and association from the illegal execution of entire data
(4) Secreted unofficial accessibility.

2 Scope of the Paper

Homomorphic-based hybrid encryption will overcome these complexities and perform the encryption-based derivations by exclusive of initial decrypting the data. The computational results stay behind encrypted, only to be read and understand by someone whose are having access to decryption key. The FHbE derives methods that consent to random computations on encrypted based data, to defeat a quantity of performance problems. This paper proposes three important objectives: First one, it described the hybrid-based encrypted approaches, methodology implications, applications, and lagging issues. Second one is the efficiency analysis of homomorphism encryption implications along with how homomorphic encryption based algorithms have composed for realistic complex problems, and the third one is descriptive analysis of past designed homomorphic-based encryption approaches with their implications and advances. And also, this paper shows the clear technical paths for upcoming new researchers to solve the complex problems to reduce the cyber security threads like as SSH and TLS attacks.

3 Traditional and Hybrid-Based Encrypted Methods, Methodology Implications, and Applications

The numerous dedicated environments such as searchable-based encryption, deterministic-based encryption, order-preserving-based encryption, partial and fully homomorphic-based encryptions consent to precise group of computations to be derived on encrypted data for implications in real life circumstances [8]. New innovation computational techniques can be generated and implicated for cloud by combining these encryption approaches. This combinational-based computational environment utilized a constrained group of functionalities with authenticated principles for generating the best securable outcome responses.

4 Present Scenarios of Homomorphic Encryption Implications

The HbE executes an excessively critical based algorithm which permits this complexity to workout such as medical based systems like as MRI machines. The most costly component of the MRI-based machine is treated as the algorithmic sequences in which the machine can analyze the magnetic-based resonance data. HbE permits this to implicate where the patient data along with the algorithm are together to be protected. The FHbE (i.e., inducing a ring homomorphism onto the encrypted data) permits for a good deal of proficient and vigorous set of computations to activate on the private data [7]. The Microsoft Corporation introduced and recently released its own homomorphic encryption library the Simple Encrypted Arithmetic Library (SEAL). The primary design standard of SEAL is to present the accessibility to the core-based concepts of HbE in one well-designed package with no exterior dependencies. The library is intended to be so easy to use that people with no prior experience in homomorphic encryption can start using it in their work [4]. And also, there are many new innovation experimentations are going to solve the complex problems in various fields like as auto tune-up of electronic voting system, MRI diagnosis-based systems (Table 1).

5 Past Designed Methodologies and Their Implications

The HbE method can be formed as asymmetric, symmetric, and hybrid based on input key, and also, HbE can be mostly classified into partial homomorphic-based encryption (PHbE) and fully homomorphic-based encryption (FHE). PHbE executes only with restricted numeral of functionalities such as implementation of either addition or multiplication on encrypted private data. But the FHbE executes both addition and subtraction functionalities on encrypted private data as many as times. In HbE, the encryption scheme can be probabilistic or deterministic based on the produce ciphertext type. Every time a plaintext can be encrypted along with outcomes which are in different ciphertext in probabilistic type of encryption, but in deterministic encryption, It can always be the outcome of the similar type of ciphertext as shown in Fig. 1. Efficiency of HE schemes is measured as good as if it can be produce the small size of ciphertext and measures the time considered to execute the decryption, encryption or recryption procedure is less. The several types of well-known PHE meth-

Table 1 Hybrid-based encrypted approaches, methodology implications, and applications

Hybrid-based encrypted methods	Purpose and usage	Applications	Advantages	Lagging issues
Homomorphic-based encryption	This approach performs technical functionality-based operations directly on encrypted privacy data exclusive access of data when it is in decrypted form [8], and it can build out the computations on encrypted privacy data. These services are utilized for defend secrecy in cloud computation and when the cloud analytics as a service [11, 12]	This environment derives the data privacy and protection once the data is encrypted in the end user client's network, and also, it permits the cloud-based service provider to execute the analytics feature actively on the encrypted data. This entire process to be outsourced in both of data analysis along with the storage to the cloud-based service provider	Through this approach, the end-user-client-based huge datasets like network-based operators, health-care providers, and mechanical process to engineering industry-based players are to be executed and analyzed. The FHbE and PHbE variations vary in the numeral of processes to be executed on encrypted based data	The main and important lagging issue of using HbE schemes in cloud is its involvement of complex-based mathematical constructions. It will take more processing time for processing the mathematical constructive-based encryption and decryption activities [10]
Identity-based encryption (IbE)	IbE derives and permits an end-point along the public-key of another end-point commencing a given identity	In this approach, anybody can send encrypted private data to the possessor of the e-mail address by using an e-mail address as a public-key	It contains the capability to decrypt the content which will lie along with the entity process in control of the matching hidden-based private or secret key of the creator of the e-mail address as long as the name space is appropriately maintained	It permits an end-point along the public-key of another end-point [13]

(continued)

Table 1 (continued)

Hybrid-based encrypted methods	Purpose and usage	Applications	Advantages	Lagging issues
Attribute-based encryption (AbE) variations [13] (1) Key policy attribute-based encryption(KP-AbE), (2) ciphertext policy attribute-based encryption, (3) attribute-based encryption scheme with non-monotonic access structures, (4) hierarchical attribute-based encryption, and (5) multi-authority attribute-based encryption(MAAbE)	AbE derives the functionalities of attribute-based encoding mechanism for further process of privacy data such as responsibility-based or access-based policies along with the end-user-based secret or private-confidential keys [8]	In this approach, the attributes are used to subjugate to produce a public-key for encrypting the data and utilized as an access-based policy to manage the user-based accessibility. The access-based policy can form as ciphertext-based policy: the access structure on the ciphertext or key- based policy: the access structure on the user's private key. The access-based structure characterized as monotonic version based or non-monotonic version based one [14] It is a good quality for community protection-based applications and used in the 3GPP standards for proximity-based services for LTE [8]	This approach derives the advantages such as to decrease the Internet-based communication transparency and to provide a fine-grained accessibility control [14] IbE and AbE permit user end-points without having network accessibility connections to prepare the protected and genuine machine-to-machine contact channels	The AbE is still holds the lagging issues such as: the data original owner wants to utilize the each official end-users public-key to encrypt the private data and the related applications are restricted in the real-life environment to be reason it use the access of monotonic based attributes to control end-users based accessibility in the machine. And also, it attempts to describe policies like user-based training and technical-based functioning efforts

(continued)

Table 1 (continued)

Hybrid-based encrypted methods	Purpose and usage	Applications	Advantages	Lagging issues
Post-quantum based cryptography (PbC)	The PbC has possibilities that may have theatrical consequences for cryptographic-based algorithms and their capability to preserve the information security	Through this approach, the assault-based algorithms are previously made up to ready for quantum-based computers to perform on it	The advantage of quantum cryptography lies in the fact that it allows the completion of various cryptographic tasks that are proven or conjectured to be impossible using only classical communication [15]	Security is a chain; it is as strong as the weakest link. Mathematical cryptography, as bad as it sometimes is, is the strongest link in most security chains
Deterministic-based encryption (DbE)	DbE is opposite and different to randomized derivation of probabilistic-based encryption approach [16], and it is a special cryptosystem which constantly generates the same ciphertext for the same plaintext even though over the part of executions of the encryption-based algorithm such as the RSA-based cryptosystem which consists the exclusive of encryption padding and a lot of block-ciphers when used in ECB form or else along with the regular initialization vector [17]	It allows the logarithmic time search mechanism on encrypted-based private data, and it permits linear time search when randomized encryption has been processed. It will apply the scanning operation to scan the total entire database [11]	This dissimilarity is vital for large outsourced databases which cannot afford to slowdown search [10]	This type of encryption will permit the server to execute higher equality checks; it can execute the selects with equality predicates, equality joins, GROUP BY, COUNT, DISTINCT, and so on [3]

(continued)

Table 1 (continued)

Hybrid-based encrypted methods	Purpose and usage	Applications	Advantages	Lagging issues
Order-preserving-based encryption(OPbE)	In this OPbE, the plaintext-based area's nth element is to be mapped, so that it can form the order between plaintexts is preserved between ciphertexts [18]	OPbE permits to perform the order-based assessments to guarantee that the ciphertexts keep the order recognized in between plaintexts. In this, the SQL-based range queries can executed well once the field has encrypted in this [19] Based on this OPeB, the server can also perform ORDER BY, MIN, MAX, SORT, and so on [16]	OPeE deriving the advantages like as: it boosts the search-based operational speed, and it permits to perform the order-based comparisons over the encrypted data while preserving the realistic level of protection. And also, it consists of some addition advantages [11] such as provide security-based provision; it permits the indexing mechanism And also, it permits the order and comparison relation on encrypted private database-based on range-based equality, MAX, MIN, and COUNT-based queries, the operations based on ORDER BY and GROUP BY clauses. These clauses are directly executed on encrypted private data And also, it increases the performance of the search operations	OPeE holds some lagging issues such as the leakage of plaintext-based information orders, and also, it will not execute the AVG or SUM operations to a specified group of values until it is decrypted [11]

(continued)

Table 1 (continued)

Hybrid-based encrypted methods	Purpose and usage	Applications	Advantages	Lagging issues
Searchable-based encryption (SbE)	In this SbE, the authenticated users will able to perform the private keyword search on encrypted private data in cloud For performing the SE algorithms, formulate efficient, secured over the encrypted files by the incorporation of the multi-platform-based fields such as indexing, cryptography, storage. The data user, owner, and cloud server are the active contributors of a secure-based search model in a cloud. Initially, the data owner will encrypts the correlated files and parallel keyword-based index files by using any known cryptographic algorithms, and then, the both encrypted privacy files and index-based files are uploaded into the cloud-based Web server	In this approach, based on type of the data, the following characteristic operations can be performed such as retrieval procedure query numbers and their type, result, numeral of members, proof-based security models, performance, encryption models, search types, evaluation parameters. SE algorithms have been designed and categorized based on their variations. In private key method, the same key is used for both encryption and keyword-based search. In public key scheme, different keys are used for encryption and keyword-based search, and also, the public key method is further classified based on keyword search types like conjunctive keyword search, fuzzy keyword search	SbE preserves the secrecy and privacy of data creators by make possible through the searching keywords directly on encrypted-based private data. [10] SbE is the superior trade-off between security and effectiveness	SbE holds the lagging issues, such as it does not suitable for fast parallel and I/O efficiency keyword search performance when compared to HbE, FbE, FHbE, and ORAM And also, the main lagging issue is the information leakage can occur by retrieving the adversaries [10]

Table 2 Past Designed Homomorphic based Encryption Methodologies and their Implications

Authors	Methodology	Analysis of methodology implications	Advantages and outcomes
Sonia Bogos, John Gaspoz and Serge Vaudenay [1]	Cryptanalysis of a homomorphic encryption scheme	This methodology describes the detail information and analyzes in-depth idea of HE design which has been projected by Zhou and Wornell. It describes and mounts the three attacks processes. The first attack is facilitated to recover a secret plaintext message transmitted to several end-users. The second attack executes a selected ciphertext-key recovery-based attack; it was executed and verified. The last attack is a related to selected plaintext based-decryption attack. It described three different strike-based attacks: First is the method projected encryption for transmitted messages, to support data sharing. It is stating that an attacker that snoops on broadcast traffic would get the sufficient information to resolve the system. Second is strike-based attacks: In the selected ciphertext attack, they write, an attacker with access to an oracle that decrypts the text can run a brute force to recover the encryption key. Third: As with the chosen ciphertext attack, the plaintext attack is a successful brute force against the encryption [1, 2]	It has revealed the attacks against the system broadcast encryption such as chosen ciphertext attack and a plaintext attack [1]
Hongchao Zhou, Gregory Wornell [5]	Efficient homomorphic encryption on integer vectors and its applications	It displays that the realistic homomorphic encryption methods are potential when we necessitate that not all encrypted computations be carry, but quite only those of attention to the target application. It describes that how to develop a homomorphic encryption method functioning straight on numeral vectors that supports three processes of primary attention in signal processing-based applications such as linear transformation, addition, and weighted inner products	This approach deriving some futures such as when used in combination; these primitives allow us to efficiently and strongly calculate the random polynomials [5]

(continued)

Table 2 (continued)

Authors	Methodology	Analysis of methodology implications	Advantages and outcomes
Lucas Barthelemy [20]	Cryptography-FHE privacy. A brief survey of Fully Homomorphic Encryption, computing on encrypted data	This methodology describes the clear idea about the FHbE. It stated that the FHbE is a gifted field in cryptography implications by deriving with attractive possessions. And also, it is still fairly partial regarding its computation abilities. It is transform complex applications for future problems	This approach gave the clear idea of cryptographic tools which are destined to be automated through the circuit implementations
Yi, Xun, Paulet, Russell, Bertino, Elisa	Fundamental concepts of homomorphic encryption	In this methodology, the applications are designed in the areas of secret information-based retrieval, secret searching on streaming-based data, privacy-preserving data mining, electronic-based voting, and cloud-based computing.	Through this approach, the content is presented in an instructional and practical style along with suitable examples to improve the researchers accepting
Ms. Parin. V. Patel, Mr Hitesh D Patel [6]	A survey of the homomorphic encryption approach for data security in cloud computing	HbC will represent the security in cloud computing. HbC method executes functionalities on encrypted privacy-based data; it will generate outcomes exclusive of decrypting that private data. And also, it presents the similar result as process performs on row data	Through this approach, the security-based problems can be overcome through this HbC algorithm, and the data privacy is controlled by this HbC algorithm. In this, the RSA algorithm is implicated
Prasanna B T, C B Akki [10]	A comparative study of homomorphic and searchable encryption schemes for cloud computing	A comparative study of these two efficient cloud cryptographic methods has been carried out	Detailed comparison description about the homomorphic-based and searchable-based encryption methods with their type of approaches, implications, and derivations

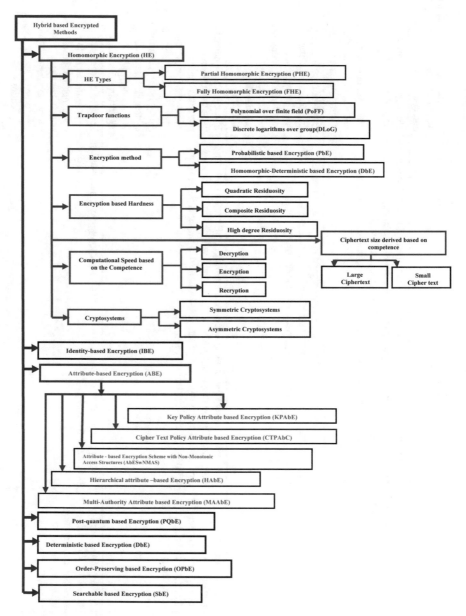

Fig. 1 Hybrid-based encrypted methods along with their types, sub-categories, and their implications

ods contain the unpadded-based RSA, ElGamal, Goldwasser-Micali, Benaloh, Paillier, Okamoto-Uchiyama, Naccache-Stern, Damgard-Jurik and Boneh-Goh-Nissim etc. Some of the known FHE methods are Gentry, Van Dijk, Smart-Vercauteren, Stehle-Steinfield, Ogura, Lyubashevsky, Gentry-Halevi, Brakerski-Vaikuntanathan,

Brakerski-Gentry-Vaikuntanathan and Chunsheng, etc. Prasanna and Akki [9] in their survey paper discussed different known PHE and FHE schemes along with their limitations and benefits [10] (Table 2).

6 Conclusion

The homomorphic-based encryption (HbE) consents to build the precise procedure on confidential dataholding by encryption; it is intended to facilitate the hybrid security key roles in the encrypted-based field, and it is an active important invention for the extensive and fast rising variety of environments such as cloud computing (CC), cyber security (CS), distributed sensor (DS) fields, and Internet of things (IoT). The numerous dedicated environments such as searchable-based encryption, deterministic-based encryption, order-preserving-based encryption, partial and fully homomorphic-based encryptions consent to precise group of computations to be derived on encrypted data for implications in real life circumstances. This paper has described the hybrid based encrypted approaches, methodology implications, applications and lagging issues. Here this paper mainly focuses on the efficiency analysis of homomorphism encryption implications along with how homomorphic encryption based algorithms has composed for realistic complex problems and described the analysis of past designed homomorphic based encryption approaches with their implications and advances. This paper shows the clear technical paths for upcoming new researchers to solve the complex problems to reduce the cyber security threads like SSH and TLS attacks.

References

1. Cryptography, "In an all Encrypted worldcharting the future of innovation", Ericsson Technology Review Cryptography in and all Encrypted world, Volume 92, 22nd December, 2015.
2. Researchers crack homomorphic encryption. https://www.theregister.co.uk/2016/08/16/researchers_crack_homomorphic_encryption/.
3. Proceedings of the 23rd ACM, 2011, CryptDB: "Protecting confidentiality with encrypted query processing," abstract available at: http://dl.acm.org/citation.cfm?id=2043566.
4. Yi, Xun, Paulet, Russell, Bertino, Elisa "Homomorphic Encryption and Applications".
5. Lucas Barthelemy, "Cryptography-FHE privacyA brief survey of Fully Homomorphic Encryption, computing on encrypted data", 29th June 2016.
6. Ms. Parin. V. Patel, Mr Hitesh D Patel, "A Survey of the Homomorphic Encryption Approach for Data Security in Cloud Computing", International Journal of Engineering Development and Research(IJEDR), 2013, ISSN: 2321-9939.
7. http://stackoverflow.com/questions/1023981/practical-applications-of-homomorphic-encryption-algorithms.
8. Hongchao Zhou, Gregory Wornell, "Efficient homomorphic encryption on integer vectors and its applications", Information Theory and Applications Workshop (ITA), 2014, Date of Conference: 9–14 Feb. 2014. https://doi.org/10.1109/ita.2014.6804228. IEEE Xplore: 24 April 2014.

9. Prasanna B.T, C.B. Akki, A Survey on Homomorphic and Searchable Encryption Security Algorithms for Cloud Computing. Communicated to Journal of Interconnection Networks, April 2014.
10. Prasanna B T, C B Akki, "A Comparative Study of Homomorphic and Searchable Encryption Schemes for Cloud Computing".
11. Nasrin Dalil, Ahmed Kayed, "Preserving Data in Cloud Computing", IJCSI International Journal of Computer Science Issues, Volume 12, Issue 2, March 2015, ISSN (Print): 1694-0814| ISSN (Online): 1694–0784.
12. Research Directorate staff, "Securing the cloud with Homomorphic encryption", The Next Wave, 2014, Vol. 20 No. 3.
13. Minu George, Dr. C.Suresh Gnanadhas, Saranya. K, "A Survey on Attribute Based Encryption Scheme in Cloud Computing", International Journal of Advanced Research in Computer and Communication Engineering, Vol. 2, Issue 11, November 2013, ISSN (Print): 2319-5940 ISSN (Online): 2278–1021.
14. Cheng-Chi Lee, Pei-Shan Chung and Min-Shiang Hwang, "A Survey on Attribute-based Encryption Schemes of Access Control in Cloud Environments, International Journal of Network Security, Vol. 15, No. 4, PP. 231–240, July 2013.
15. Quantum Cryptography, http://www.whitec0de.com/quantum-cryptography/.
16. Raluca Ada Popa, Catherine M. S. Redfield, Nickolai Zeldovich, and Hari Balakrishnan, "CryptDB: Protecting Confidentiality with Encrypted Query Processing ", In: Proceedings of the 23rd ACM Symposium on Operating Systems Principles (SOSP), 2011.
17. http://www.Wikipedia.com/Deterministic encryption, May 2014.
18. Dongxi Liu and Shenlu Wang, "Programmable Order-Preserving Secure Index for Encrypted Database Query", Proceeding in IEEE Fifth International Conference on Cloud Computing, 2012.
19. Santi Martınez, Josep M. Miret, Rosana Tom'as and Magda Valls, "Security Analysis of Order Preserving Symmetric Cryptography", Applied Mathematics & Information Sciences, 2013, Vol. 7, No. 4.
20. https://www.microsoft.com/en-us/research/project/homomorphic-encryption/.   Yi,   Xun, Paulet, Russell, Bertino, Elisa "Homomorphic Encryption and Applications".

Performance and Analysis of Human Attention Using Single-Channel Wireless EEG Sensor for Medical Application

Sravanth Kumar Ramakuri, Anudeep Peddi, K. S. Nishanth Rao,
Bharat Gupta and Sanchita Ghosh

Abstract Specialists are dealing with the improvement of EEG-based human–computer interface for upgrading the personal satisfaction in restorative and additionally non-medicinal applications using the blink of eyes. Such innovation can be consolidated to brain science, anesthesiology, gaming, security framework, and for continuous patients checking. It is easy to use the Neurosky Mindwave headset gadgets, which are for the most part used to identify and measure electrical action of the client's temple and transmit the gathered information remotely, to a computer. Subsequent to preparing EEG signal, it is classify into different recurrence groups for highlight extraction. This paper for the most part deals with extricating the component of EEG sign in OpenViBE. Here, the characteristics and specification of EEG-based HCIs for real-time applications are presented. Furthermore, the discussion about the mental or behavioral state of the person (eyeblink, meditation, attention levels) through the NeuroSky Mindwave (MW001) device using OpenViBE has been done.

Keywords EEG sensor · BCI · Signal processing · OpenViBE software

S. K. Ramakuri (✉) · A. Peddi
Department of Electronics and Communication Engineering,
VNRVJIET, Hyderabad, India
e-mail: sravanthkumar_r@vnrvjiet.in

A. Peddi
e-mail: anudeep_p@vnrvjiet.in

K. S. Nishanth Rao
Department of Electronics and Communication Engineering,
MLRIT, Hyderabad, India

B. Gupta
Department of Information and Technology, Institute of Engineering and Managment, Kolkata,
India
e-mail: bharat@nitp.ac.in

S. Ghosh
Department of Electronics and Communication Engineering,
NIT, Patna, India
e-mail: bij_arn@yahoo.com

© Springer Nature Singapore Pte Ltd. 2019   355
R. S. Bapi et al. (eds.), *First International Conference on Artificial Intelligence
and Cognitive Computing* , Advances in Intelligent Systems and Computing 815,
https://doi.org/10.1007/978-981-13-1580-0_34

1 Introduction

In BCI framework, the client sends the data through the cerebrum signals on the recipient side these signs will be gotten and apply for target applications. The application then changes over these signs into charges that cause developments. The BCI has been utilized for a broadening assortment of such applications. The beginning applications have been gone for offering incapacitated individuals some assistance with utilizing machines. Nowadays, numerous scientists consider that remote BCI frameworks are a critical step toward getting BCI applications out of research facilities. Late advances in implanted framework, sensor innovation, computerized signal handling strategies and remote communication innovation have made ongoing mind action observing conceivable [1, 2]. The data (cerebrum sign) is transmitted either wired or remotely to the computer framework utilizing a different disseminated electroencephalogram (EEG) cathodes or sensors put over the human head [3]. We have described wired and remote gadgets in light of writing work, late advances, and their applications. These gadgets enhance the personal satisfaction incapacitated or rationally debilitated patients in a healing center and home [4].

In this paper, we focus on the different type of EEG systems and their applications. We have also mentioned the state-of-the-art available wireless EEG devices and their features. Initially, researchers proposed wired BCI system to diagnose disease [4] which is very uncomfortable for the patients. Currently, wireless BCI system helps people's life more comfortable. These systems are able to offer improvements in entertainments, games, medical engineering, rehabilitation, and daily life because such systems have obvious advantages. They are simple, convenient, and flexible to use because of wireless.

The NeuroSkyMindWave headset is utilized to gather crude sign from the human mind. The got to data from the Neurosky headset are then plotted through the application program interface of Neurosky Inc. utilizing a preparing-based system. We appoint a subject to perform facial signals an Whatever remains of paper is as per the following Literature Survey on EEG sign and its qualities are given in area II. Area gives the proposed work, and Sect. 4 gives Results and Discussion on feature extraction through behavior state monitoring utilizing NeuroskyMindwave Device 2 System Model.

2 Literature Survey

In 1924, Hans Berger found that electrical signals can be measured from the scalp of the human brain and published his first paper which has ever since established electroencephalography (EEG) as a basic tool for clinical diagnosis and brain research [5]. Presently, several research sectors around the world are focusing on BCI investigation and its performance. Some illustrations of EEG-dependent applications contain wheelchairs, control of cursors, typing skills, robots, and diagnosis of disease.

Table 1 Frequency band classification based on brain activity [8]

Brainwave type	Frequency range (Hz)	Mental states and conditions
Delta	0–4	Dreamless, sleep, unconscious
Theta	4–8	Recall, fantasy, imaginary, dream, creative
Alpha	8–13	Relaxed, but not drowsy, tranquil, conscious
Low Beta	13–15	relaxed yet focused, integrated
Midrange Beta	16–20	Thinking, aware of self and surroundings
High Beta	20–30	Alertness, agitation
Gamma	30–100	Motor functions

Fig. 1 User handles NeuroskyMindwave with OpenViBE [9]

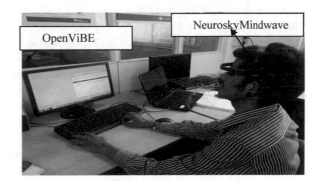

EEG is the representation of the electrical sign of the cerebrum movement from the scalp. The principal neural movement of the human mind distributed in [1] utilizing a basic galvanometer. EEG is for the most part portrayed in recurrence band shift of plenty tuple and recurrence of the wave speak to different cerebrum states which rely on upon interior mind conduct state and outside reenactment [2]. As of late, clinical utilization of EEG 21 cathodes is utilized to recognize five crucial waves.

By and large, brainwaves are characterized in five recurrence groups, i.e., Delta, Theta, Alpha, Beta, and Gamma. Delta signal (0.1–3.5 Hz) partners with a profound rest condition. Theta (4–8 Hz) signal comprised of with dreams, fanciful, sitting without moving, sleepiness. Alpha (8–12 Hz) speaks to loose, reflection, and example shows up in alertness. Beta (> 13–30 Hz) ready, working dynamic, caught up with, deduction and fixation, and Gamma (30–100 Hz) is connected with the rhythms for abnormal state data handling as appeared in Table 1. Recently, incredible improvement is going on in remote EEG gadgets for patient observing, gaming, etc. Earlier, scientists were utilizing wired EEG gadgets for patient checking, however, now through remote EEG gadgets [6]. These EEG gadgets have been developed by Neurosky, OCZ innovation, InteraXon, PLX gadgets, and Emotive frameworks 1 [2, 3] (Fig. 1) [7].

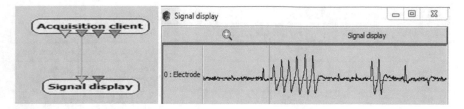

Fig. 2 Acquisition of EEG signal in OpenViBE software

3 Methodology

A. *Utilized Resources: Hardware*
The main hardware for used this study NeuroSky Mind-Wave Device is a single-channel EEG headset and it costs only around $100. The device consists of eight main parts ear clip, ear arm, battery area, power switch, adjustable headband, sensor tip, sensor arm, and thinks gear chipset. The main principle of this operation is very simple; in this device, we used two sensors to detect and filter the EEG signals. First sensor tip detects electrical signals from the forehead of the brain; it detects the signal from the forehead through the electrode positioning (FP1) according to 10–20 system of the EEG. The second sensor ear clip is useful to a ground reference in this chip think gear module allows filter out the electrical noise. Signals from this sensor receiving via electrode placed in left earlobe, where the data obtained from the 512 Hz of sampling rate with each sample; there is a 16-bit ADC resolution, and we are observing given Fig. 2. Users handle Neurosky device with the computer [10].

B. *Utilized Resources: Software*
More simplify signal processing techniques and data analysis are possible through the OpenViBE software [11]. It is a computing language using software programming designed for data analysis, computation, and visualization [9]. In this paper, OpenViBE is used to process the EEG signal. Normally, EEG machines electrodes are placed on the head of the subject and transmit EEG signal to computer with huge external noise, artifacts, and interference but in OpenViBE software is easily tackled these issues [5, 12–16].

Nowadays, more simplify signal processing and data analysis through the Open-ViBE software (2010) are a computing language using software programming designed for data analysis, computation, and visualization [17]. OpenViBE allows 2D and 3D visualization of data through the topology. In this scenario OpenViBE in processing EEG signal, Normally EEG machines electrodes placed on the head of the subject and transmitted data to computers these signals huge external noise, artifacts and interference, these issues OpenViBE software easily tackled. First we set up the NeuroskyMindwave device to the computer for acquiring brain signal [7].

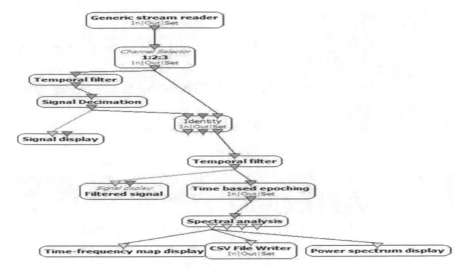

Fig. 3 Algorithm for PSD using generic stream reader

4 Result and Discussion

First we connect the Neurosky device setup with OpenViBE software. OpenViBE is a mathematical and graphical programming language. In this simple design, we can easily drop and drag inserting options in this software. First initiate into acquisition server of OpenViBE software applying Neurosky device. Sample per count is 32, connection port is 1024, and sampling frequency of the device is 512. Then opening the OpenViBE [6] designer for programming we design a simple program in Fig. 2 for obtaining raw signal of a user. NeuroskyMindwave device is a single electrode (FP1) on the forehead from where we observe the raw signal on Fig. 2. In this raw signal, we observe the change occurring due to the eyes closed and open of a user in accurate signal.

Normally, EEG signal will be characterized through the features in different frequency bands. We divide the raw signal into frequency bands and observe the NeuroskyMindwave device giving proper result in the presence of EOG artifacts through the device eye movements/blinking, below 4 Hz (shown in Fig. 3). Next we design algorithm using generic stream reader through OpenViBE software.

First we acquire signal from acquisition client. It is client-based server, and in this scenario, we are using a client NeuroskyMindwave device which gives sampling frequency 512 Hz and data block size is 32 per samples per second. In next step normally, we remove noise in EEG signals using filter. In our design, we are using band-pass filter for extracting frequency bands as it eliminates noise of the signal. Then we are extracting signal epochs at every second 512 samples where each block has 32 samples ($32/512 = 1/16$).

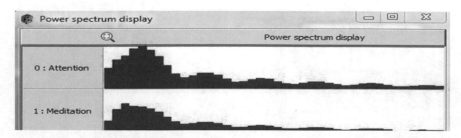

Fig. 4 Power spectral density of filtered EEG signal

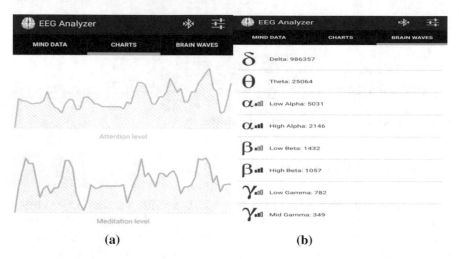

Fig. 5 **a** Attention improvement using human–machine interface, **b** frequency band spectrum level

Next we use this signal for spectral analysis. Here applying amplitude in spectral analysis settings we forward the signal to spectral density. In this scenario, we give subjects to handle different events. First subject is said to relax with eyes closed, and we observe power spectrum shown in Fig. 4. When subject in eyes close and relax, we observe its effect in delta waves. Normally, delta wave function is associated with dreams and meditation.

Next we are observing same experiment with NeuroskyMindwave device to operate smartphone (One plus 3T) using EEG analyzer application. EEG Analyzer application gives structured information operating EEG devices.

We are observing from Fig. 5a, b that when subject eye closed in deep mediation, level increased in packets of Delta (0.5–3 Hz) frequency and attention of subject will be increased. With our proposed algorithm, we would be able to find the behavior state of the user. Mainly it gives behavior state like attention or meditation levels we can manage human behavior state through our gadgets (Smart phones, Devices), Motor skills like hand/Eye Condition and conscious thoughts.

5 Conclusion

We have proposed an algorithm for the component extraction from EEG signal in OpenViBE environment. Initially, we watch the crude EEG information for a specific time period in OpenViBE. Next we have stored the information through a Neurosky-Mindwave gadget in generic stream file design. Then we proposed human attention and meditation levels calculated using activities observation through single-channel wireless EEG device, and it gives proper observations.

In the future work, we will observe different subjects to generate a more comprehensive model for different emotions.

Declaration We have taken necessary permission to use the dataset/images/other materials from respective parties. We, authors undertake to take full responsibility if any issue arising in future.

References

1. B. I. Morshed, and A. Khan, "A brief Review of Brain Signal Monitoring Technologies for BCI Applications: challenges and Prospects", Journal of Bioengineering and Biomedical Science, vol. 4, no. 1, pp. 1–10, May 2014.
2. J. P. Donoghue, "Connecting Cortex to Machines: Recent Advances in Brain Interface", Nature Neuroscience, vol. 5, pp. 1085–1088, November 2002.
3. Y. Lieu, O. Sourina, and M. K. Nguyen, "Real-time EEG-based Human Emotion Recognition and Visualization", international Conference on Cyberworlds, pp. 262–269, October 2010.
4. S Mangalagowri, S. G., and P. Cyril Prasanna Raj. "EEG feature extraction and classification using feed forward back propagation algorithm for emotion detection." Electrical, Electronics, Communication, Computer and Optimization Techniques (ICEECCOT), 016 International Conference on. IEEE, 2016.
5. Jasper, H., 'Report of committee on methods of clinical exam in EEG', Electroencephalogr. Clin. Neurophysiol., 10, 1958, 370–375.
6. Suleiman, Abdul-Bary Raouf, and Toka Abdul-Hameed Fatehi. "Features extraction techniques of EEG signal for BCI applications." Faculty of Computer and Information Engineering Department College of Electronics Engineering, University of Mosul, Iraq (2007).
7. Sravanth Kumar Ramakuri, Vivek Kumar, and Bharat Gupta, "Feature Extraction from EEG Signal through one electrode device for medical application", 1st International Conference on Next Generation Computing Technologies (NGCT 2015), Dehradun, 4–5, September 2015.
8. Murugappan, M., et al. "EEG feature extraction for classifying emotions using FCM and FKM." International journal of Computers and Communications 1.2 (2007): 21–25.
9. Millán, J.R (2004) 'On the need for online learning in brain-computer interfaces', Proc. 2004 Int. Joint Conf. Neural networks, Vol. 4, pp. 2877–2882.
10. Garcia, G. N., Ebrahimi, T. and Vesin, J. M. (2003) 'Support vector EEG classification in the fourier and time-frequency correlation domains', Proceedings of IEEE EMBS Conference on Neural Engineering, pp 591–594.
11. Pfurtscheller, G., Neuper, C., Schlogl, A. and Lugger, K. (1998) 'Separability of EEG signals recorded during right and left motor imagery using adaptive autoregressive parameters', IEEE Transactions on Rehabilitation Engineering, Vol. 6, no. 3, pp. 316–325.
12. Salazar-Varas, R., and D. Gutiérrez. "Feature extraction for multi-class BCI using EEG coherence." Neural Engineering (NER), 2015 7th International IEEE/EMBS Conference on. IEEE, 2015.

13. Barrett, G., Blumhardt, L., Halliday, L., Halliday, A. M., and Kriss, A., 'A paradox in the lateralization of the visual evoked responses', Nature, 261, 253–255.1976.
14. Effects of electrode placement', http://www.focused-technology.com/electrod.htm, California. Accessed on 22/2/2014.
15. Mindwave,http://developer.neurosky.com/docs/lib/exe/fetch.php?media=mindwave_user_gui de_en.pdf. Accessed on 4/3/2015.
16. Emotiv,https://emotiv.com/.../EmotivEPOCSpecifications2014.pdf. Accessed on 4/3/2015.
17. S. Debener, F. Minow, R. Emkes, K. Gandras, and M. D. Vos," How about taking Low-cost, Small, and Wireless EEG for a Walk", Psychophysiology.
18. Palaniappan, R., Paramesran, R., Nishida, S. and Saiwaki, N. (2002) 'A new brain computer interface design using Fuzzy ART MAP', IEEE Transactions on Neural Systems and Rehabilitation Engineering, Vol. 10, pp. 140–148.
19. Wojciech SAŁABUN 'Processing and spectral analysis of the raw EEG signal from the Mind-Wave' PRZEGLAD ELECKTROTECHNICZNY, ISSN 0033-2097, R. 90 NR 2/2014.
20. M. Murugappan, M. Rizon, R. Nagarajan, S. Yaacob, I. Zunaidi, and D. Hazry EEG feature Extraction for classifying Emotions using FCM and FKM International Journal Of Computers And Communications, issue 2, Vol. 1, 2007.

Smart Heartbeat Monitoring System Using Machine Learning

K. Nirosha, B. Durga Sri and Sheikh Gouse

Abstract The increase in quality for wearable technologies has opened the door for an Internet of things (IoT) resolution to attention. One among the foremost current attention issues nowadays is that the poor survival rate of out-of-hospital abrupt internal organ arrests. The target of this study is to gift a multisensory system mistreatment IoT which will collect physical activity heart rates and vital sign. For this study, we have a tendency to enforce an embedded sensory system with an occasional Energy Bluetooth communication module to gather ECG and vital sign knowledge employing a smartphone in a very common setting. This study introduces the employment of signal process and machine learning techniques for detector knowledge analytics for abrupt a seize systole and or heart failure prediction. Big data analytics, legendary in the organization for its use in controlling and managing huge data sets, is applied with success to forecast, avoidance, management, and cure for disorder. Big data analytical visualization tools are useful for health care and forecasting heart attacks.

Keywords IoT · Big data analytics · Data mining · Machine learning
Internet of things (IoT) · ECG · Heart rate · Smartphone · Heart failure

1 Introduction

The increase in quality for wearable technologies has opened the door for an Internet of things (IoT) answer to health care [1]. One in all the foremost prevailing healthcare

K. Nirosha (✉) · B. Durga Sri · S. Gouse
Department of Information Technology, MLR Institute of Technology,
Hyderabad, India
e-mail: nirosha.kunduru@gmail.com

B. Durga Sri
e-mail: durga.sree14@gmail.com

S. Gouse
e-mail: gouse.sheikh@gmail.com

© Springer Nature Singapore Pte Ltd. 2019
R. S. Bapi et al. (eds.), *First International Conference on Artificial Intelligence and Cognitive Computing* , Advances in Intelligent Systems and Computing 815,
https://doi.org/10.1007/978-981-13-1580-0_35

issues nowadays is that the poor survival rate of out-of-hospital explosive viscous arrests. All existing systems for predicting a seize systole within the senior primarily contemplate the center rate parameters. By 2050, it is calculable that over one in five individuals are going to be ancient sixty-five or over. Heart diseases within the senior appeared to be a really common incidence, and roughly, tierce to common fraction of the senior population experiences coronary failure or a seize systole repeatedly on a yearly basis. Our analysis tries to deal with this downside by specializing in ECG signals' analysis and detection that may eventually cause a risk prediction. Since abnormal ECG patterns will cause a coronary failure, our system uses the identification of Associate in nursing abnormal heart rate to alert the user concerning a possible coronary failure.

In associate degree aging society, heart attacks have vast consequences since they have a tendency to cause tremendous issues as associated with deterioration within the quality of life and a rise within the price of aid. Though there has been an excellent deal of analysis on automatic heart failure detection, the world of risk of heart failure prediction remains lacking in study and investigation. The necessity to spot all the doable patterns which will result in a heart failure is extremely difficult.

Big data, [2] an inspiration currently many years previous, is changing into the first technique to harness knowledge, and a lot of healthcare sector discover opportunities and forecast the client behaviors. Big data is the knowledge that exceeds the process capability of standard information systems. The data is simply too massive, quick, or does not work the structures of standard information architectures. Big data characteristics may be represented by "6Vs."

Therefore, our focus is on heart risk prediction instead of detection. The target of this study is to gift a multisensory system that may collect heart rates and body temperatures. For this study, we tend to enforced associate embedded sensory system with a coffee Energy Bluetooth communication module to gather ECG and blood heat information employing a smartphone in a very common setting. This study introduces the employment of signal process and machine learning techniques for device information analytics for sudden as seize systole and or heart failure prediction. Our planned system is going to be helpful not solely to the senior, however, conjointly includes a scope in characteristic cardiopathy among youngsters, adults, stroke patients, therapeutic rehabilitation patients, and human behavior analysis.

2   Literature Work

There has been lots of research done that we will relate to: It's vital for United States to be told regarding connected work to ascertain what aspects we will improve on. Health could be a quite common topic for analysis, and plenty of corporations have taken advantage of that by planning systems that connect patients with doctors around the world. "Patients like Me" have launched its first online community in 2006 [3], and its main goal was to pay attention to patients to spot outcome measures, symptoms, and coverings. It is principally centered on serving to patients answer the question:

Symptoms of heart attacks
• Chest pain or discomfort. This involves uncomfortable pressure, squeezing, fullness, or pain in the center or left side of the chest that can be mild or strong. This discomfort or pain often lasts more than a few minutes or goes away and comes back.
• Upper body discomfort in one or both arms, the back, neck, jaw, or upper part of the stomach.
• Fullness, indigestion, or choking feeling (may feel like "heartburn")
• Shortness of breath, which may occur with or before chest discomfort.
• Nausea or vomiting.
• Light-headedness or sudden dizziness, or breaking out in a cold sweat.
• Sleep problems, fatigue (tiredness), and lack of energy.
• Rapid or irregular heart beats

Fig. 1 Symptoms of heart attacks

"Given my standing, what's the simplest outcome I will hope to realize, [4] and the way do I receive there". They answered patient queries in many forms like having patients with similar conditions connect with one another and share their experiences (Fig. 1).

Everyday IoT systems manufacture content concerning health and goodness. Despite the fact that an outsized variety of corporations give health services, none of them have the feature of providing hardware devices that may be employed by patients to observe their everyday activities and alert them once required. There square measures lots of heart monitors out there that give users with their EKG signals in order that they will keep track of their condition, however, none of that WHO alert the users upon emergencies [5]. "**Qardiocore**" could be a terribly well known and top quality cardiac monitor that tracks a user's complete heart health and displays it on smartphones. The device yields terribly in terms of showing real-time graphs of EKG and graph. However, it solely permits users to share information with their doctors upon receiving it in Associate in nursing offline manner, and it does not offer them the choice of alerting them in real time once their heart is at a heavy condition, and it definitely does not predict heart attacks. Neither of the higher than systems have the potential of predicting cardiac failure.

Ciccone et al. tested a practicability study of integrate care managers to the healthcare system to support general practitioners and specialists of patients, diabetes, and cardiac failure. This resulted during a stable improvement within the clinical service and achieved higher management of unwellness. Ultimately, integrate care manager system provides a positive effect on patient health, and therefore, their results may be attributed to the solid cooperation between the care manager and patient.

3 Methodology and Data Analysis

3.1 Architecture of the System

From Fig. 2, it shows the fundamental flow of the system's design [6]. All the parts are mentioned intimately within the next section, however, in the main the IoT device perpetually collects information from the users and sends it via Bluetooth to the applying as shown within Fig. 2. The applying is wherever all the process and information analysis happen. As mentioned, the user has the choice to look at his/her period of time plots which can offer him/her a basic plan of his/her body's standing. The user does not ought to keep track of his/her information to create certain that she/he is okay since the application's job is to alert the user upon associate emergency [7]. Finally, once the applying senses associate abnormality, it either alerts the user or sends associate attentive to the emergency contact betting on, however, serious the condition is.

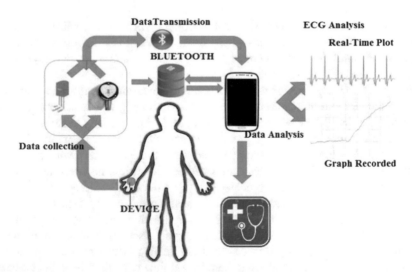

Fig. 2 Architecture of heart attack prediction system

3.2 ECG Analysis

The heart contains of a muscle referred to as heart muscle that smoothly drives the circulation of blood flow throughout the body [8]. In traditional system, a stream of electrical current flows through the complete heart and its propagation isn't random and spreads over the complete structure of the center in a very coordinated pattern and ends up in a good flow of blood in and out of the center. This leads to a measurable amendment in voltage on the some surface. The resultant amplified signal is thought as in cardiogram [9]. A broad variety of things has an effect on the electrocardiogram, not restricted to irregularity in internal organ muscles, metabolic irregularity of the heart muscle, and center of the heart [10]. The electrocardiogram measurements show that the center follows a particular pattern of progression and also the sub-waves are known because the $P, Q, R, S,$ and also the T sub-waves and every of those have a time period in humans. As a results of the electrical activity of the center cells, this flows among the body and potential variations square measure fixed on the surface of the skin, which might be measured ill-treatment, that is that the ECG [11]. The graphical recording of that body surface potentially generates the cardiogram (Fig. 3).

3.3 Proposed Algorithm

We have applied following five ordinarily used classifiers for prediction on the basing on their performance [12]. These classifiers area unit as follows:

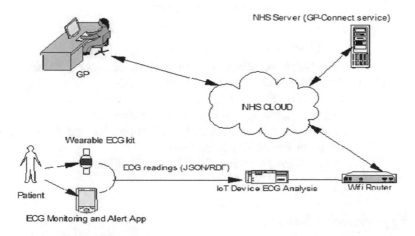

Fig. 3 ECG real-time monitoring IOT network

1. Theorem network naïve mathematician (NB)
2. Support vector machine SMO
3. C4.5 call tree J48
4. K-nearest neighbor 1 Bk
   Clustering and nearest neighbor strategies area unit ideally suited to use with numeric knowledge.
   However, knowledge typically uses exploitation categorical values, i.e., names or symbols.
   In this scenario, it is going to be higher to use a probabilistic technique [13], like the naive mathematician classifier (NBC).

Naive mathematician Classifier:
A naive mathematician classifier could be a program that predicts a category worth given a collection of set of attributes.
   For each class label,

1. Calculate chances for every attribute, conditional on the category worth.
2. Apply the product rule to get a joint probability for the attributes.
3. Apply Bayes rule to derive conditional chances for the category variable.

Once this has been done for all category values, then the output category will be the best likelihood. The basis of naive mathematician algorithmic program is theorem or as an alternative referred to as Bayes' rule or Bayes' law [14, 15]. It offers us a way to calculate the probability, i.e., the likelihood of an occasion supported previous data obtainable on the events. Additionally, theorem is expressed with the following equation:

$$P(A/B) = \frac{P(B/A)P(A)}{P(B)}$$

where

$P(A/B)$	conditional probability of event A given the event B is true
$P(A)$ and $P(B)$	Probabilities of event A and B, respectively
$P(B/A)$	Probability of event B given the event A is true
A	is **proposition, and** B is called the **evidence.**
$P(A)$	is called the **prior** probability of proposition and $P(B)$ is called the **prior** probability of evidence.
$P(A/B)$	is called the **posterior,** $P(B/A)$ is the **likelihood**.

4 Experimental Results

Figures 4, 5, 6, 7, 8, and 9.

Relation: heart-statlog

No.	age Numeric	sex Numeric	chest Numeric	resting Num	serum_ch Num	fasting_ N	resting	maximum	exercise_ind Nu	oldpeak Numeric	slope Numeric	number_of_maj Numeric	thal Numeric	class Nominal
1	70.0	1.0	4.0	130.0	322.0	0.0	2.0	109.0	0.0	2.4	2.0	3.0	3.0	present
2	67.0	0.0	3.0	115.0	564.0	0.0	2.0	160.0	0.0	1.6	2.0	0.0	7.0	absent
3	57.0	1.0	2.0	124.0	261.0	0.0	0.0	141.0	0.0	0.3	1.0	0.0	7.0	present
4	64.0	1.0	4.0	128.0	263.0	0.0	0.0	105.0	1.0	0.2	2.0	1.0	7.0	absent
5	74.0	0.0	2.0	120.0	269.0	0.0	2.0	121.0	1.0	0.2	1.0	1.0	3.0	absent
6	65.0	1.0	4.0	120.0	177.0	0.0	0.0	140.0	0.0	0.4	1.0	0.0	7.0	absent
7	56.0	1.0	3.0	130.0	256.0	1.0	2.0	142.0	1.0	0.6	2.0	1.0	6.0	present
8	59.0	1.0	4.0	110.0	239.0	0.0	2.0	142.0	1.0	1.2	2.0	1.0	7.0	present
9	60.0	1.0	4.0	140.0	293.0	0.0	2.0	170.0	0.0	1.2	2.0	2.0	7.0	present
10	63.0	0.0	4.0	150.0	407.0	0.0	2.0	154.0	0.0	4.0	2.0	3.0	7.0	present
11	59.0	1.0	4.0	135.0	234.0	0.0	0.0	161.0	0.0	0.5	2.0	0.0	7.0	absent
12	53.0	1.0	4.0	142.0	226.0	0.0	2.0	111.0	1.0	0.0	1.0	0.0	7.0	absent
13	44.0	1.0	3.0	140.0	235.0	0.0	2.0	180.0	0.0	0.0	1.0	0.0	3.0	absent
14	61.0	1.0	1.0	134.0	234.0	0.0	0.0	145.0	0.0	2.6	2.0	2.0	3.0	present
15	57.0	0.0	4.0	128.0	303.0	0.0	2.0	159.0	0.0	0.0	1.0	1.0	3.0	absent
16	71.0	0.0	4.0	112.0	149.0	0.0	0.0	125.0	0.0	1.6	2.0	0.0	3.0	absent
17	46.0	1.0	4.0	140.0	311.0	0.0	0.0	120.0	1.0	1.8	2.0	2.0	7.0	present
18	53.0	1.0	4.0	140.0	203.0	1.0	2.0	155.0	1.0	3.1	3.0	0.0	7.0	present
19	64.0	1.0	1.0	110.0	211.0	0.0	2.0	144.0	1.0	1.8	2.0	0.0	3.0	absent
20	40.0	1.0	1.0	140.0	199.0	0.0	0.0	178.0	1.0	1.4	1.0	0.0	7.0	absent
21	67.0	1.0	4.0	120.0	229.0	0.0	2.0	129.0	1.0	2.6	2.0	2.0	7.0	present
22	48.0	1.0	2.0	130.0	245.0	0.0	2.0	180.0	0.0	0.2	2.0	0.0	3.0	absent
23	43.0	1.0	4.0	115.0	303.0	0.0	0.0	181.0	0.0	1.2	2.0	0.0	3.0	absent
24	47.0	1.0	4.0	112.0	204.0	0.0	0.0	143.0	0.0	0.1	1.0	0.0	3.0	absent

Fig. 4 Data sets

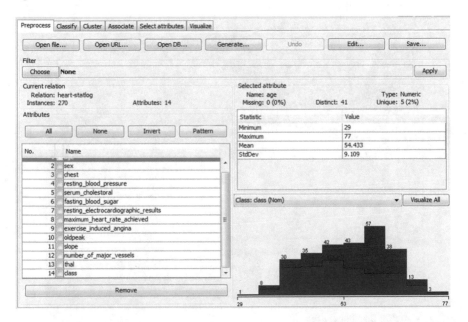

Fig. 5 Attribute analysis

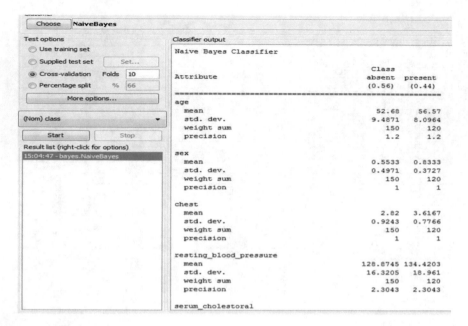

Fig. 6 Classifier output

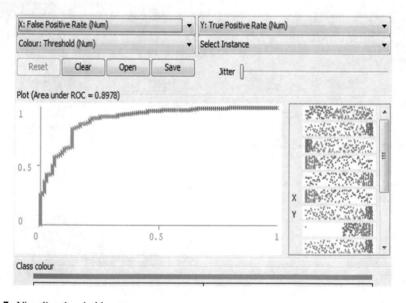

Fig. 7 Visualize threshold curve

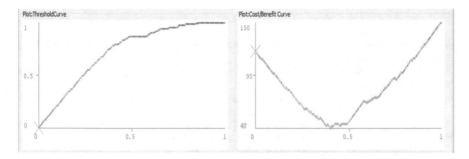

Fig. 8 Cost/benefit analysis

Fig. 9 Final ECG result to mobile

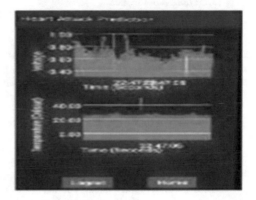

5 Conclusion

In this we have designed an embedded IoT system to continuously monitor the heartbeat, body temperature cholesterol, chest pain using ECG analysis through mobile and alerting the time-to-time symptoms to the device. In case if any abnormality occurs, it reports to respective persons' mobile and doctor. The system learns the symptoms from the previous database and alerts the person with in a stipulated time and to prevent from the heart risk. In future, we plan to design our system to control the cholesterol, smoking, alcohol through application and it predicts the lifetime of the heart.

References

1. K Nirosha, Durga sree. B, Dr. Sheikh Gouse. (2016), iHOME: Bio-Health Intelligent Mobile System Using IoT, International Journal of Innovations in Engineering and Technology (IJIET) ISSN 2319–158.
2. B. Durga Sri, K.Nirosha, M. Padmaja, (2017), HEALTHCARE ANALYSIS USING HADOOP, International Journal Of Current Engineering And Scientific Research (IJCESR), 2394–0697.

3. United States, Heart Association and Stroke Association, "Heart Disease and Stroke Statistics—At-a-Glance," 2015.
4. Qardiocore,https://www.getqardio.com/qardiocore-wearable-ecg-ekg-monitor-iphone/, [Lat Accessed: 22 May 2017].
5. United States, Heart Association and Stroke Association, "Heart Disease and Stroke Statistics—At-a-Glance," 2015.
6. Clifford, Gari D. Advanced Methods and Tools for ECG Data Analysis. Norwood, MA, USA: Artech House, 2006. ProQuest ebrary. Web. 24 November 2015.
7. N.J. Holter, "New methods for heart studies,"Science, vol. 134, p. 1214, 1961.
8. Al Mamoon I, Sani AS, Islam AM, Yee OC, Kobayashi F, Komaki S (2013) A proposal of body implementable early heart attack detection system, 1–4.
9. Wanaskar UH, Ghadge P, Girmev V, Deshmukh P, Kokane K (2016) Intelligent Heart attack prediction system using big data. International Journal of Advanced Research in Computer and Communication Engineering 5: 723–725.
10. Lee B, Jeong E (2014) A design of a patient-customized healthcare system based on the hadoop with text mining (PHSHT) for an efficient disease management and prediction. International Journal of Software Engineering and Its Applications 8: 131–150.
11. Ebrahimzadeh, E, Pooyan, M, & Bijar, A 2014,'A novel approach to predict sudden cardiac death (SCD) using nonlinear and time-frequency analyses from HRV signals', *Plos One*, 9, 2, p. e81896, MEDLINE with Full Text, EBSCO*host*, viewed 30 June 2016.
12. De Chazal, P.; O'Dwyer, M.; Reilly, R.B., "Automatic classification of heartbeats using ECG morphology and heartbeat interval features," in Biomedical Engineering, IEEE Transactions on, vol.51, no.7, pp. 1196–1206, July 2004 https://doi.org/10.1109/tbme.2004.827359.
13. J. -V. Lee and Y. -D. C. a. K. T. Chieng, "Smart Elderly Home Monitoring System with an Android Phone," International Journal of Smart Home, vol. 7, pp. 17–32, May 2013.
14. Prerana THM, Shivaprakash NC, Swetha N (2015) Prediction of heart disease using machine learning algorithms- NaÃ¯ve Bayes, Introduction to PAC Algorithm, Comparison of Algorithms and HDPS. International Journal of Science and Engineering 3: 90–99.
15. Kumar KG, Arvind R, Keerthan PB, Kumar SA, Dass PA (2014) Wireless methodology of heart attack detection. International Journal for Scientific Research and Development 2: 673–676.

Compact Clusters on Topic-Based Data Streams

E. Padmalatha and S. Sailekya

Abstract The regular clustering algorithms follow a batch mode of clustering. The existing clustering algorithms that use the incremental approach are lagging with drawbacks like inefficient memory utilization, data loss. The proposed system resolves all the issues by incremental clustering algorithm to cluster data incrementally. The clustering algorithm uses heuristic measures. Every incoming data is compared with the existing clusters. If it matches with any of the existing clusters, it finds a place to reside. Otherwise, it creates a new cluster. There are a few prerequisites. The raw data (unstructured) has to be brought into a structured format to initiate the clustering process. Each cluster is represented in the form of a vector called a data cluster vector. Finally, the proposed method is proved to be better than the existing method, and misclassification is completely removed.

Keywords Data cluster vector · Incremental clustering

1 Introduction

Clustering is an unsupervised learning method in which there is no target value to be predicted; the goal is to find common patterns or grouping similar examples [3]. It is a main task of exploratory data mining, and a common technique for statistical data analysis, used in many fields, including machine learning, pattern recognition, image analysis, information retrieval, bioinformatics, data compression, and computer graphics. In the clustering algorithms prior to this followed a batch mode process of clustering. The drawback here was that the number of clusters was fixed, thereby increasing the ambiguity in each cluster [5]. Although there are some cluster-

E. Padmalatha (✉)
Chaitanya Bharathi Institute of Technology, Hyderabad, India
e-mail: padmalatha@cbit.ac.in

S. Sailekya
B.V. Raju Institute of Technology, Medak, India
e-mail: lekya.sheral@gmail.com

© Springer Nature Singapore Pte Ltd. 2019
R. S. Bapi et al. (eds.), *First International Conference on Artificial Intelligence and Cognitive Computing* , Advances in Intelligent Systems and Computing 815,
https://doi.org/10.1007/978-981-13-1580-0_36

ing algorithms that cluster incrementally, they are not very effective in some aspects like memory utilization. Birch [1] clusters the data based on CF-tree instead of the original large dataset. Broadly discards the data that it considers it to be "unimportant." CluStream [2] uses two phases for clustering—online and offline phase.

The proposed system in this paper uses sessions which enable the data to be transferred without any usage of external databases. A session is associated with a data visitor. There is only one user access admin to this system. Stream data is the input that is fed initially to the system. This data in raw form is unstructured. It is first brought into a structured format.

It consists of the following steps.

- Representation of a sentence in the stream data as a tuple (tv_i) by calculating Tf-Idf score.
- Implementation of k-means clustering algorithm to create initial clusters.
- The corresponding data cluster vector (DCVs) are then initialized. DCV maintain statistics of the stream data.
- When new stream data is sent to the system, an incremental stream data clustering algorithm will cluster the sentences and update the clusters.

2 Methodology

Implementation is the realization of an application or execution of plan, idea, model, design, specification, standard, algorithm, or policy. The various modules necessary for the implementation of the proposed method are specified.

Figure 1 illustrates that the admin, after logging into the system, will be allowed to send the stream data. This stream data is brought into a structured format by the system by calculating the Tf-Idf scores for each word in a sentence. Each sentence is represented by a tv_i value. These values are then sent into the k- means clustering algorithm to create initial clusters. Each cluster is represented in a vector format by initializing the DCV. Incremental clustering is now brought into picture where it clusters the data incrementally.

2.1 Stream Data

A data stream is an ordered sequence of instances that in many applications of data stream mining [4] can be read only once or a small number of times using limited computing and storage capabilities. It is usually in large size. In the proposed system, the stream data is taken from the University Of California Irvin (UCI) [7]. Two different domain-related stream data are considered in which one stream data is about description of the film and other stream data is related to staff review of Swiss hotel at Chicago.

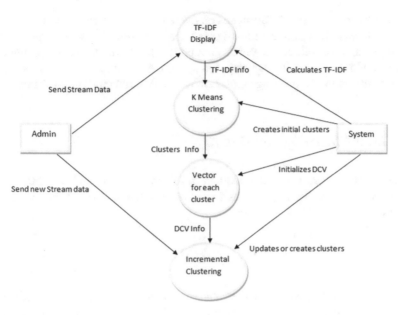

Fig. 1 Process of proposed method

2.2 Tf-Idf Module

- The first step is to the convert the unstructured data into structured data. This is done by calculating the Tf-Idf score for each word. Tf-Idf stands for term frequency-inverse document frequency, and the Tf-Idf weight is a weight often used in information retrieval and text mining. This weight is a statistical measure used to evaluate how important a word is to a document in a collection or corpus. The importance increases proportionally to the number of times a word appears in the document but is offset by the frequency of the word in the corpus. Variations of the Tf-Idf weighting scheme are often used by search engines as a central tool in scoring and ranking a document's relevance given a user query.
- **Term frequency(Tf)**
  $Tf(t)$ = (Number of times term t appears in a sentence)/(Total number of terms in that sentence).
- **Inverse document frequency (Idf)**
  $Idf(t) = \log$(Total number of sentences/Number of sentences with term t in it).

The product of Tf and Idf values are calculated by the above-mentioned formula, and the Tf-Idf value is obtained by multiplying Tf and Idf value. tv_i values are calculated by squaring and adding the Tf_Idf values of each word in that sentence. Each sentence is now represented by a numeric tv_i value. Once the Tf-Idf values are computed, tv_i values for every sentence are calculated and are represented in the form of a matrix. The attributes on the vertical line represent the non-redundant words of a sentence, and the attributes on the horizontal line represent the sequence number of the sentences, and the matrix contains the Tf-Idf scores. Once the tv_i values are obtained, these values are sent as an input for the k-means clustering algorithm to create initial clusters. In the proposed system, k value is considered as two. To perform incremental clustering algorithm, we need initial clusters for mandatory.

2.3   Data Cluster Vector Module

These initial clusters are represented in the form of a vector. It consists of four attributes-—sum_v, wsum_v, cv, and n. **sum_v** is the sum of normalized textual vectors.

$$\mathbf{sum\_v} = \sum_{i=1}^{n} tvi/||tvi|| \tag{1}$$

This is obtained by adding up the values on the vertical axis, i.e.., the tf-idf values for a particular word from all the sentences and dividing by the square root of a value which is obtained by adding the squares of values on the vertical axis. Similarly, it is calculated for all the words.

$$\mathbf{wsum\_v} = \sum_{i=1}^{n} wi.tvi \tag{2}$$

wsum_v is the sum of weighted textual vectors. This is obtained by adding up the values on the vertical axis, i.e., the tf-idf values for a particular word from all the sentences. Standard weight is 1.

$$\mathbf{cv} = \mathbf{wsum\_v}/\mathbf{n} \tag{3}$$

cv is the vector of the cluster centroid. For a cluster, a centroid value is necessary, which is calculated by taking wsum_v of a word by diving it by the number of sentences in the cluster and calculated in a similar fashion for all the words. **n** is the number of sentences in a cluster.

2.4   Incremental Clustering Module

In the initial system will have k clusters as the new stream is given as the input, every sentence in the new stream data is compared with every other existing cluster. For a particular sentence S, Tf-Idf scores are calculated, and then, it is represented by a tv_i value. We make the sentence words and all the existing initial clusters words

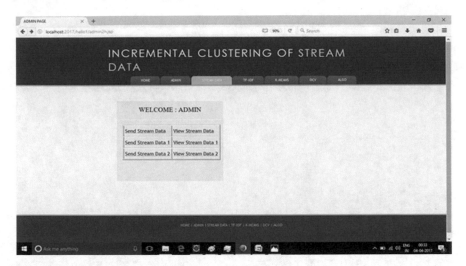

Fig. 2 Stream data page

to be same, and also, the words should be in same order. Now, actual incremental clustering algorithm starts. We take each cluster centroid vector and the particular sentence Tf-Idf scores in a vector, and cosine similarity is found out by substituting both vectors values in the below-mentioned formula. This process is repeated for all initial clusters. We now have all clusters cosine similarities. By performing selection sort method, we obtain the cluster having the maximum cosine similarity value (MaxSim(S)) and that cluster is taken. We calculate the minimum bounding similarity (MBS) of a cluster which is having maximum cosine similarity value, by substituting in the below-mentioned formula. The formula explains, taking the cluster which is having maximum cosine similarity value, attributes sum_v, wsum_v, n and all clusters wsum_v's. If this MBS is found to be greater than the maximum cosine similarity value (MaxSim(S)), a new cluster is created. DCV of the new cluster is initialized. Otherwise, it is directly updated into that cluster having maximum cosine similarity value. DCV of the corresponding cluster is updated. In this way, we send each sentence of stream data into the incremental clustering algorithm and all sentences are correctly placed in the clusters.

For all the sentences of the stream data, initial clusters are the clusters which we created through k-means [6] algorithm, and if they are updated, updated information of those clusters is taken as initial cluster set.

Result and Analysis Stream Data Page

Figure 2 tells that once the user logs in, a session is created. Two links are provided in this page: One accepts input in the form of stream data, and the second link finds the TF-IDF [8] scores for each word in the stream data to represent in a matrix format.

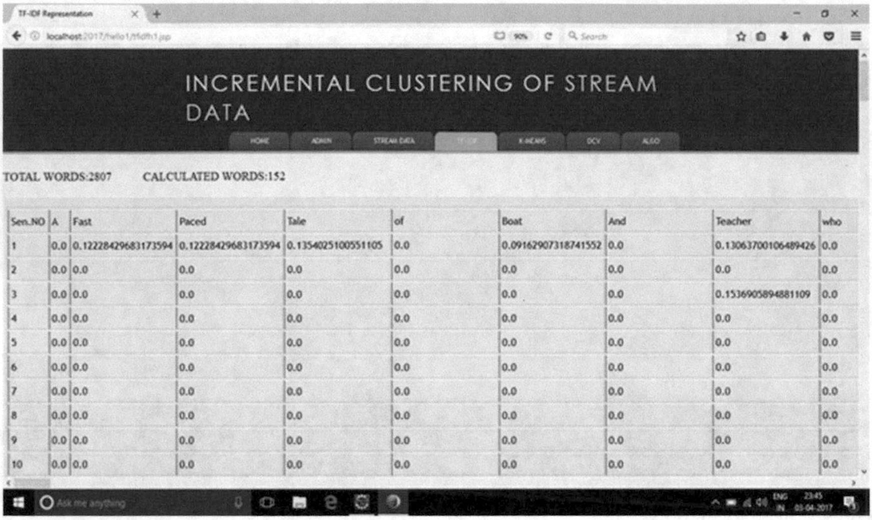

Fig. 3 TF-IDF representation

2.5 TF-IDF Representation

After calculating Tf-Idf [8–10], scores for each word in the sentences present in the stream data are shown in Fig. 3. tv_i scores for every sentence are found out by squaring and adding up the Tf-Idf values. Sentence count, word's count is maintained is also displayed in Fig. 3. This is a two-dimensional matrix that consists of non-redundant array of words. The horizontal axis represents the words, and the vertical axis represents serial number of the sentence.

3 Implementation

The essential requirements for developing the hardware of this system include a minimum of Pentium IV 3.5 GHz, a hard disk of 40 GB, and a ram of 1 GB. At the same time, for developing the software of this system, we require an operating system of minimum Windows 7. For the front end development, we have used cascaded style sheets. Coding language used is Java, and the Web server used is Tomcat. Database used is MySQL. An integrated development environment (IDE) used is Eclipse Mars.

The proposed algorithm for the implementation of incremental clustering is as follows

Input: a Cluster Set C_Set/step1
While! Stream.end() do/step2
Sentence S = Stream.next();/step3

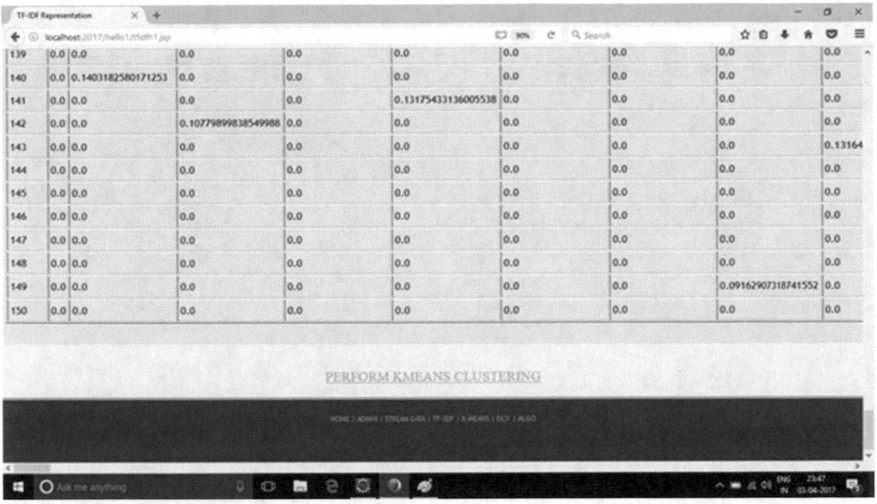

Fig. 4 Perform K-means clustering

Choose Cp in C_Setwhose centroid is closest to S;/step4
If MaxSim(S) < MBS/step5
Create a new cluster Cnew = {S};/step6
Else update Cp with S;/step7

Figure 4 shows that for the entire stream data Tf-Idf [9], values are calculated. The next step is to perform k-means [6] clustering to create initial clusters. Each cluster is showcased by showing the tv_i values of all the sentences in the cluster. Creating initial clusters is shown in Fig. 5 The above figure gives the details of tv_i values; initial clusters are created based on tv_i values. The entire stream data is divided into two clusters, and each cluster contains values which are closely related. In the next step, these clusters are represented in the vector format. In the proposed algorithm, Step 1 is obtained here by creating initial clusters C_Set.

3.1 Data Cluster Vector Initialization of Stream Data 1

Data cluster vector initialization of stream data 1 is shown in Fig. 6. Data cluster vector initialization is the form of a vector that consists of four attributes—'sum_v'(sum of the normalized textual vectors), 'wsum_v'(sum of the weighted textual vectors), 'cv'(vector of cluster centroid), and 'n'(number of sentences in the cluster).

Figure 7 shows the stream data 1 which is related to the same domain of initial clusters (clusters formed by k-means clustering) is sent to the system. The link view stream data 1 shows the sentences contained in stream data 1. Incremental Clustering of Stream Data 1. In the proposed algorithm, step 2 and step 3 are performed where it

Fig. 5 Creating initial clusters

Fig. 6 Data cluster vector initialization

takes each sentence of stream data 1 in the loop, and after completion of processing sentence, the next sentence is taken. Figure 8 shows that the clustering process is performed incrementally for every sentence. Representation is given by sentence tv_i value, all clusters cosine similarities, maximum cosine similarity, MBS for that maximum cosine similarity cluster is found out. For sentence 1, we can see that maximum cosine similarity value is less than MBS, and a new cluster is created because it does not belong to any of the existing clusters. For sentence 2, the sentence

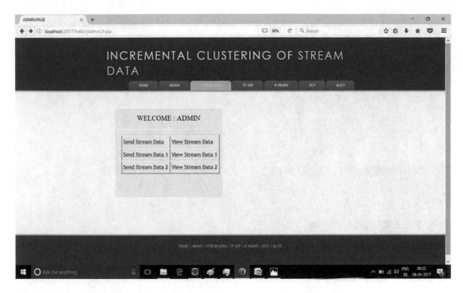

Fig. 7 Send stream data 1

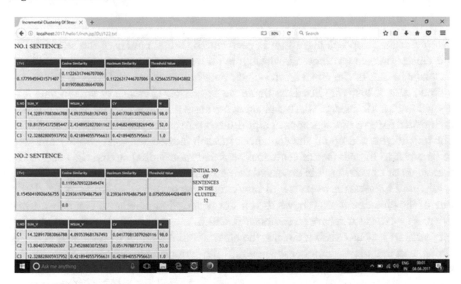

Fig. 8 Clustering for sentence 1 and 2

is relating to second cluster so it has been updated to the second cluster and it can be observed that DCV of that cluster has been updated through the initial number of sentences count in the cluster shown in the above Fig. 8. Clustering for subsequent sentences is shown in Fig. 9.

Figure 9 shows that for sentence 6, the sentence is relating to first cluster so it has been updated to the first cluster and we can see DCV of that cluster has been updated

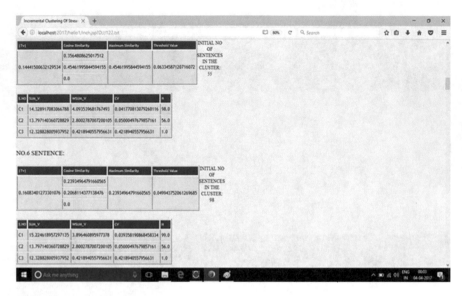

Fig. 9 Clustering for subsequent sentences

(Step 7 of the proposed algorithm is performed). Initial counts of the sentences in the cluster present are shown in the figure above. Once a new sentence has found its place in one of the clusters, the count gets incremented, and TCV's values are also updates. Clustering For Last Sentence of the last sentence of stream data 1 is shown in Fig. 10. In Fig. 10, five clusters are shown from which three new clusters are created (Step 6 of the proposed algorithm is performed) and remaining are initial clusters. If you add up all the sentences in each cluster, the total would come out to be equal to the number of sentences sent and more initial stream data count. By formation of clustering, it is observed that no data is going to misclassified.

Figure 11 shows the overview of how every new sentence is getting updated into one of the clusters. It also shows the cluster description whether it is a new cluster or existing cluster whenever a sentence is sent. If it is a new cluster, cluster number increases. If it is an existing cluster, the cluster initial number of sentences count is shown and that cluster number is shown. If the same stream data 1 is resent for the second time. It is now observed that in the previous pass of the same data, it created a new cluster at that time. Now no new cluster is formed, only the initial clusters are there because the sentence has now found its place in one of the existing, updated clusters. Figure 12 shows the updated clusters. Both the initial clusters have some number of sentences which are closely related to each other. As we had sent the new stream data which is related to the same domain only, the number of new clusters creation is reduced. These updated initial clusters will be the input of the next pass of stream data.

Fig. 10 Clustering for last sentence

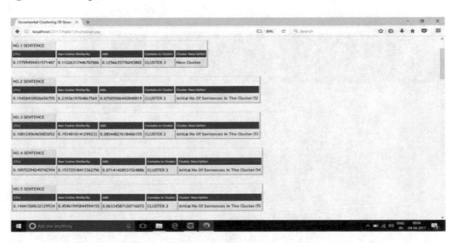

Fig. 11 Overview of stream data 1

3.2 Stream Data 2 Related to Other Domain

Figure 13 shows the stream data 2 which is not related to the same domain of initial clusters (clusters formed by k-means clustering) is sent to the system. This algorithm not confined to a particular domain. It accepts data from the other domain as well. Incremental Clustering of Stream Data 2. Step 2 and step 3 are performed where it takes each sentence of stream data 2 in the loop and after completion of processing sentence, the next sentence will be considered. Clustering for sentence 1 and sentence

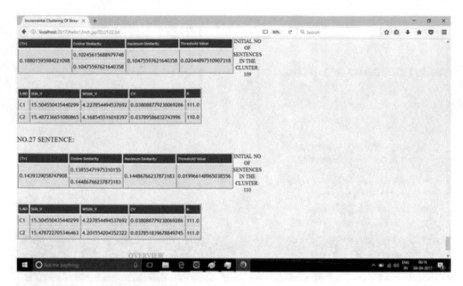

Fig. 12 Updated initial clusters

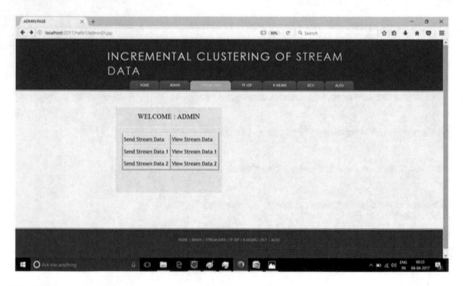

Fig. 13 Send stream data 2

2 of stream data 2 is shown in Fig. 14. Figure 15 shows that the number of clusters keeps increasing because the data is not from the same domain. It does not find its place in the existing clusters. Hence, for sentence 1 and sentence 2, new clusters are created.

Figure 16 tells that all the updated clusters and also new clusters are shown. Since the fresh data does not belong to the same domain, more number of new clusters is

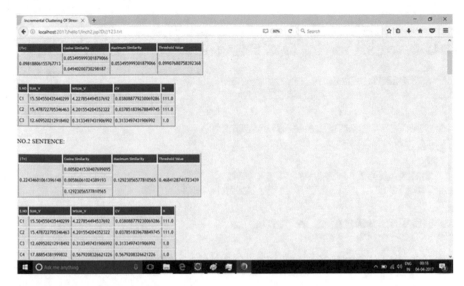

Fig. 14 Clustering for sentence 1 and sentence 2

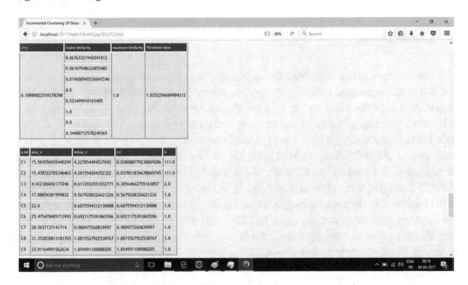

Fig. 15 Clustering for last sentence

created. They are seven new clusters which are created out of which one new cluster got updated with two sentences having in it. Figure 16 shows the overview of the clusters that have newly created. We can see that all new clusters are created through cluster description field.

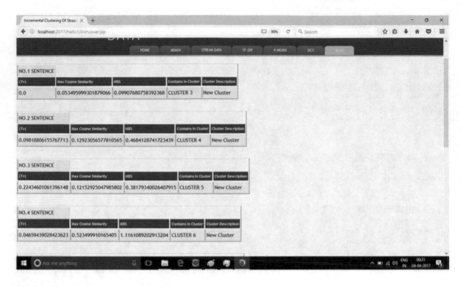

Fig. 16 Overview of stream data 2

4 Conclusion

Incremental clustering algorithm is not specific to any particular domain but is effective when implemented in the same domain. There is no loss of data. All the data that is sent gets categorized into any of the existing clusters. No data is discarded. Even if it does not belong to any of the existing clusters, new clusters are created. All the data is kept in sessions. The data stays until the session is kept active. No concept of storing the data in database is used here. Therefore, memory space utilization is kept at minimum. As part of future work, if the number of clusters keeps increasing, the clusters that have very little content and have not been updated for a while could be safe to delete. No database is used to store the data it uses sessions instead. Therefore, the memory storage is not required. Unlike the clustering algorithms that discard the data that they feel is "unimportant," there is no chance of data loss in this system. It increases the distinctness, thereby reducing the ambiguity.

References

1. T. Zhang, R. Ramakrishnan, and M. Livny, "BIRCH: An efficient data clustering method for very large databases," in Proc. ACM SIGMOD Int. Conf. Manage. Data, 1996, pp. 103–114.
2. P. S. Bradley, U. M. Fayyad, and C. Reina, "Scaling clustering algorithms to large databases," in Proc. Knowl. Discovery Data Mining, 1998, pp. 9–15.
3. C. C. Aggarwal, J. Han, J. Wang, and P. S. Yu, "A framework for clustering evolving data streams," in Proc. 29th Int. Conf. Very Large Data Bases, 2003, pp. 81–92.

4. S. Zhong, "Efficient streaming text clustering," Neural Netw., vol. 18, nos. 5/6, pp. 790–798, 2005.
5. C. C. Aggarwal and P. S. Yu, "On clustering massive text and categorical data streams," Knowl. Inf. Syst., vol. 24, no. 2, pp. 171–196, 2010.
6. An Efficient k-means Clustering Algorithm: Analysis and Implementation by Tapas Kanungo, David M. Moun, Nathan S. Netanyahu, Christine D. Piatko, Ruth Silverman and Angela Y. Wu.
7. UCI Machine Learning Repository [http://archive.ics.uci.edu/ml]. Irvine, CA: University of California, School of Information and Computer Science.
8. Sparck J.K. A statistical interpretation of term specify and its application in retrieval. J. Doc., 28:11–20, 1972.
9. Manning C.D., Raghavan P., and Schütze H. Introduction to Information Retrieval. Cambridge, UK, Cambridge University Press, 2008.
10. Salton G. and Buckley C. Term-weighting approaches in automatic text retrieval. Inf. Process. Manage., 24(4):513–523, 1988.

Author Profiling Approach for Location Prediction

G. Srikanth Reddy, T. Murali Mohan and T. Raghunadha Reddy

Abstract Author Profiling is a type of text categorization technique which is used to predict the profiling characteristics of the authors like gender, age, and location by observing their writing styles in the text. Various researchers proposed several types of techniques to predict the demographic characteristics like gender and age of the authors from different types of datasets. Few researchers show interest on location prediction of the texts. In this work, we concentrated on prediction of the location of the authors. Most of the existing approaches in Author Profiling concentrated on extraction of different types of stylistic features to discriminate the writing style of the text. In this work, the experimentation carried out with content-based features like most frequent terms and a document representation technique profile specific document weighted approach is used to predict the location of the authors. In this approach, a new term weight measure is proposed to compute the term weights of the features. These term weights of most frequent terms were used to compute the document weights specific to every location profile group. These document weights were exploited to represent the document vectors for generating the classification model. The obtained results were good when compared with existing approaches for location prediction.

Keywords Author profiling · Profile specific document weighted approach
Location prediction · Term weight measure · Document weight measure

G. Srikanth Reddy (✉) · T. Raghunadha Reddy
Department of IT, Vardhaman College of Engineering, Hyderabad, India
e-mail: srikanthreddygopu@gmail.com

T. Raghunadha Reddy
e-mail: raghu.sas@gmail.com

T. Murali Mohan
Department of CSE, Swarnandhra Institute of Engineering and Technology,
Narsapuram, Andhra Pradesh, India
e-mail: drtmm512@gmail.com

© Springer Nature Singapore Pte Ltd. 2019
R. S. Bapi et al. (eds.), *First International Conference on Artificial Intelligence and Cognitive Computing* , Advances in Intelligent Systems and Computing 815,
https://doi.org/10.1007/978-981-13-1580-0_37

1 Introduction

The World Wide Web is growing exponentially day by day mainly through blogs, reviews, twitter, and other social media content. It was very difficult for the information analyst to analyze the text when the text was uploaded without any details. In this context, the researchers were looking for a solution to predict the details of the anonymous text. The Author Profiling is one such technique, which is used to predict the demographic characteristics of the authors such as gender, age, and location of the texts.

The Author Profiling techniques were used in several applications such as business, social media, and security. In business point of view, the business analysts take decisions about their business by analyzing the reviews of the products given by the users. It was difficult for analysts when the reviews do not contain any known details. Here, the Author Profiling techniques were used to predict the gender, age, and location of authors of reviews. In social media, the users created profiles with false details and they posted threatening messages and harassing messages to other users. The Author Profiling techniques were used to predict the correct details of the users by analyzing these threatening texts and harassing messages. The terrorist organizations send threatening mails to government agencies to convey information without specifying their correct details. In this context, Author Profiling is used to predict the details of the mail like from which country the mail came, the gender and age group of the person.

This paper is organized into six sections. The related work in Author Profiling for predicting age, gender, and location profiles were explained in Sect. 2. Section 3 describes the dataset characteristics and evaluation measures. The profile specific document weighted (PDW) approach was explained in Sect. 4. The experimental results of PDW approach were presented in Sect. 5. Section 6 concludes this paper with conclusions.

2 Literature Review

Most of the researchers proposed different types of stylistic features like character, word, syntactic, content, structural, readability, and information retrieval features for predicting age, gender, and location of the authors in Author Profiling [1]. Collection of dataset for experimentation in Author Profiling is a difficult task. Koppel et al. collected [2] 566 documents from the British National Corpus (BNC) for gender prediction. They experimented with 1081 features and obtained an accuracy of 77.3%.

Argamon et al. collected [3] a blog posts of 19,320 blog authors and obtained 76.1% accuracy for gender dimension by using content-based and stylistic features. They observed that the style-based features were most useful compared to content-based features to discriminate the gender. In their another work [4], they experimented with 37,478 blog posts and achieved an accuracy of 80.1% using 1502 features of content-based and stylistic features.

Dominique Estival used [5] a dataset of 1033 authors e-mails of 9836. They extracted 689 features of word-based, character-level, and structural features. They used many classifiers for this analysis but random forest algorithm gave an overall accuracy of 0.8422 for native language prediction. The bagging algorithm obtained an accuracy of 0.7998 for education dimension prediction by using the function words as features. For country dimension, the SMO machine learning algorithm obtained an accuracy of 0.8113 by using all features.

Lim et al. determined [6] common or rare words in the total corpus based on TFIDF scores of words. Mechti et al. grouped [7] words into classes according to their similarity based on TFIDF scores of words. Maharjan et al. exploited [8] TFIDF measure to compute the important word n-grams. TFIDF scores of the word n-grams were used to remove the n-grams that were not been used by at least two authors.

3 Dataset Characteristics and Evaluation Measures

The dataset was collected for location prediction from hotel reviews Website www. tripadvisor.com. The researchers faced many problems in collection of reviews and preparation of data suitable to Author Profiling tasks. In this work, we take some special precautions such as the reviews written in English language only, the reviews who gave the details of location in their profile and the reviews which contain at least five sentences considered in the collection of dataset. The location dataset contains 4000 hotel reviews of ten different countries authors, and each country contains 400 reviews to balance the corpus countrywide.

The approaches for Author Profiling used accuracy, recall, precision, and F1-score measures to evaluate the efficiency of their approach for predicting demographic characteristics of the authors. In this work, the accuracy measure is used to check the efficiency of our approach for location prediction. The accuracy is the ratio of the number of reviews predicted their location certainly to total number of reviews in the test dataset.

4 Profile Specific Document Weighted (PDW) Approach

The PDW approach for location prediction is represented in Fig. 1.

The PDW approach was proposed in [9] for representation of a document. In this approach, $\{T_1, T_2, \ldots T_n\}$ is a set of vocabulary terms and $\{D_1, D_2, \ldots D_m\}$ is a set of documents in the dataset. The dataset used in this work for location prediction contains ten countries documents, such as United States of America (USA), Australia (AU), China (CH), Brazil (BZ), India (IN), Germany (GM), Japan (JP), Russia (RS), Pakistan (PK), and United Kingdom (UK).

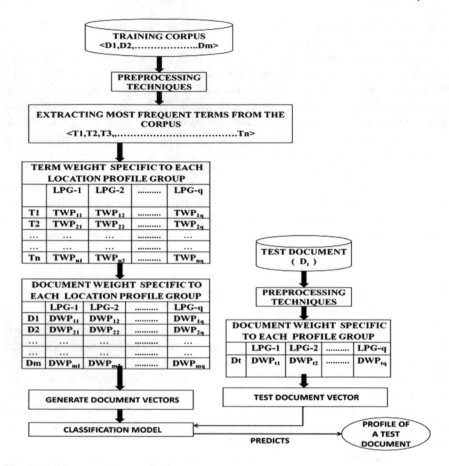

Fig. 1 PDW approach for location prediction

{LPG-1, LPG-2, …LPG-q} is the set of location profile groups of different countries. TWP$_{nq}$ is the weight of the term T_n in location profile group LPG-q. DWP$_{mq}$ is the weight of document D_m in location profile group LPG-q. In this approach first, the term weights are calculated specific to each location profile group using a term weight measure. Then, document weights were computed specific to every location profile group using document weight measure and weights of the terms. The document vectors were represented with document weights. Different classifiers are used in experimentation to generate the classification model using these document vectors.

In this approach, the usage of term weight measure and document weight measure influences the accuracy of location prediction.

4.1 Term Weight Measure

Various researchers proposed different types of term weight measures in different research areas. The term weight measures are classified into two classes such as supervised term weight measure and unsupervised term weight measures based on the class membership information is used or not. In this work, a new supervised term weight measure is proposed to compute the weights of the terms. The proposed supervised term weight measure is represented in Eq. (1).

$$W(t_i, d_k \in \text{LPG}-q) = \frac{tf(t_i, d_k)}{\text{DF}_k} * \frac{\sum_{x=1, d_x \in \text{LPG}-q}^{m} tf(t_i, d_x)}{1 + \sum_{y=1, d_y \notin \text{LPG}-q}^{n} tf(t_i, d_y)}$$

$$* \frac{\sum_{x=1, d_x \in \text{LPG}-q}^{m} \text{DC}(t_i, d_x)}{1 + \sum_{y=1, d_y \notin \text{LPG}-q}^{n} \text{DC}(t_i, d_y)} \tag{1}$$

In this measure, $tf(t_i, d_k)$ is the term frequency in document d_k, DF_k is the total number of terms in a document d_k.

$\sum_{x=1, d_x \in \text{LPG}-q}^{m} tf(t_i, d_x)$ gives the total count of the term t_i in all the documents of location profile group LPG-q.

$\sum_{y=1, d_y \notin \text{LPG}-q}^{n} tf(t_i, d_y)$ gives the total count of the term t_i in all the documents of location profile groups except LPG-q.

$\sum_{x=1, d_x \in \text{LPG}-q}^{m} \text{DC}(t_i, d_x)$ gives the number of documents in profile group LPG-q contains the term t_i.

$\sum_{y=1, d_y \notin \text{LPG}-q}^{n} \text{DC}(t_i, d_y)$ gives the number of documents in all profile group except LPG-q contains the term t_i.

The main principle of this term weight measure is: It assigns more weight to the terms which are having more frequency in interested document and contained in more number of documents in interested location profile group.

4.2 Document Weight Measure

In this work, a document weight measure is used proposed by Raghunadha Reddy et al. [10]. The document weight measure determines the weight of a document by considering different information of terms in a document. Equation (2) represents the document weight measure used in our experiment.

$$W(d_k, \text{LPG}-q) = \sum_{t_i \in d_k, d_k \in \text{LPG}-q} \text{TFIDF}(t_i, d_k) * W(t_i, \text{LPG}-q) \tag{2}$$

This measure used two types of information of terms such as TFIDF (Term Frequency and Inverse Document Frequency measure) weight of a term and the term weight calculated by term weight measure to compute the weight of a document. In this measure, $w(d_k, \text{LPG-}q)$ is the weight of document d_k in location profile group LPG-q.

5 Experimental Results

Table 1 shows the PDW approach accuracies for location prediction. In general, the researchers used different types of stylistic features to differentiate the writing styles of the authors. In our previous works, it was observed that the stylistic features are not more suitable to predict the location of the author certainly. By examining the dataset of location of different countries, it was observed that the differences in writing style of authors were identified in the terms used by the different countries authors. Based on this analysis, we selected 8000 most frequent terms for predicting the location of the author. In this work, the experimentation started with 1000 terms then increased to 8000 terms with an increase of 1000 terms in every iteration. It was observed that the obtained results were good for location prediction when compared with existing approaches for location prediction in Author Profiling. Different classification algorithms such as simple logistic (SL), logistic (LOG), IBK, bagging (BAG), naïve Bayes multinomial (NBM), and random forest (RF) were used to generate the classification model.

When compared with all classifiers, the random forest classifier achieved highest accuracy of 82.32% for location prediction. The naïve Bayes multinomial classifier achieved an accuracy of 80.38% for location prediction. It was observed when the number of terms was increased then the accuracies of location prediction were also increased in all classifiers.

Table 1 Accuracy of PDW approach for location prediction

Classifiers/number of terms	SL	IBK	LOG	BAG	NBM	RF
1000	60.59	65.70	68.07	70.21	71.17	72.71
2000	61.41	66.54	69.34	71.95	74.24	73.47
3000	63.26	69.36	71.24	72.67	75.84	76.80
4000	64.83	70.79	72.12	73.81	76.11	78.45
5000	65.51	71.85	73.87	74.85	77.49	79.31
6000	66.80	72.20	74.64	76.20	78.67	80.91
7000	68.39	73.14	75.25	77.14	79.62	81.65
8000	69.47	74.91	77.18	79.27	80.38	82.32

6 Conclusions

In this work, a PDW approach was used to predict the location profile of the authors from hotel reviews dataset. The random forest classifier obtained highest accuracy of 82.32% for location prediction when most frequent 8000 terms were used as features. It was planned to increase the accuracy of location prediction by considering the semantic relationship between terms in a document using deep learning concepts.

References

1. T. Raghunadha Reddy, B. VishnuVardhan, and P. Vijaypal Reddy, "A Survey on Authorship Profiling Techniques", International Journal of Applied Engineering Research, Volume 11, Number 5 (2016), pp 3092–3102.
2. Koppel, M., Argamon, S., Shimoni, A.R.: Automatically categorizing written texts by author gender. Literary and Linguistic Computing 17(4), pp. 401–412 (2002).
3. Argamon, S., Koppel, M., Pennebaker, J. W., and Schler, J. (2009). Automatically profiling the author of an anonymous text. Communications of the ACM, 52(2):119 (2009).
4. Schler, J., Koppel, M., Argamon, S., and Penebaker, J. "Effects of Age and Gender on Blogging". AAAI Spring Symposium on Computational Approaches to Analysing Weblogs (AAAI-CAAW), AAAI Technical report SS-06-03, (2006).
5. Estival, D., Gaustad, T., Pham, S.B., Radford, W. and Hutchinson, B., "Author profiling for english emails", in Proceedings of the 10th Conference of the Pacific Association for Computational Linguistics (PACLING'07)., (2007), 263–272.
6. Wee-Yong Lim, Jonathan Goh and Vrizlynn L. L. Thing, "Content-centric age and gender profiling", Proceedings of CLEF 2013 Evaluation Labs, 2013.
7. Seifeddine Mechti, Maher Jaoua, Lamia Hadrich Belguith, "Author Profiling Using Style-based Features", Proceedings of CLEF 2013 Evaluation Labs, 2013.
8. Suraj Maharjan and Thamar Solorio, "Using Wide Range of Features for Author Profiling", Proceedings of CLEF 2015 Evaluation Labs, 2015.
9. Raghunadha Reddy T, Vishnu Vardhan B, Vijayapal Reddy P, "Profile specific Document Weighted approach using a New Term Weighting Measure for Author Profiling", International Journal of Intelligent Engineering and Systems, Nov 2016, 9(4), pp. 136–146. https://doi.org/10.22266/ijies2016.1231.15.
10. Raghunadha Reddy T, Vishnu Vardhan B, Vijayapal Reddy P, "Author profile prediction using pivoted unique term normalization", Indian Journal of Science and Technology, Vol 9, Issue 46, Dec 2016.

Machine Learning and Mining for Social Media Analytics

G. Sowmya, K. Navya and G. Divya Jyothi

Abstract The expansion in the informal communication locales facilitates the way toward publicizing the results of associations and taking feelings from the clients. The associations which are showcasing their items need to comprehend what individuals are discussing their items. This can be accomplished by utilizing famous long-range informal communication locales. It has enhanced the heading of social media examination, which incorporates social network analysis, machine learning, and data mining. Presently a-days on account of the expansion in the interpersonal organizations, individuals began to utilize these sites (social organizing destinations) to share the data. Twitter is exceptionally prevalent for information mining in all the online networking locales as a result of its prominence and use by acclaimed individuals. In this paper, we will exhibit how the showcasing associations are publicizing their items or taking the criticism for their items utilizing Web-based social networking systems like Twitter. In this procedure, the framework will gather messages (tweets) from the locales and in light of that we will investigate the item's input. Inputs of any item will be shown as tweets regarding positive, negative, or nonpartisan. The machine learning algorithms (Naive Bayes, maximum entropy) are utilized to choose the outcomes. We likewise play out a pre-preparing stage to enhance the information precision. The point of our paper is to arrange the suppositions from the messages from the long-range interpersonal communication.

Keywords Machine learning · Data mining · Naïve Bayes · Maximum entropy

G. Sowmya (✉) · K. Navya · G. Divya Jyothi
Department of CSE, MLR Institute of Technology, Hyderabad, India
e-mail: sowmya.g@mlrinstitutions.ac.in

K. Navya
e-mail: navyareddy.karangula@gmail.com

G. Divya Jyothi
e-mail: divyag.1605@gmail.com

© Springer Nature Singapore Pte Ltd. 2019   397
R. S. Bapi et al. (eds.), *First International Conference on Artificial Intelligence and Cognitive Computing* , Advances in Intelligent Systems and Computing 815,
https://doi.org/10.1007/978-981-13-1580-0_38

1 Introduction

In this paper, we will utilize sentiment analysis which is vital for information mining. Assessment analysis is the computational treatment of suppositions. The vast majority of the general population are utilizing Twitter for sharing the data. By utilizing conclusion examination and the Twitter as a specialized device, we can gather the tweets on any inclining theme and order the outcomes as positive, negative, and nonpartisan [1]. In this paper, we will utilize an instrument called nostalgic examination device, which contains three capacities: the sentiment among Twitter messages, discovering which are certain, negative, and unbiased tweets from data assets [2]. Here we will concentrate on dissecting messages or tweets from the online networking locales [3].

2 Related Work

2.1 Social Media Analysis

The term online networking refers to the social relationship among the general population, contrast groups, locales, etc. Everybody of them will assume double parts, going about as hub and a social performer. In an advanced world, Web is the primary concern for everybody. Social media has hugeness in many fields like science, correspondence thinks about, financial aspects, data science, authoritative examinations, and social brain research [4, 5].

2.2 Twitter

Twitter is an online networking system which is utilized to enhance the connections among the general population by sharing the musings or messages called "tweets" [6]. The message length in Twitter is restricted to 140 characters; however, on November 7, 2017, the cutoff was stretched out to 280 characters for all dialects aside from Japanese, Korean, and Chinese. The clients who are enlisted can post the messages, yet the general population who are not enrolled can just read them. The clients can utilize this administration through the Twitter site or portable application programming. Twitter was made by Jack Dorsey, Noah Glass, Biz Stone, and Evan Williams in March 2006 and propelled in July of that year [7, 8].

The Twitter clients can choose the general population with whom they need to communicate. The clients will get the warnings when their companions have shared a tweet. A client who is trailed by another client require not to tailing them back. Twitter is named as a smaller-scale blogging administration. Small-scale blogging implies that it enables the clients to send messages, photographs, sounds, emoticons, or some other media. Among all the microblogging destinations (Twitter, Plurk, Tumblr, identi.ca, and others), Twitter contains many number of tweets and ended up noticeably famous in less time. And furthermore, the clients of Twitter fluctuate from standard individuals to celebrities, organization CEOs, government officials, and even presidents in this manner which gives a colossal base to information mining. On account of its ubiquity and information volume, we have chosen Twitter as the hotspot for this framework [9].

2.3 How Twitter Is Used in Social Network Analysis

An interpersonal interaction benefit is an online administration that is utilized to create informal community among the general population who are intrigued to share musings, exercises, data, or genuine associations. As interpersonal organization benefit stage is developing fast on the Internet, along these lines it gives adequate data to informal community examination. Among all the person-to-person communication sites, Twitter is the prevalent informal communication benefit [10]. Twitter prompts great position for informal community investigation since it has an exceptional confinement that it will accept just 140 characters in each tweet. A few analysts utilized Twitter for spatio-worldly data. Sakaki, Okazaki, and Matsuo researched the working idea of Twitter and presented an occasion warning framework which screens tweets and conveys notice [11]. Pak and Paroubek utilized Twitter for assessment mining and conclusion examination.

3 Framework

This model gathers opinions (tweets) from long-range informal communication site and gives a perspective of business insight. In this system, we have two layers in the sentiment analysis. They are data processing layer and sentiment analysis layer. The main layer manages information accumulation and information mining, while the second layer utilizes an application to introduce the aftereffect of information mining.

3.1 Collecting Data and Preprocessing

To begin with, we will accumulate the rundown of tweets or remarks on various items physically and then after experience the site of informal community locales to gather tweets. All the collected tweets can be stored in database for further processing. In the examination procedure, words and their implications will be considered. As indicated

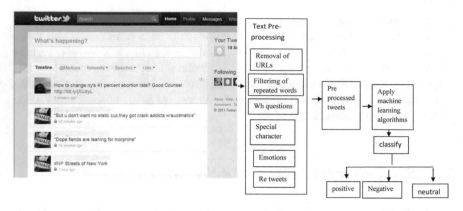

Fig. 1 Architecture for sentiment analysis system

by social semantic examination, tweets about tattles or disconnected information will be disposed of and subsequently real substance is precisely removed (Fig. 1).

The framework design comprises three sections, i.e., gathering messages from long-range interpersonal communication destinations, preprocessing the messages, and applying calculations. To begin with, we need to choose the messages from a content document or exceed expectations record to preprocess the messages. In pre-preparing layer, the information which is not imperative like rehashed messages (tweets), rehashed letters, URLs, feeling symbols, WH questions, exceptional images will be disposed of. In the wake of preprocessing, we have to choose a calculation to get the surveys on items like telephones, iPods, silver screens, electronic media.

3.2 Algorithms Used in Machine Learning

Machine learning is a piece of counterfeit consciousness. It is a logical control which manages the advancement of calculations which enable the PC to deliver the outcomes in view of information exhibit in database. The data will provide the relations between the variables. The primary point of machine learning is to naturally figure out how to perceive and settle on better choices in light of information [12].

3.3 Naive Bayes Algorithm

The Bayesian classification shows an immediate learning strategy and additionally a measurable technique for characterization. Expect a probabilistic model and it enables us to catch the danger about the model fairly by deciding the probabilities of the results. By using this algorithm, we can understand and evaluate many learning algorithms

$$C* = \text{argmac } P_{\text{NB}}(c|d)$$

$$P_{\text{NB}}(c|d) := (P(c)) \sum_{i=1}^{m} P(f|c)^{n1(d)}/P(d)$$

3.3.1 Proposed Naive Bayes Classifier

Information: messages $M = \{m1, m2, m3 \ldots mn\}$,
  Database: Bayes Table BT
  Yield: great messages $F = (f1, f2 \ldots)$,
  Non-great messages $N = \{n1, n2, n3 \ldots\}$,
  Impartial messages $nu = \{nu1, nu2, nu3 \ldots\}$
  $M = \{m1, m2, m3 \ldots\}$

Stage 1: Divide the tweet message into words $mi = \{w1, w2, w3 \ldots\}, i = 1, 2, \ldots n$
Stage 2: If wi BT
return +ve extremity and −ve extremity.
Stage 3: We can ascertain the general extremity of a word by utilizing the accompanying equation log(+ve extremity)- log(−ve extremity).
Stage 4: Repeat the stage 2 until for every one of the words till end of the words.
Stage 5: To get the aggregate extremity of a message, we have to include the polarities of all expressions of a message.
Stage 6: The message might be sure or negative and this will be chosen by the extremity.
Stage 7: We have to rehash the stage 1 for every one of the messages until the point that M is NULL.

3.4 Maximum Entropy Algorithm

This is a machine learning algorithm used for grouping and predicting. The maximum entropy techniques are used in part of speech tagging, ambiguity resolution, and parse selection. The advantage of maximum entropy algorithm is its flexibility. At the same time, it is very cost-effective. Indeed, even the advanced maximum entropy models require numerous assets and a lot of information with a specific end goal to precisely assess the model's parameters. These models contain hundreds of thousands of parameters. This algorithm is not only expensive but also sensitive to roundoff the errors. For estimating the parameters, it requires highly efficient, accurate, scalable methods. Here, we consider various calculations for assessing the parameters of ME displays, including generalized iterative scaling and improved iterative scaling, and broadly useful streamlining systems.

$$P_{ME}(c/d_1\lambda) = \exp\left[\sum_i \lambda_i f_i[c, d]\right]$$

$$\sum_{ei} \exp\left[\sum_i \lambda_i f_i[c, d]\right]$$

3.4.1 Proposed Maximum Entropy Classifier

Information: Messages $M = \{m1, m2, m3\ldots, mn\}$,
  Database: Entropy Table ET Output:
  Great messages $F = \{f1, f2\ldots\}$,
  Non-great messages $N = \{n1, n2, n3\ldots\}$
  Nonpartisan messages nu = {nu1, nu2, nu3...}
  $M = \{m1, m2, m3\ldots\}$

Stage 1: Divide a tweet message into words $mi = \{w1, w2, w3\ldots\}, i = 1, 2, \ldots, n$
Stage 2: If wi NT.
return +ve extremity and −ve extremity.
Stage 3: We can ascertain the general extremity of a word by utilizing the accompanying equation ((+ve polarity) * log(1/+ve extremity))−((−ve polarity) * log(1/−ve extremity)).
Stage 4: Repeat the stage 2 until for every one of the words till end of the words
Stage 5: To get the aggregate extremity of a message, we have to include the polarities of all expressions of a message.
Stage 6: The message might be certain or negative and this will be chosen by the extremity.
Stage 7: We have to rehash the stage 1 for every one of the messages until the point that M is NULL.

4 Runtime Execution

In the pre-preparing module, we need to dispose of the information which is not vital like repeated tweets, URLs, feeling symbols, WH questions, unique images. The message size will be measured after preprocessing. The outcomes are as per the following. In the preprocessing in the event that we consider, the first message estimate is 100% in this way, and subsequent to evacuating the repeated tweets, the recorded measure continuously diminishes to 96.9%. After URL expulsion, the document estimate diminishes. The file size will be still decreased after removing the emotion icons, WH questions, special. In this way, after pre-preprocessing the messages, it step by step declines to 92.5%. Precision: The greatest entropy calculation is contrasted and another instruments of wistful examination like sentiment 140, twittratr, tweetfeel. We should give same messages to both most extreme entropy and other wistful devices (Figs. 2 and 3).

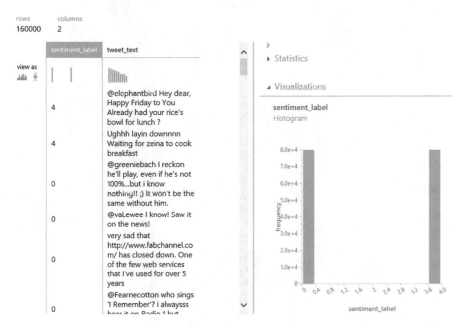

Fig. 2 Preprocess the Twitter reviews

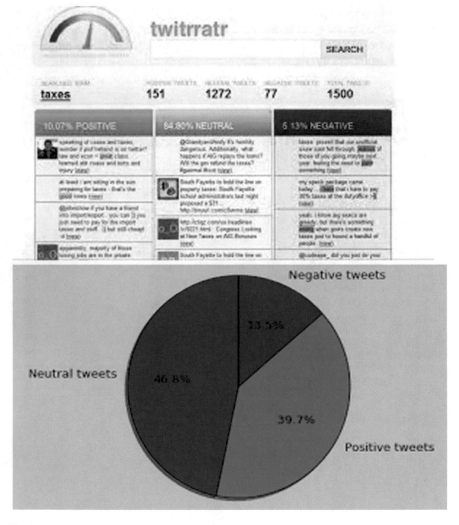

Fig. 3 Messages classified as positive, negative, and neutral

5 Conclusion

In this examination paper, we have proposed a model for taking the client suppositions on any item or theme. One of the elements of the proposed framework is to preprocess the tweets to find the client assessments. After preprocessing stage, the data will be stored in database. This stored data will be used for machine learning algorithms. For this assurance, we are utilizing machine learning technique.

References

1. Becker, H., Naaman, M., Gravano, L.: Beyond Trending Topics: Real-World Event Identification on Twitter. ICWSM, 11, 438–441, 2011.
2. Bethard, S., Yu, H., Thornton, A., Hatzivassiloglou, V.,Jurafsky. D.: Automatic Extraction of Opinion Propositions and their Holders. In: Proceedings of the AAAI Spring Symposium on Exploring Attitude and Affect in Text, 2004.
3. Au Yeung, C. M., and Iwata, T.: Strength of social influence in trust networks in product review sites. In Proceedings of the fourth ACM international conference on Web search and data mining (pp. 495–504). ACM, 2011.
4. Aggarwal, N., Liu, H.: Blogosphere: Research Issues, Tools, Applications. ACM SIGKDD Explorations. Vol. 10, issue 1, 20, 2008.
5. Aggarwal, C.: An introduction to social network data analytics. Springer US, 2011.
6. Adedoyin-Olowe, M., Gaber, M., Stahl, F.: A Methodology for Temporal Analysis of Evolving Concepts in Twitter. In: Proceedings of the 2013 ICAISC, International Conference on Artificial Intelligence and Soft Computing. 2013.
7. Aiello, L. M., Petkos, G., Martin, C. et al.: Sensing trending topics in Twitter. 2013.
8. Bakshy, E., Hofman, J. M., Mason, W. A., Watts, D. J.: Identifying influencers on twitter. In Fourth ACM International Conference on Web Seach and Data Mining (WSDM), 2011.
9. Becker, H., Chen, F., Iter, D., Naaman, M., Gravano, L.:Automatic Identification and Presentation of Twitter Content for Planned Events. In ICWSM, 2011.
10. Becker, H., Iter, D., Naaman, M., Gravano, L.: Identifying content for planned events across social media sites. In Proceedings of the fifth ACM international conference on Web search and data mining (pp. 533–542). ACM, 2012.
11. Bekkerman, R., McCallum, A.: Disambiguating web appearances of people in a social network. In Proceedings of the 14th international conference on World Wide Web (pp. 463–470). ACM, 2005.
12. Asur, S., and Huberman, B.: "Predicting the future with social network." Web Intelligence and Intelligent Agent Technology (WIIAT), 2010 IEEE/WIC/ACM International Conference on. Vol. 1. IEEE, 2010.

Accent Issues in Continuous Speech Recognition System

Sreedhar Bhukya

Abstract Speech is the output of a time-varying excitation excited by a time-varying system. It generates pulses with fundamental frequency F0. This time-varying impulse trained as one of the features, characterized by fundamental frequency F0 and its formant frequencies. These features vary from one speaker to another speaker and from gender to gender also. In this paper, the accent issues in continuous speech recognition system are considered. Variations in F0 and formant frequencies are the main features that characterize variation in a speaker. The variation becomes very less within speaker, medium within the same accent, and very high among different accent. This variation in information can be exploited to recognize gender type and to improve performance of speech recognition system through modeling separate models based on gender type information. Five sentences are selected for training. Each of the sentences is spoken and recorded by five female speakers and five male speakers. The speech corpus will be preprocessed to identify the voiced and unvoiced region. The voiced region is the only region which carries information about F0. From each voiced segment, F0 is computed. Each forms the feature space labeled with the speaker identification: i.e., male or female. This information is used to parameterize the model for male and female. K-means algorithm is used during training as well as testing. Testing is conducted in two ways: speaker dependent testing and speaker independent testing. SPHINX-III software by Carnegie Mellon University has been used to measure the accuracy of speech recognition of data taking into account the case of gender separation which has been used in this research.

Keywords Speech recognition (SR) · Linear prediction coding (LPC) · Accent

S. Bhukya (✉)
Speech and Vision Laboratory, IIIT-Hyderabad, Hyderabad, India
e-mail: sr2naik@gmail.com

© Springer Nature Singapore Pte Ltd. 2019
R. S. Bapi et al. (eds.), *First International Conference on Artificial Intelligence and Cognitive Computing* , Advances in Intelligent Systems and Computing 815,
https://doi.org/10.1007/978-981-13-1580-0_39

1 Introduction

As speech recognition is a complex task, it is still difficult to find the complete solution, because every human being has his/her own different characteristics of voice and accent, this in itself has become one of the main problems in the field of speech recognition [1]. It is worth pointing out the fact that this research investigates an approach for identifying genders from spoken data and builds separated models for accent that can enhance the performance of speech recognition by reducing the search space in lexicon.

Studies in gender classification using voice shall give insights into how humans process voice information. What may be important is that gender information is conveyed by the F0, type of sound, and the size of vocal tract. It can be assumed that modeling the vocal tract using linear prediction coding (LPC) and cepstrum information [2]. Furthermore, small errors in gender classification can be allowed as sometimes it is even hard for a person to identify the gender of a speaker. Studies by Susan Schötz at Lund University [3] have showed that a feature-based system with a trained decision tree can successfully classify male and female voices automatically. However, her studies have been most concentrated on age separation.

In the context of speech recognition, accent and gender separation can improve the performance of speech recognition by limiting the search space to speakers from the same accent and gender. In the gender classification accuracy, it has been noticed that perhaps the most important variability across speakers besides gender is a role played by accent. Therefore, it is probably that any recognition system attempting to be robust to a wide variety of speakers and languages should make some effort to account for different accents that the system might encounter.

In this paper, one approach is to use gender dependent features, such as the pitch and formants. The pitch information was used in [3–5] for the problem of gender separation. However, fundamental frequency and formant frequencies estimation rely considerably on the speech quality and accent. Although the quality of speech used in this study was not free from noise, tried to improve the gender model by using K-means algorithm to get high accuracy.

2 Implementation

Gender separation is the process of separating the speaker's gender type directly from the acoustic information. This is possible using gender invariant feature separation and learning from variations within gender and accent. Furthermore, speech is produced as a result of convolution of excitation of the vocal cord with the vocal tract system coupled with the nasal tract. It is decided to make a practical study of accent issues in continuous speech recognition system, trying to find out the most correct results by implementing autocorrelation technique. This practical study was carried out through several stages that will be described as follows.

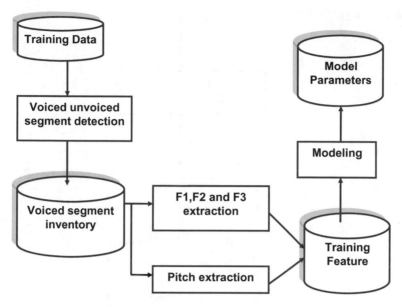

Fig. 1 Matching of voiced data segment detection

Stage 1: Data Collection

Training Data:
Five female and five male speaker subjects are selected to record five properly selected sets of sentences. The selected recorded sentences for the training purpose are: *welcome, where Mike is, I believe you are fine, Have fun with him, Thanks to God*.

Testing Data:
For the testing purpose, five females and five males have been selected to speak one sentence and each has been recorded. In fact, every speaker has given a sentence different from other speaker. It has to be observed that the group of testing is actually independent from that of the training data.

Stage 2: Feature Extraction
In this stage, we are going to build the process that can identify the gender type on the basis of individual information included in speech waves through extracting the features, viz., fundamental frequency F0 from the collected training data making separate models depending on the type of features and optimal parameters for each gender. The fundamental frequency shows high variation from one speaker to another. The variation becomes higher when the comparison is among speakers of different gender.

This process is represented in Fig. 1. Technically, when the voice characteristics of utterance are checked, there will be a wide range of probabilities that exist for parametrically representing the speech signal for the gender separation task.

Fig. 2 Matching of features

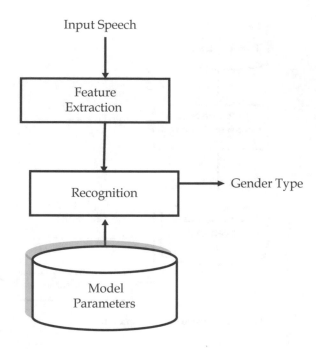

Stage 3: Feature Matching and Decision Making

In this stage, extracted the same feature type for testing data as done during training stage. Then applied the concept of pattern recognition to classify objects of interest into one of the desired gender types. The objects of interest are called sequences of vectors that are extracted from an input speech. Since the classification procedure in our case is applied on extracted features, it can be also referred to as feature matching (Fig. 2).

The K-means approach [6–8] is used because of its flexibility and ability to give high recognition accuracy. K-means can be simply defined as a process of mapping vectors from a large vector space to a finite number of regions in that space. Each region is called a cluster and can be represented by its center called a codeword. The collection of all code words is called a codebook. The gender separation system will compare the codebooks of the tested speaker with the codebooks of the trained speaker. In other words, during recognition, among the models, the best model that maximizes the joint probability of the given observation will be selected as recognized model. The best-matching result will be the desired gender type and this can be verified as the decision making logic. Finally, the gender type which is modeled by the recognized model will be given as an output of our system [9–17].

3 Data Analysis and Observation

In this section, the data analysis of the work is discussed. Figure 3 shows feature extraction (F0) in which the utterance *welcome* of both genders is analyzed showing that female on the right and male on the left. It shows that female speakers have higher pitch than male. Regarding the technique of feature extraction, each subfigure has its description in Fig. 3.

Speech Recognition Accuracy Results:

Speech recognition accuracy has been measured under Sphinx system and the results are as follows:

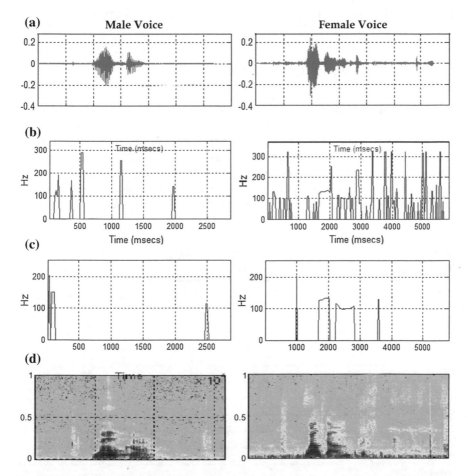

Fig. 3 **a** Speech single for male speaker speech signal for female speaker, **b** Pitch estimated by autocorrelation method, **c** Pitch estimated by cepstrum method, **d** Spectrogram, *Note* X axe represents the time in *m* sec and *Y* axe represents frequency in Hz

When the data is mixed from both genders, the accuracy resulted in 58%, and we consider this as low results of accuracy. But when we separate the gender type, an increase in accuracy is obtained. This appears obviously through the results achieved by the same gender of female which is 84%, while the accuracy results of the same gender of male are 78%. In the step followed, we test the accuracy within the same gender and same speaker and the accuracy resulted is 100%.

In addition, for the training data of male and testing data of female, the accuracy of speech recognition resulted in 45%, while training data of female and testing data of male is resulted in 34%. Moreover, the results of same accents (Indian Accent) appear to be somewhat different for both genders. Their accuracy is 70%. And when they are separated, the accuracy of male is 72% whereas the female is 90%. To sum up, we conclude that the gender separation and accent plays an important role in increasing the rate of accuracy results of speech recognition (Table 1).

4 Conclusions and Future Work

Speech is considered as the essential form of communication between humans. It plays a central role in the interaction between human and machine, and between machine and machine. The automatic speech recognition is aimed to extract the sequence of spoken words from a recorded speech signal and so it does not include the task of speech understanding, which can be seen as an even more elaborated problem. Because of the fact that the goal of speech recognition is still far away from the optimal solution for higher accuracy, the gender separation system has been

Table 1 Speech recognition accuracy of different accent, same accent, and same speaker

Training data		Testing data		Accuracy %
Male	Female	Male	Female	
Different accent				
☑	☑	☑	☑	58
	☑		☑	84
☑		☑		78
☑			☑	34
	☑	☑		45
Same accent				
☑	☑	☑	☑	70
	☑		☑	90
☑		☑		72
Same speaker				
☑		☑		100
	☑		☑	100

proposed to enhance the performance of speech recognition through building separate gender accent models by limiting search from whole space of acoustic models that can further lead to improve accuracy of speech recognition.

Although the speech data used in this study are collected from different nationalities with different accents (Arabs, Russians, Americans, and Indians) and recorded in different sampling rate and channels. High accuracy of gender recognition is obtained by using the technique which has been mentioned so far.

When applying K-means algorithm as pattern recognition for the extraction of features, the results of the experiments for estimated pitch value through autocorrelation and cepstrum have shown 100% accuracy. Consequently, we conclude that pitch features are more suitable and strongly advised to distinguish the gender type.

Future work

In future work, we are planning to expand the range of evaluation set. We are also planning to investigate the performance of the system to make it applicable, usable as well as useful for the study purpose in speaker separation and speaker verification.

Summary

In both the gender separation accuracy and the speech recognition accuracy of Sphinx transcriptions in our work, we have noticed that the accent variability among speakers plays a crucial role in speech recognition accuracy besides gender. Therefore, it seems that any recognition system attempting to be robust to a wide variety of speakers and languages should make some effort to account for different accents that the system might encounter.

References

1. Breazeal, C. and Aryanda, L. (2000),'Recognition of affective communicative intent in robot-directed speech,' in' Proceedings of Humanoids 2000.
2. http://www.ece.auckland.ac.nz/p4p_2005/archive/reports2003/pdfs/p60_hlai015.pdf.
3. http://www.ling.lu.se/persons/Suzi/downloads/RF_paper_SusanneS2004.pdf.
4. S. Davis and P. Mermelstein. Comparison of parametric representations for monosyllabic word recognition in continuously spoken sentences. IEEE Transactions on Acoustics Speech and Signal Processing, 28:357–366, Aug 1980.
5. Parris E. S., Carey M. I., Language Independent Gender Identification, Proceedings of IEEE ICASSP, pp 685–688, 1996.
6. Linde, Y., A. Buzo, and R.M. Gray, "An Algorithm for Vector Quantizer Design,"IEEE Trans. on Communication, 1980, COM-28(1), pp. 84–95.
7. Hartigan, J.A., Clustering Algorithm, 1975, New York, J. Wiley.
8. Gersho, A., "On the Structure of Vector Quantization," IEEE Trans. on Information Theory, 1982, IT-28, pp. 256–261.
9. Richard P. Lappmann, Speech recognition by Machines and Humans, SPEECH Comm., pp. 1–15, 1997.
10. Santosh K. Gaikwad, Bharti W. Gawali and Pravin Yannawar, A Review on speech recognition technique, International Journal of Computer Application, Vol. 10(3), pp. 16–24, 2010.
11. M. Prabha, P. Viveka and Bharatha sreeja, Advanced gender recognition system using speech signal, IJSET, Vol.6(4), pp. 118–120, 2016.

12. Chetana Prakash and Suryakanth V Gangasetty, Fourier- Bessel based cepstral coefficient features for text-indipendent speaker identification, IICA, pp. 913–930, 2-11.
13. Musaed Alhussein, Zalfiqar Ali, Muhammad Imran and Wadood Abdul, Automatic gender detection based on characteristics of vocal folds for mobile healthcare system, Hindawi, pp. 1–12, 2016.
14. Suma Swamy and K. V Ramakrishnan, An efficient speech recognition system, CSEIJ, Vol3(4), pp. 21–27, 2013.
15. Preeti Saini and Parneet Kaur, Automatic speech recognition: A review, IJETT, Vol (2), pp. 132–136, 2013.
16. Bhupinder Singh, Neha Kapur and Puneet Kaur, Speech recognition with Hidden Markow model: A review, IJARCSSE, Vol. 2(3), pp. 400–403, 2012.
17. M.A Anusuya and S.K Katti, Speech recognition by Machine: A review, IJCSIS, Vol6(3), pp. 181–205, 2009.

Effective Handling Personal Electronic Health Records Using Metadata Over Cloud Computing

E. V. N. Jyothi and B. Rajani

Abstract Cloud computing plays an essential role in managing the personal clinical data, i.e., data storage and access. In overall, improving patient outcomes is a critical phase, and accordingly, personal clinical data are challenging to collect and manage from other sources due to their disseminated in nature, i.e., patient data located at different sources of places such as hospitals, clinics, test centers, and doctor's office; in overall, data are in the form of structured and unstructured formats like images, text, chart, or hard copies of documents. But when the situation is critical it is challenging to collect personal clinical data from several sources in different formats which cause no evidence-based diagnosis and treatment. In response to such scenarios, we propose a methodology that accomplishes personal health data by exploiting metadata for the organization and easy retrieval of clinical data and cloud storage for easy access and sharing with doctors to apply the continuity of care and evidence-based treatment which improves human lives.

Keywords Personal electronic health records (PEHRs) · Dublin core metadata
Cloud storage · Standard medical codes

1 Introduction

In general, people with good health are rich than wealthy people; in this connection, health care is considered essential as we experience the increase in chronic diseases such as heart disease, cancer, diabetes, and asthma which require continuous treatment, reduce the value of life, and increase overall medical expenses. According to the National Centers for Disease Control (NCDC), 17.5 million people die each

E. V. N. Jyothi (✉)
PACE Institute of Technology & Sciences, Ongole, India
e-mail: jyothiendluri@gmail.com

B. Rajani
Keshav Memorial Institute of Technology, Hyderabad, India
e-mail: rajani.badi@gmail.com

© Springer Nature Singapore Pte Ltd. 2019
R. S. Bapi et al. (eds.), *First International Conference on Artificial Intelligence and Cognitive Computing* , Advances in Intelligent Systems and Computing 815,
https://doi.org/10.1007/978-981-13-1580-0_40

year in India from cardiovascular diseases, amounting to a staggering 31% of all deaths [1]. As per the World Health Organization, non-communicable diseases or chronic diseases, such as cancer, heart ailments, respiratory diseases and diabetes, kill 38 million people globally every year. [2]. However, many of these diseases can be prevented and managed through early detection, physical activities, a balanced diet, and treatment therapy. Recently, there has been more focus on preventive care or monitoring and control of the symptoms. Nowadays, there are many mobile health applications and sensors such as blood pressure sensors, electrocardiogram sensors, blood glucose measuring devices which are used by the patients who monitor and control their health. These apps and sensors produce personal health data that can be used for treatment purposes if managed and handled properly, since it can be considered patient-generated data.

There are other types of personal health data that are available from various sources such as hospitals, doctor's offices, clinics, radiology centers, or any other caregivers. Health above documents is deemed as an electronic personal health record (EPHR). According to AHIMA [32], PHR can be defined as an electronic, lifelong resource of health information needed by individuals to make health decisions. However, it is not easy to collect all the relevant personal health data because of the facts that they are in different data types, available from different sources, and stored in different media and devices. To overcome such difficulties, it is desirable to have personal health data in one place where users have full control over their clinical data. To be useful, the clinical data should be shareable when needed for the diagnosis and treatment. Without proper clinical information, such as medical history, allergies, current medication, adverse reaction, medical mistakes could occur when making medical decisions due to insufficient information. Even if a patient has complete medical history and all the necessary clinical data, if it is not shared properly among caregivers at the time of need, discontinuity in care may occur. To meet the needs of such scenario, a personal health record system should have the following properties robust and private storage, easy retrieval and maintenance, secure, shareable and be able to handle emergency situations.

There are two types of personal health record systems—untethered and tethered. Untethered PHR is an independent PHR where patients have full control over their health records and to collect, manage, and share them. On the other hand, the tethered PHRs is linked to a specific healthcare providers' EHR system where the users typically gain easy access to their records through secure portals and see their clinical information such as test results, immunization record, family history and also secure messaging with their collaborating clinicians. However, the participating patients need to share the cost, and the information they provide may not be complete since the information sources are from one provider only. In our previous work [16], we identified six barriers (usability, ownership, interoperability, privacy and security, portability, and motivation) that cause the slow adoption of PHRs. We propose an untethered PHR that utilizes personal cloud storage, offers simplicity in organizing various kinds of clinical data by utilizing metadata and easy access to emergency clinical data to paramedics or clinicians in case of emergency.

The Government of India (GoI) is also in the process of Standardizing Codification for Health Procedures in India and framing Metadata and Data Standards (MDDS) for health sector. MDDS will facilitate interoperability of e-Governance applications by providing a common data information model for various stakeholders in the health system—national and state, public and private, to begin sharing meaningful information with each other in a timely manner, but since it is not specifically designed for clinical data, there are some limitations in its expressive power in healthcare domain. In this paper, we simplified the categorization of clinical data by human body part for easy retrieval of clinical data using MDDS so users can manage their clinical data without in-depth knowledge about clinical information. As a proof of our concept, we developed a system called My Clinical E-Record System (MCERS) to help users to store, organize, retrieve, and share their clinical data with caregivers when needed including emergency situations. In an emergency situation, clinicians, e.g., physicians, paramedics, nurses, can access patient's data using their license numbers, patient name, and their date of birth only. Emergency information consists of current medication list, known allergies, and side effects. By having complete medical history, the MCERS' users may be able to reduce medical errors and improve patients' outcome. It also ensures the continuity of care by sharing personal clinical data among healthcare providers when needed.

In this paper, we began our discussion by analyzing the problem as it relates to the users' needs and expectations for collecting and maintaining their personal health records. The remainder of this paper is organized as follows: Sect. 2 discusses related work, and in Sect. 3, need for electronic health record and data standards for image, multimedia, waveform, and document is discussed. In Sect. 4, we discuss how to solve the data organization and retrieval using MDDS to facilitate better and more accurate data retrieval. Section 5 talks about cloud storage. In Sect. 6, finally we conclude our study and discuss our future work in Sect. 7.

2 Literature Review

Zhang et al. [3] developed an application to apply metadata efficiently to clinical trial data. The authors chose Microsoft Excel due to the wealth of built-in features (e.g., spell-checking, sorting, filtering, finding, replacing, importing, and exporting data capabilities) which contribute to the ease of use, power, and flexibility of the overall metadata application. They focused on the analysis process in a drug development environment such as adverse clinical events (ACEs), electrocardiogram (ECG), laboratory (LAB), and vital signs (VITAL) where the raw data are stored in the clinical trial database and then manipulated the data.

Teitz et al. [4] developed a Web site called HealthCyberMap with the goal of mapping Internet health information resources in novel ways for enhanced retrieval and navigation. They are using Protege-2000 to model and populate a qualified DC RDF metadatabase. They also extended the DC elements by adding quality and location elements of the resource. Also, the W3C RDFPic project extends the DC schema by adding its elements such as camera, film, lens, and film development date for describing and retrieving digitized photos [4].

Ariel Ekblaw et al. [5] built a system (RedRec) to enable patients to access their medical health records across health providers (e.g., pediatrician, university physician, dentist, employer health plan provider, specialists). Their system applies novel, blockchain smart contracts to create a decentralized content-management system for healthcare data across providers. The RedRec governs medical records access while providing the patient the ability to share, review, and post new records via the flexible interface. The raw medical record content is kept securely in providers' existing data storage. But when the patient wants to retrieve data from their provider's database, their database Gatekeeper checks authentication and then if it is approved, the Gatekeeper retrieves the relevant data for the requester and allows sync with the local database.

Dogac et al. [12] proposed archetypes to overcome the interoperability problem in exchanging clinical data. They provided guidelines on how ebXML registries can be used as an efficient medium for annotating, storing, discovering, and retrieving archetypes. They also used ebXMLWeb services to retrieve data from clinical information systems. "An archetype is defined as "a reusable," formal expression of a distinct, domain-level concept such as "blood pressure," "physical examination," or laboratory results, expressed in the form of constraints on data whose instances conform to some reference model" [12]. However, most existing research in the health domain focused on a single data type but our study is a comprehensive study that covers many different clinical and nonclinical documents such as images (e.g., x-ray, scanned document, ultrasound), text (e.g., CDA, CCR, CCD), observed symptoms noted by patients, clinical sensors' data which can help to organize these various data in a way that can help in storing and retrieving such data in an efficient way.

Fearon et al. [13] defined metadata as "structured information that describes, explains, locates, or otherwise makes it easier to retrieve, use, or manage an information resource." However, metadata standard has not been employed by many repositories, and most of the metadata was descriptive rather than administrative or for preservation [14].

3   Clinical Data

In this section, we describe the types, format, and sources of the clinical data and the techniques on how to store and retrieve them.

A.   Dimension data from portable medical devices, sensors, or mobile application

A clinical sensor is a device that responds to a physical stimulus and transmits a resulting impulse for interpretation or recording. Some sensors are designed to work outside of the human body while others can be implanted within the human body.

In this paper, we are referring to the sensors for the homecare setting, such as blood oxygen monitors, thermometers for body temperature, heart rate, sensor glucose (SG), blood pressure. There can be non-textual data generated from sensors such as electrocardiogram measurement device. The clinical sensors play major roles in health care, including early detection of diseases, diagnosis, disease monitoring, and treatment monitoring [24]. Another method to collect measurement data is mobile health applications. For instance, most smartphones (e.g., Android, iPhone, Samsung Galaxy) offer health and fitness apps that help users to monitor their daily activities and health (e.g., track diet and nutrition calories, track vital signs, track fitness progress, share health data with their doctor electronically). The data collected from these applications can be sent as a message or an email attachment to which the users want to share it with. MDDS will facilitate interoperability of e-Governance applications by providing a common data information model for various stakeholders in the health system—national and state, public and private, to begin sharing meaningful information with each other in a timely manner.

B. Observed Symptoms

Patients sometimes experience particular symptoms (e.g., chest pain, nausea, vomiting, shortness of breath). If the patient notices, they should be recorded and shared with their physician for proper treatment. If those are not shared with their physician, due to incomplete information, misdiagnosis could occur. When recording, the observed symptoms should be described in standardized code such as MDDS for semantic interoperability as the same symptoms can be described in multiple ways. Without codified description, there can be discrepancies of the perceptions regarding symptoms between patient, nurses, or physicians.

C. Images

Most of the medical imaging machines produce standard image format called digital imaging and communications in medicine (DICOM). There are two types of clinical data images: images that are based on DICOM standard (e.g., x-rays, computed tomography (CT), magnetic resonance (MR), and ultrasound devices) and scanned documents. DICOM is defined as the international standard for medical images and related information (ISO 12052). The DICOM format combines images and metadata that describes the medical imaging procedure. Accessing data in DICOM files becomes as easy as working with TIFF or JPEG images. On the other hand, the scanned documents (e.g., PDF/JPEG) are difficult to retrieve because the content is not searchable. For example, some physicians write notes on clinical forms while diagnosing their patients and then type them on the computer or just scan them and upload them to the patient records. Either way is time-consuming, difficult to retrieve promptly, and consumes relatively large storage space. Also, the patient may have more than one doctor or had been treated by many health providers, which in turn fragments his/her records. So when the patients obtain their records, they mostly got them either printed out or sent as an email attachment. This, in turn, makes it difficult to retrieve scanned documents because its content cannot be retrieved by computers. To alleviate such issues, we have utilized metadata to describe such medical documents, so computerized retrieval and systematic organization are possible.

D. Clinical Document

PEHR data may be collected from healthcare providers. There are three types of clinical document formats—continuity of care record (CCR), clinical document architecture (CDA), and continuity of care document (CCD)—that allow healthcare providers to exchange clinical information summary about a patient. But CCR was excluded from the of the 2014 Edition of EHR certification as a valid way to send the summary of care documents. Hence, the content from a CCR was merged into a CDA format and called continuity of care document (CCD). Currently, with Meaningful Use Stage 2 and the 2014 Edition of EHR certification, consolidated CDA includes CCD as one of its document types. When using untethered personal health record system, patients are responsible for collecting clinical data from their healthcare providers or their patient-generated measurement data and keep it in their storage such as personal cloud space. For example, CDA, CCR, and CCD can be obtained from healthcare providers; X-rays can be obtained from the radiology department and laboratory test results from test.

Laboratory or doctor's office: Patients can share their health records with their clinicians by either electronically transmitting or granting access to their storage through the PHR whenever possible. If electronic sharing is not allowed, the patient may download the file and make hard copies or store them in USB, CD, or other mediums for sharing clinical data needs to be stored in one place for easy sharing and retrieval. Well-managed personal cloud space could outlive the lifetime of PHRs since the discontinuity of the service does not affect the data stored in the cloud space. In our approach, we separate the clinical data from applications to make the data independent of the application. Also, the users can have alternative applications for their clinical data. Such independence motivates the user to adopt PHR and use PHRs with flexibility. Our proposed concept is illustrated in Fig. 1. In the figure, the clinical data are separated from the application for data independence.

4 Metadata

Metadata benefits personal health record management in many ways. Among those are

- **Uniformity in definitions**: Properly defined tags provide structured information about the clinical data users stored.
- **Explaining the relationships**: Metadata can be used to clarify the relationships among the clinical data by defining categories and associated relationships in the category. We have defined the usage of each tag in the DC for clinical data organization as well as easy retrieval. When the data are uploaded or modified, so is the

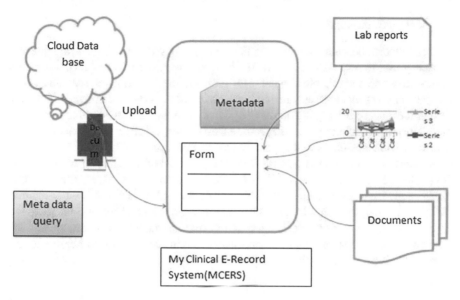

Fig. 1 Overview of the relationships among metadata, clinical data, cloud storage, PHR application, and clinical data sources

corresponding metadata. There are two different methods of storing metadata. In the first method, metadata can be embedded in the data (e.g., in the header of the digital file). The advantages of this option are ensuring that the metadata will not be lost, no need for linking data and metadata, and both the object and metadata can be updated together. In the second method, metadata can be stored separately where the metadata is stored in a database and linked to the objects. The advantage of this option is that it can simplify the management of metadata and can expedite the retrieval of the data [13]. In our approach, we employed the latter method to accelerate the retrieval of clinical data and to enhance expressive power. However, in this method, there can be inconsistencies between metadata and clinical data when transitioning to a new platform, integration between different systems, or sharing data across multiple systems [15]. There are many metadata formats that have been accepted internationally including Dublin Core (DC), Federal Geographic Data Committee (FGDC), Encoded Archival Description (EAD), and Government Information Locator Service (GILS), to name a few [17]. One of the major issues in using metadata among healthcare systems is the lack of interoperability [16]. The Government of India (GoI) is also in the process of Standardizing Codification for Health Procedures in India and framing Metadata and Data Standards (MDDS) for health sector. MDDS will facilitate interoperability of e-Governance applications by providing a common data information model for various stakeholders in the health system—national and state, public and private, to begin sharing meaningful information with each other in a timely manner.

A. Metadata management

In the healthcare domain, making a medical decision based on incomplete or incorrect data may lead to mistakes in treatments. Hence, it is important for decision-makers to have accurate and reliable clinical information at the time of need. Metadata management ensures that the data are associated with the datasets and utilized efficiently throughout and across organizations [18]. Data governance is needed for successful metadata implementation so it can provide trustworthy, timely, and relevant information to decision-makers as well as personal users. For successful implementation, data governance must be aligned with the intended purposes of the users or organizations [19]. The Government of India (GoI) is also in the process of Standardizing Codification for Health Procedures in India and framing Metadata and Data Standards (MDDS) for health sector. MDDS will facilitate interoperability of e-Governance applications by providing a common data information model for various stakeholders in the health system—national and state, public and private, to begin sharing meaningful information with each other in a timely manner.

5 Personal Cloud Storage

Table 1.
  Table 1 MDDS metadata content for doctor visit summary

Table 1 Metadata schema for EPHR

Entity	Description
Title	The title of the uploaded document
Creator	The author of the document
Subject	Chief complaint/reason of visit(pick lists)
Description	Abstract
Relation	One of the body parts (thorax, abdomen, heart extremities, integumentary, head, urinary, or reproductive); this element is linked to the subject element
Date	Date of visit, Laboratory test-ray
Type	X-ray, scan/CT, ultrasound, MR, ECG
Format	PDFs, Text, JPEGs, TIFFs, etc.
Identifier	Document ID
Language	English and other languages
Coverage	Geographical and time-related information
Rights	Access rights(secured or unsecured)
Source	Data source

```
<? xml version="1.0" encoding="utf-8"?>
<Metadata  xmlns:xsi=http://www.w3.org/2001/XMLSchema-instancexmlns:M
DDS=http://serl.org/dc/elements/1,1
  Xmlns:mr=http://ClinicalRec.org/MyclinicalRecords.aspx">
  <Report id="sample">
  <Title>Doctor Visit Summary</Title>
  <Creator>Dr.Singh</Creator>
  <Subject>Teath</Subject>
  <Description>Pain in the morning</Description>
  …
  …
  <
```

And operated by cloud storage service provider [23]. Cloud storage services have many advantages such as cost savings, ease of use, ability to share data, accessibility, and sustainability. Personal cloud storage (PCS) is getting more popular because of the aforementioned convenience. Any cloud storage service provider may be used—SugarSync, Carbonite, IDrive, Dropbox, Google Drive, etc., [22]—for storing personal clinical data as long as they provide required functionality and security. In our proof of concept, MCRS, we are currently using DropboxTM as storage. The contents of each storage are organized by directories and described by MDDS metadata for interoperability. In case of the data that have embedded metadata, we create another layer of metadata so entire files can be located through our MDDS metadata content.

6 Our Proof of Concept—My Clinical E-Record System (MCERS)

As mentioned in the introduction, there are a number of obstacles in collecting and maintaining personal health data. In an attempt to remove such obstacles, we developed a Web-based system called My Clinical e-Record System (MCERS) that can help users to upload, organize, and retrieve relevant health data. Some of the features of MCERS are:

- Users can search within their health records not only by keyword but also by any of the metadata elements. They also can search by two elements such as subject and date in order to filter data by showing more relevant data. Also, they can find a group of records based on a date range they specify.
- MCERS allows users to share their document with their physicians.
- MCERS helps users to create metadata for any documents and upload them to their cloud storage. It also helps to retrieve those documents easily and can point to its location if more information is needed.
- To overcome the ownership barrier, we separate the clinical data from applications which will give the users more freedom by not limiting themselves to one provider

or application. Also, their data are saved on their own storage and we do not have to store it in our system.

- To overcome the interoperability barrier, we used MDDS standard to describe any clinical data using our tag definition.
- The easy access to users' health data and the ability to contribute to their record enhances users' motivation to use PHR.
- MCERS enables emergency clinical data access by emergency crew only by their license number which is the UIDAI Aadhaar number, patient name, and date of birth.

Using MCERS, we use patient's Dropbox access token to allow the connection between Dropbox and MCRS so patients can have their own storage and can access their storage through MCERS. This way, users keep their own data without binding to any specific application. MCERS contains no clinical data as they are stored in the patient's cloud storage. Patients need accounts for Dropbox and MCRS separately.

Managing health data MCERS categorizes clinical data based on human body parts. There are eight categories—abdomen, heart, head, thorax, extremities, integumentary, urinary, and reproductive. Any clinical data will be stored and linked based on these categories using the relation and subject tag elements of the MDDS metadata. We also kept the category to a minimum so it can be simple enough to be used by patients. Users can specify the category, i.e., the human body part of interest, when searching for relevant clinical data, so it can show only the clinical documents (e.g., doctor visit summary, x-ray) that are related to that part. When using the relationships between the resource (MDDS subject) and target resource (MDDC relation), it is possible to combine the result to a greater scope; e.g., instead of eyes and ears, it can be categorized by head. This can be done by predefining each part of the human body and associating it with its related category in the system. Also, we have constrained the MDDS subject to a small core set that can be selected from a pick list (all possible parts of the human body) to best describe the subjects. So, when the users select the subject element, the MDDS relation field will be populated automatically with the associated part of its related category. For example, when a user searches by keyword (e.g., head) and chooses the element (e.g., relation), the search results will be filtered and show only all clinical documents that are relevant to the head (e.g., eyes, ears, brain, mouth tooth/teeth, nose, chin) instead of showing all documents. Also, the user can filter the search by date if they need to specify a period of time to find clinical documents.

7 Conclusion

As the medical industry is going through a paradigm shift from clinician-centered to patient-centered, readily available complete personal medical history became the crucial part to ensure the three major goals in medical industry: evidence-based treatment, continuity of care, and prevention of medical mistakes. In this paper, we propose a methodology that accomplishes personal health data by exploiting metadata for the organization and easy retrieval of clinical data and cloud storage for easy access and sharing with doctors to apply the continuity of care and evidence-based treatment which improves human lives.

8 Result and Discussion

Each PHR can upload health records to Dropbox and can access whenever required. Fig 2 shows number of health records storage vs access in Dropbox.

Downloading PHR from Dropbox using metainformation is studied (Fig. 3).

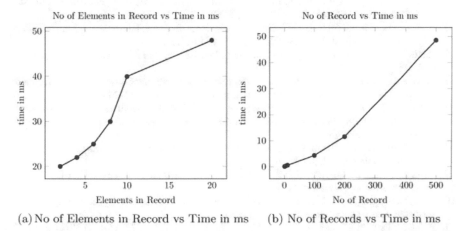

(a) No of Elements in Record vs Time in ms (b) No of Records vs Time in ms

Fig. 2 PHR versus time

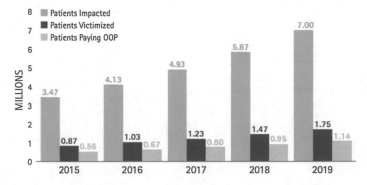

Fig. 3 PHR security analysis

Above graph shows security analysis in cloud storage in during 2015–2019, nearly 25 m of PHR have leaked and 6 m of PHR among 25 m are privacy victim identified and during this 4 m of PHR are caused with leakage theft.

References

1. Liu, L.S., Shih, P.C.& Hayes, G.R. Barriers to the adoption and use of personal health record systems. Proceedings of the 2011 conference 363–370 (2011).
2. Price, M. et al. Conditions potentially sensitive to a Personal Health Record (PHR) intervention, a systematic review. BMC Medical Informatics and Decision Making 15, 32 (2015).
3. Dogac, A. et al. Exploiting ebXML registry semantic constructs for handling archetype metadata in healthcare informatics. International Journal of Metadata, Semantics and Ontologies 1, 21–36 (2006). [13] FEARON, J.D. & LAITIN, D.D. Ethnicity, Insurgency, and Civil War. American Political Science Review 97, 75 (2003).
4. Sweet, L.E. & Moulaison, H.L. Electronic Health Records Data and Metadata: Challenges for Big Data in the United States. Big Data 1, 245– 251 (2013).
5. AHIMA. "Rules for Handling and Maintaining Metadata in the EHR" Journal of AHIMA 84, no.5 (May 2013): 50–54.
6. Dhivya, P., Roobini, S. & Sindhuja, A. Symptoms Based Treatment Based on Personal Health Record Using Cloud Computing. Procedia Computer Science 47, 22–29 (2015).
7. Patel, C., Gomadam, K., Khan, S. & Garg, V. TrialX: Using semantic technologies to match patients to relevant clinical trials based on their Personal Health Records. Journal of Web Semantics 8, 342–347 (2010).
8. Appelboom, G. et al. Smart wearable body sensors for patient selfassessment and monitoring. Archives of public health 72, 28 (2014).
9. Zhang, J., Chen, D., Services, E.D.P.C. & Cynwyd, B. Posters Metadata Application On Clinical Trial Data In Drug Development. 1–3.
10. Teitz, T., Stupack, D.G. & Lahti, J.M. Halting neuroblastoma metastasis by controlling integrin-mediated death. Cell Cycle 5, 681–685 (2006).
11. Ariel Ekblaw, Asaf Azaria, Thiago Vieira, Andrew Lippman. (2016). MedRec: Medical Data Management on the Blockchain. PubPub, [https://www.pubpub.org/pub/medrec version: 57e013615dbf3f3300152554].

Enhancing Prediction Accuracy of Default of Credit Using Ensemble Techniques

B. Emil Richard Singh and E. Sivasankar

Abstract Credit rating of an institution or individual provides a suggestive financial picture and strength of the individual or the institution. It gives the lender the ability to visualize the potentiality to the extent to which credit could be availed by the institution or the individual. Default prediction on the sum of all attributes such as payment history is a common instrument used for generation of credit rating. This research is aimed at comparing the predictive accuracy of ensemble of base classifiers using techniques of bagging, boosting, and random forest in the prediction of default of credit card clients and suggesting the technique with the highest accuracy. Customers' default payment in Taiwan dataset is used to build the model. ML classification algorithms such as K-nearest neighbor, Naive Bayesian, decision tree, and support vector machines are applied to create the base model on the dataset. Bagging, boosting, and random forest are applied on the dataset to generate model for prediction. The accuracy of each of the models for various degrees is tabulated. Information gain feature filter method is used to identify features with maximum entropy. The features with high entropy suggested by information gain together with ensemble techniques are used to build the new model. The accuracy of the new model is then tabulated. Boosting ensemble technique is found to have the best accuracy of prediction.

Keywords Ensemble techniques · Bagging · Boosting · Random forest
Feature selection · Machine learning · Credit default dataset

B. Emil Richard Singh (✉) · E. Sivasankar
Department of Computer Science and Engineering, National Institute of Technology,
Tiruchirappalli, Tamil Nadu, India
e-mail: emilrichard@gmail.com

E. Sivasankar
e-mail: sivasankar@nitt.edu

© Springer Nature Singapore Pte Ltd. 2019
R. S. Bapi et al. (eds.), *First International Conference on Artificial Intelligence and Cognitive Computing* , Advances in Intelligent Systems and Computing 815,
https://doi.org/10.1007/978-981-13-1580-0_41

427

1   Introduction

As financial instruments and market grow more dynamic and economy continues to evolve, credit analyst also needs to change the way they assess risk. Every company tries to build a credit risk model based on their experience in handling newer request for credit. With the eruption of data storage facility and the ability to capture every miniscule of every transaction, companies have an edge in utilizing this data knowledge to build a more robust risk model than what they could have done a decade ago. Machine learning algorithms were able to utilize this opportunity in narrowing the gap between data analytics and prediction. The system to be designed should be able to predict more true positives and negatives rather than giving a false prediction. The accuracy of prediction by machine learning algorithms such as SVM, NB, KNN, and DT as an ensemble together with feature selection techniques will provide a convoluted way to predict default in credit problem. The following is a brief view of various sections presented in this paper. Section 2 presents an overview of work carried out using ensemble techniques. Section 3 presents an insight into ensemble techniques bagging, boosting, random forest, and the most common base classifiers. Section 4 outlines credit default dataset, experiments conducted on the dataset and tabulates and picturizes results. Section 5 concludes on the best technique on predictive accuracy for credit default.

2   Literature Survey

The articles have been published on the prediction of credit default using the default of credit card client's dataset. In [1], bagging is explained for the first time as generation of n bootstrap replicates, where $n <$ sample space N whose majority vote predicts the class label. It explains that bagging could bring in optimum results even in unstable procedures or could bring down the performance of stable procedures. The paper tells us that that the number of bootstrap bag replicates required is less if the values in the dataset are numerical and the required replicates increase as the number of classes in the dataset increases. In [2], weighted majority voting aggregation rule is used to find an optimum upper bound of the number of classifiers required for the highest accuracy of an ensemble. In [3], the dataset is preprocessed into clusters with homogeneity using the technique optimal associate binning and then ensemble classifier techniques are applied. The above preprocessing technique is found to increase the accuracy of prediction of the ensemble classifier. In [4], balancing outliers and default values in separate bins with weights assigned to each bin and preprocessing with logistic regression for feature selection are found to increase the efficiency of bagging ensemble. In [5], bagging is found to perform better among the ensemble methods for credit data. In [6], random subspace feature selection with bagging is used to remove redundant features and reduce noise in the data. It is found that random subspace bagging decision tree gives better accuracy of prediction than bagging decision tree. In [7], C4.5 decision tree is shown as the best method for ensemble methods and it also confirms that Bayes and K-nearest neighbor classifiers have the poorest performance among ensemble methods. [8] proposes composite ensembles, AdaBoost, and bagging together with feature selection methods random subspace and rotation forest as the suitable method for credit problem. The paper shows that the order of application of the methods (feature selection algorithm followed by AdaBoost) is vital for obtaining better accuracy in results. In [9], randomization is shown to have equal prediction accuracy as bagging but not as accurate as boosting technique.

The comparative study made is as follows: (i) Classification algorithms such as KNN, BN, DT, and SVM are applied to predict the class label on the dataset. Appropriate normalization techniques were also applied to enhance the correctness of prediction; (ii) ensemble techniques such as bagging, boosting, and random forest are applied on the dataset, and the accuracy with increase in bags, iterations, and trees was tabulated; (iii) Step (ii) is repeated with the top eight highest entropy features selected using feature selection filter method information gain; and (iv) ensemble classification technique with minimum features, iterations, and maximum accuracy is found.

3   Methodology

3.1   Ensemble of Classifier

In ensemble technique, we use an aggregate of machine learning algorithms to derive better predictive performance than which could have been obtained from any of the constituent individual learning algorithms separately. An ensemble in itself is a supervised learning algorithm, which can be trained and then used to make predictions. The trained ensemble, hence, represents a single model for prediction. The new ensemble model tends to have better accuracy in prediction than the model based on single ML algorithm, when there is significant diversity among the ensemble models. Hence, diversity among the models is an important criterion for improving the prediction accuracy. However, combining strong classifier with a weak classifier could also result in dumb down model.

Ensemble size and the number of constituent classifiers of an ensemble involved in generation of the model play a vital role in performance and accuracy of prediction. The ideal ensemble size needs to be found beyond which the model could not be improved further. The above theory is called as the law of diminishing returns in ensemble construction. This theory also shows that using the same number of independent component classifiers as class labels gives the maximum accuracy.

Bagging: Bootstrap aggregating or bagging is applying the same learning algorithm to train each learner on a different set of data. It was invented by Breiman in the year 1996. N' subsets of data or bags of data are drawn randomly with the replacement from the training data N, wherein $N' < N$. The N' subset of data are chosen in parallel using uniform distribution. Random draw with replacement gives the possibility of duplicates in the subset generation. Each of the N' subsets of data is used to train a model M. Test data X is applied to each of the M models for predicting Y. The mean of Y or majority vote of each of the learner models M gives the required ensemble of prediction. The bagging models' sample mean accuracy of prediction for any point X is given by

$$f_{\text{bag}}(X) = \frac{1}{N} \sum_{n-1}^{N} f^n(X) \tag{1}$$

where N is the number of bags.

Boosting is similar to bagging, except that the iterations are sequential, and in every iteration, the new classifier tries to improve the accuracy of prediction of the previous iteration classifier. The first subset of data $N1'$ is chosen from the training data N to train the model M. Test data X is applied to the model for predicting Y. The incorrectly learned data of $N1'$ is biased for the selection as the next subset for prediction in the next iteration. The new model has better accuracy in areas, where the first model has failed. This iteration is repeated several times to learn those data points that are inaccurately learnt in the previous iteration. The models weighted on prediction accuracy are then voted for majority for final prediction. Boosting as such tries to reduce both the bias and variance of the data (Fig. 1).

$$f_{\text{bag}}(X) = \sum_{n=1}^{N} \alpha^n f^n(X) \tag{2}$$

where α^n is the weight considering the error in the last iteration (Fig. 2).

Random forest is predicted by creating N' decision trees. N' subsets of data are drawn randomly with the replacement from the training data N, wherein $N' < N$. m input features of M are split using information gain or the Gini index creating the nodes of the tree. As the number of decision trees grown increases, the degree of

BAGGING

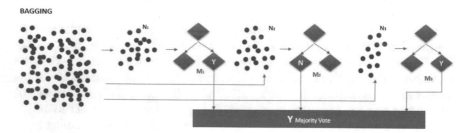

Fig. 1 Bagging ensemble technique

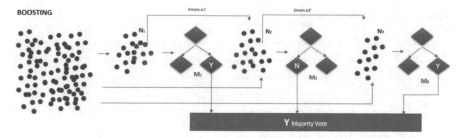

Fig. 2 Boosting ensemble technique

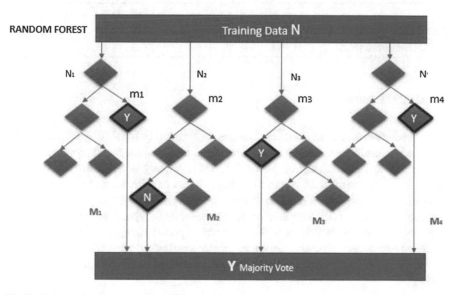

Fig. 3 Random forest ensemble technique

precision of the prediction also increases. The final prediction is done by majority vote of each class predicted by each tree created (Fig. 3).

3.2 Data Mining Techniques

Data mining tools and algorithms provide means to uncover important patterns in the data acquired from various sources like transactional, logs, and historical data [10]. Some of the classifiers used for the comparison in this paper include K-nearest neighbor (KNN) classifier [11], Naive Bayesian classifier (NB) [12], decision trees (DTs) [13], and support vector machines (SVMs) [14].

Table 1 Dataset description

Property	Count
Number of attributes	25
Number of records	30,000
Number of churn customers	6636
Number of non-churn customers	23364

Table 2 Churn prediction confusion matrix

	Predict	
	Churn (C)	Non-churn (NC)
Churn (C)	T P	F N
Actual non-churn (NC)	F P	T N

4 Experiments

Class prediction is carried out using four classifiers namely KNN, BN, DT, and SVM. Ensemble techniques such as bagging, boosting, and random forest are then applied for enhancing the accuracy of prediction. The accuracy of each of the ensemble methods for the number of bags in bagging, number of iterations in boosting, and number of trees created in random forest is tabulated. Information gain (IG) feature selection technique is then applied on the dataset. Newly identified features are then used on the ensemble techniques. The accuracy of prediction with features suggested by IG feature selection technique on ensemble methods is then tabulated and analyzed. Class prediction and accuracy measure are implemented in R programming packages.

4.1 Dataset Description

The credit default dataset used is accessed from UCI library [15]. The classifier performances are evaluated on this dataset. Table 1 presents the important information about the dataset taken for evaluation.

4.2 Performance Measures

The accuracy is used as a measure to identify the correctness of the prediction of the ensemble techniques. The accuracy is evaluated by the proportion of correct number of predictions to the total number of instances. 2×2 contingency matrix as in Table 2 is assumed for each classification algorithm. The accuracy is given by

$$\text{Accuracy} = \frac{\text{TP} + \text{TN}}{\text{TP} + \text{FN} + \text{FP} + \text{TN}} \tag{3}$$

Table 3 Accuracy (%) of base classifiers

Base classifiers	Accuracy (%)
KNN	78.81
NB	66.32
DT	81.74
SVM	81.48

Table 4 Accuracy average (%) of ensemble techniques with the number of bags or trees chosen

Number of models	Bagging	Boosting	Random forest
01–10	77.48	81.73	77.73
11–20	80.52	81.77	80.62
21–30	80.88	81.64	81.14
31–40	81.11	81.55	81.34
41 and above	81.39	81.55	81.31

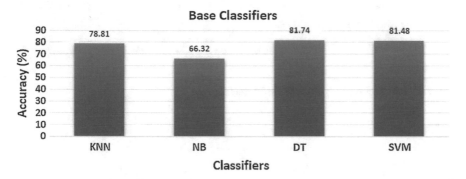

Fig. 4 Accuracy of base classifiers on credit card default dataset

4.3 Result and Analysis

The accuracy percentage of base classifiers such as KNN, NB, DT, and SVM on credit default dataset after normalization using min–max method is presented in Table 3 and graphically represented using Fig. 4. The prediction accuracy of DT and SVM is found to be in close proximity to each other. Table 4 presents the accuracy average of the ensemble techniques for different ranges of the number of bags in bagging, iterations in boosting, and trees built in random forest. It could be inferred that the accuracy average of ensemble techniques peaked in different ranges of bags, iterations, or trees created. Accordingly, bagging ensemble achieved maximum accuracy when the number of bags chosen is above 40; boosting ensemble could achieve maximum accuracy when the number of iterations crossed 11, and thereafter beyond 20 iterations, the accuracy is found to decrease; random forest ensemble achieved maximum accuracy when the number of trees created is in the range of 31–40 and its performance is in between bagging and boosting techniques.

Table 5 Accuracy (%) of ensemble techniques with feature selection

Ensemble technique	Number of models	Accuracy (%)
Bagging	60	79.13
Boosting	12	82.03
Random forest	40	80.95

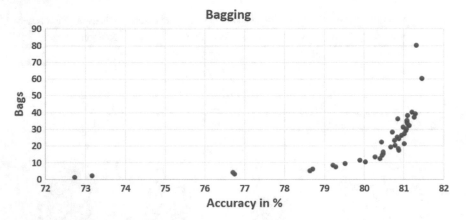

Fig. 5 Accuracy of bagging on credit card default dataset

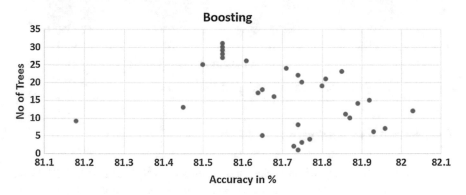

Fig. 6 Accuracy of boosting on credit card default dataset

Table 5 presents the accuracy percentage of ensemble techniques on credit default dataset, with the top eight attributes suggested by the IG feature selection method. Table 4 is graphically represented using Figs. 5, 6, and 7. Table 5 is graphically represented using Fig. 8.

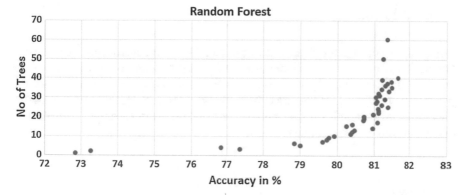

Fig. 7 Accuracy of random forest on credit card default dataset

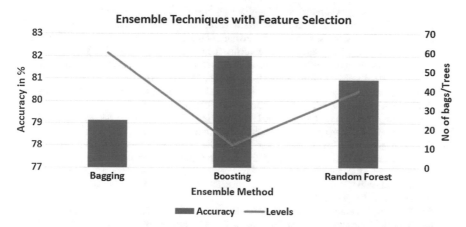

Fig. 8 Comparison of ensemble techniques with feature selection on credit card default dataset the best prediction 81.46 with the number of bags at 60. Boosting technique achieved the best prediction accuracy 82.03 at 12th iteration. Random forest achieved the best accuracy of 81.68 with the number of trees at 40. With the top eight features selected using the IG feature selection technique, the accuracy of prediction of boosting technique remained at 82.03 at 12 levels of iteration. The other techniques had a marginal drop in their prediction accuracy. It is found that feature selection algorithm was not able to provide considerate increase in prediction accuracy. However, owing to less number of attributes, the speed of classification increased. The C5.0 decision tree is able to perform better when used in boosting than when used with random forest ensemble technique. To conclude, boosting ensemble technique had the most accurate prediction accuracy before and after the application of feature selection technique with minimum number of iterations. Study on the accuracy of ensemble of classifiers with preprocessing. Feature extraction techniques such as LDA and PCA could be a trigger for futuristic work

5   Conclusion

Financial institutions look forward to the individual's credit repayment history to conclude and reason out a valid credit score for the individual. Building a risk model using machine learning algorithm is a modern-day challenge. Here in this work, the prediction of ensemble of classifiers is studied with feature selection. It is found that the ensemble of classification techniques has better prediction accuracy than the base classifiers. Bagging technique reached.

References

1. Leo Breiman. Bagging predictors. Machine learning, 24(2):123–140, 1996.
2. Hamed R Bonab and Fazli Can. A theoretical framework on the ideal number of classifiers for online ensembles in data streams. In Proceedings of the 25th ACM International on Conference on Information and Knowledge Management, pages 2053–2056. ACM, 2016.
3. Nan-Chen Hsieh and LunPing Hung. A data driven ensemble classifier for credit scoring analysis. Expert systems with Applications, 37(1):534–545, 2010.
4. Steven Finlay. Multiple classifier architectures and their application to credit risk assessment. European Journal of Operational Research, 210(2):368–378, 2011.
5. Gang Wang, Jinxing Hao, Jian Ma, and Hongbing Jiang. A comparative assessment of ensemble learning for credit scoring. Expert systems with applications, 38(1):223–230, 2011.
6. Gang Wang, Jian Ma, Lihua Huang, and Kaiquan Xu. Two credit scoring models based on dual strategy ensemble trees. Knowledge-Based Systems, 26:61–68, 2012.
7. AI Marqu´es, Vicente Garc´ıa, and Javier Salvador Sanchez. Exploring the behaviour of base classifiers in credit scoring ensembles. Expert Systems with Applications, 39(11):10244–10250, 2012.
8. AI Marqu´es, Vicente Garc´ıa, and Javier Salvador Sanchez. Two-level classifier ensembles for credit risk assessment. Expert Systems with Applications, 39(12):10916–10922, 2012.
9. Thomas G Dietterich. An experimental comparison of three methods for constructing ensembles of decision trees: Bagging, boosting, and random- ization. Machine learning, 40(2):139–157, 2000.
10. Jiawei Han, Jian Pei, and Micheline Kamber. Data mining: concepts and techniques. Elsevier, 2011.
11. Keinosuke Fukunaga. Introduction to statistical pattern recognition. Academic press, 2013.
12. Harry Zhang. The optimality of naive bayes. AA, 1(2):3, 2004.
13. S Rasoul Safavian and David Landgrebe. A survey of decision tree classifier methodology. IEEE transactions on systems, man, and cybernetics, 21(3):660–674, 1991.
14. Nello Cristianini and John Shawe-Taylor. An introduction to support vector machines and other kernel-based learning methods. Cambridge university press, 2000.
15. default of credit card clients data set. https://archive.ics.uci.edu/ml/datasets/default+of+credit+card+clients.

A Comparison Review on Comb–Needle Model for Random Wireless Sensor Networks

M. Shanmukhi, J. Amudahavel, A. Vasanthi and G. Naga Sathish

Abstract In this paper, the purpose of random deployment wireless sensor network (WSN) is presented. Then compared the extended comb–needle model (ECNM) and cluster-based ECNM (CECNM) for random deployment of WSN is explained with illustrations. The mathematical analysis of ECNM and cluster-based ECNM are explained in terms of communication and storage overhead is carried out. The simulation environment is presented and concluded with the performance evaluation of ECNM and CECNM in terms of parameters such as communication cost, energy consumption, throughput, delay, packets delivered, and packet loss. The communication cost of the ECNM model is $O(n)$ where $n = \sqrt{N}$. N denotes the total number of sensor nodes in the network, and the communication cost of the CECNM model is $O(n/Nc)$.

Keywords Wireless sensor networks · Comb–needle model · ECNM · CECNM

1 Introduction

As of now, random wireless sensor networks are widely used in civilian and military applications [1]. Right from the inception of WSN, there has been research on enhancing lifetime of the network as nodes in WSN are resource constrained. Energy efficiency is given importance in all aspects of WSN. The many to one kind of communication in WSN which enables sensor nodes to gather data (sense data) and disseminate it towards base station causes energy consumption and communication overhead. In order to reduce energy consumption due to communication overhead,

M. Shanmukhi (✉) · A. Vasanthi · G. Naga Sathish
Department of CSE, BVRIT HYDERABAD College of Engineering for Women,
Bachupally 500090, Hyderabad, India
e-mail: shanmukhi.m@gmail.com

J. Amudahavel
Department of CSE, KLEF, Vaddeswaram, Andhra Pradesh, India
e-mail: info.amudhavel@gmail.com

© Springer Nature Singapore Pte Ltd. 2019
R. S. Bapi et al. (eds.), *First International Conference on Artificial Intelligence and Cognitive Computing* , Advances in Intelligent Systems and Computing 815,
https://doi.org/10.1007/978-981-13-1580-0_42

437

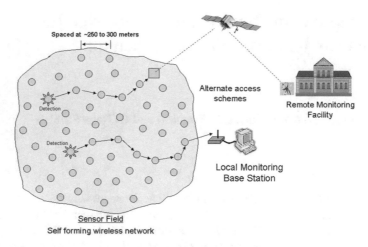

Fig. 1 Sample architecture of random WSN

many researchers contributed to devise techniques that could aggregate data so as to reduce communication overhead and improve energy efficiency and longevity of the grid network. Practically, most of the WSNs for military and other applications are deployed randomly, but not as a grid (Grid) network. Hence, this paper reviews the extension to CNM for randomly deployed WSN. Simulation-based experiments are carried out for extended CNM in NS2. The extended comb–needle model (ECNM) for random networks for improving efficiency of data aggregation is the main focus of this paper. The simulation results of ECNM and ECNM showed significant improvement in terms of reduction of communication cost, energy efficiency and reduction of overall network overhead.

1.1 Why Random Wireless Sensor Networks?

Wireless sensor networks (WSNs) are widely used in different domains such as disaster management, general surveillance, ocean pollution detection, health care domain, home automation, monitoring wildlife habitat, military applications where the network operates from a hostile environment [1]. In some applications, it is found that huge number of sensor nodes is deployed, and the sensor nodes are connected to a base station or sink. In such applications, if 1000 s of nodes are need to be deployed in hostile environments, where a man cannot go into those areas to deploy a grid network. So in most of the applications, random deployment is inevitable like Fig. 1.

As mentioned in our previous research papers [2–4], energy conservation is most important concept in WSN. Sensor nodes consume energy for the activities such as sensing data from surroundings, processing data if required and sending it to base

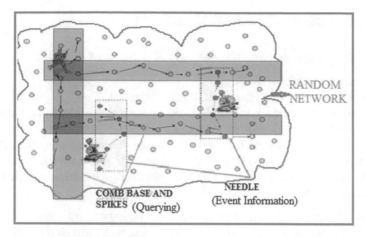

Fig. 2 Comb–needle model for random networks

station. It is understood that sensing and data processing need less energy when compared with the activity known as data transmission. Therefore, it is essential to have measures that can result in reduced transmission cost. The aim of this is to reduce the energy consumption in WSN. Many researchers came up with different ideas and perceptions on the process of energy consumption [5]. In this paper, proper deployment of sensor nodes, data aggregation and routing are focused in order to reduce energy consumption in WSN. It is achieved by extending CNM into random networks and further improved with clustering.

As shown in Fig. 1, the sensor nodes are deployed such that there are no voids in the sensing area. They are supposed to sense data and send to local monitoring base station. A major issue in WSN is energy consumption as the nodes have only limited energy resources [6]. Data aggregation techniques are helpful in energy conservation, various data aggregation techniques are elaborated in [7, 8].

2 Extended Comb–Needle Model (ECNM)

The original CNM proposed by Liu et al. [9] which was meant for grid networks. The CNM is extended for random deployments and presented in this section. The extension is done by redefining the notion of comb and needle for random networks. The ECNM makes use of the blend of push and pull approaches for information discovery and dissemination. The process of querying by the base station for events is regarded as combing, whereas the detection of events as needles [10].

This is like combing for needle in a haystack or pool of sand. The proposed CNM for random networks is shown in Fig. 2.

2.1 Assumptions of the ECNM and CECNM

The following are the assumptions with respect to proposed ECNM:

1. Sensor nodes are location-aware.
2. Sensor nodes are randomly and uniformly deployed.
3. Base station is located at the top left corner of sensing region. And there might be multiple base stations in the network.
4. Comb formation and establishment of route are done when query is made.
5. Once an event is detected or a needle is found, the nodes present on the spikes of the comb are communicated regarding the event occurrence. A viable path is chosen to notify the base station about the event.

The ECNM and CECNM models are based on two important approaches known as push and pull. Base station broadcasts the query, and then the path for dissemination of query is established in the form of a comb (with spikes). Any node that finds an event occurrence can push it to neighbours towards spikes in two directions such as downward and upward. Then, it appears like a needle [10].

2.2 Analysis of ECNM

This section provides the mathematical analysis of extended CNM for random networks.

Mathematical Notations:

n	x-axis
m	y-axis
(x, y)	Coordinates
(x_n, y_m)	boundaries of (x, y) coordinates
s	Inter-spike spacing
a to c	Spike 1 of the comb
b to d	Spike 2 of the comb
r	Range of the strip in comb base and spikes

Consider a random network where the number of nodes is represented by n, and they are deployed at (x_i, y_i) coordinates where $x_i <= x_n$ and $y_i <= y_m$. The deployment is considered at (x, y) and boundary is provided as (x_n, y_m). Here is an assumption that says that there is uniform distribution of sensor nodes in the network. A comb structure is the possible result of a generated query. Let us consider that there is query node located at (x, y). When query is sent in vertical direction query node (x, y_m) to (x, y), then the direction of query takes place in horizontal lines from nodes such as $(x, y+s)$, $(x, y+2s)$ to $(x_n, y+s)$, $(x_n, y+2s)$, where s is inter-spike spacing or combing degree. As a result, the routing structure appears like a comb as illustrated in Fig. 3.

The notion of the comb base, spikes and needles are redefined in random network ECNM [129] is as follows:

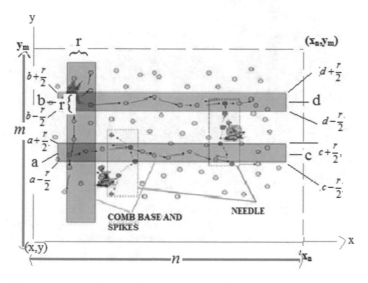

Fig. 3 Definition of comb base and spike for ECNM

$$(x, y + s) = a \tag{1}$$

$$(x, y + 2s) = b \tag{2}$$

$$\left(x_{n,y} + s\right) = c \tag{3}$$

$$\left(x_{n,y} + 2s\right) = d \tag{4}$$

The horizontal lines are formed from Eqs. (1) to (3) and from (2) to (4) provided above, while vertical lines are formed from (x, y) to (x, y_m). The estimation of distance is computed as follows.

$$s = (y_m - y)/3 \tag{5}$$

In order to describe the query dissemination in horizontal direction as shown in Fig. 3, the points considered are $(x, y+s)$ and $(x, y+2s)$. The comb structure has been developed. Then, there is a need to specify the range (width) of the comb and its spikes of CNM in the random network by using the certain criteria, where range of the comb base and spike is represented by r. The vertical range for random networks is shown below:

$$(x + r, y) \text{ to } (x + r, y_m)$$

Based on Eqs. (1) and (3), for spike a to c, the horizontal range is as follows.

$$\left(a + \frac{r}{2}, a - \frac{r}{2}\right) \text{ to } \left(c + \frac{r}{2}, c - \frac{r}{2}\right)$$

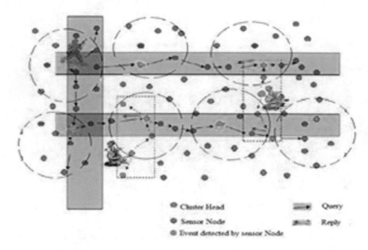

Fig. 4 Cluster-based ECNM

In the same fashion based on Eqs. (2) and (4), the horizontal range for spike b to k is as follows.

$$\left(b + \frac{r}{2}, b - \frac{r}{2}\right) \text{ to } \left(d + \frac{r}{2}, d - \frac{r}{2}\right)$$

Euclidean distance is used by sensor nodes while passing query in order to find shortest path. Euclidian distance is computed as follows.

$$\xi_{\text{dist}} = (p, q)(u, v) = \sqrt{(p - u)^2 + (q - v)^2} \tag{6}$$

As shown in Fig. 4, the nodes such as e, f and g are neighbours in the given spike. According to the Euclidian distance, there is link established between e and f.

As per the analysis made above, the communication cost is $O(n) + O(n/2)$ in ECNM, i.e., (as discussed in [2] $C_l = l\text{-}1$ *hops, where* $l = n$. *The comb spike and needle are reached to finite level*).

From the above analysis, the communication is computed as

$$C_{\text{optimal}} = O(n) \tag{7}$$

3 Analysis of CECNM

The research carried out on the cluster-based random network with CNM considers the network model presented in Fig. 2, where clustering has been added to ECNM for more efficient data aggregation. The formation of a comb and a needle is depicted in Fig. 4. The cluster-based ECNM is implemented, and its performance is observed in terms of the identified parameters provided later in the paper.

Assumptions • Wireless sensor network consists of N stationary and location-aware (using GPS or some other localization method) sensor nodes deployed randomly in the square field region of size $n \times m$.

• Base station is in top left corner (multiple base stations may be available in network, each sensor node with soldier is considered as base station).

• All the sensor nodes have the equal transmission power, in other words, all communication links are assumed to be symmetric.

• There are no transmission errors.

• Sensor node acts as an aggregator, if the size of the data is bigger than the certain limit.

Clustering is done using ISODATA [11] scheme.

3.1 Applying ISODATA Clustering Algorithm

Iterative self-organizing data analysis technique (ISODATA) algorithm is originally explored in [11]. This is a variant of popular clustering algorithm known as K-Means. The resultant clusters produced by K-Means are subjected to splitting and merging. This makes the new variant aforementioned. In other words, it is understood that ISODATA is nothing but K-means with heuristics additionally added for splitting and merging of clusters. The heuristics are used for optimal partitioning and guiding the process using a threshold for controlling the process. When the cluster exceeds the given threshold, splitting is performed. In the same fashion, merging is performed when two cluster centroids have distance which is less than another threshold that guides merging process using heuristics of the ISODATA. ISODATA is one of the unsupervised learning techniques studied and adapted to realize extended cluster-based CNM on random networks.

ISODATA is used in many real-time applications with respect to clustering. In this research, it is used for specific purpose. With respect to random networks, sensor nodes are initialized in 2D plane prior to applying ISODATA. Initial cluster vector is provided to ISODATA. Then, the algorithm considers the positions of nodes and performs clustering. Many clusters are formed, and each node is associated with closest cluster. These two steps are crucial in the algorithm and they are repeated iteratively until all clusters are formed as intended. In the same fashion, clusters are merged based on the criteria aforementioned. The splitting and merging processes are used by the algorithm with known heuristics for making a network with clusters formed ideally.

Table 1 Simulation parameters

Parameter	Value
MAC protocol	802.11
Number of nodes	40–50
Node deployment	Random
Radio propagation model	Two-ray ground
Radio transmission range	200 m
No. of clusters	Variable

4 Simulation Environment

The ECNM and CECNM are implemented using NS2 simulations. The simulations demonstrate the hypothesis pertaining to the models to reduce energy consumption and reduction of communication cost for efficient data aggregation in WSN. The important observations made in the simulations related to ECNM include throughput, packets delivered, packet loss, delay, energy consumption and communication cost. These observations are presented in this section, and they are compared with the state-of-the-art model known as simple random network model.

The performance of ECNM is evaluated using NS2 simulations besides comparing it with that of CECNM. Simulation of WSN is made with 50 sensor nodes and a sink node. The sensor nodes are deployed randomly in 20 × 10 area. The sensor nodes are assumed to have uniform and independent distribution across the simulation area.

Table 1 shows the NS2 environment used for simulations that demonstrate the hypothesis of proposed ECNM and CECNM. Different performance metrics are used to evaluate the results of simulations. They are described in the ensuing subsection.

4.1 Performance Evaluation of ECNM

The simulation results show that extended comb–needle model in random networks and CECNM for random networks provide efficient means for data gathering and information dissemination. Two network models are compared with performance metrics such as communication cost, energy consumption, delay, packet loss, packets delivered and throughput. The network models are random network with CNM and random network without CNM and cluster-based random network with CNM and without CNM. The performance of random network with and without CNM is shown in Fig. 5.

The results reveal that the implementation of random network with CNM has performance improvement over its counterpart without CNM. It is true with all performance metrics considered. The CNM when employed with random network is able to reduce communication cost by 29.26%, energy consumption by 48.75%, delay by 63.15% and packet loss by 84.81%. The CNM with random network has

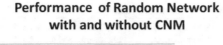

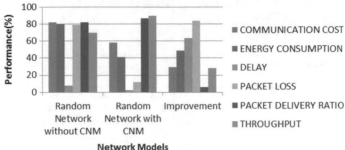

Fig. 5 Performance of random network with and without CNM

Table 2 Performance of random network with and without CNM

Parameter (Units)	Random network without CNM	Random network with CNM	Improvement (%)
Communication cost (Packets)	82	58	29.26
Energy consumption (Joules)	80	41	48.75
Delay (ms)	7.6	2.8	63.15
Packet loss (Packets)	79	12	84.81
Packets delivered (Packets)	82	87	6.09
Throughput (Packets)	70	90	28.57

improved packets delivered by 6.09% and throughput by 28.57%. The results showed insights on the performance of two networks in Table 2.

4.2 Performance Evaluation of CECNM

Two network models are used in simulation study of the ECCNM. They are cluster-based random network with CNM and cluster-based random network without CNM. The results reveal that the performance of cluster-based random network with CNM is more when compared with its counterpart without employing CNM in all aspects considered.

The cluster-based random network with CNM reduces communication cost by 37.5%, energy consumption by 19.35%, delay by 0.5% and packet loss by 50%. Besides, it improves packets delivered by 5.55% and throughput by 5.37%. These

Table 3 Performance of cluster-based random network with and without CNM

Parameter (Units)	Random network with clustering without CNM	Random network with clustering and CNM	Improvement (%)
Communication cost (Packets)	48	30	37.5
Energy consumption (Joules)	31	25	19.35
Delay (ms)	1	0.5	50
Packet loss (Packets)	10	5	50
Packets delivered (Packets)	90	95	5.55
Throughput (Packets)	93	98	5.37

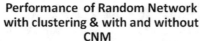

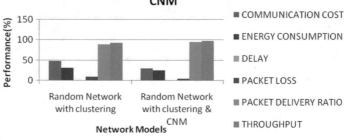

Fig. 6 Performance of random network with clustering and with and without CNM

statistics reveal the fact that the ECCNM has performance improvement over the cluster-based random network without CNM in Table 3.

The performance of the two network models is visualized in Fig. 6. From the results, it is evident that the two network models are same with one difference. The former does not use CNM while the latter employs it. The results reveal that there is comparable performance improvement with the cluster-based random network which exploits CNM. The rationale behind this is the ability of CNM to utilize best features of push and pull models besides its notion of comb and needle which has efficient query processing.

As shown in Fig. 7, it is clear that there is an improvement in performance when CNM is employed with random network and cluster-based random network. With cluster-based random network, CNM has improved its performance in reducing communication cost and energy consumption.

It showed difference with respect to throughput as well. The other parameters, such as PD, packet loss, and delay, the performance difference is there with respect to random network with CNM.

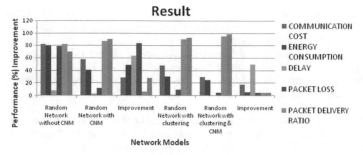

Fig. 7 Improvement when CNM and clustering applied in random network

5 Conclusion

This paper reviews the proposed extended comb–needle model which is based on CNM and supports data aggregation. In this model, sensor nodes are deployed randomly. The review describes that the models are able to reduce communication cost and energy consumption. The ECNM is studied with random network to demonstrate the functionality of aggregation technique that consumes less energy thereby increasing the life of network and extended cluster-based CNM which makes use of random networks. The network model is used with clustering and CNM. The CNM exploits push and pull models of communication strategies while the clustering makes the network more efficient. The performance of the CECNM is compared with the cluster-based random network without CNM. NS2 simulations are made to have observations of ECNM for efficient data aggregation besides performance measures like delay, energy consumption, communication cost, packets delivered and throughput. The results revealed that the proposed ECNM outperforms simple random network as it improves communication efficiency and conservation of energy, and it is understood that the cluster-based random network with CNM has comparable performance over cluster-based random network without CNM. The same work can further extend with mobile sink in future.

References

1. M. Cardei, J. Wu, M. Lu, M. Pervaiz, "Maximum network lifetime in wireless sensor networks with adjustable sensing ranges", IEEE *International Conference on Wireless and Mobile Computing, Networking And Communications, (WiMob'2005)*, vol. 3, IEEE, pp. 438–445, 2005.
2. M. Shanmukhi, O.B.V. Ramanaiah, "Cluster-based Comb-Needle Model for Energy efficient Data Aggregation in Wireless Sensor Networks", 2nd *International conference on Applications and Innovations in Mobile Computing* (AIMoC).12th–14th, pp. 42–47, February-2015, IEEE.
3. M. Shanmukhi, O.B.V. Ramanaiah, "Extended Comb Needle Model for Energy Efficient Data Aggregation in Random Wireless Sensor Networks", Indian *Journal of Science and Technology*, Vol 9(22), https://doi.org/10.17485/ijst/2016/v9i22/89953, June 2016.

4. M. Shanmukhi, O.B.V. Ramanaiah, "An Energy Efficient Data Aggregation for Random Sensor Networks", *Transactions on Networks and Communications*, Vol 4, No. 5,2016.
5. R. Subramanian, F. Fekri, "Sleep scheduling and lifetime maximization in sensor networks: fundamental limits and optimal solutions", *Proceedings of the 5th International Conference on Information Processing in Sensor Networks, ACM,* pp. 218–225, 2006.
6. F. Cuomo, A. Abbagnale, E. Cipollne, "Cross-layer network formation for energy-efficient IEEE 802.15. 4/zigbee wireless sensor networks", *Journal AdHoc Networks*, Vol.11, issue.2, pp. 672–686, 2013.
7. M. Shanmukhi, O.B.V. Ramanaiah, "A Survey on Energy Efficient Data Aggregation Protocols for Wireless Sensor Networks", published in *International Journal of Applied Engineering Research*, ISSN 0973-4562, Volume 11, Number 10 pp. 6990–7002, 2016.
8. Koustuv Dasgupta, Konstantinos Kalpakis and Parag Namjoshi, "An Efficient Clustering–based Heuristic for Data Gathering and Aggregation in Sensor Networks", *IEEE Wireless Communications and Networking,* Year: 2003, Volume: 3 Pages: 1948–1953, Vol. 3, IEEE, pp. 1–6, 2003.
9. X. Liu, Q. Huang, and Y. Zhang, "Combs, Needles, Haystacks: Balancing Push and Pull for Discovery in Large- Scale Sensor Networks," *Proceedings of ACM Conferenceon Embedded Networked Sensor Systems (SenSys '04)*, pp. 122–133, Nov. 2004.
10. Xin Liu, Qingfeng Huang and Ying Zhang," Balancing Push and Pull for Efficient Information Discovery in Large-Scale Sensor Networks", IEEE *Transactions on Mobile Computing*, Vol. 6, No. 3, pp. 241–251, 2007.
11. Ball, G.H., Hall, D.J., "ISODATA: A Novel Method of Data Analysis and Classification", *Technical Report, Stanford University, Stanford, CA,* 1965.

Automatic Classification of Bing Answers User Verbatim Feedback

Annam Naresh and Soudamini Sreepada

Abstract Microsoft Windows products receive huge amount of feedback across different channels. The amount of feedback received monthly around 11 K, which is humanly inefficient to analyze. When user issues a technical query (e.g., connect projector in win 10) on Bing search engine, it shows an answer (e.g., steps to connect to projector in Windows 10) to user query. These answers are created by content author's team. The triggering team shows the relevant answer for a user query. When users are not satisfied with the answer, they might provide feedback. It is very crucial to understand user feedback to improve user experience. The existing approach to analyze user feedback is to go through each piece of feedback and assign it to right team for resolution. This approach is laborious, expensive and does not scale well. We proposed an approach, which understands user query, answer, and feedback and automatically categorize the verbatim feedback into one of the following three categories: *authors, triggering, irrelevant (Junk)*. The classified feedback is routed to the respective team. We trained a supervised machine learning classifier to perform feedback classification. We have extracted different features from query, answer, and user feedback. Our features composed of bag-of-n-grams extracted from verbatim feedback and deep semantic structured model (DSSM) score between query and answer title. We have achieved 82% classification accuracy using support vector machine (SVM) algorithm. This classifier has been improved over the time. Our approach reduced huge amount of manual work. The proposed solution also helped in reduction of dissatisfaction ration (internal success measure) by 2%, which indicates the enhancement in overall user experience with tech answers.

Keywords User verbatim feedback · Classification · Deep semantic similarity model

A. Naresh (✉) · S. Sreepada
Microsoft R&D India, Hyderabad, Telangana 500032, India
e-mail: annaresh@microsoft.com

S. Sreepada
e-mail: ssreepad@microsoft.com

© Springer Nature Singapore Pte Ltd. 2019   449
R. S. Bapi et al. (eds.), *First International Conference on Artificial Intelligence and Cognitive Computing* , Advances in Intelligent Systems and Computing 815,
https://doi.org/10.1007/978-981-13-1580-0_43

1 Introduction

In today's big data era, all organizations and product development teams started utilizing customer feedback data very effectively to make important decisions. Sometimes user feedback will completely change the way we think about the product or service. Analyzing user feedback and deriving actionable insights are very crucial to improve the overall user experience. User feedback data analysis helps in prioritizing actions. The analyzed feedback needs to be categorized so that relevant teams can track and work on the problems identified in their feature areas. Microsoft offers wide variety of software suites like Windows OS, Office. Microsoft provides help for all of its products through many channels. There are support Web pages for all of its products (e.g., support.microsoft.com). Microsoft's search engine provides the answers for queries related to technical help (e.g., how to do remote desktop on Windows 10) on top of Web results. Typically, these answers consist of sequential steps that solve user query.

1.1 Query and User Feedback Workflow

When a user enters the query on search engine, and it provides results in the form of hyperlinks or Web URLs. For technical related queries, search engine shows up an instant answer/short snippet on top of Web results. If the returned result has required information, user will be satisfied (SAT), else user will be dissatisfied (DSAT) and provides verbatim feedback. The complete steps for user query and feedback flow are shown in Fig. 1. The instant answers (answer title and answer content) and instructions provided in support pages are created by product content team authors. Retrieving an appropriate answer for a given user search query is handled by the triggering group (Dev team). User verbatim feedback is always associated with the query and answer. User feedback talks about authors (content) issue: when content of an answer is inadequate or incomplete to finish the desired task; triggering issue: if inappropriate answer is shown to user (also known as user intent mismatch); or sometimes junk verbatim.

In this paper, we proposed an intelligent and automatic classification of user verbatim feedback. The remainder of the paper is organized as follows. Section 2 describes the existing approach. The details of proposed approach are explained in Sect. 3. In Sect. 4, the experimental setup and results of experimental studies have been discussed. We conclude the paper and scope of our approach in Sect. 5.

Fig. 1 Typical workflow of
user query and feedback

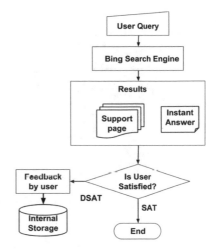

2 EXISTING Approach: Internal Bug Bash

The existing approach to analyze user verbatim is to conduct an internal bug bash periodically and go through each user feedback and understand what user has provided for answers. But this approach takes 20 developers human effort and 3 days' complete time to analyze three months of feedback data. It takes 540 working hours for complete analysis. It is even difficult to collate individual results. This approach is not scalable and takes time to react for user feedback immediately.

Why It is Essential to Solve: The feedback analysis is often left as manual or semi-manual. It is a very important problem to solve as answers show up on top of search engine results page (SERP). Hence, answers or experiences shown to users should be as accurate as possible! The end goal of search engine is to provide right help available to users.

3 Proposed Approach

We employed a machine learning-based solution for automatic classification of user feedback. This decision has been motivated by practical considerations: first, the feedback data is often very short and incoherent; secondly, we believe that an appropriately chosen machine learning technique will be able to draw its own conclusions from the distribution of lexical elements in a piece of feedback.

We formulate automatic grouping of user verbatim feedback as a multi-class text classification. We proposed an approach that a supervised machine learning algorithm is learnt to classify the user feedback associated with query and answer into one of the three categories: *authors, triggering,* and *irrelevant*. We applied text analytics on user verbatim to generate bag-of-words (*n*-grams) features. In addition to this,

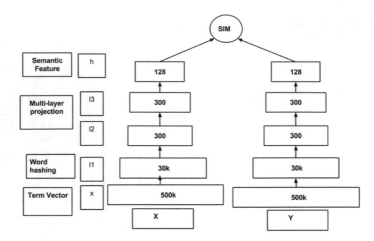

Fig. 2 The schematic diagram of DSSM model

we encoded the information between query and answer title using deep semantic structured model (DSSM). The DSSM score is used as a strong input signal to the classification model.

3.1 Deep Semantic Structured Model

Deep structured semantic model, or more general deep semantic similarity model, is developed by the Microsoft Research Deep Learning Technology Center (DLTC) [1]. It is a deep neural network (DNN) modeling technique for representing text strings (sentences, queries, predicates, entity mentions, etc.) in a continuous semantic space and modeling semantic similarity between two text strings. Two strings X and Y are mapped to feature vectors in latent semantic space via DNN.

DSSM Overview DSSM maps the high-dimensional sparse word-occurrence vector into low-dimensional dense semantic vector. It even maps a query and a relevant document into a common semantic space in which a similarity measure can be calculated so that even there is no shared term between the query and the document, the similarity score can still be positive (e.g., about my pc and my computer details). We explain the different layers of DSSM model architecture in the following subsections, and these are represented in Fig. 2.

Input and Word Hashing Layer: The input layer represents simple term occurrence vector of a string based on the vocabulary. The word hashing layer reduces the dimensionality of the bag-of-words term vectors. It is based on letter n-gram concept. Given a word (e.g., good), first starting and ending marks to the word (e.g., #good#) is applied. Then, the word is broken into letter n-grams (e.g., letter trigrams: #go, goo, ood, od#). Finally, the word is represented using a vector of letter n-grams.

Multi-Layer Projection and Semantic Layer: Successive nonlinear transformations are applied on output of word hashing layer. This will learn very lesser dimensions and yet captures the semantic information. The final layer is 128-dimensional representation for input term vector. Finally, cosine similarity between these semantic features is computed as per Eq. (1).

$$\text{DSSMScore}\,(X, Y) = \text{Cosine}(h_X, h_Y) = \frac{h_X^T h_Y}{\|h_X\| \|h_Y\|}, \tag{1}$$

where h_X and h_Y represent semantic representation of strings X and Y, respectively. The range of DSSMScore (X, Y) score is normalized between 0 and 1. Higher score denotes the more similarity.

Learning DSSM: The click-through logs (from repository of search engine logs) consists of a list of queries and their clicked documents. We assume that a query is relevant to the documents that are clicked on for that query. The essential part of the DSSM is to maximize the conditional likelihood of the clicked documents given the queries. A DNN is trained using SoftMax function as loss function. DSSM model is trained on huge search click logs. It can be customized to own corpus as well. In our experiments, a pretrained DSSM has been used to compute the similarity between query and answer tittle. The complete details of training DSSM are available in [1].

3.2 Supervised Classifiers

We encode user verbatim using bag-of-n-gram features, and these are added with DSSM score. These combined features along with class labels (assigned with one of the following categories: content, triggering, irrelevant) are used in training logistic regression, support vector machine, and random forest supervised classifiers. The internals of these classifiers can be found in [2].

3.3 Post-Classification

Our predictive machine learning solution is applied to classify the user feedback and assign it one of three categories: authors, triggering, and irrelevant. This classified feedback is routed to respective teams, authors, and triggering. It reduces the service-level agreement (SLA) in taking the appropriate action by respective teams. This way it is very helpful in tracking the issues team wise.

4 Experimental Studies

In this section, we describe dataset details and experimental results. We also discuss insights gained and business impact of this work.

4.1 Datasets

Our dataset consists of user feedback from different channels such as instant answers and Support.Microsoft.Com (SMC) pages. The feedback is logged in internal data storage infrastructure. Each feedback is associated with query, answer information, and other metainformation. The nature of these verbatim feedbacks is noisy and unstructured. The summary of dataset is shown in Table 1. These numbers are for en-US market only. Table 2 shows sample annotated user feedback. Let us consider a user query is 'Windows 10 search for files' and returned answer talks about documents search in Windows 10. But one of the instructions says go-to universal search box. It is not clearly mentioned about universal search box in the answer. The feedback is given by user for this, "didn't know where is the universal search box." An icon in the instruction would have been helpful for the user in finishing his/her task.

Table 1 Statistics of dataset

Month	#User verbatim
Oct 2016	10649
Nov 2016	11661
Dec 2016	10700
Jan 2017	10200

Table 2 Sample annotated data

Query	Answer title	User feedback	Category
Windows 10 search for files	Where are my documents in Windows 10?	Method 1: did not know where the search box is	Authors
What time is it Cortana	Chat with Cortana	Asked time and it didn't tell me, only provided suggestions of what to ask about	Triggering
Windows hello in windows 10	Learn Windows Hello and set it up	sr791 xyzzy outlook.com	Irrelevant

4.2 Train–Validation–Test Split

Since there is no ground truth available for the user feedback data that has been logged in our internal storage. We have annotated 10,000 examples of October 2016 month data. This set has been used as labeled dataset. We split this dataset into training and validation set in the ratio 70 and 30%, respectively. The trained classifiers are evaluated on validation dataset. The next month data has been used as test data. The performance measure used in all our experiments is average classification accuracy on test data. Since there is no ground truth available for test data, we randomly chose 10% of the test data for manual judgement. We retrained the model using October month data and November data with confidently labeled examples by classifier. This is also known as self-training in the literature of semi-supervised learning. A classifier is first trained with the small amount of labeled data. The classifier is then used to classify the test data. Typically, the most confidently labeled test examples, together with their predicted labels, are augmented to the initial training set. This will be repeated for every month. Hence, the predictive ability of model improves over the time.

4.3 Data Cleaning

It is known that user verbatim text is of free-form text. So, we applied following NLP pre-processing techniques:

- Bing speller checker: Bing spell checker API is used to correct spelling mistakes in user verbatim based on context of neighboring words.
- Case normalization: User text is normalized to lower case.
- Stop word removal: The most frequent words often do not carry much meaning. Examples: the, a, of, for, in.
- Stemming (Lemmatization): Converts multiple related words to a single canonical form (e.g., playing, play to play).
- Remove special characters: Removed special characters, emails, and URLs.
- We do not remove numbers as our feedback text contains version numbers of products (e.g., Windows 10, Office 16, Office 2013).

4.4 Feature Engineering

Feature engineering is the most important step in determining the performance of any machine learning model. We extracted features from user verbatim, query, and answer. We have extracted bag-of-n-grams. Our vocabulary size is 33 K. We have extracted bag-of-n-grams features with TF-IDF as weighting function. Best features have been chosen based on mutual information (MI). We have used up to 3-grams in

Table 3 Accuracy comparison (in %) of different classifiers

Classifier	Oct	Nov	Dec and Jan
Logistic regression	77	78.12	NA
Random forest	76.2	78.5	NA
Support Vector Machine (SVM)	79	80.25	83.5

NA: we have not trained LR and Random Forest models for further months as they didn't perform well on previous months

our all experiments. We did not extract *n*-grams features from query and answer title as they are interrelated. Instead, we computed the DSSM score between query and answer title. If DSSM score is not available by pretrained model, Jaccard similarity between query and answer title is used. This feature provides a good input signal in identifying the triggering issues.

4.5 Results

We trained logistic regression, support vector machine, and random forest classifiers in our experimental studies. The parameters of classifiers are tuned using validation set. The results of our experimental studies are presented in Table 3. It is observed that support vector machine (SVM) classifier has performed better than other classifiers. Support vector machines (SVMs) have a good track record in text classification, they can be trained using many features, and both training and classification for linear SVMs are fast with optimized learning algorithms. SVM model worked well for the following plausible reasons:

- High-dimensional data and few irrelevant features.
- Text classification problems are linearly separable in higher dimensions.

SVM-based classifier has been retrained for further months. As it is a recurrent task, we retrained our model based on the confident classifications of previous month results to enhance the classification performance. We noted the classification accuracy 83.5% (+4.5% improvement over initial model) on December and January month data using our retrained model.

The confusion matrix of SVM classifier is shown in Table 4. The model is able to distinguish content and triggering issues from irrelevant (junk) category very well. Currently, the classier is identifying content and triggering issues with an average classification accuracy of 80%. The misclassification results of SVM classier are presented in Table 5.

Table 4 Confusion matrix of SVM classifier

Actual predicted	Content (%)	Triggering (%)	Irrelevant (%)
Content	80	10	10
Triggering	16	83.33	1
Irrelevant	6.25	6.25	87.5

Table 5 Analysis of classification results

Query	Answer title	User verbatim	Actual category	Predicted category
Microsoft Outlook connecting issues	How to fix mailbox connection issues in Outlook	#1 Takes me to a dead URL	Content	Content
Windows 10 startup programs folder	Change startup apps in Windows 10	I asked for the folder. This does not help me add new programs to the list	Content	Triggering
How do i sync my photos in Windows 10?	About sync settings on Windows 10 devices	Trying to find out how to sync photos	Triggering	Triggering
Can a Microsoft Office 365 legal template personal subscription altered	Cancel your Office 365 Home or Personal subscription	My question was can the office 365 legal template be altered (in other words, move the vertical lines and vertical numbers moved or altered)	Triggering	Content

4.6 Top Insights Gained

- Most of the Cortana-related answers are having issues with content and triggering gaps.
- Some of the answers contain go-to hub as part of instructions. And user feedback corresponding to that is What is hub? How it looks like. Adding hub icon in the instructions would help users.

4.7 Business Impact

Our machine learning solution has reduced lot of human efforts and helped in estimating SLA with author's team and triggering team. Our proposed solution has helped in reduction of DSAT ration (= DSAT/(DSAT + SAT), where DSAT and SAT are quantification of user dissatisfaction and satisfaction, respectively) by 2%, and it indicates the enhancement in overall user experience with our answers.

5   Conclusion

Understanding the user feedback and taking appropriate decisions are very important for any product team to improve the user experience. In this paper, we addressed the issues in manual analysis of user feedback and we developed an automatic intelligent machine learning algorithm in conjunction with DSSM model to classify user feedback. The proposed approach helped in reducing huge manual efforts and resulted in reduction of DSAT ratio by 2%. This indicates overall user satisfaction is increased with the answers. Our proposed solution can be leveraged easily by many other products feedback pipeline.

References

1. Po-Sen Huang, Xiaodong He, Jianfeng Gao, Li Deng, Alex Acero, Larry Heck: Learn- ing deep structured semantic models for web search using clickthrough data. In: Pro- ceedings of the 22nd ACM Conference on Information and Knowledge Management (2013).
2. Rich Caruana, Alexandru Niculescu-Mizil: An empirical comparison of supervised learning algorithms. In: Proceedings of the 23rd International Conference on Ma- chine Learning (ICML) ACM, pp. 161–168 (2006).

Big Data Analytics on Aadhaar Card Dataset in Hadoop Ecosystem

D. Durga Bhavani, K. Rajeswari and Nenavath Srinivas Naik

Abstract Aadhaar is a unique identity issued to Indian citizens by Unique Identification Authority of India (UIDAI). The Aadhaar enrollment data are in a CSV file format with the details of citizens such as state, district, gender, age, Aadhaar generated, enrollment rejected, date of enrollment. This paper proposes the inconsistencies and fluctuation in enrollments based on demographics, ages, time, and the reaction of state governments and habitants of states and country. The main idea is to infer the fluctuations in enrollment due to the effect of demonetization, essential linking of Aadhaar and PAN, child enrollment, and other government policies.

Keywords Aadhaar card · Big data · Hadoop · Hive

1 Introduction

The provision of Aadhaar was launched on January 28, 2009 by the Planning Commission of India, promising that the data would be protected and security would be maintained. Hence it is now the worlds largest biometric. The benefits of Aadhaar include the reduction of 24% sale in subsidized LPG by excluding ghost beneficiaries saving 127 k crores. It also includes removal of duplicate ration cards, voter id, easy verification for passport, biometric attendance in government offices, unemployment benefit, linking land records with Aadhaar, PRAGATI crop insurance scheme, linking Aadhaar number with SIM number, and Aadhaar linking with banking services, etc.

D. Durga Bhavani (✉) · K. Rajeswari · N. Srinivas Naik (✉)
Department of Computer Science and Engineering, CVR College of Engineering (Autonomous),
Hyderabad, India
e-mail: drddurgabhavani@gmail.com

N. Srinivas Naik
e-mail: srinuphdcs@gmail.com

K. Rajeswari
e-mail: rajeswarikatta10@gmail.com

© Springer Nature Singapore Pte Ltd. 2019
R. S. Bapi et al. (eds.), *First International Conference on Artificial Intelligence and Cognitive Computing* , Advances in Intelligent Systems and Computing 815,
https://doi.org/10.1007/978-981-13-1580-0_44

After the Kargil War, a security and defense expert stated reasons of defects at Kargil. One of the reasons included an absence of identity proof and accurate data of battalions in the area opposite to Kargil War during 1998 which led intrusions in April/May 1999 launched by Pakistan. Hence, it was recommended that every citizen in the village borders must be issued an identity. A Group of Ministers (GoM) led by L. K. Advani in May 2001 announced the requirement of national identity in 2003 which is a multi-purpose identity card would be soon rolled out.

The UIDAI was formally launched in 2009 by the UPA government. Manmohan Singh linked Aadhaar to Direct Benefit Transfer Scheme (DBT) to eliminate leaks in the system on November 26, 2012. A compulsion for a passport that links has to be made for the verification process, and it is accessible from July 5, 2014. On June, 18, 2015, high-level review meetings were conducted, and the media publicized that the DBT on LPG subsidy given to Aadhaar cardholders by leaching off shadow dealers and the government made 14–15% more profit on LPG. On June, 16, 2017, Aadhaar made mandatory to open a bank account, and the existing bank accounts must be linked with Aadhaar number by December 31, 2017 [1].

The UIDAI provides a catalog of downloadable datasets collected at national level. The dataset of interest is enrollments processed in detail. Since the dataset is in gigabytes, here the challenge is how to process this data. The rest of this paper is organized as follows. Section 2 describes literature survey. The proposed work is presented in Sect. 3. Experimental analysis is presented in Sect. 4. Finally, Sect. 5 describes conclusion and future work.

2 Literature Survey

2.1 Analysis of Social Security Number Notification Component of the Social Security Number, Privacy Attitudes, and Notification Experiment

This investigation was done by US Census Bureau regarding the connectivity between social security number and administrative records. This connection between SSN and administrative records resulted in less-respondent burden and less memory errors. The analysis determines whether SSN requests and utilization of notification for administrative record influence census response, form completeness among forms returned, and response to the SSN request item. The results stated that the presence of no SSN in a household or low coverage area for an SSN request dropped the rate of response to the Census Bureau. The findings suggest that the use of specific notification rather than general notification hiked the response rate. But combination of specific notification and SSN lessened the response rate. The authors suggest that the response to SSN request must be minimized for privacy concerns. When more complete data are required, notification of the administrative record must include in a cover letter with a census form [2].

2.2 The Swedish Personal Identity Number: Possibilities and Pitfalls in Healthcare and Medical Research

Swedish healthcare services and national health registers are subject to the nearness of a unique identifier. Around 75,638 people have changed PIN by January 2008. The most popular explanations for the difference in PIN are a wrong account of the date of birth or sex among workers or infants. While phenomenal, variation in sex dependably prompts a variation of PIN as it is sex-particular and even date of birth. The popular explanations behind reutilization of the PIN are that when migrants are appointed with a new PIN, the PIN might have been used by another person beforehand. This is a vital deficiency of specific PIN blends alluding to 1960 births. A few moral problems can be elevated professional and on the utilization of PIN in restorative study. They recommend that coordinating of registers over the PIN and coordinating of national well-being registers without the express endorsement of the distinct patient is the advantage [7].

2.3 An Anatomization of Aadhaar Card Dataset a Big Data Challenge

Big data analytics can be utilized for changing raw information into essential data which supports in the professional examination and structures a choice which can be an emotionally supportive network for the administrators in the association. The authors broke down the Aadhaar card informational index in contradiction to various research questions. For instance, add up Aadhaar cards enrolled by state, rejected by the state, add up to the number of Aadhaar cards by gender, and the aggregate number of Aadhaar cards by age [5].

2.4 Aadhaar Data Analysis Using SAP Lumira

SAP Lumira is known as a visual intelligence tool to create and visualize information on the dataset. The tool imagines data and develops accounts to provide graphical details of the data. The reason for selecting this dataset was it allows geographical analysis. SAP Lumira enables prepare and share mechanism [4].

3 Proposed Work

Hadoop is a distributed storage, and using Hive, we submit queries. The queries are applied to huge datasets. The datasets are so large that high-end, expensive,

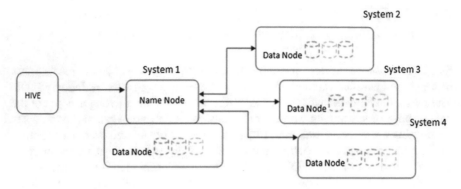

Fig. 1 Cluster setup

traditional databases would fail to perform operations. The Hadoop cluster consists of NameNode and DataNodes. The two primary nodes are NameNode (the master) and various DataNodes (slaves). The NameNode deals with the file system as it stores meta-data for all documents and indexes in the file system. The client machine can access file system through communicating with the NameNode and DataNodes. DataNodes are the slaves to NameNode, and client can recover blocks whenever they require. At least, two copies replicate the data in DataNodes exchanged and copied between those nodes [3, 6, 9, 10].

3.1 Methodology

To manage large chunks of data, Hadoop is installed in a cluster of commodity nodes to perform operations on fully distributed storage. The Hive system converts the query into mappers and reducers operations which works for the given data in the cluster [8]. Steps to initiate query analysis on big data, i.e., the Aadhaar dataset, in this paper are described in the following steps.

1. Download the .csv files from the UIDAI Web site.
2. Using Java, create a new column to store the data year wise in CSV file format which is given for each day in each month from August 2016 to July 2017.
3. Install Java 1.8.0 71, Hadoop 1.2.1 version, and Ubuntu operating system in the cluster nodes.
4. Configure NameNode in system 1 and DataNodes in systems 2, 3, and 4 as shown in Fig. 1.
5. Install and run Hive on the NameNode.
6. Upload the dataset and append it using joins on the NameNode.
7. Run required Hive queries on the NameNode.
8. The final output is extracted from the HDFS.
9. Finally, analyze the data.

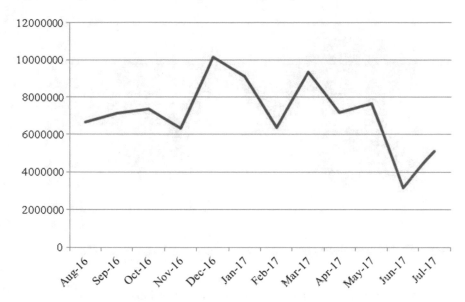

Fig. 2 Month-wise enrollments

4 Experimental Analysis

In our empirical analysis, Aadhaar dataset of 10.8 GB in .csv format is used. Hive execution mode is used for analyzing Aadhaar dataset.

Research queries (RQs) described below are to analyze Aadhaar dataset.

– RQ1. Find out the total number of cards approved month wise.
– RQ2. Find out the total number of cards approved age wise.
– RQ3. Find out the total number of Aadhaar applicants by gender in South India.

RQ1. Find out the total number of cards approved month wise.

Queries that are used in Hive shell are as follows:

1. Group data by month (date of enrollment).
2. For each month, aggregate group using SUM (Aadhaar generated).

From the results of the first query as shown in Fig. 2, the most significant numbers of enrollments done in December 2016 as enrollment for Aadhaar have increased from 7 to 8 per day post-demonetization in November 2016 against 5–6 lakhs, which also grew the Aadhaar payment system. Another hike in increase in enrollments was in May 2017 as the news headlines reported from that linking of Aadhaar and PAN card for income tax returns was mandatory from July 2017.

RQ2. Find age-wise Aadhaar cards from different states.

Queries that are used in Hive shell are as follows:

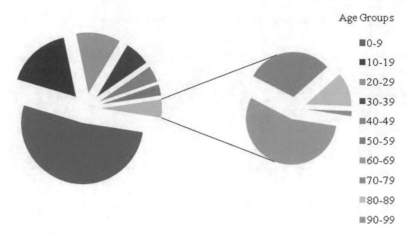

Fig. 3 Age-wise enrollments

1. Create a table with age range from a lower to upper limit.
2. Group ages within the given ranges.
3. Aggregate those groups with SUM (Aadhaar generated).

From the results of the second query, 81% of the enrollments are done below the age of 30 and with the most enrollments are within the age range of 0–9 years as shown in Fig. 3. UIDAI focused on child enrollment during 2016 and 2017, besides persuading those who have not registered. India ranks 141 in the world regarding young population concerning the countries, and by 2020, almost 64% of the countries population will be in working age group. Aadhaar has to be updated once the child crosses five years of age and again after fifteen years of age. Thus, it resulted in highest enrollments in the age bands of 0–9 and 10–19.

RQ3. Identify Aadhaar enrollments by gender in South India.

Queries that are used in Hive shell are as follows:

1. Identify the Aadhaar cards accepted with gender as a male in each state and store it in the table
2. Identify the Aadhaar cards accepted with gender as female in each state and store it in the table
3. Join the above two tables.

Percentage of population enrolled from August 2016 to July 2017 is as shown in Fig. 4:

1. Andhra Pradesh, 0.20%
2. Telangana, 0.18%
3. Karnataka, 7.75%
4. Kerala, 4.79%
5. Tamil Nadu, 10.44%

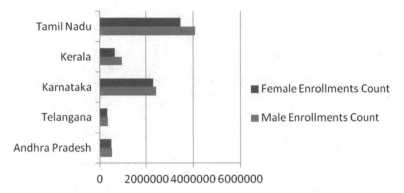

Fig. 4 Gender- and state-wise enrollments

From the results of the third query, we observed that Tamil Nadu state has the highest number of Aadhaar applicants from August 2016 to July 2017. Tamil Nadu has National Population Register (NPR), under which each family is not registered as an individual. But each time a family comes for registration, the male member will be missing as he will be working in some other place. Only 88% of the adults in Tamil Nadu had Aadhaar card by the end of 2015 which was the least in prominent states of India which resulted in higher enrollments in the dataset between the given months.

5 Conclusion and Future Work

The Aadhaar dataset is analyzed for various queries. Hadoop framework is used for storing, processing and analyzing 10.8 GB of Aadhaar data. Hive query language is used for analyzing and extracting required data. More than 1.1 billion out of 1.3 billion population has an Aadhaar by October 2017. This paper points the inconsistencies and fluctuations in enrollment regarding demographics, timeline, age, and the reaction of the state governments and citizens of states. Using Hive, we analyze the acceptance and rejection of enrollments with invalid entries. Analysis on data can be used for highly specific purpose via Hive, Hbase, Giraph. Using Giraph, Hadoop can implement thousands of reading/writes per second and might be used directly from the enrollment offices to upload directly into the primary HDFS storage for enrollments of data. Using Hbase, semi-structured or unstructured data can be maintained.

As part of future research work, the proposed work can be further extended to analyze or manage Aadhaar dataset for frauds, errors, and duplication of Aadhaar, etc.

References

1. 'Aadhaar' most sophisticated ID programme in the world: World Bank, Daiji World (2017).
2. Linda, B.:Analysis of the Social Security Number Validation Component of the Social Security Number, Privacy Attitudes, and Notification Experiment. In: Report of Census 2000 Testing, Experimentation, and Evaluation Program (2003).
3. Madhavi, V.: Parallel Processing of cluster by Map Reduce. In: International Journal of Distributed and Parallel Systems (IJDPS), vol. 3, no.1, (2012).
4. Chandrashekar, M.: Aadhaar Data Analysis using SAP Lumira. In: Data Geek Chal- lenge, https://blogs.sap.com/2013/08/20/.
5. Mohit, D., Nanhay, S.: An Anatomization of Aadhaar card dataset- a big data challenge. In: Procedia computer science. vol. 85, pp. 773–739 (2016).
6. Assuncao, M., Calheiros, R., Bianchi, S., Netto, M., Buyya, R.: Big data computing and clouds: trends and future directions. In: Journal of Parallel and Distributed Computing (JPDC), volumes 79, pp. 3–15 (2015).
7. Ludvigsson, J.F., Otterblad-Olausson, P., Pettersson, B.U., Ekbom, A.: The Swedish personal identity number: possibilities and pitfalls in healthcare and medical re- search. Europian Journal of Epidemiol, vol. 24, pp. 659–667 (2009).
8. White, T.: Hadoop: The Definitive Guide. In: OReilly Media (2015).
9. Chih-Fong, T., Wei-Chao, L., Shih-Wen, K.: Big data mining with parallel com- puting: A comparison of distributed and MapReduce methodologies. In: Journal of Systems and Software, vol. 122, pp 83–92 (2016).
10. Naik, N.S., Negi, A., and Sastry, V.N.: Improving straggler task performance in a heterogeneous MapReduce framework using reinforcement learning. In: International Journal of Big Data Intelligence, vol. 5, no. 4, pp. 201–215 (2018).

Improving Performance of MapReduce in Hadoop by Using Cluster Environment and Key-Value Pair Localization

Boddu Ravi Prasad and K. Anil Reddy

Abstract Nowadays, enormous amount of data is being generated every minute. Data is in different forms, i.e., structured, semi-structured, and unstructured. Many industries are using Hadoop to analyze this data to improve their businesses. In Hadoop environment, MapReduce is being widely used to analyze the data. MapReduce uses distributed approach to process the data. It indicates data transfers between different nodes in a cluster and between different clusters also. Due to this, the performance of Hadoop framework is degrading. Many researchers have proposed different techniques to improve the performance of MapReduce in Hadoop framework. In this paper, Improved K-Mediods algorithm to form clusters and key-value pair's locality mechanism in the shuffle phase to improve the performance of MapReduce are used. This work is tested in multi-cluster Hadoop environment.

Keywords Big data · Hadoop · MapReduce · Key-Value pair locality

1 Introduction

In the present tech-savvy world, for every activity of human or machine, data is generated. This leads to huge amount of data which can be seen in different formats, i.e., structured, semi-structured, and unstructured formats. The data is treated as big data. The traditional databases are unable to process this data because of huge volumes and different varieties. After analyzing in proper manner, if used, this big data is a value-add to the organizations which in turn can create new opportunities.

B. R. Prasad (✉) · K. A. Reddy
CSE Department, Marri Laxman Reddy Institute of Technology and Management,
Dundigal, Hyderabad, Telangana, India
e-mail: rprasad.boddu@gmail.com

K. A. Reddy
e-mail: anilreddyk@mlritm.ac.in

© Springer Nature Singapore Pte Ltd. 2019 467
R. S. Bapi et al. (eds.), *First International Conference on Artificial Intelligence and Cognitive Computing* , Advances in Intelligent Systems and Computing 815,
https://doi.org/10.1007/978-981-13-1580-0_45

Big data has five characteristics:

Volume: Because of latest technical developments, huge amount of data is generated from every sector. Industry requires latest mechanism to store and process this data. Implementation of security features may degrade system performance which is not acceptable.

Velocity: Due to huge volume of data, velocity of processing of data and implementation of security mechanisms decreases which is not acceptable in critical applications like healthcare. To increase the velocity, system has to follow parallel processing and distributed systems.

Variety: Big data is collected in structured, semi-structured, and unstructured forms. Different mechanisms are to be implemented to process different varieties of data.

Veracity: This talks about uncertainty of data. This is formed mainly because of unstructured data which comes from social media, stock exchanges, etc. Processing and providing security to uncertain data is a very difficult task. Veracity describes consistency and trustworthiness of data.

Value: It provides the value of outputs for gains from large data sets.

In Hadoop environment, MapReduce is being widely used to analyze the data. MapReduce uses distributed approach to process the data. It indicates data transfers between different nodes in a cluster and between different clusters also. Due to this, the performance of Hadoop framework is degrading. Many researchers have proposed different techniques to improve the performance of MapReduce in Hadoop framework.

This paper is organized as follows: Sect. 2 describes existing system; Sect. 3 describes cluster formation and key-value pair localization; Sect. 4 gives the analysis of results; and concluding remarks are given in Sect. 5.

2 Existing System

Hadoop framework uses two modules, namely HDFS and MapReduce [8]. HDFS distributes the data into different data nodes. Data analysis is carried out by MapReduce. The execution of above two modules is carried out as follows:

NameNode, Secondary NameNode and JobTracker are the master services in a HDFS [2] framework. DataNode and TaskTracker are the slave services. All the master and slave services can communicate each other. If NameNode is the master service, its corresponding slave service is DataNode and they can talk to each other. Client sends the request to NameNode to store the data or the request file. After receiving the request, NameNode takes the metadata of the request file and sends the response to store the file in specific DataNodes which are having free space. Client then puts the files in respective DataNodes.

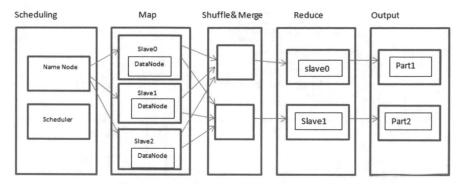

Fig. 1 Workflow of MapReduce

To overcome the problem of losing the data because of the failure of DataNodes which are the commodity hardware, by default, HDFS is given with three replications. DataNodes which have the replications of the same file communicate with each other and sends the response back to the client. All the DataNodes send proper block report and heartbeat to the NameNode for every short period of time. NameNode stores the detail in its metadata. All the data is stored in HDFS in this manner. If NameNode does not receive block report and heartbeat of a particular DataNode, it treats as that DataNode is not active thus removes the metadata of that DataNode. If the metadata is lost, entire cluster is inaccessible. To overcome this problem, it is advised to have highly reliable hardware for NameNode. NameNode is the single point of failover. DataNodes represent actual storage and NameNodes as the metadata.

Processing of data is done by MapReduce. JobTracker is the service to process the data by using the program. JobTracker sends a request to NameNode to get the details of the request file and then gets the response from NameNode. JobTracker sends the request to TaskTracker which is nearer to it to process the data. JobTracker sends the computation (program) to all the TaskTrackers to process all the data. This process is called as Map. The number of mappers in a cluster is equal to the number of input splits. This is taken care by JobTracker. Every TaskTracker sends the block report and heartbeat to JobTracker in every 3 s. If a TaskTracker is down for any reason, the same task is given to other TaskTracker which is having the same file (replication) by JobTracker. JobTracker is also called single point of failover. This also requires highly reliable hardware. Shuffler takes care of merging all the outputs from all the mappers. Reducer takes care of combining all the outputs from all the mappers. Number of output files is equal to the number of Reducers. The workflow of MapReduce is shown in Fig. 1.

Many algorithms are proposed to improve the performance of Hadoop framework. Performance [5] of MapReduce can be improved by forming efficient clusters by using K-means [4] and K-Mediods [7] algorithms. By applying deep data locality [6], the execution of shuffle phase can be reduced.

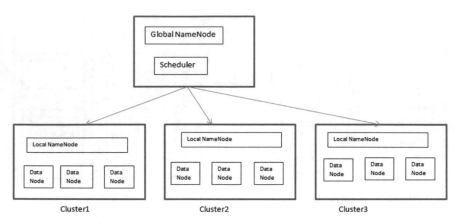

Fig. 2 Cluster formation

3 Cluster Formation and Key-Value Pair Localization

In the existing approach, entire data is distributed in different DataNodes. In the proposed approach by using Modified K-Mediods algorithm, the data can be divided into smaller data sets of similar type [3]. These data sets are treated as clusters. As shown in Fig. 2 for each cluster one Local NameNode and multiple DataNodes are maintained. For the entire network, one Global NameNode is maintained. When a request comes from the client, initially it goes to the Global NameNode. The scheduler who is present at Global NameNode decides to accept the request or not. If the request is accepted, Global NameNode transfers it to the appropriate Local NameNode depending on the data present in corresponding cluster.

After the completion of mapper phase, key-value pairs are generated. In the shuffling phase, the key-value pairs which are present in different DataNodes of same cluster or of different clusters are brought to the blocks in the same nodes. In shuffle phase, shuffle operation is performed in single node which leads less data movement in the system. Due to this, load on network decreases and shuffle phase completes in less amount of time. Reducer performs its operation in the blocks which are present in single node. The above mechanism decreases data movement in the environment.

For example, in the above Fig. 3, if key-value pairs are present in block2, block3, block4, block10, block14, then all the key-value pairs in different blocks are shifted to block2, block3 which are present in single node. Shuffle operation is performed on block2 and block3 which are present in single node. Reducer operation is carried on the output of shuffle phase.

The combination of cluster mechanism and key-value locality mechanism improves the performance of MapReduce.

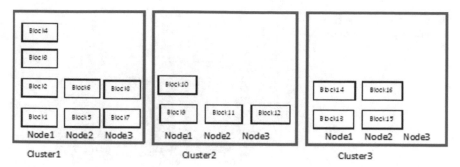

Fig. 3 Key-value pair localization

Fig. 4 Comparision in terms of total execution time

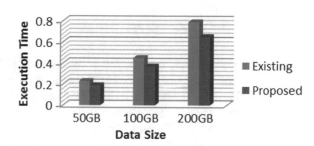

4 Experimental Setup and Analysis of Results

4.1 Experimental Setup

The proposed work is tested on the multi-cluster environment, where 13 systems were configured. One machine was used as Global NameNode, and remaining systems were divided into three clusters. Each cluster consists of one Local NameNode and three DataNodes. In this environment, Hadoop 2.6.0 [1] is installed on Ubuntu 14.04 operating system.

4.2 Analysis of Results

To perform the experiment, the data from one shopping mall was used. The experiment was carried out on different sizes of data; 50, 100, 200 GB.

Figure 4 shows the total execution time of existing modified K-Mediods and the proposed approach. Both the approaches completed their execution almost at the same time when the data is 50 GB. But for the data size of 200 GB, the proposed approach completed the processing of data in less amount of time.

Figure 5 shows shuffling phase execution time where the proposed approach required more time for all the sizes of data.

Fig. 5 Comparision in terms of shuffling phase execution time

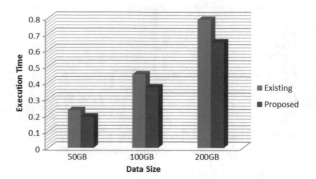

5 Conclusion and Future Scope

For all business entities, information is the backbone. These business entities are spending lot of money in processing the data. Hadoop platform is most suitable to analyze the data. It uses HDFS and MapReduce. To improve the performance of Hadoop environment, MapReduce is to be optimized. In this paper to improve the performance of MapReduce, the modified K-Mediods algorithm and key-value data locality in shuffle phase are used. By applying the above two approaches, the processing time of Hadoop environment is decreased. Present study is checked in multi-cluster environment. This can be extended to multi-cluster environment in wide area networks.

References

1. Hadoop Implementation, http://Hadoop.apache.org/docs.
2. HDFS Permissions Guide: http://Hadoop.apache.org/docs/stable/Hadoop-projectdist/Hadoop-h dfs/HdfsPermissionsGuide.html[09-01-2014].
3. Parth Gohil, Bakul Panchal, J. S. Dhobi "A Novel Approach to Improve the Performance of Hadoop in Handling of Small Files" in IEEE International Conference on Electrical, Computer And Communication Technologies PP.1–5 DOI :https://doi.org/10.1109/icecct.2015.7226044 (2015).
4. Pramod Bide & Rajashree Shedge "Improved Document Clustering using K-means Algorithm" in IEEE International Conference on Electrical, Computer And Communication Technologies PP.1–5 https://doi.org/10.1109/icecct.2015.7226065 (2015).
5. Swathi Prabhu, Anisha P Rodrigues, Guru Prasad M S, Nagesh H R "Performance Enhancement of Hadoop MapReduce Framework for Analyzing Big Data" in IEEE International Conference one Electrical,Computer And Communication Technologies, ISBN: 978-1-4799-6084-2, PP.1–8 (5-7 March 2015).
6. Sungchul Lee,, Ju-Yeon Jo, Yoohwan Kim "Performance Improvement of MapReduce Process by Promoting Deep Data Locality " in IEEE International Conference on Data Science and Advanced Analytics PP.292–301 (2016) https://doi.org/10.1109/dsaa.2016.38.
7. Subhash Chandra, Deepak Motwani "An Approach to Enhance the Performance of Hadoop MapReduce Framework for Big Data " International Conference on Micro-Electronics and Telecommunication Engineering PP. 178–182 (2016) https://doi.org/10.1109/icmete.2016.64.
8. Borthaku, D. The hadoop distributed file system: Architecture and design. Retrieved from http://lucene.apache.org/Hadoop.

Migration of Big Data Analysis from Hadoop's MapReduce to Spark

J. Pradeep Kumar, Sheikh Gouse and P. Amarendra Reddy

Abstract Hadoop was the first major platform for big data, later leading to the development of the better known Spark by the Apache Software Foundation. For a long time of use the inquires about experienced which concentrated on Hadoop empower examines to have a decent review of its points of interest, disadvantages and confinements keeping in mind the end goal to enhance the arrangement started presented. Despite the fact that in 2012 Hadoop 2.0 introduced a number of improvements, particularly in regards to accessibility of the Hadoop Distributed File System (HDFS) and bunch asset administration amid MapReduce work execution through the YARN design, the impacts of these changes are yet to be investigated. This paper aims to address the disadvantages and limitations of Hadoop and what these residual extents of enhancements. Spark has outperformed Hadoop to become the most powerful open source software for dealing with big data. It handles most of its operations in-memory, copying them from the passed on physical limited into the far speedier and more reliable RAM memory. This decreases the amount of dull composed work and scrutinizing to and from direct, clumsy mechanical hard drives that occurred under Hadoop's MapReduce structure. Spark provides a number of innovations that increase the estimation of huge information and allows new utilize cases. Spark gives us a complete structure for constructing, overseeing, and actualizing enormous information preparing necessities.

J. Pradeep Kumar (✉) · S. Gouse · P. Amarendra Reddy
Department of Information Technology, MLR Institute of Technology,
Dundigal, Hyderabad, Telangana, India
e-mail: Pradeep.jakkulla@gmail.com

S. Gouse
e-mail: gouse.sheikh@gmail.com

P. Amarendra Reddy
e-mail: amarpanyala88@gmail.com

© Springer Nature Singapore Pte Ltd. 2019
R. S. Bapi et al. (eds.), *First International Conference on Artificial Intelligence and Cognitive Computing*, Advances in Intelligent Systems and Computing 815,
https://doi.org/10.1007/978-981-13-1580-0_46

1 Introduction

Data is developing at a rapid pace in terms of both its volume and the speed with which it can be processed. This semi-organized data could potentially provide knowledge whether broke down. This large amount of data can be prepared in parallel by MapReduce [1]. In 2004, Google Inc. produce a paper titled "Google's MapReduce". Queries were split and circulated crosswise over numerous nodes that were handled in parallel. This was called the Map phase. At that point outcomes were consolidated by the Reduce phase and an output was created. This prompted the development of the Hadoop open source system for conveyed figuring over numbers of nodes. Its major drawback was the time it took to process iterative datasets. Spark provided the solution to this problem with its extremely high calculation speed. Simply put, it stores data in a fault tolerant model in resilient distributed datasets (RDD) [1, 2].

Despite the fact that Spark works up to 100 times faster than Hadoop in particular conditions, it doesn't have its own specific appropriated storage system. Disseminated stockpiling is essential to a significant part of the present growth of big data as it allows immense multi-petabyte datasets to be secured over an endless number of PC hard drives, rather than requiring monstrously large custom equipment which holds everything on one device. These structures are versatile and more drives can be added to the framework as the dataset grows [3, 7].

MapReduce returns a large part of the data back to the physical storage medium after each operation. This was initially done to ensure that a full recovery could be introduced in barrier something turns out gravely —as data held electronically in RAM is more unusual than that neatly held on disks. However, Spark arranges data in RDDs, which can be recovered after a failure [1, 4].

Spark's functionality for handling advanced data processing tasks, such as real-time stream processing and machine learning is way ahead of what is possible with Hadoop alone. This, along with the increase in speed provided by in-memory operations, is the real reason, in my opinion, for its growth in popularity. Real-time processing means that data can be fed into an analytical application the moment it is captured, and insights immediately fed back to the user through a dashboard, to allow action to be taken. This sort of processing is increasingly being used in all sorts of big data applications, for example, in recommendation engines used by retailers, or for monitoring the performance of industrial machinery in the manufacturing industry [1, 5].

Spark's usefulness in taking care of cutting edge information preparing undertakings, for example, ongoing stream handling and machine learning is streets ahead of what is conceivable with Hadoop alone. The increase in speed provided by in-memory functions is the real reason for its increasing popularity. Real-time processing implies that information can be placed into a big data analytics application the minute it is captured, and experiences instantly sustained back to the client through a dashboard, to enable move to be made. This kind of handling is increasingly being utilized as a part of a wide range of big data applications, such as in proposal motors utilized by retailers, or for observing the operation of mechanical apparatus in the manufacturing industry [6, 9].

2 Hadoop Distributed File System (HDFS)

To process any data, the client submits both data and a program. HDFS stores that data and MapReduce then shapes it. Hadoop functions take the input data, break into 128 Mb units and then moves the pieces to different center points. Once each of the squares of data is secured on data centers, the customer can process the data [4]. Resource Manager by then designs the program on particular center points. Once each one of the center points methodology the data, and the yield is returned to the HDFS [1, 3, 5]. The components of the HDFS and MapReduce are shown in Table 1 and the limitations of Hadoop are shown in Table 2.

3 Limitations of Hadoop for Big Data Analytics

See Table 2.

4 Spark

This is an open source cluster-computing framework; a general purpose and lightning fast cluster-computing system [1, 2], and a tool for running Spark applications [10]. It is 100 times faster than Hadoop and 10 times faster than accessing data from disks. It is written in Scala, Java, Python, and R. It runs in Standalone, YARN, and Mesos cluster managers [8, 9]. It is a powerful open source engine that provides real-time stream processing, interactive processing, graph processing, and in-memory processing as well as high-speed batch processing. It is also ease to use and has a standard interface. The spark components are shown in Fig. 1. Spark streaming is shown in Fig. 2 and the architecture is shown in Fig. 3 [10].

4.1 Spark Components

From Fig. 1, the components are [1, 9]:
Spark Core
This is the kernel of Spark and provides an execution platform for all Spark applications.
Spark SQL

Table 1 Components of HDFS and MapReduce

Type	Description	Components		Features
HDFS (Hadoop distributed file system)	Primary data storage Java-based file system. Runs on commodity hardware. Interacts with shell commands	Name node (Master)	Stores metadata (**blocks**, location, rack) runs file system (namespace, client's access and executes as naming, closing, opening files, and directories) [9]	Robust Cost Effective Faster Data Processing Flexibility Fault tolerance
		Date node (Slave)	Storing actual data. Performs read and write operation. Replica block (data and recording the block's metadata) Performs checksums for data Performs handshaking by verifying (namespace ID and software version) Mismatch it goes down automatically. Manages data storage **Operations**: Block replica creation, deletion, and replication [9]	
MapReduce	Provides data processing software framework for easily writing applications. Programs run in parallel Each phase has **key-value pairs**	Map phase	Converts data it into another set. Broken down into key and value pairs	Simplicity Scalability Speed Fault tolerance
		Reduce phase	Takes output from the Map Combines and modifies those data tuples based on the key	

Table 2 Limitations of Hadoop

Limitations	Issue	Solution
Small files	Not suited for small data default block size 124 MB [3] [9]	HAR files Sequence files HBase
Low processing speed	Data on-disk	In-memory (RAM)
Batch processing only	Does not process streamed data	Stream processing
No real-time processing	Not suitable	Real-time processing
No iteration	Does not support cyclic data flow	Cyclic data flow
Latency	Key value pair	Resilient distributed dataset
Not easy to use	Code for each and every operation	Built-in code
Security	Kerberos authentication good security but difficult to manage	Password authentication
No abstraction	Does not have	Resilient distributed dataset abstraction
Vulnerable by nature	Entirely written in Java	Scala Python
Lengthy code	Larger	Shorter
No caching	Inefficient	Cache data in memory
Uncertainty	Ensures that data job is complete but no guarantee	Guarantee
Processing algorithms	Graph processing slow	Graph processing high
Fault tolerance	Replication	No replication
Limitations	Hadoop MapReduce	Spark Flink Apache Flume, MillWheel, and Google's own Cloud Dataflow [10]

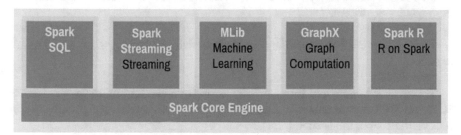

Fig. 1 Spark components

This enables users to run SQL/HQL queries on top of it.

Spark Streaming

This is an interactive and data analytics application used with real-streaming data [2].

Fig. 2 Spark streaming

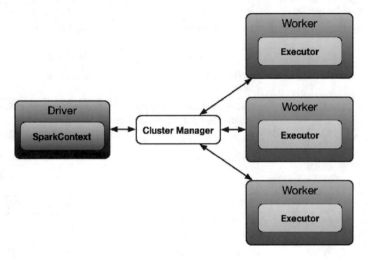

Fig. 3 Spark architecture

Spark MLlib

This is the machine learning library which provides both efficiencies and the high-quality algorithm.

Spark GraphX

This is the graph data processing engine that enables the processing of graph data at scale.

SparkR

This is an R bundle that forms a lightweight front end tool to utilize spark from R. It enables data researchers to breakdown substantial datasets and intuitively run jobs on them from the R shell [3, 4].

1. **Spark architecture components** [1]

    1. Resilient distributed datasets (RDD)
    2. Directed acyclic graph (DAG).

1. **RDD**:

R provides fault tolerance through a lineage graph. D is the data present on multiple nodes in a cluster.

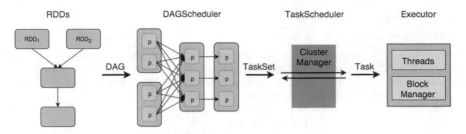

Fig. 4 DAG processing

D is the collection of partitioned data with primitive values [1].

RDDs are immutable but can generate a new RDD by transforming an existing RDD as shown in Fig. 4.

RDDs creating

1. Parallelized collections: Create parallelized collections by invoking the parallelized method in the driver program.
2. External datasets: calls the text file method one can create RDDs.
3. Existing RDDs: Apply transformation operation on existing RDDs to create a new RDD.

RDDs operations

1. Transformation: A new RDD is created from the existing one. It sends the dataset to function and returns a new dataset.
2. Action: The end result is returned to the driver and then sent to external storage.

2. **DAG**

Direct means coordinate action which changes a data segment state from A to B as shown in Fig. 4.

Acyclic means non-cyclic; there is no return to the start position.

Graph is a sequence of calculations performed on data where every node is an RDD segment and edge is a change on top of data [7].

4.2 Components

From Fig. 4, architecture the components are given in details from Table 3 [1, 2, 9].

1. Master Daemon—(Master/Driver Process).
2. Worker Daemon—(Slave Process).

Table 3 Architecture components

Driver (Master node)	Executor	Cluster manager
Midpoint and the start point for the Shell. Changing over client program into the task. Scheduling undertaking on the executor	It plays out every one of the information preparing. Runs the undertaking that makes up the application and returns the outcome to the driver. Provides in-memory stockpiling to Resilient distributed databases that are stored by the user	Require an outside administration for gaining assets and distribution of jobs
Executes the main function of the operation	Reads and writes data to external sources	Allocation and deallocation of various physical resources
Components like DAGScheduler, TaskScheduler, BackendScheduler, and Block Manager	Stores the calculation that comes about data in-memory, cache or on hard disk drives	Types: Standalone Apache, Mesos, Hadoop's, YARN, and Kubernetes
Keeps running on the master node of the Spark cluster, plans the job execution, and consults with the cluster manager	Interacts with the storage systems.	Choice of cluster manager depends on application goal
Makes an interpretation of the RDDs into the execution and splits the diagram into different stages	Distributed agent responsible for the execution of tasks	Standalone is the easiest to use
Stores metadata about all the resilient distributed databases and their allotments	Every application has its own executor process. Run for the entire lifetime (Static Allocation of Executors)	Standalone spark cluster at run

5 Results

The difference between features of MapReduce and Spark are shown in Table 4 [9]. A comparison of MapReduce and Spark in terms of page rank is shown in Fig. 5 [9, 10]. A comparison of the number of iterations with running time in seconds for both Hadoop and Spark is shown in Fig. 6. The main differences between MapReduce and Spark are shown in Table 4 [9, 10].

5.1 Differences between MapReduce and Spark

See Table 4 and Figs. 5 and 6.

Table 4 Differences between MapReduce and Spark

Features	MapReduce	Spark
Speed	Fast	100 × faster
Processing	Batch	Real-time
Storage	Disk	RAM
Language	JAVA	Scala
Latency	High	No
Analytics	Simple MapReduce	Streaming, machine learning, and complex analytics
Coding	More lines	Fewer lines
Cores	8–16	4
Memory	24 GB	8–100 GB
Disks	4–6 one-TB disk	4–8
Network	1 GB Ethernet all-all	10 GB/more
Cost	Cheaper	Costly
Compatibility	Data types and data sources	Same as MapReduce
Failure tolerance	More tolerant	Tolerant
Security	More security features	Still in its infancy

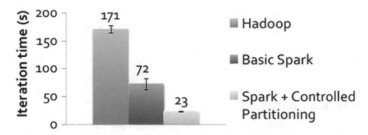

Fig. 5 Page rank between Hadoop and Spark

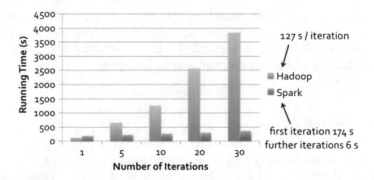

Fig. 6 Iterations in Hadoop and Spark

MapReduce code:

Map Code:

String line = value.toString().replaceAll("[^\\w\\s]|('s|ly|ed|ing|ness) ", " ").toLowerCase();

Reduce code:

```
                for (IntWritable value : values) {
                        sum = sum + value.get();
                }
[cloudera@quickstart Desktop]$ hadoop fs -cat mrWordCount_sorted/part-r-00000 | head
the  | 155355
and  | 122639
of   | 79506
a    | 73622
to   | 71718
it   | 51427
in   | 48202
i    | 48037
that | 40835
was  | 39604
```

Fig. 7 Result of MapReduce word count

```
[cloudera@quickstart Desktop]$ hadoop fs -cat /user/cloudera/sparkWordCount/part-00000 | head
(the,155355)
(and,122639)
(of,79506)
(a,73622)
(to,71718)
(it,51427)
(in,48202)
(i,48037)
(that,40835)
(was,39604)
```

Fig. 8 Result of Spark word count

5.2 Analysis of Results

The given task involved counting the occurrence of particular words and then sorting the counts in descending order. Special characters were removed, along with the endings s, ly, ed, ing, and ness, and all words were converted to lowercase. It almost took 58 lines to implement the WordCount program using MapReduce (see Fig. 7) but the same WordCount was implemented in just 3 lines using Spark (see Fig. 8). Also compared the clock item different systems in Fig. 9.

6 Conclusions

Hadoop incorporates not only a capacity segment known as the Hadoop Distributed File System (HDFS), but also a handling part called MapReduce. As a result it is not necessary to use Spark to complete preparations. On the other hand, you can similarly use Spark without Hadoop. Spark does not possess its own particular record admin-istration framework, however, so it should be incorporated with one if not HDFS,

System	Wall-clock time (seconds)
MATLAB	15443
Mahout	4206
GraphLab	291
MLlib	481

Fig. 9 Comparison of clock item different systems

at that point another cloud-based information stage. Hadoop is naturally resilient to system faults or failures since data are written to disk after every operation. Spark has similar built-in resiliency by virtue of the fact that its data objects are stored in resilient distributed datasets (RDDs) distributed across the data cluster. Our experimental results show that the proposed framework not only enables the possibility of distributing Spark workflow throughout multiple clusters, but also provides significant improvement in performance compared to single cluster environments by optimizing the utilization of multi-cluster computing resources.

References

1. Z. Liu, H. Zhang and L. Wang, "Hierarchical Spark: A Multi-Cluster Big Data Computing Framework," 2017 IEEE 10th International Conference on Cloud Computing (CLOUD), Honolulu, CA, 2017, pp. 90–97. https://doi.org/10.1109/cloud.2017.20.
2. A. V. Hazarika, G. J. S. R. Ram and E. Jain, "Performance Comparison of Hadoop and Spark Engine," 2017 International Conference on I-SMAC (IoT in Social, Mobile, Analytics and Cloud) (I-SMAC), Palladam, 2017, pp. 671–674. https://doi.org/10.1109/i-smac.2017.8058263.
3. C. Aatish and T. Tashley, "Categorizing the .mu Domain Using MapReduce," 2017 IEEE 2nd International Conference on Signal and Image Processing (ICSIP), Singapore, Singapore, 2017, pp. 470–474. https://doi.org/10.1109/siprocess.2017.8124586.
4. L. E. V. Peña, L. R. Mazahua, G. A. Hernández, B. A. O. Zepahua, S. G. P. Camarena and I. M. Cano, "Big Data Visualization: Review of Techniques and Datasets," 2017 6th International Conference on Software Process Improvement (CIMPS), Zacatecas, Mexico, 2017, pp. 1–9. https://doi.org/10.1109/cimps.2017.8169944.
5. Raswitha Bandi, Dr. Sheikh Gouse, Dr. J. Amudhvel, (2017) A Comparative Analysis For Big Data Challenges And Big Data Issues Using Information Security Encryption Techniques1, 2 in International Journal of Pure and Applied Mathematics Vol.8, 183–189.
6. Dr. Sheikh Gouse, Raswitha Bandi, Dr. P. Armanedar Reddy. (2017) Comprehensive Survey on Big Data Analytics and Tools presented in the National Seminar on "International Conference on Recent Trends in Computer Science and Technology (ICRTCST-2017)" R.V.S College of Engineering and Technology Edalbera, Bhilai Pahari, Jamshedpur-831012, 21/04/2017.
7. Manjunath R., Tejus, Channabasava R. K and Balaji S., "A Big Data MapReduce Hadoop distribution architecture for processing input splits to solve the small data problem," 2016

2nd International Conference on Applied and Theoretical Computing and Communication Technology (iCATccT), Bangalore, 2016, pp. 480–487.

8. N. Ramakrishnaiah and S. K. Reddy, "Performance Analysis of Matrix and Graph Computations Using Data Compression Techniques in MPI and Hadoop MapReduce in Big Data Framework," *r*, Chennai, 2017, pp. 54–62. M. Zhang, F. Liu, Y. Lu and Z. Chen, "Workload Driven Comparison and Optimization of Hive and Spark SQL," 2017 4th International Conference on Information Science and Control Engineering (ICISCE), Changsha, 2017, pp. 777–782.

9. Mayer-Schönberger V, Cukier K (2013) Big Data: A Revolution That Will Transform How We Live, Work, and Think. Eamon Dolan/Houghton Mifflin Harcourt.

10. R. C. Maheshwar and D. Haritha, "Survey on High Performance Analytics of Big Data with Apache Spark," 2016 International Conference on Advanced Communication Control and Computing Technologies (ICACCCT), Ramanathapuram, 2016, pp. 721–725.

Healthcare Monitoring Using Internet of Things

Sripada Soumya and Sandeep Kumar

Abstract In recent years, technology has developed and created human life an easy and luxury one. People can use a lot of electronic gadgets in their day-to-day life to upgrade their lifestyle. We can also use them for our health care with the help of smart techniques. IoT is a smart system consisting of a different type of sensors and is a real-world application connected with each other through the Internet via wired or wireless network structure. In India, the cause of death is many diseases probably. In that, some diseases may be curable. The chance of death will be reduced when the problem identified is instant. For an example, heart attacks, the patient cannot be able to describe his situation. For that, we proposed a healthcare monitoring system, in which the tracking place of the patient and his health conditions will be monitored. In a hospital, the healthcare monitoring system is very essential to observe the patient's physiological parameters. The smart hospital system (SHS) is two method phenomena, first to track the people by automatic identification and biomedical devices in the hospital, second deal with the patient's monitoring. Wireless sensor network (WSN), radio frequency identification (RFID), ultra-high frequency (UHF), global system for mobile communication (GSM) and the smart mobile are the technologies used in the implementation of smart healthcare systems. The second is a monitoring system that will collect the data, in real time. We can also get environmental parameters and also the patient's physiological conditions with the help of ultra-low-power hybrid sensing network (HSN) composed of UHF, RFID functionalities.

Keywords Radio frequency identification (RFID) · Wireless sensor network (WSN) · Smart hospital system (SHS) · Ultra-high frequency (UHF)

S. Soumya (✉) · S. Kumar
Department of Electronics and Communications, Sreyas Engineering College, Hyderabad, India
e-mail: sowmyasripada11@gmail.com

S. Kumar
e-mail: drsandeep@sreyas.ac.in

© Springer Nature Singapore Pte Ltd. 2019
R. S. Bapi et al. (eds.), *First International Conference on Artificial Intelligence and Cognitive Computing* , Advances in Intelligent Systems and Computing 815,
https://doi.org/10.1007/978-981-13-1580-0_47

1 Introduction

Internet of things (IoT), the innovative technology that will connect the living things and the nonliving things with the help of the Web and the communication technology. In a hospital, healthcare monitoring system is essential to observe the patient's physiological parameters. Consider an example of a pregnant woman, her foetal parameters such as blood pressure (BP), heart rate and its movements can be easily obtained. Continuously monitoring the patient's situations with the implantable and wearable body sensor networks can improve the early detection of emergency conditions. People with various degrees of attention and physical disabilities can be easily understood by the healthcare monitoring system to give a fast treatment. Not only the old people and disabled people but also for the families where both the parents need to work will be benefitted from these systems by providing care services for their babies and children. This paper will give IoT awareness, smart methods to get the patients address and their health report with biomedical devices in the hospitals and nursing institutes. Smart Hospital System (SHS) [18], means the getting of particular health parameters on different and complementary techniques used such as WSN, RFID [5, 10] and smart mobile. Ultra-high frequency (UHF), global system for mobile communication (GSM) [13], radio frequency identification (RFID), wireless sensor network (WSN) and smart mobile represent the most important technologies used for the implementation of smart healthcare systems.

2 Technologies Used

RFID is a technology with low-cost, low-power consumption. RFID-based sensors in healthcare devices enable zero-power and easy to implement the monitoring and gives the patient's physiological parameters. RFID is operating within the region 15–25 m. A typical wireless body area network (WBAN) [4] is consisting of a number of miniatures, low-power sensing devices. WBAN system will have wireless transceivers. They are more advantageous because of low power supply small-sized, light in weight, environmental-friendly and also long-lasting. The ZigBee [5, 8] is a wireless mesh loop system. In ZigBee, many devices can connect at a time. The range of the ZigBee is 100 m and has low cost. It can transfer the data 250 kbps with high reliability. Body sensor network systems (BSNs) [11] are used for medical monitoring and communication in emergency situations. BSN will access the medical data, memory enhancement. It will offer the flexibility to human. Vitals signs are observed by the BSN to measure body's basic parameters such as blood pressure which is helpful for monitoring health of a person. Fig. 1 explains the body sensor network and the flow of information.

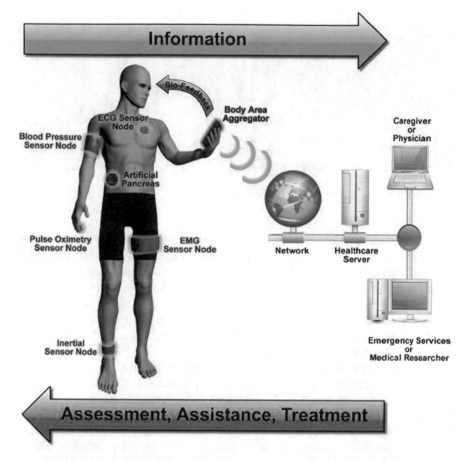

Fig. 1 Healthcare with body sensor network [1]

3 Applications

The accessibility of mobile phones is increasing day by day, and mobile communication has a broad scope of a global system. GSM systems provide a great chance to use to trigger improvement [2] and enhance individuals. By using sensors, the information which is recorded and contrasted will be configurable and also predefined so that will be sent to the microcontroller which is already accessed by a specific doctor who takes care of the patient; regardless of emergency, using GSM and GPS devices [13], SMS would be sent to phone number of the particular doctor's with accurate values and patient's data about current position. Figure 2a provides the exact location of the patient.

Heartbeat sensor is used to give a digital output when we place a finger on it. The LED flashes for each heartbeat when the heartbeat detector is in working condition.

(a)   (b)

Fig. 2 **a** detection of position [13], **b** heart rate sensor [15]

The output of this will be sent to the microcontroller in order to measure the beats per minute (BPM) rate. The principle involved in this is light modulation by blood flow through finger at every pulse. IR sensor mainly having an IR LED and a photodiode, this pair is known as photocoupler. The IR LED emits the IR radiation, and then this will be sensed by the photodiode. The resistance of the photodiode will depend on the amount of IR radiation which is falling on it. So the voltage drop also depends on the voltage comparator. In this, LM358 voltage comparator is used to sense the change in voltage and then generate the output.

Blood pressure nothing but the pressure of the blood in the arteries as it is pumped throughout the body by the heart. During the heart beats, it will send blood to all the body parts and also receive from the part of the body. A sphygmomanometer is used to check the blood pressure. Here, we use a wrist blood pressure monitor device. Figure 3a explains the way of detecting the pulse by using the pulse rate sensor. Figure 3b shows the reading of blood pressure. A wrist-worn wearable medical device is used for monitoring cardiac/respiratory condition [12] of the patients. The physiological parameters such as electrocardiograms, heart rate, blood pressure, body temperature. The technology provides us, with the mobile applications that will show the information about the list of the patients. It will contain all the physical parameters such as blood pressure, temperature and humidity details. The heart rate is also shown for the particular patients.

Figure 4a shows the list of patients in the respective rooms [18], and their actual position of a particular patient in detail. Figure 4b gives details about the emergency situation. This will be highlighted when an emergency condition occurs. These are the time variables which will vary with respect to the time such that clear updates will be gathered.

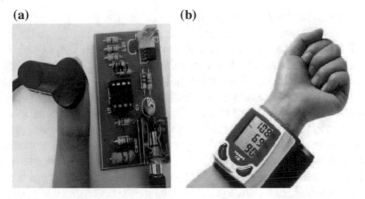

Fig. 3 **a** Pulse rate sensor [12], **b** blood pressure measuring wrist device [12]

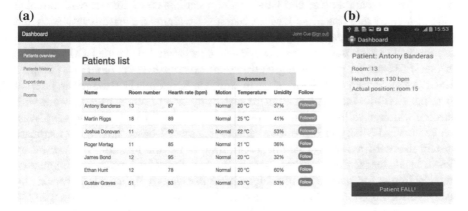

Fig. 4 **a** Patients list [18], **b** screenshots of the medical app [18]

4 Literature Analysis

Vivek Pardeshi, Saurabh Sagar, Swapnil Murmurwar, Pankaj Hage [3] explains health monitoring systems using IoT and Raspberry Pi—a review system by using Raspberry Pi as a sensor node and a controller and use GSM technology which is very simple, power efficient. Taiyang Wu, Jean-Michel Redout´e [4] proposed an wireless body area network implementation using wearable sensor node with solar energy harvesting and Bluetooth low-energy (BLE) transmission. Solar energy used for accessing long-term continuous medical monitoring. Shubaham Puri, Tanupriya Choudhury, Nirbhay Kashyap, Praveen Kumar [5] explains specialization of IoT applications in healthcare industries based on the RFID tags used in watches. In that system, vitals can be wirelessly monitored using a smartwatch. Himadri Nath Saha, Supratim Auddy, Subrata Pal, Shubham Kumar, Shivesh Pandey, Rocky Singh, Amrendra Kumar Singh, Priyanshu Sharan1, Debmalya Ghosh, Sanhita Saha [6] explains the health monitoring using Internet of things (IoT), for the trace patient's health with the assistance of sensors and Internet which observes the change in the heartbeat, and it will be represented in mechanical mode. S. Pinto, J. Cabral and T. Gomes [7] proposed We-Care: An IoT-based health care system for elderly people by the monitor and register patients vital information by using sensors which provide mechanisms to trigger alarms in emergency situations. Konstantinos Karamitsios, Theofanis Orphanoudakis [8] described efficient IoT data aggregation for connected health applications based on the different IoT protocol stacks for sensor data aggregation to remote healthcare status monitoring. Marjan Gusev, Ana Guseva [9] proposed state of the art of cloud solutions based on ECG sensors with the help of automated analysis of received data, monitoring of specific ECG-related data to store it and transfer relevant information to a medical institution. Mert Bal and Reza Abrishambaf [10] proposed a system for monitoring hand hygiene compliance based on Internet of things, uses passive RFID wearable tags and hand hygiene stations consisting of wall-mounted soap dispensers and touchless faucets. Sammy Krachunov, Christopher Beach and Alexander J.Casson, James Pope, Xenofon Fafoutis, Robert J. Piechocki and Ian Craddock [11] proposed heart rate sensing using a painted electrode ECG wearable. In that, ECG sensor can be integrated into an existing IoT wearable and compare the device's accuracy. Vikas Vippalapalli, Snigdha Ananthula [12] described Internet of things (IoT)-based smart healthcare system by using body sensor network with aurdino which have lightweight wearable sensor nodes for real-time sensing. Warish Patel, Viral Mistry [13] proposed developing an IoT-based intelligent medicare system for real-time remote health monitoring, to get the location of a person by GSM. Jusak Jusak, Heri Pratikno, Vergie Hadiana Putra [14], proposed Internet of medical things for cardiac monitoring: paving the way to 5G mobile networks, by using sensor node on Arduino microcontroller board that based on ATmega328P and Raspberry Pi version 3 Model B. K. Sundara Velrani, G. Geetha [15] proposed sensor-based healthcare information system, and in this human vital signs are sensed for heartbeat and temperature (Table 1).

5 Result

The Raspberry is a single board computer of the accrediting card. The features of Raspberry Pi 2, which we used is having 900 MHz quad-core, ARM Cortex A7 CPU, 1 GB RAM (Fig. 5).

According to our experiments, the Raspberry Pi 2 is connected with the temperature and humidity sensor as shown in Fig. 6. The Python script is used in this project as the software code. The features of Raspberry Pi 2 are 4USB port, 1.5 W of power in idle mode, 6.7 W power in maximum under stress, 40GPIO pins, display interface, full HDM1 port, ethernet port, combined 3.5 audio jack and composite video, camera interface. There we can get the temperature of the environment and its humidity details with respect to the time. So that accuracy is carried out for the temperature which is the most important parameter to get the patients details. Figure 6 shows

Table 1 Comparison analysis

S. No	Author name	Year	Methodology	Remarks
1	Vivek Pardeshi et al. [3]	2017	Raspberry Pi as a sensor node and GSM	Simple, power-efficient technology
2	Taiyang Wu, Jean-Michel Redout´e [4]	2017	Wearable sensor node and Bluetooth low energy (BLE)	Solar energy used for long-term continuous medical monitoring based on WBAN
3	Shubaham Purri et al. [5]	2017	RFID tags used in watches	Vitals can be wirelessly monitored using smartwatch
4	Himadri Nath Saha et al. [6]	2017	Trace health with the sensors and internet	Change in a heartbeat can be represented mechanically
5	S. Pinto et al. [7]	2017	Monitor patients vital information by using sensors	It will provide mechanisms to trigger alarms in emergency situations
6	Konstantinos Karamitsios et al. [8]	2017	Different IoT protocol stacks for sensor data aggregation	Remote healthcare status should be monitoring
7	Marjan Gusev, Ana Guseva [9]	2017	Automated analysis of data, monitoring of ECG-related data	Process and store data then transfers information to a medical institution
8	Mert Bal and Reza Abrishambaf [10]	2017	RFID wearable tags and wall-mounted soap dispensers and touchless faucets	Hand washing stations at patient's bed and the entrance of the patient's room in order to detect hand hygiene events
9	Sammy Krachunov Christopher Beach et al. [11]	2017	ECG sensor	ECG sensor can be integrated into an IoT wearable and compare the device's accuracy
10	Vikas Vippalapalli1 et al. [12]	2016	Body sensor network with Arduino	Less weight wearable sensor nodes for real-time sensing
11	Warish Patel et al. [13]	2016	GPS, temperature sensor pulse sensor	To get the location of the person by GPS
12	Jusak Jusak, Heri Pratikno et al. [14]	2016	Arduino micro-controller, Raspberry Pi	Average throughput and bandwidth utilization in the mobile device can be achieved
13	K.Sundara Velrani et al. [15]	2016	Bluetooth communication	Human vital signs are sensed for heartbeat and temperature

(continued)

Table 1 (continued)

S. No	Author name	Year	Methodology	Remarks
14	Prabhu Ravikala Vittal & Natarajan Sriraam et al. [16]	2016	Photoplethysmography (PPG) for measuring the heart rate	Designed patch sensor is low cost and monitors the heart rate continuously
15	Yong Lian et al. [17]	2015	CTDA system architecture, wireless sensor	CTDA architecture has great potential for self-powered wireless sensor
16	Luca Catarinucci, Danilo De Donno et al. [18]	2015	Ultra-low-power HSN integrating UHF RFID functionalities	SHS is used to collect the patients' physiological parameters via HSN
17	Sourav Kumar Dhar et al. [19]	2015	Scheduling technique is used to differentiate the sensors	Gives body temperature with large data size and low sample rate

(a)   **(b)**

Fig. 5 **a** Raspberry Pi 2, **b** temperature interface

the screenshot of temperature and humidity. There is a temperature graph in Fig. 7, making an idea regarding the temperature of the sensors.

6   Conclusion

Health is an important thing for the human beings. For the patients, health should be monitored in all the time. It is impossible to say the doctor to care for only single patients even there are caretakers of patients they may not have a sufficient knowledge to monitor them. The general parameters such as blood pressure, heartbeat rate and temperature are needed to check frequently. For the continuous monitoring, here we are using the several devices such as wristwatch-based blood pressure device and

Fig. 6 Temperature and humidity at different times

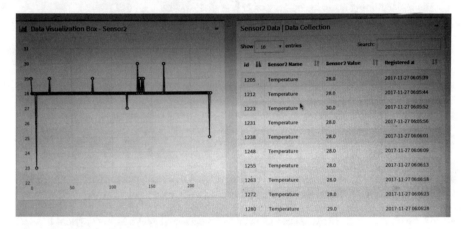

Fig. 7 Temperature graph

pulse rate monitor. At sometimes, it is required to get the position of the patient, by using GSM module. A combined report about all the patients will be shown based on the particular application. All these devices are based on the different systems such as RFID, BSN, WCN, GSM. Many things are developed based on the IOT to take care of the patients in emergency conditions.

Declaration We have taken necessary permission to use the dataset/images/other materials from respective parties. We, authors, undertake to take full responsibility if any issue arising in future.

References

1. Sandeep Kumar & Hemlata Dalmia, "A Study on Internet of Things Applications and Related Issues", International Journal of Applied and Advanced Scientific Research, Volume 2, Issue 2, pp 273–277, 2017.
2. Karthi Jeyabalan, "Home Healthcare and Remote Patient monitoring" Internet of Things and Data Analytics Handbook, First Edition. Edited by Hwaiyu Geng, pp: 675–682, 2017.
3. Vivek Pardeshi, Saurabh Sagar, Swapnil Murmurwar, Pankaj Hage, "Health Monitoring Systems using IoT and Raspberry Pi", International Conference on Innovative Mechanisms for Industry Applications, pp: 134–137, 2017.
4. Taiyang Wu, Jean-Michel Redoute, "An Autonomous Wireless Body Area Network Implementation towards IoT Connected Healthcare Applications", pp: 2169–3536 (c), 2016 IEEE Conference.
5. Shubaham Purri, Tanupriya Choudhury, Nirbhay Kashyap, Praveen Kumar, "specialization of IoT applications in healthcare industries"International Conference On Big Data Analytics and Computational Intelligence (ICBDACI), pp: 252–256, 2017.
6. Himadri Nath Saha, Supratim Auddy, Subrata Pal, Shubham Kumar, Shivesh Pandey, Rocky Singh, Amrendra Kumar Singh, Priyanshu Sharan1, Debmalya Ghosh, Sanhita Saha, The Health Monitoring using the Internet of Things (IoT), pp. 66–73, 2017 IEEE.
7. S. Pinto, J. Cabral, and T. Gomes, "We-Care: An IoT-based Health Care System" pp:1378–1383 on IEEE 2017.
8. Konstantinos Karamitsios, Theofanis Orphanoudakis," Efficient IoT data aggregation for connected health applications", 2017 IEEE Symposium on Computers and Communications (ISCC).
9. Marjan Gusev, Ana Guseva," State-Of-The-Art of Cloud Solutions Based on ECG Sensors", pp: 501–506, IEEE Eurocon 2017, Ohrid, R. Macedonia.
10. Mert Bal and Reza Abrishambaf, " A System for Monitoring Hand Hygiene Compliance based-on Internet-of-Things" pp:1348–1353, 2017 IEEE.
11. Sammy Krachunov, Christopher Beach and Alexander J. Casson, James Pope, Xenofon Fafoutis, Robert J. Piechocki and Ian Craddock," Energy Efficient Heart Rate Sensing using a Painted Electrode ECG Wearable",2017 IEEE.
12. Vikas Vippalapalli, Snigdha Ananthula, "Internet of things (IoT) based smart healthcare System", International Conference on Signal Processing, Communication, Power and Embedded System (SCOPES), pp:1229–1233, 2016.
13. Warish Patel, Viral Mistry, "Developing an IoT based intelligent Medicare system for Real-Time Remote Health monitoring", 8th International Conference on Computational Intelligence and Communication Networks. pp: 641–645, 2016.
14. Jusak Jusak, Heri Pratikno, Vergie Hadiana Putra," Internet of Medical Things for Cardiac Monitoring: Paving the Way to 5G Mobile Networks" On IEEE International Conference on Communication, Networks, and Satellite (COMNETSAT), pp: 75–79, 2016.
15. K.Sundara Velrani, G.Geetha," Sensor Based Healthcare Information System" IEEE International Conference on Technological Innovations in ICT For Agriculture and Rural Development", pp:86–92, 2016.
16. Prabhu Ravikala Vittal & Natarajan Sriraam, "A Pilot Study on Patch Sensor Based Photo Plethysmography (PPG) for Heart Rate Measurements", 2016.
17. Yong Lian, "Challenges in the Design of Self-Powered Wearable Wireless Sensors for Healthcare Internet-of-Things" 2015 IEEE.
18. Luca Catarinucci, Danilo De Donno, Luca Mainetti, Luca Palano, Luigi Patrono, Maria Laura Stefanizzi, and Luciano Tarricone, ''An IoT-Aware Architecture for Smart Healthcare Systems" IEEE Internet of Things Journal, pp:1–11, 2015.
19. Sourav Kumar Dhar, Suman Sankar Bhunia, Nandini Mukherjee,"Interference-Aware Scheduling of Sensors in IoT Enabled Health-care Monitoring System" 2014 Fourth International Conference on Emerging Applications of Information Technology, pp:152–157.

Enhanced Homography-Based Sports Image Components Analysis System

Abhay Atrish, Navjot Singh and Vinod Kumar

Abstract Sports data analysis is an active research area in recent years to automatically extract all the significant information from the sports data. One of the significant tasks is to trace the 3D calibrated sports image to a 2D soccer reference field and map the player's position in the reference plane. This paper proposed an automated, efficient, and more accurate system for field registration on a reference plane, player's localization, and team recognition. A feedback-based enhanced homography transformation is used for field registration. Also, the player's team and the referee are recognized based on their jersey color and marked on the reference plane with their respective team color. The various algorithms constituting vision-based features such as color, shape, size are used to implement the proposed system. Extensive experiments are conducted on different datasets to demonstrate the effectiveness and efficiency of the system in terms of achieving the accuracy to determine the ground truth values of the field, players, and their corresponding team. Additionally, the proposed system's (comprising enhanced homography) results are juxtaposed with normal homography-based system to outline the improvement.

Keywords Computer vision · Pattern recognition · Field registration
Homography · Image processing · Sports video analysis

A. Atrish (✉) · N. Singh
National Institute of Technology Uttarakhand, Srinagar, Uttarakhand, India
e-mail: abhay.cse12@nituk.ac.in

N. Singh
e-mail: navjot.singh.09@nituk.ac.in

V. Kumar
Indian Institute of Technology Roorkee, Roorkee, India
e-mail: vinodfee@iitr.ac.in

© Springer Nature Singapore Pte Ltd. 2019
R. S. Bapi et al. (eds.), *First International Conference on Artificial Intelligence and Cognitive Computing* , Advances in Intelligent Systems and Computing 815,
https://doi.org/10.1007/978-981-13-1580-0_48

(a)   **(b)**

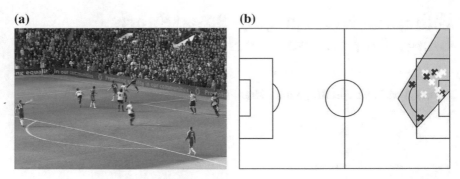

Fig. 1 **a** Soccer image with 3D projection. **b** Mapping soccer image on a 2D reference soccer plane

1 Introduction

The sports image and video data are increasing drastically, and Internet is flooded with it. Sometimes there is a need to extract some significant information [1, 2] from these images and videos; extracting the relevant information manually from the sports data is a tedious and time-consuming task. Thus, this process needs to be automated to extract the information effectively and more precisely. A variety of information like player recognition, score identification, event detection, trajectory analysis can be extracted from the sports data, and a lot of research is going on to come up with more efficient and effective way to extract this information. One such important aspect of the sports data, that this paper has focused on, is to draw the field region of the image on a rectangular reference frame and represent all the important information like player's position and their team, referee position on it. This paper primarily focuses on soccer, as it is a widely popular and easily relatable to the other outdoor games in terms of field structure and composition. The proposed system uses the low-level vision-based features like color, texture, shape, size to extract the information from the image [2, 3]. The main aim of this automated system is to transform and plot the original sport's image to a rectangular reference soccer field and mark all the player's position on it along with their teams as shown in Fig. 1a and b. The sports images are captured by the camera from certain distance and height, and at a certain angle covering some specific area of the field. This original image is 3D calibrated which is transformed into a rectangular reference plane [4] using homography. [4–7] have discussed the homography and suggested some mathematical solutions to estimate it correctly. A variety of players detection strategies are discussed in [8–11], where [8, 9] uses a blob-based template matching algorithm, [10] implements AdaBoost classifier, and [11] uses connected component analysis to detect the players. [2, 5, 12] have used the player's jersey color to recognize their team.

The remaining paper is structured as follows: Sect. 2 gives an overview of the proposed system, whereas Sect. 3 delineates the complete proposed system including playfield area detection, homography estimation, field registration, player localization, and team recognition. Finally, Sect. 4 discusses the experimental results, and Sect. 5 concludes the paper.

2 System Overview

The complete automated process is divided into following consecutive steps implementing different vision-based algorithms to accomplish various tasks: (1) determine the playfield area, (2) homography improvisation and estimation, (3) field registration from original soccer image to reference plane, (4) players registration on the reference field, and (5) team recognition, as shown in Fig. 2. Initially, the playfield area is extracted from the image using a dominant color algorithm. Refs. [1, 10, 13] have discussed various ways to extract the playfield area by using HSI or RGB color space. Subsequently, the field lines need to be extracted from the playfield area, and their length and angle are also evaluated. In order to detect field lines, [8] have used the top-hat transform to detect lines, it failed to provide line segment equation, its length, and angle, whereas [1, 7, 11, 13] have used Hough transform to detect the fields lines effectively. In addition to that [1, 13] have recognized the penalty box based on the two or more parallel vertical lines found in the field area. In the proposed system, some key field lines are traced on the reference field, based on the fixed vertices of the field. Further, the homography estimation is performed to transform one plane to another, and the field line equations are used as feedback to enhance the homography transformation by minimizing the mapping model error [1]. Moreover, the players are detected and localized in the original image, [8] have suggested the blob detection algorithm, [10] have built an AdaBoost classifier to detect the players, and [11] have used connected component analysis. Our system uses the blob detection algorithm over other approaches as the former only focuses on detecting and localizing the player, whereas the other focuses primarily on the extraction of the players. The positions of the players are mapped to the reference plane using the homography transformation. Finally, the player's team and the referee are recognized based on their jersey color, and their respective position on the reference plane is represented with their team (jersey) color. The proposed system has shown significant results over a large set of images as discussed in Sect. 4.

3 Detailed Proposed System

This section elaborates the step-by-step execution of the proposed system which involves (1) playfield area detection, (2) enhanced homography estimation, (3) field registration, (4) player detection, localization, and mapping, and (5) team recognition.

Fig. 2 End-to-end flow of
the proposed system

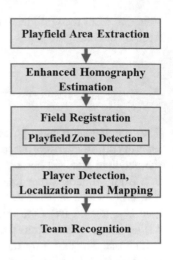

3.1 Playfield Area Extraction

As discussed in our previous work [9], usually the soccer fields are covered with grass, resulting in the green color of the playfield area. The actual shade of green color may vary due to various constraints like lighting, weather, grass patterns. Thus, a range of dominating colors are extracted from the playfield area by applying dominant color algorithm [9, 13]. The dominant color algorithm uses an RGB color space-based histogram to define a range of dominant color for each color space components. Firstly, for an image I the color peak I_{peak} for each color space component is determined from the histogram G, then a range $[I_{\text{min}}, I_{\text{max}}]$ is determined around their respective peaks such that:

$$G[I_{\text{min}}] \geq k * G[I_{\text{peak}}] \tag{1}$$

$$G[I_{\text{min}} - 1] < k * G[I_{\text{peak}}] \tag{2}$$

$$G[I_{\text{min}} - 1] \geq k * H[I_{\text{peak}}] \tag{3}$$

$$G[I_{\text{min}} + 1] < k * G[I_{\text{peak}}] \tag{4}$$

$$I_{\text{min}} \leq I_{\text{peak}} \tag{5}$$

$$I_{\text{min}} \geq I_{\text{peak}} \tag{6}$$

where I_{min} and I_{max} are the minimum and maximum indices of the range, k is a constant whose value is fixed to 0.2 based on the experimentation results. k represents the deviation in the dominant color shade from its peak value. After applying the dominant color algorithm, morphological operations are applied to remove blobs, lines, and other noise. Figure 3 depicts the playfield area extraction from the original image (Fig. 1a).

Fig. 3 Extracted playfield
area from Fig. 1a using the
dominant color algorithm

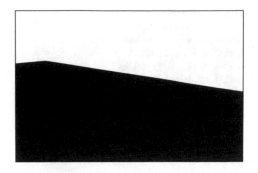

3.2   Enhanced Homography Estimation

The sports image is a 2D image with 3D calibration as the image is taken by a camera
from certain angle and distance. Here, the primary purpose is to map the original
3D planar image to a 2D rectangular soccer field or reference plane. Figure 1a
shows the original field image, and Fig. 1b shows 2D reference plane with original field's components marked on it. In order to project the complete field region
and the player's position from the original image to the reference soccer plane, a
perspective homography transformation H is determined between the 3D camera
plane and the reference plane. Refs. [4, 6, 7] have discussed various approaches for
homography estimation; [4] have mathematically evaluated the homography transformation, [6] has implemented homography with hypothesis validation and [7] has
used homography with Kalman filter to overcome misdetection. The homography is
nonlinear, planar, 3×3 homogeneous matrix and exhibits eight independent entries,
which pairwise represents shearing, translation, scaling, and perspective distortions.
Moreover, the camera is assumed to be pinhole and kept at some static position. The
coordinates of the actual image c_o are projected on the reference plane as c_r such that
$c_r = Hc_0$. The coordinates of vertices and intersection points help in determining the
transformation matrix H:

$$[x', y', z']^T = H[x, y, z]^T \tag{7}$$

where x, y, z are the coordinates of sports image, and x', y', z' are the coordinates
on the reference frame. Once H is evaluated, all the reference point, i.e., vertices
are mapped on the reference plane, later, H value is cross-checked by a feedback-
oriented approach which remaps the line segment $A'x + B'y + C' = 0$ between the
known reference points back on the original frame as $Ax + By + C = 0$, such that,

$$[A, B, C]^T = L[A', B', C']^T \tag{8}$$

where $L = H^T$, H is the homography point transformation and L is the homography
line transformation. For all coordinates (p, q), $H_{(p, q)} = K_{(p, q)}^T$, replot the endpoints

Fig. 4 Different field
regions captured by the
cinematographer as follows:
1. midfield area, 2. left
penalty box, 3. right penalty
box, 4. top-left corner, 5.
top-right corner, 6.
bottom-left corner, and 7.
bottom-right corner

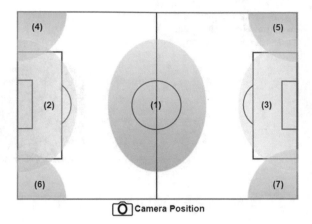

of each line onto the playfield model. Further, the actual lines and the traced lines
are matched and the model mapping error [1] is assessed as:

$$E_{pq} = \sum_r \delta_{pqr} \qquad (9)$$

where δ_{pqr} calculates the error in mapping the field by measuring the distance of
each replotted line to the closest line of the playfield model. Finally, the enhanced
homography $\hat{H}$ is evaluated by minimizing the model mapping error. Thus, the
enhanced homography transformation helps in outlining the region more precisely.

3.3 Enhanced Homography-Based Field Registration

Playfield Zone Detection: In order to transform one plane into another plane, the
playfield region is detected in the original image. Figure 4 depicts the various possible
regions which cinematographer may capture (assume camera as pinhole size); these
regions are majorly categorized as midfield area, penalty box, and the area between
the midfield and penalty box. Out of these, midfield area is easy to detect and portray,
as it contains a single long vertical line with an ellipse attached to it, whereas area
between midfield and penalty box is complex to outline due to limited reference
points. Moreover, the penalty box is traced based on the two or more vertical parallel
lines and vertices. This region is further sub-categorized into (1) left penalty box, (2)
right penalty box, (3) top-left corner, (4) top-right corner, (5) bottom-left corner, and
(6) bottom-right corner, as shown in Fig. 4. The field lines, their intersection points,
and their angles with respect to horizontal axis play a pivotal role in the categorization
of the image region.

The midfield consists of a single long vertical line and an ellipse attached to it,
and the range of vertical line angle is between 80° and 100° and surrounded by the

Fig. 5 Line detection and categorization: Yellow, red, and green lines represent vertical lines, horizontal lines, and an arc, respectively

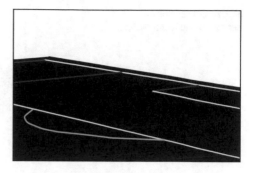

dominant color region on its both sides. A penalty box is detected if two or three parallel vertical lines are found. The longest line among these lines is the boundary line, which along with the relative position of other two lines or dominant color region helps in determining the right or left penalty box. A corner is located, if the vertical boundary line intersects the top or bottom horizontal boundary line and the type of corners are decided based on the relative position of boundary lines and the enclosed field area. For example, if the vertical line angle is in the range from 100° to 150° and horizontal line angle is in the range from 0° to 30°, or the field area in on the left side of the vertical line and below the horizontal line, then it is a top-right corner as shown in Fig. 5. Hough transform is used to detect these lines and help in determining their angle, length, and mathematical equation as well.

Field Registration: For field registration, many papers [4–7] have used homography or a predefined template to plot the area covered in the soccer image. The playfield region traced on the reference plane using normal homography is not very accurate, and in order to obtain a more exact result, the proposed system considered a feedback-oriented enhanced homography as discussed in the previous section. The intersection points like the field corners, the vertices of goal rectangular box and the penalty box, and the intersection point of midfield line with the horizontal boundary lines can be easily mapped in the reference plane. Now, the lines enclosed by these vertices and intersection points are sketched, and remaining lines are outlined on the reference plane using the enhanced homography transformation as depicted in Fig. 5. Hence, the resultant reference field is sketched more precisely with the feedback-based enhanced homography.

3.4 Player Detection, Localization, and Mapping

After homography estimation and field registration, the player's blob is detected and localized on the original image and further mapped on the reference plane. As discussed before, a variety of player detection and localization approaches have been discussed in [6, 8], and [9] which are primarily based on color differencing,

Fig. 6 Player's blobs
detection and localization

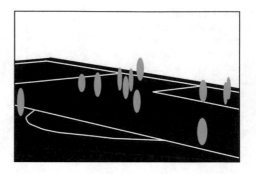

blob detection algorithm, and connected component algorithm. Our proposed system has used the blob detection algorithm to detect and localize the players by detecting the blobs created in the playfield region after applying the dominant color algorithm. All the extracted blobs are matched with predefined elliptical templates of a certain range of height–width ratio to identify all the players present in the playfield area. Once the player's blobs are identified, their positions are localized on the original field area as represented in Fig. 6 and the coordinate value of the center of the blobs is determined. Furthermore, the blob's coordinates are marked as a cross symbol on the reference plane using the enhanced homography transformation which helps in the precisely determining the player's location in the reference plane.

3.5   Team Recognition

Once all the players are localized, their teams are decided based on their jersey color. For all the player's blob segmented in the previous section, their jersey color is examined to deduce the player's team. The k-mean clustering is applied to the above half of the player's blob, and the color cluster with the maximum region covered is the player's jersey region. Hence, the player with similar color blobs belongs to the same team. There are two teams in the soccer game, but sometimes a third color blob is also observed which belongs to the referee. Once the player's team is finalized, the player's position on the reference plane is marked with the same team color as illustrated in Fig. 7.

4   Experimental Evaluation

The proposed system aims to trace the field region along with the player's position and team on a reference rectangular field. The experimental results analyze to what extent the final position of the players and the field lines have been correctly achieved with respect to their ground truth value. The whole system has been developed in

Fig. 7 Green-shaded region shows the area covered in the Fig. 1a, cross symbols represent the player's position, and their color depicts their respective team

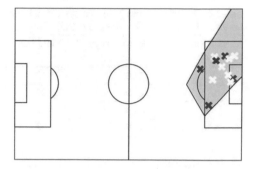

MATLAB, and the experiments are conducted on a system of 4 dual-core CPUs, 2.8 GHz processor, 16 GB RAM, and GeForce 8800 GPU. The experiments are performed on three datasets of 1000, 5000, and 10,000 soccer images, respectively, with an average image resolution of 960 × 640, along with an overall dataset of 16,000 images. These datasets consist of a well-proportioned mixture of a variety of images like the midfield view, different penalty box views, corner view, in-field view, complete field view. These datasets help in demonstrating the consistency in the result and effectiveness of the proposed system in portraying the image information on the reference plane. The following parameters are taken into consideration for measuring the consistency and effectiveness of the proposed system:

(1) Ground truth % accuracy in representing field region and lines = Total number of lines correctly plotted/Total number of lines in soccer images * 100
(2) Ground truth % accuracy in representing player's location = Total number of players correctly located/Total number of player in soccer images * 100
(3) Overall ground truth % accuracy of the proposed system = Total number of field's components correctly mapped/Total number of field's components present in the soccer images * 100

Table 1 discusses the ground truth % accuracy in mapping the field lines, player's position, and the overall accuracy of the system for three different datasets. The table depicts the consistency in the results for determining the field lines, player's location, and the overall accuracy. The overall ground truth accuracy in determining the field lines, player's location, and all the field components are 96.52, 98.51, and 97.77%, respectively. Additionally, the results of the enhanced homography transformation incorporated system are compared with the general homography transformation on a complete dataset of 16,000 images as shown in Table 2. The results have shown a significant improvement in the accuracy which denotes the effectiveness of the proposed system. Hence, our proposed system shows outstanding results with high transformation and localization accuracy.

Table 1 Accuracy of the proposed system in mapping the field lines, player's location, and both on three datasets of 1000, 5000, and 10,000 images

General truth % accuracy	No. of images	Field lines	Player's location	Overall system
	1000	95.73	97.82	97.11
	5000	96.24	98.44	97.73
	10,000	96.86	98.71	97.95
Overall accuracy	16,000	96.52	98.51	97.77

Table 2 Comparison of accuracy of the proposed system with normal homography and with enhanced homography on a dataset 16,000 images

General truth % accuracy	Field lines	Player's location	Overall system
Homography	93.03	94.72	93.96
Enhanced homography	96.52	98.51	97.77

5 Conclusion

In this paper, we have presented a more effective and accurate system for field registration, player localization, and team recognition. This system first enhances the homography plane transformation with the feedback provided by the field lines between the reference points to attain better accuracy. Additionally, the complete playfield region is outlined on the reference plane using the enhanced homography. Later, the players are detected and localized in the original image and marked on the reference plane. Finally, the player's team are detected and represented with their team color on the reference plane. In the experimental analysis, the proposed system shows promising results in terms of (1) the ground truth accuracy in determining the field's line and player's position on the reference plane, and (2) the overall accuracy of the enhanced homography over the general homography. The proposed system can be easily generalized and implemented to cater to soccer videos as well as to other sports domains. The other possible extension of the work are as follows: (1) Implement and validate the system for sports videos, (2) the system has the capability to track the trajectory of ball or players on the reference plane, and (3) extract other important information like event detection, goal detection, trajectory analysis, team strategy analysis.

References

1. Ekin, A.: Automatic Soccer Video Analysis and Summarization. IEEE Transactions on Image Processing. 12.7, 796–807 (2003)
2. D'Orazio, T., Leo, M.: A review of vision-based systems for soccer video analysis. Pattern recognition. 43.8, 2911–2926 (2010)
3. Ping, Q., Weitao, Z.: Sport Video Intelligence Analysis Using Mid-Level and Low-Level Vision Information. In: IEEE International Conference on Future Computer Science and Education, pp. 165–168. IEEE (2011)
4. Gerke, S., Linnemann, A., Müller, K.: Soccer player recognition using spatial constellation features and jersey number recognition. Computer Vision and Image Understanding. 159, 105–115 (2017)
5. Lu, W. L., Ting, J. A., Little, J. J., Murphy, K. P.: Learning to track and identify players from broadcast sports videos. IEEE transactions on pattern analysis and machine intelligence. 35.7, 1704–1716 (2013)
6. Assfalg, J., Bertini, M., Colombo, C., Del Bimbo, A., Nunziati, W.: Semantic annotation of soccer videos: automatic highlights identification. Computer Vision and Image Understanding. 92.2-3, 285–305 (2003)
7. Linnemann, A., Gerke, S., Kriener, S., Ndjiki-Nya, P.: Temporally consistent soccer field registration. In: 20th IEEE International Conference on Image Processing, pp. 1316–1320. IEEE (2013)
8. Sun, L., Liu, G.: Field lines and players detection and recognition in soccer video. In: IEEE International Conference on Acoustics, Speech and Signal Processing, pp. 1237–1240. IEEE (2009)
9. Atrish, A., Singh, N., Kumar, K., Kumar, V.: An Automated Hierarchical Framework for Player Recognition in Sports Image. In: ACM International Conference on Video and Image Processing. ACM (2017)
10. Liu, J., Tong, X., Li, W., Wang, T., Zhang, Y., Wang, H.: Automatic player detection, labeling and tracking in broadcast soccer video. Pattern Recognition Letters. 30.2, 103–113 (2009)
11. Khatoonabadi, S. H., Rahmati, M.: Automatic soccer players tracking in goal scenes by camera motion elimination. Image and Vision Computing. 27.4, 469–479 (2009)
12. Fu, T. S., Chen, H. T., Chou, C. L., Tsai, W. J., Lee, S. Y.: Screen-strategy analysis in broadcast basketball video using player tracking. In: IEEE Visual Communications and Image Processing, pp. 1–4. IEEE (2011)
13. Huang, C. L., Shih, H. C., Chao, C. Y.: Semantic analysis of soccer video using dynamic Bayesian network. IEEE Transactions on Multimedia. 8.4, 749–760 (2006)

DCT- and DWT-Based Intellectual Property Right Protection in Digital Images

Singh Arun Kumar, Singh Juhi and Singh Harsh Vikram

Abstract Keeping in mind the end goal to shield copyright material from illicit duplication, many advances have been introduced. In this paper, a new technique is presented using discrete cosine transform and discrete wavelet transform. The purpose of this technique is to achieve all its requirement like high impartibility and security. The main aim of this paper is to provide application in medical field. The new method will be utilized for concealing a mystery picture inside a cover picture utilizing one mystery key to generate a stego-image. The proposed strategy does not require the first cover picture to remove the installed mystery picture. The near examination between the explained procedure and the other existing systems in time and frequency has demonstrated the predominance of the proposed procedure in terms of PSNR.

Keywords DWT · DCT · PSNR · Image compression · Watermarking
Data hiding

1 Introduction

Steganography is the process of embedding text or image information into cover image. Data hiding or steganography is the science of hidden communication which has got much attention from the research community recently. There are many objec-

S. A. Kumar (✉) · S. Juhi
Amity University, Gurgaon, Haryana, India
e-mail: arunsingh86@gmail.com

S. Juhi
e-mail: juhisingh17@gmail.com

S. A. Kumar
SNU Ranchi, Ranchi, India

S. H. Vikram
KNIT Sultanpur, Sultanpur, Uttar Pradesh, India
e-mail: harshvikram@gmail.com

© Springer Nature Singapore Pte Ltd. 2019
R. S. Bapi et al. (eds.), *First International Conference on Artificial Intelligence and Cognitive Computing* , Advances in Intelligent Systems and Computing 815,
https://doi.org/10.1007/978-981-13-1580-0_49

Table 1 Process of calculation of scaling factor

S. No.	Cover image	Average value of DCT coefficient of secret image (A)	Average value of 2L DWT coefficient of cover image (B)	Scaling factor(A/B)
1	Image1	24.6815	0.0124	1.9930e+03
2	Image2		0.0005	4.8934e+04
3	Image3		0.0206	1.1969e+03
4	Image4		0.0043	5.7714e+03
5	Image5		0.0025	1.0043e+04
6	Image6		0.0162	1.5221e+03
7	Image7		0.0012	2.0814e+04

tives of steganography, in which few main objectives are imperceptibility (or unde-tectability), security, embedding payload, and robustness (required against common attacks) [1].

In this paper, we introduced a very simple method in frequency domain based on DCT and DWT. Frequency domain offers good undetectability as compared to time domain approach. DCT is being used for compression of secret message image. The purpose of doing compression is to reduce the size of payload. In the next step of procedure, 2-level DWT of cover image was calculated, and vertical detail matrix (LH) is being used for data hiding. Advantage of taking LH matrix is to improve the robustness [2].

2 Proposed Method

There are various methods for data hiding in spatial domain in which most common and easiest method is LSB hiding [3]. If changes in the cover image after data hiding are negligible, then distortions brought by the insertion process remain imperceptible. To minimize the changes due to insertion process, the DCT coefficients are divided by the scaling factor. Scaling factor for various images is given in Table 1.

The filter used in proposed technique for data compression having following coefficients

$$\text{Filter Coefficient} = \begin{bmatrix} 1\ 1\ 1\ 1\ 1\ 1\ 1\ 0; \\ 1\ 1\ 1\ 1\ 1\ 1\ 0\ 0; \\ 1\ 1\ 1\ 1\ 1\ 0\ 0\ 0; \\ 1\ 1\ 1\ 1\ 0\ 0\ 0\ 0; \\ 1\ 1\ 1\ 0\ 0\ 0\ 0\ 0; \\ 1\ 1\ 0\ 0\ 0\ 0\ 0\ 0; \\ 1\ 0\ 0\ 0\ 0\ 0\ 0\ 0; \\ 0\ 0\ 0\ 0\ 0\ 0\ 0\ 0; \end{bmatrix}$$

From the above filter coefficients, it is clear that only 21 coefficients of 8*8 matrix are taken. Rest 43 coefficients are zero. So in hiding algorithm, only 21 elements of DCT coefficient will replace DWT of cover image in each 8*8 grid.

3   Literature Review

Ajaya Shrestha proposed a method based on LSB hiding. In this method, the author has selected grayscale secret image for data hiding. Each pixel of this image is first XORed with the secret key. This key will be used at the time of data extraction. In the next step, the author calculated the DWT of cover image and selected any two bands (i.e., HL/HH or LH/HH or HL/LH) from the DWT coefficient. After that, he decomposed the given secret image pixel into higher and lower nibble. These lower and higher four bits of secret information get placed into the lower four bits of two wavelet coefficients of one of the band selected. Main drawback of this procedure is low PSNR [4].

N Sathisha proposed a method based on calculating the LWT of cover image and payload. Then mantissa part of cover image is replaced by the mantissa part of payload. The inverse LWT is applied to generate stego-image. PSNR value calculated in this procedure was 58.52 dB [5].

Asna Furqan proposed a method using SVD and DWT which performed very well under various attacks. The common attacks were taken under consideration like cropping, image compression, and other signal processing operations [6].

Y. Taouil proposed new steganographic methods based on Haar discrete wavelet transform. The coefficients of H, V, D bands of transformed cover image could hold the binary bits of secret message. H, V, D bands of transformed image have shown high degree of impeccability if random pixel was chosen for data hiding. Mean square error and the peak signal to noise ratio were calculated to evaluate the quality of the hiding process. Maximum value of peak signal to noise ratio calculated in this method was 48.952 dB [7].

Samaneh Shafee proposed a method based on compressive sensing (CS) with considering human visual system (HVS) features in data hiding area. HVS visible spectrum is not been able detect small changes in cover image which determines

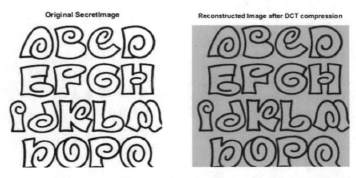

Fig. 1 Original data images and image reconstructed after compression

Fig. 2 Original cover image taken

the security factor of image. This property of HVS is used in certain area of spatial domain region of image where secret messages can be embedded without being seen into a cover image [8].

4   The Proposed Algorithm

4.1   Data Hiding Steps

1. Take secret image as data (Fig. 1)
2. Compress the above data using DCT transform (up to 52 dB) (Fig. 1)
3. Take mean of DCT coefficients of step 2
4. Select cover image (Fig. 2)

Fig. 3 Stego-image after data hiding

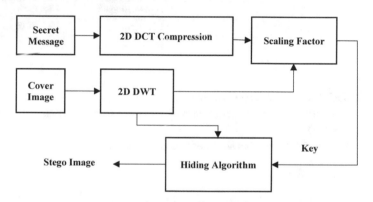

Fig. 4 Algorithm for data hiding

5. Calculate 2-level DWT of cover image and select vertical detail matrix for data hiding
6. Take mean of vertical detail matrix in above step
7. Take ratio of quantities in step 6 and step 3 (scaling factor)
8. Take DCT coefficient matrix in step 2 and divide this matrix with the ratio calculated in the above step
9. Hide the above found matrix directly in vertical detail matrix of step 5 (Figs. 3 and 4).

4.2 Data Extraction Steps

1. Take stego-image
2. Calculate 2-level DWT of the stego-image

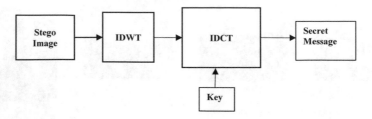

Fig. 5 Algorithm for data extraction

Table 2 Experimental result of various cover images

S. No.	Cover image	PSNR(dB)
1	CoverImage1	105.8065
2	CoverImage2	132.8678
3	CoverImage3	140.9324
4	CoverImage4	109.4143
5	CoverImage5	130.4868
6	CoverImage6	104.7068
7	CoverImage7	120.2325

3. Extract DCT coefficient from vertical detail matrix
4. Multiply the above found coefficient with the ratio found in data hiding steps 7 (scaling factor)
5. Take IDCT of the above extracted image (Fig. 5).

5 Noise Analysis and Experimental Result

Mean square error (MSE) and peak signal to noise ratio are the two parameters for quality measure, and their corresponding equations are also given below [9].

Consider two images, $x\,(a, b)$ and $y\,(a, b)$ of $M \times N$ dimensions. The formula for mean square error is [10]

$$\text{MSE} = \frac{1}{MN} \sum_a \sum_b [x(a, b) - y(a, b)]^2$$

$$\text{PSNR} = 10\log\left(\frac{255}{\sqrt{\text{MSE}}}\right)$$

There are seven images selected for this paper, and the result in given below (Table 2).

6 Conclusions

Our implemented technique has proved a high degree of impartibility as PSNR is very high. Introduced method could be used in many fields like medical applications where watermarking is found as an instrument for protection of medical information, secure e-voting system, broadcast monitoring, owner identification, proof of ownership, and content authentication [11].

References

1. Kumar, Sushil, "A Comparative Study of Image steganography in Wavelet Domain", IJCSMC, Vol. 2, Issue. 2, February 2013, pg.91–101.
2. Ahmed A. Abdelwahab, "A Discrete Wavelet Transform Based Technique for Image Data Hiding" 25th National Radio Science Conference (NRSC 2008) March 18-20, 2008, Faculty of Engineering, Tanta Univ., Egypt.
3. Chikouche Sofyane Ladgham, "An Improved Approach for LSB Based Image Steganography using AES Algorithm", IEEE 5th International Conference on Electrical Engineering – Boumerdes (ICEE-B) October 29–31, 2017, Boumerdes, Algeria.
4. Ajaya Shrestha, "Color Image Steganography Technique using Daubechies Discrete Wavelet Transform" 9th International Conference on Software, Knowledge, Information Management and Applications (SKIMA), 2015.
5. Sathisha, N, "Image Steganography Based on Mantissa Replacement using LWT", International Journal of Advanced Research in Computer Engineering & Technology (IJARCET) Volume 4 Issue 2, February 2015.
6. Furqan Asna, "Study and Analysis of Robust DWT-SVD Domain Based Digital Image Watermarking Technique Using MATLAB", 2015 IEEE International Conference on Computational Intelligence & Communication Technology.
7. Taouil Y., "High Imperceptibility Image Steganography Methods Based on HAAR DWT" International Journal of Computer Applications (0975 – 8887) Volume 138-No.10, March 2016.
8. Shafee Samaneh, "A Secure Steganography Algorithm Using Compressive Sensing Based on HVS Feature", IEEE 2017 Seventh International Conference on Emerging Security Technologies (EST).
9. Mohammad Hadi, "A Image Steganography Scheme Based On DWT Using Lattice Vector Quantization and Reed-Soloman Encoding",2015 2nd International Conference on KBEI, Tehran Iran.
10. https://www.ncbi.nlm.nih.gov/pmc/articles/PMC4391065/.
11. Mohamed M. Abd-Eldayem, "A Proposed Security Technique Based on Watermarking and Encryption for Digital Imaging and Communications in Medicine", Egyptian Informatics Journal (2013).

A Study on Multi-agent Systems in Cognitive Radio

Sandeep Sharma, Sunil Karforma and Sripati Mukhopadhyay

Abstract Cognitive radio (CR) is getting huge attention with its spectrum optimising ability and demand of inherent cognitive abilities at wireless nodes in 5G communication (comn). It can utilise the unused part of spectrum, which facilitates sharing of spectrum amongst Primary User (PU) and Secondary User (SU). Users sense the radio environment and take appropriate actions, making multi-agent system (MAS) as natural choice for CR. This paper presents state-of-the-art survey on use of MAS in CR research and proposals.

Keywords CR · PU · SU · Dynamic spectrum access (DSA) · Spectrum · MAS
Auction

1 Introduction

CR detects free channel (ch) available in its vicinity. It then changes own parameters allowing more concurrent wireless comns in same band [1]. It is a great enabler for evolution from wired to wireless world [2]. We are heading towards a crisis of spectrum [3]. CR if designed in lines of MAS has potential to overcome this problem.

S. Sharma (✉) · S. Karforma
The University of Burdwan, Burdwan, West Bengal, India
e-mail: sandeepsharma050265@gmail.com

S. Karforma
e-mail: sunilkarforma@yahoo.com

S. Mukhopadhyay
The Academy of Technology, Kolkata, West Bengal, India
e-mail: dr.sripatim@gmail.com

© Springer Nature Singapore Pte Ltd. 2019
R. S. Bapi et al. (eds.), *First International Conference on Artificial Intelligence and Cognitive Computing* , Advances in Intelligent Systems and Computing 815,
https://doi.org/10.1007/978-981-13-1580-0_50

Fig. 1 Spectrum
management functionalities

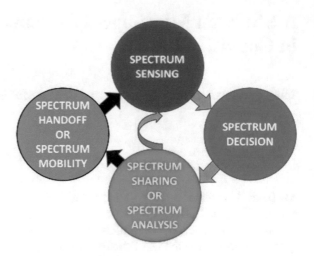

1.1 Introduction to CR

CR was introduced in 1998 by Mitola [4, 5] as a smart radio which senses the radio
environment. On detection of idle parts of spectrum or holes in its environment,
CR makes use of them opportunistically [6]. FCC first used the term CR in March
2000 [7]. In 2005, Simon Haykin treated CR as brain-empowered wireless comn
[8]. It was followed by many proposals for CR. Reference [9] gave power and bandwidth allocation in cognitive radio network (CRN). Optimisation formulations in
[10] provided optimal solutions for necessary resource allocations. Reference [11]
used innovative method for implementing CR based on human behaviour. Reference
[12] used machine learning techniques in CR. Utility of cooperation was deliber-
ated in [13]. SUs basically utilise the bandwidth of PUs smartly [14–16]. Figure 1
gives out four main functionalities of the CR cycle [17]. For spectrum sensing (SS),
some studies proposed RF energy detection [18, 19], while others proposed empty
spectrum as combination of signals from PU, AWGN and signal gain [20, 21]. Many
other works oriented themselves towards matched filter detection [22, 23]. Some
used Game Theory [24]. In many approaches [25–28], SUs formed coalitions and
cooperated to sense spectrum to identify and access free chs. MAC-based solutions
have been recently developed for DSA [29–34]. Most of MAS solutions address
issues of SS and sharing [44]. Spectrum handoff is new area of research. In reactive
handoff approach [33, 34], SUs switch the spectrum on sensing a PU demanding the
same band. Target ch is then selected instantaneously. In proactive approach [35–43],
SUs can envisage ch status and thus can switch chs even before a PU demands or
approaches ch.

Table 1 Analogies between an agent and a CR node [45]

Agent	CR terminal/node or SU
A virtual entity	A real entity
Interacts and perceives environment and acts accordingly. It can communicate with other agents	Interacts with radio environment and detects holes. It can also communicate with other nodes
Environment awareness via past observations	Learning is an important part
An entity which functions autonomously. It uses sensors for observing and makes use of actuators to act. Overall, its activities are directed towards achieving specific goals	Has ability to perform resource management (RM) for radio resources. It has set of goals. It can also direct its activities towards achieving these goals
Actuators are used for performing actions	Decision regarding selection of the bands/chs
They can be very simple or complex	They are invariably extremely complex
They can function independently	They may or may not function independently
Autonomy	Autonomy
Cooperation is used for interaction	Beaconing is used for interaction
Function together to achieve shared goals	Function together to share the spectrum efficiently
Knowledge base consists of agents present locally and in neighbourhood	Knowledge base consists of neighbouring PUs' usage

1.2   Introduction to MAS

An agent can be either hardware or software based. It can perform a particular task automatically on behalf of either an application or a user [39]. Agents can be of two types, e.g. intelligent and mobile [40, 41]. An MAS consists of multiple interacting computing elements (agents) [42]. MASs contain environment, objects, agents and different relations between them [43]. To be intelligent, an agent must have some properties, e.g. reactivity, pro-activity and sociability [44, 45]. Further, MAS is an ever-interacting group of agents which have common environment, goals and plans [46].

2   Similarities Between CR and MAS

Analogies between agent and CR node are being brought out as Table 1:

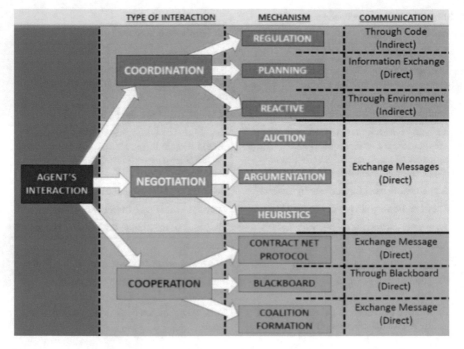

Fig. 2 Agents' interaction mechanisms, their classification and comn types [44, 47]

3 Interaction Between Agents

Agents' interaction types, mechanisms and comn types related to each are as given in Fig. 2 [44, 47]:

3.1 Coordination

It ensures that agent's activities preserve some desired relationships, e.g. sequence and complementarily. Most popular coordination mechanisms are regulation, planning and reactive cooperation [48].

3.2 Cooperation

Agents work together to maximise their utilities and reach a common goal. Blackboard [49], Contract Net Protocol (CNP) [50] and Coalition Formation [51] are some very useful cooperation-based protocols.

Table 2 Auctions in classical sense vis-à-vis auctions in CRNs

Auctions—classic sense	Auctions—CRNs
Goods or items to sell	Idle chs
Buyer or bidder	SU
Seller or provider	PU
Auctioneer or auctioning agency	Regulator or controller or broker

Fig. 3 Types of auctions

3.3 Negotiation

It enables agents to reach agreement about service by an agent for another and resolves conflicts between agents. Auctions [52], argumentation [53] and heuristics [54] are most popular negotiation protocols.

Auctions. Auction has several stakeholders. Table 2 compares traditional auction vis-à-vis auctions in CRN.

Traditionally, four types of auctions are used as shown in Fig. 3. Figure 4 summarises spectrum management of CRN using MAS [44, 47]. A hybrid variant of auction is double auctions [52]. In almost all traditional settings except Vickrey, auctioneer desires to maximise price at which good is allocated, while bidders desire to minimise price. Auctions resolve conflicts between agents. Major problems of auctions are possibility of fraud and situations when correct winner cannot be determined [53, 54].

Argumentation. Argumentation is an attempt to agree what to believe by supporting arguments and resolving inconsistency [55]. The mechanism is very complex and needs knowledge of past actions and agents' arguments.

Heuristics. Using heuristics, seller agents can learn from previous bids and information exchanges of malicious buyers, thus avoiding future trades with malicious buyers. It is extremely difficult to identify efficient heuristic [56, 57].

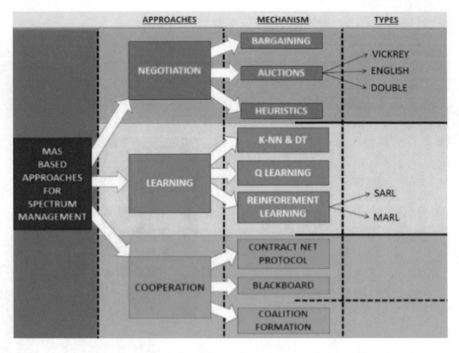

Fig. 4 MAS-based spectrum management for CRN

4 MAS Applications in CRNs

Application of MAS for addressing the CR problems can be broadly divided as in the Fig. 4.

4.1 Negotiation-Based Approaches

Negotiation-based approaches are used for spectrum allocation. PUs and SUs negotiate reaching agreement.

Bargaining. It is a negotiation mechanism based on selling/buying concept and is used in CR to solve sensing issue [58] or for fair spectrum allocation [39, 40]. Reference [59] proposed a bargaining-based pair-wise cooperative SS scheme. A solution for spectrum allotment of distributed type via bargaining locally was presented in [40].

Heuristics. Heuristics-based solutions in CR context are provided to solve complex problems in spectrum allocation issue. For example in [41], the problem of allocating the chs amongst CR nodes is handled.

Auctions. Auctions are popular negotiation processes used in CRN in order to solve and optimise spectrum utilisation. In [53, 60–70, 94], all users submit their bids to a centralised manager. Auctioneer allocates the resources in a way to maximise its utility. The utility function changes according to type of auction. A spectrum management policy is suggested in [60] relying on Vickrey. A pricing- and billing-based solution is proposed in [61] presenting an auction sequence mechanism, which allows buyers to express their urgency, requirements, buying capacity and choices. Vickrey auction [60] seems to be more suitable to execute within a specified time as compared to English auction used in [61]. In [63], double auction is proposed to resolve spectrum issues amongst PUs and SUs. Unlike normal spectrum sharing techniques for sharing spectrum, in [64] SUs can make independent and simultaneous decisions. They can also make decisions about bid based on self-interest considerations. Another auction approach is proposed in [66] where SU can make a bid for the spectrum that it needs and PU may assign same to SU without degrading PU's own QOS. Spectrum trading model proposed in [67] uses agent as an additional part in the trading process that is different from SUs and PUs. Reference [68] proved in some situations spectrum is more effectively utilised in the scenarios when many SUs gain access to a solitary ch. In most situations, the users' behaviours are not truthful, so the centralised authority or regulator or broker cannot optimise the utility of the CRN [69]. Reference [70] gives secondary spectrum commerce scenario where a PU can sell its spectrum which is not used or underutilised. Such spectrum is sold in the form of spectrum space–time unit which is finely grained. Reference [71] explains how a single round auction can be used when we aim at meeting applications which demand reaction immediately, as using multiple rounds of auctions will amount to very significant and prohibitive delays in terms of time for obvious reasons. Little work had been done in FPSB, omitting any discussion on same here. However, Emna Trigui et al. [94] suggested a novel auction technique using MAS based on FPSB. It is both fast and workable and does not require multiple rounds of biding.

4.2 Learning-Based Approaches

Learning methods have been only recently used to solve multiple issues in CRN, e.g. [71–73] for SS, [74, 75] for spectrum sharing and [76] for spectrum access and control. Two types of learning algorithms are generally used, i.e. Reinforcement Learning (RL) and Q-learning. Reference [76] gave out the use of two types of RL used with MAS in CR. However, [75] gave many other learning algorithms, e.g. K-NN or decision tree (DT) for classification of data. Reference [77] gives ch selection while [67] brought out two types of RL. Reference [78] gave a novel approach for Multi-agent Reinforcement Learning (MARL). In [79], unique learning arrangement is given out for each agent. Reference [80] gave decentralised Q-learning. Problem of aggregated interference is discussed in [81]. The issue of real-time RM for CRs is discussed in [82] on entering a zone with poor QOS. Reference [83] described system-dependent learning techniques.

RL. RL allows agents to learn from past states (sharing pattern, neighbourhood movement, etc.) in order to improve subsequent actions and moves. Approach in [71] proved that allowing SUs learning via interaction can improve overall spectrum usage especially for SUs' transmissions and ch switching capabilities. Two types of RL approaches were discussed in [68], namely Single Agent Reinforcement Learning (SARL) and MARL. In general, as proved in [68], RL provides high network-wide performance with respect to dynamic chs selection. Reference [78] describes a new approach based on MARL in CRNs with ad hoc decentralised control which gives network capable of converging into fair sharing of spectrum. Reference [79] gave a learning mechanism for each agent providing reward to every agent enabling it to take right decision and chooses action which is most appropriate.

Q-learning. In this, we do not need to emulate its vicinity and is being used more for spectrum access as it allows agents learn to act in optimal manner though it is essentially RL. If two or more SUs choose same ch for data transmission, collision occurs in [76], similar to Aloha without collision avoidance. However, SUs here aim at learning of how to avoid collisions, according to their experience. Q-learning was exploited in [77] to ensure ch selection in CRNs. Both [76] and [77] deal with RM giving out a method for modelling negotiations which cuts down overhead using a unique and new method aimed at accessing of spectrum. In order to address the issue of interference caused to PUs, an algorithm using multi-agent learning was given in [80] as a solution named as "decentralised Q-learning". A real-time algorithm based on decentralised Q-learning is given in [81] in order to address the issue of aggregate interference.

K-NN and DT. Reference [82] addresses real-time RM of CR using MAS and K-NN/DTs. Dependence of learning technique on system characteristics is given in [83].

4.3 Cooperation-Based Approaches

CNP. CNP-based approach in [84] relied on Call for Proposal (CFP) concept. PU agent is considered as a "manager", and SUs are considered as "contractors" enhancing overall spectrum allocation and ensures user satisfaction. References [77, 85] proposed how agents of SUs and PUs can be made to cooperate and also coexist. This allows maximising the utility function of other agents without adding cost of unnecessary messages as agents are cooperative and selfless in nature. In [24, 86], a model is given employing agents for trading spectrum in CRNs with help of an agent which acts as broker. References [87] developed allocation of spectrum which can be effective in environments which are dynamic in nature, using cooperation.

Coalition Formation. Reference [15] uses Coalition Formation for sharing unlicensed spectrum bands by equipping SUs with agents and make them communicate with neighbours forming coalitions in the bands which are not licensed. It thus resolves most conflicts arising during access of spectrum since agents cooperate with each other to share the spectrum. Reference [88] uses MAS for dynamic spectrum sharing whereby every time SU is in need of spectrum, agent representing it cooperates and communicates for the purpose of sharing the spectrum with the agents which represent PU. In [89], embedded intelligent agents coordinate to exploit resources while overcoming interference with PUs. Reference [71] proposes modular architecture for information collection regarding radio environment which is stored as knowledge base shared and hence accessed by all the agents. Sharing of knowledge increases the collective and individual gain of MAS. Reference [16, 45, 90] proposes that on getting an appropriate offer SUs must decide immediately depending on spectrum availability. In such case, SUs need not wait for response from every single PU. Reference [13] proposes how PU can be made immune of interference. This is achieved by regulating the power being transmitted.

Blackboard. Blackboard approach in [91] helps to solve some related network problems such as intrusion detection and jamming. In [92], author's main goal is to reduce interference, handoff delay and blocking probability.

5   Spectrum Mobility and Handoff

References [93, 94] give a new scheme for handoff of spectrum which enables switching to best spectrum by CR with help of MAS negotiation. All CR nodes aim to get maximum profits for themselves while negotiating. A proposal was presented to solve the problems of spectrum mobility, sharing and handoff in CRN using MAS. A novel scheme for auction which is both practical and implementable was given by [94]. While performing handovers, interference can be minimised just by controlling the handover in [65, 82].

6   Conclusion

MAS can be used in many diverse ways and many different scenarios in the CR, e.g. SUs cooperating within themselves, PUs and SUs cooperating within themselves and even ones using agents acting as broker for purpose of negotiation. There are ceaselessly many other models using MAS. The paper brings out in detail CR, its main functionalities and the DSA. Of special interest is use of MAS in CRNs based on auctions mechanisms due to its envisaged great practical utility in CRNs. Learning methods have been only recently used, and they are still evolving. However, cooperative approaches are not widely applied. Little work had been done in spectrum mobility and handoff which calls for attention of researcher community worldwide. MAS clearly emerges as the approach which perfectly matches the requirements of CRNs in accessing the spectrum as it ensures the users autonomy since agents which are embedded can then perform their own spectrum management in accordance to requirements in dynamic and decentralised way. There is lot of scope for researchers to improve the auction specially in modelling the interaction between the users. It can be finally concluded that use of MASs in CRNs can provide very efficient solutions to many CR issues.

References

1. Wikipedia, https://en.wikipedia.org/wiki/Cognitive_radio/
2. Mitola, J.: Software radios: Survey, critical evaluation and future directions. In: IEEE Aerospace and Electronic Systems Magazine, vol. 8, No. 4, pp. 25–36. (1993)
3. Cisco visual NW index: forecast & methodology, In: Cisco White Paper, pp. 1–14 (2012)
4. Mitola, J., Maguire, G.: CR: Making software radios more personal. In: IEEE Personal Communications vol. 6, No. 4, pp. 13–18. (1999)
5. Mitola, J.: CR: an integrated agent architecture for software defined radio. In: Dissertation for Doctor of Technology, Teleinformatics, KTH, Stockholm (2000)
6. Mitola, J.: CR Architecture: The Engg Foundations of Radio XML Link. In: Wiley (2006)
7. NPRM (Notice of Proposed Rule Making), FCC (2000)
8. Haykin, S.: CR: Brain-empowered wireless comns. In: IEEE Selected Areas Communication, vol. 23, No. 2, pp. 201–220. (2005)
9. Wang, L., Xu, W., He, Z., Lin J.: Algorithms for optimal resource allocation in heterogeneous CRNs. In: 2nd IEEE International Conference on Power Electronics and Intelligent Transportation System. Globecom, pp. 1–5, Shenzen (2009)
10. Abbas, N., Nasser, Y., Ahmad, K.: Recent advances on artificial intelligence and learning techniques in CRNs. In: EURASIP J., pp 1–20. (2015)
11. Xing, Y., Chandramouli, R.: Human behavior inspired CRN design. In: IEEE Comns, pp. 122–127. (2008)
12. Clancy, C., Hecker, J., Stuntebeck, E., OShea, T.: Applications of machine learning to CRNs. In: IEEE Wireless Comns, vol. 14, No. 4, pp. 47–52. (2004)
13. Li, Y., Wang, X., Guizani, M.: Resource pricing with primary service guarantees in CRNs: A Stackelberg game approach. In: IEEE GLOBECOM, Honolulu (2009)
14. Xing, Y., Chandramouli, R., Mangold, S.: DSA in open spectrum wireless networks, IEEE J. on Selected Areas on Comns vol. 24, No. 3, pp. 626–637. (2006)
15. Usama, M.: Utilization of Coop MASs for Spectrum Sharing in CRNs. In: Theses (2011)
16. Usama, M., Boulahia, L., Gaïti, D.: A Cooperative Multiagent Based Spectrum Sharing. In: IEEE Sixth Advanced International Conference on Telecomns, pp. 1—7. Barcelona (2010)
17. Akyildiz, I., Lee, W., Chowdhury, K.: CRAHNs: CR Ad Hoc Networks. In: Elevier J. Ad Hoc Networks, pp. 810–836. http://www.omidi.iut.ac.ir (2009)
18. Oh, D., Lee, Y.: Energy Detection Based SS for Sensing Error Minimization in CRNs. In: International J. of Comn Networks and Information Security, pp. 1–5. (2009)
19. Yaqin, Z., Shuying, L., Nan, Z., Zhilu, W.: A Novel Energy Detection Algorithm for SS in CR. In: Information Technology J., vol. 9, pp. 1659–1664. (2010)
20. Ghasemi, A., Sousa, E.: Collaborative SS for opportunistic access in fading environment. In: IEEE Symposia on New Frontiers in DSA Networks, pp. 131–136. Baltimore (2005)
21. Sahin, M., Guvenc, I., Arslan, H.: Uplink user signal separation for OFDMA-based CRs. In: EURASIP, J. on Wireless Comns and Networking. (2010)
22. Bouzegzi, A., Ciblat, P., Jallon, P.: Matched Filter Based Algorithm for Blind Recognition of OFDM Systems. In: IEEE VTC-Fall, pp. 1–5. Calgary, Canada (2008)
23. Kapoor, S., Rao, S., Singh, G.: Opportunisitic SS by Employing Matched Filter in CRN. In: IEEE International Conference on Comn Sys & NW Tech, pp. 580–583. Katra (2011)
24. Wang, B., Wu, Y., Liu, K.: Game theory for CRNs: An overview. In: Elsevier J. Computer Networking, pp. 2537–2561. https://www.elsevier.com (2010)
25. Rajasekharan, J., Eriksson J., Koivunen, V.: Coop game-theoretic modeling for SS in CRs. In: IEEE Conf Rec of 44th Asilomar Conf on Sigs, Sys & Comps, pp. 165–169. CA (2010)

26. Saad, W., Zhu, H., Basar T., Debbah, M., Hjorungnes, A.: Coalition Formation Games for Collaborative SS. In: IEEE Transaction on Vehicular Technology, pp. 276–297. (2011)
27. Rajasekharan, J., Eriksson, J., Koivunen, V.: Cooperative Game-Theoretic Approach to Spectrum Sharing in CRs. In: Computer Science and Game Theory. (2011)
28. Liang, M., Qi, Z.: An Improved Game-theoretic Spectrum Sharing in CR Systems. In: 3rd International Conf on Comns and Mobile Computing, pp. 270–273. Qingdao (2011)
29. Domenico, A.D., Strinati, E.C., Benedetto, M.G.D.: A Survey on MAC Strategies for CRNs. In: IEEE Communications Surveys & Tutorials, vol. 14, No 1, pp. 21–44. (2012)
30. Zhang, X., Hang, S.: CREAM-MAC: CR-EnAbled Multi-Ch MAC Protocol Over DSA NWs, IEEE J. of Selected Topics in Signal Processing, pp. 110–123. (2011)
31. Wang, C.W., Wang, L.C..: Analysis of Reactive Spectrum Handoff in CRNs. In: IEEE J. on Selected Areas in Communication, vol. 30, No 10, pp. 2016–2028. (2012)
32. Wang, C.W., Wang, L.C., Adachi, F.: Modeling and Analysis for Reactive-decision Spectrum Handoff in CRNs. In: IEEE Globecom, pp. 1–6. (2010)
33. Wang, C.W., Wang, L.C.: Modeling and Analysis for Proactive-decision Spectrum Handoff in CRNs. In: IEEE International Conference on Comn (ICC), pp. 1–6. Dresden (2009)
34. Y. Song and J. Xie. Common Hopping based Proactive Spectrum Handoff in CR Ad Hoc Networks. In: IEEE Global Telecom. Conf. GLOBECOM, pp. 1–5. Miami (2010)
35. Duan J., Li. Y.: An optimal Spectrum Handoff scheme for CR mobile Adhoc Networks. In: Advances in Electrical and Computer Engineering J., vol. 11, No. 3, pp. 11–16. (2011)
36. Wang, L., Cao, L., Zheng, H.: Proactive ch access in dynamic spectrum networks. In: Elsevier J. of Physical Communication, pp. 103–111. (2008)
37. Song, Y., Xie, J.: ProSpect: A Proactive Spectrum Handoff Framework for CR Adhoc Networks without Common Cont Ch. In: IEEE trans on mob Comp, pp. 1127–1139. (2012)
38. Trigui, E., Esseghir, M., Boulahia, L.M.: On Using MASs In CRNs: A Survey. In: International J. of Wireless & Mobile Networks (IJWMN), vol. 4, No 6, pp. 1–16. (2012)
39. Liu, H., MacKenzie, A.B, Krishnamachari, B.: Bargaining to Improve Ch Sharing between Selfish CRs. In: IEEE GLOBECOM - Global Telecom. Conf., pp. 1–7. Honolulu (2009)
40. Cao, L., Zheng, H.: Distributed spectrum allocation via local bargaining. In: IEEE Comns Society Conference on Sensor and Ad Hoc Comns and Networks, pp. 475–486. (2005)
41. Rao, V.S., Prasad, R.V., Yadati, C., Niemegeers, I.G.M.: Distributed heuristics for allocating spectrum in CR adhoc networks. In: IEEE GLOBECOM, pp. 1–6. Miami (2010)
42. Wooldridge, M.: An Introduction to MASs. In: Wiley & Sons. West Sussex (2002)
43. Ferber J.: MAS: An Introduction to Distributed AI. In: Addison Wesley Longman. (1999)
44. Weiss G.: A modern approach to distributed AI. In: MIT press, USA (2000)
45. Yau K.L.A., Komisarczuk P., Teal P.D.: Achieving Context Awareness & Intelligence in Distributed CRNs: A Payoff Propagation Approach. In: IEEE Workshops of International Conf. on Advanced Information Networking and Applications, pp. 210–215. Bipolis (2011)
46. Huhns, M.N.: Multi Agent Systems. In: Tutorial at the EASSS (1999)
47. Espinasse, B., Fournier, S., Freitas, F.: Agent and ontology based info gathering on restricted web domains with AGATHE. In: ACM symposium on Applied Computing. pp. 2381—2386. Fortaleza (2008)
48. Ahmed, M., Kolar, V., Petrova, M., Mahonen, P., Hailes, S.: A component-based architecture for CR RM. In: IEEE CRWNCOM, pp. 1–6. Hannover (2009)
49. Jurado, F, Redondo, M.A., Ortega, M.: Blackboard architecture to integrate components and agents in heterogeneous distributed eLearning systems: An application for learning to program. In J. of Systems and Software, vol. 85, No. 7, pp. 1621–1636. (2012)
50. Mir, U., Aknine, S., Boulahia, L.M., Gaïti, D.: Agents' Coordination in Ad-hoc Networks. In: 8th ACS/IEEE conference AICCSA, pp. 1–8. Hammamet (2010)
51. Gruszczyk, W., Kwasnicka, H.: Coalition Formation in MASs-an evolutionary approach. In: IEEE International Multiconf. on CS and IT (IMCSIT), pp. 125–130. Wisia (2008)
52. Amraoui, A., Benmammar, B. Krief, F., Bendimerad, F.T.: Auction-based Agent Negotiation in CR Ad Hoc Networks. In: Fourth ADHOCNETS, Paris (2012)
53. Cassady, R.: Auctions & Auctioneering. In: Univ of California Press, California, (1967)

54. Adomavicius, G., Gupta, A.: Toward comprehensive real-time bidder support in iterative combinational auctions. In: Info. Systems Research, vol. 16, No. 2, pp. 169–185(2005)
55. Rahwan, I., Ramchurn, S.D., Jennings, J.R.: Argumentation-based negotiation. In: The Knowledge Engineering Review, Cambridge Univ. Press, pp. 343–375. Cambridge (2003)
56. Amanna, A., Ali, D., Fitch, D.G., Reed, J.H.: Hybrid Experiential-Heuristic CR Engine Architecture and Implementation. In: J. of Computer NW and Comns, pp. 1–15 (2012)
57. Hu, D., Mao, S.: Cooperative relay interference alignment for video over Cognitive Radio Networks. In: IEEE INFOCOM, pp. 2014–2022. Orlando, (2012)
58. Xie, J., Howitt, I., Raja, A.: CR RM using MASs, In: CCNC, pp. 1123–1127. (2007)
59. Pan, M., Fang, Y.: Bargaining based pairwise cooperative SS for CRNs. In: IEEE Military Communications Conference MILCOM, pp. 1–7. San Diego(2008)
60. Chang, H.B., Chen, K.C..: Auction-based spectrum management of CRNs. In: IEEE Transactions on Vehicular Technology, vol. 59, No. 4, pp. 1923–1935. (2010)
61. Kloeck, C., Jaekel, H., Jondral, F. K.: Multi agent wireless System for Dynamic and Local Combined Pricing, allocation and Billing. In: J. of comn, vol.1, No. 1, pp. 48–59. (2006)
62. Niyato, D., Hossain, E.: Competitive pricing for spectrum sharing in CRNs: Dynamic game, inefficiency of Nash equilibrium and collusion. In: IEEE J. on Selected Areas in Communications, vol. 26, No.1, pp. 192–202. (2008)
63. Teng, Y., Zhang, Y., Dai, C., Yang, F., Song, M.: Dynamic spectrum sharing through double auction mechanism in CRN. In: IEEE Wireless Communications and Networking Conference (WCNC), pp. 90–95. Cancun (2011)
64. Mohammadian, H., Abolhassan, B.: Auction-based Spectrum Sharing for Multiple Primary and Secondary Users in CRNs, IEEE Sarnoff Symposium, pp. 1–6. Princeton (2010)
65. Kasbekar, G.S., Sarkar, S.: Spectrum auction framework for access allocation in CRNs, IEEE/ACM Transactions on Networking, vol. 18, No. 6, pp. 1841–1854. (2010)
66. Wang, X., Li, Z., Xu, P., Xu, Y., Gao, X., Chen, H.: Spectrum sharing in CRNs- An Auction based Approach. In: IEEE Transactions on Systems, Man, and Cybernetics, Part B (Cybernetics) - Special issue on game theory, vol. 40, No. 3, pp. 587–596. (2010)
67. Qian, L., Ye, F., Gao, L., Gan, X., Chu, T., Tian, X., Wang, X., Guizani, M.: Spectrum Trading in CRNs: An Agent-Based Model under Demand Uncertainty. In: IEEE Transactions on Comns, vol. 59, No. 11, pp. 3192–3203. (2011)
68. Yau, K.L.A., Komisarczuk, P., Paul, D.T.: Enhancing Network Performance in Distributed CRNs using Single Agent and Multi Agent Reinforcement Learning. In: IEEE Local Computer Networks (LCN), pp. 152–159, Denver (2010)
69. Usama, M., Boulahia, L.M., Gaïti, D.: Multiagent Based Spectrum Sharing Using Petri Nets. In: Springer, Trends in Prac. Appls of Agents and MASs, pp. 537–546. Berlin (2010)
70. Jia, J., Zhang, O, Zhang, Q., Liu, M.: Revenue generation for truthful spectrum auction in Dynamic Spectrum Access. In: Proceedings of the tenth ACM international symposium on Mobile ad-hoc networking and computing. ACM, pp. 3–12. New Orleans (2009)
71. Wu, C., Chowdhury, K., Felice, M.D., Meleis, W.: Spectrum management of CR using Multi Agent Reinforcement Learning. In: Proceedings of the 9th International Conference on Autonomous Agents and MASs, pp. 1705—1712. Toronto (2010)
72. Lunden, J., Koivunen, V., Kulkarni, S.R., Poor, H.V.: Reinforcement Learning based distributed multiagent sensing policy for CRNs. In: IEEE International Symposium on Dynamic Spectrum Access Network (DySPAN), pp. 642–646. Aachen (2011)
73. Chen, Z., Qiu, R.C.: Cooperative SS using Q-learning with experimental validation. In: IEEE Southeastcon, pp. 405–408. (2011)
74. Fu F., Schaar, M..: Dynamic Spectrum Sharing Using Learning for Delay-Sensitive Applications. In: IEEE International Conference on Comns, pp. 2825–2829. Beijing (2008)
75. Venkatraman, P., Hamdaoui, B., Guizani, M.: Opportunistic BW sharing through RL. In: IEEE Transactions on Vehicular Technology, vol. 59, No. 6, pp. 3148–3153. (2010)
76. Li, H.: Multiagent Q-Learning for Aloha-Like Spectrum Access in CR Systems. In: EURASIP, J. on Wireless Comns and Networking. Springer, Heidelberg (2010)

77. Li, H.: Multi-agent Q-learning of ch Selection in multi-user CR Systems: A two by two case. In: IEEE International Conference on System, Management & Cybernetics, pp. 1893–1898. San Antonio (2009)
78. Chen, B., Hoang, A.T., Liang, Y.C.: CR Ch Allocation Using Auction Mechanisms. In: Vehicular Technology Conference (VTC Spring), pp. 1564–1568. Singapore (2008)
79. Gafar, M., Elnourani, A.: Cognitive Radio and Game Theory: Overview and Simulation. In: Thesis work, Blekinge Institute of Technology, Karlskrona (2008)
80. Hossain, E., Niyato, D., Han, Z.: DSA & management in CRNs. In: Cambridge University Press, Cambridge (2009)
81. Serrano, A.G., Giupponi, L.: Distributed Q-learning for aggregated interference control in CR networks. In: IEEE Transactions on Vehicular Technology, pp. 1823–1834. (2010)
82. Amraoui, A., Benidriss, F.Z, Benmammar, B., Krief, F., Bendimerad, F.T.: Toward Cognitive Radio Resource Management Based on Multi-Agent Systems for Improvement of Real-Time Application Performance. In: Proceedings of the 5th International Conference on New Technologies, Mobility and Security (NTMS), pp 1–4. Istanbul (2012)
83. Bkassiny, M., Li, Y., Jayaweera, S.K.: A survey on machine-learning techniques in CRs. In: IEEE Communications Surveys & Tutorials, vol. 15, No. 3, pp. 1136–1159, (2013)
84. Mir, U., Merghem-Boulahia, L., Gaïti, D.: COMAS: A Cooperative Multiagent Architecture for Spectrum Sharing. In: Springer, EURASIP, J. on Wireless Communications and Networking, 15 pages. www.springer.com (2010)
85. Mir, U., Merghem-Boulahia, L., Esseghir, M., Gaïti, D.: "A Continuous Time Markov Model for Unlicensed Spectrum Access. In: IEEE 7th International Conference on Wireless and Mobile Computing, Networking and Communications, pp. 68–73. Wuhan (2011)
86. Zhao, Q., Qin, S., Wu, Z.: Self-Organize Network Architecture for Multi-Agent CR Systems. In: International Conference on Cyber-Enabled Distributed Computing and Knowledge Discovery, pp. 515–518, Beijing (2011)
87. Mir, U., Merghem-Boulahia, L., Gaïti, D.: Dynamic Spectrum Sharing in CRNs: a Solution based on MASs. In: International J. on Advances in Telecom, pp. 203–214 (2010)
88. Mir, U., Merghem-Boulahia, L., Esseghir, M., Gaïti, D.: Dynamic spectrum sharing for CRNs using multiagent system. In: Proceedings of IEEE Consumer Communications and Networking Conference, pp. 658–663. Las Vegas (2011)
89. A. Ahmed, O. Sohaib, W. Hussain: An Agent Based Architecture for Cognitive Spectrum Management. In: Australian J. of Basic and Applied Sciences, pp. 682–689. (2011)
90. Tan, Y., Sengupta, S., Subbalakshmi, K.P.,: Competitive spectrum trading in DSA markets: A price war. In: Proceedings of IEEE Global Telecommunications Conference GLOBECOM, pp. 1–5. Miami (2010)
91. Reddy, Y.B., Bullmaster, C.: Cross-Layer Design in Wireless Cognitive Networks. In: Ninth International Conference on Parallel and Distributed Computing, Applications and Technologies (PDCAT), pp. 462–467. Otago (2008)
92. Raiyn, J.: Toward Cognitive Radio handover management based on social agent technology for spectrum efficiency performance improvement of cellular systems. In: IEEE 19th International Symposium on Personal, Indoor and Mobile Radio Communications PIMRC, pp. 1–5. (2008)
93. Trigui, E., Esseghir, M., Merghem-Boulahia, L.: Spectrum Access during Cognitive Radio Mobiles' Handoff. In: 7th International Conference on Wireless and Mobile Communications (ICWMC), pp. 221–224. Luxembourg (2011)
94. Trigui, E., Esseghir, M., Merghem-Boulahia, L.: MAS based auction for channel selection in mobile cognitive radio networks. In: J. of ACSIJ, vol. 4, No.6. www.acsij.org (2015)

Mutual Information-Based Intrusion Detection System Using Multilayer Neural Network

V. Maheshwar Reddy, I. Ravi Prakash Reddy and K. Adi Narayana Reddy

Abstract Intrusion detection system detects the malicious activities on the network. The mutual information is used to select the features which are important to classify the user. In this paper, we propose simplified mutual information-based feature selection (SMIFS) algorithm to select the subset of features to train the multilayer neural network which classifies the new users as normal user or attacker. The accuracy of the proposed model is more than the existing proposals.

Keywords Intrusion · Detection · System · Mutual information · Neural networks · Feature selection

1 Introduction

The use of distributed network is increasing tremendously in e-commerce, banking, marketing, etc., due to the rapid growth in applications and technologies of network. At the same time, it also brings a lot of security issues. The security issues lead to the different vulnerabilities. The traditional firewalls only filter the packets based on rules. It cannot handle network-based attacks, so we require another layer of security. One such layer of security tool is provided by intrusion detection system (IDS) and is proposed by Anderson in 1980. Many researches proposed different detection and prevention systems using IDS. Such techniques include genetic algorithm, neural network, support vector machine, rule-based systems.

V. Maheshwar Reddy (✉) · K. Adi Narayana Reddy
ACE Engineering College, Telangana, India
e-mail: mahesh.vancha@gmail.com

K. Adi Narayana Reddy
e-mail: aadi.iitkgp@gmail.com

I. Ravi Prakash Reddy
G. Narayanamma Institute of Technology and Science, Telangana, India
e-mail: irpreddy@gnits.ac.in

© Springer Nature Singapore Pte Ltd. 2019   529
R. S. Bapi et al. (eds.), *First International Conference on Artificial Intelligence and Cognitive Computing* , Advances in Intelligent Systems and Computing 815,
https://doi.org/10.1007/978-981-13-1580-0_51

The main challenges of IDS are to capture, analyze, and classify the behavior of users on network efficiently. To classify the behavior of users on network signature-based IDS and anomaly-based detection systems are used. The signature-based technique tests both network and system activity for known instances of misuse using signature matching algorithms. This method detects known attacks efficiently but fails for unknown attacks, and it generates false negative alarm. This false alarms waste the time and computational resources, and it leads to inability of the system. Anomaly-based detection systems construct the user behavior using statistical or machine learning algorithms to examine network traffic or system calls and processes. But, normal behavior is not well defined in a large and dynamic system.

In addition to detection, it was observed that the data is collected from various applications and places in different ways. The size of the collected data is very large, and the number of features is also high. The data may be noisy. The selection of features from high-dimensional data is one of the main objectives of preprocessing. The feature extraction improves the effectiveness and accuracy of the system. In this paper, we propose a two-level model to detect the intrusions. In first level, we select optimal relevant features using simplified mutual information-based feature selection (SMIFS), and in the second level, these extracted features are used to train the model using multilayer neural network.

The paper is organized as: The related work is presented in Sects. 2 and 3 and describes the proposed work; the results are presented in Sects. 4 and 5 and conclude the paper.

2   Related Work

The intrusion detection system proposals were implemented on standard dataset KDD CUP 99. The number of features in this dataset is 41. Many researchers proposed feature selection methods to reduce the dimensionality of the dataset.

2.1   Feature Selection

Feature selection eliminates the irrelevant and redundant features from feature set and selects the most optimal subset of features that produce more accuracy to classify different classes of attacks.

The feature selection techniques are classified as filter-based and wrapper-based techniques [12]. The filter-based technique utilizes the independent measures like mutual information, information gain, distance measure, or consistency measures as a criterion for estimating the relation between the features. The wrapper methods make use of a particular learning algorithm to measure the importance of each feature. In this paper, we focus on filter-based technique mutual information to extract the features.

Saurabh Mukherjee and Neelam Sharma [9] proposed feature vitality-based reduction method (FVBRM) to reduce the number of features in standard KDD'99 dataset. It is compared of correlation-based feature selection (CFS), information gain (IG), and gain ratio (GR). It is trained on Naïve Bayes classifier and produced 97.78 percentage of accuracy but U2R true positive rate is very low.

2.2 Mutual Information

The concept of mutual information is introduced by Shannon in 1948 in his information theory. It describes the symmetric relationship between two different random features, and produces a zero if mutual information (MI) between observed random variables are independent and a nonzero if the variables are dependent.

Mutual information (MI) is one of the statistical measures to find the relationship between two random variables. For given two continuous random variables $X = \{x_1, x_2, \ldots, x_n\}$ and $Y = \{y_1, y_2, \ldots, y_n\}$ with n sample values, the MI between X and Y is presented in Eq. (1).

$$I(X, Y) = E(X) + E(Y) - E(X, Y) = E(X) - E(X/Y) = E(Y) - E(Y/X) \quad (1)$$

where $E(X)$ and $E(Y)$ are entropies of X, Y. These information entropies are calculated as $E(X) = \int_x p(x) \log p(x) \mathrm{d}x$ and $E(Y) = \int_y p(y) \log p(y) \mathrm{d}x$ and the joint entropy of X and Y is defined as

$$E(X, Y) = \int_{x,y} p(x, y) \log p(x, y) \mathrm{d}x \, \mathrm{d}y$$

The mutual information between two continuous random variables calculates the amount of knowledge on a feature X given by feature Y (and vice versa) that is presented in Eq. (2).

$$MI(X, Y) = \int_x \int_y p(x, y) \log \frac{p(x, y)}{p(x).p(y)} \mathrm{d}x \, \mathrm{d}y \quad (2)$$

where $p(x)$ and $p(y)$ are marginal density functions, and $p(x, y)$ is a joint probability density function.

Mutual information between two discrete random variables with marginal probability $P(a)$ and joint probability mass function $P(x, y)$ is defined by replacing the integration notation with the summation notation as shown in (3).

$$MI(X, Y) = \sum_{x \varepsilon X} \sum_{y \varepsilon Y} p(x, y). \log \frac{p(x, y)}{p(x).p(y)} \quad (3)$$

As stated above, a feature is relevant to the class if it contains related information about the class; otherwise, it is irrelevant. The feature selection algorithms [3–7] proposed based on mutual information. Battitis [4] proposed the first feature selection algorithm based on mutual information MIFS. It extracts the features based on their relevance to class. In MIFS, first feature is selected based on the maximum amount of mutual information between target class and the feature. To extract next features, a greedy algorithm is applied. Here β is a random value between 0 and 1.

$$G_{\mathrm{MI}} = \max\left\{ I(c, f_i) - \beta \sum_{f_i \varepsilon F, s \varepsilon S} I(f_i, f_s) \right\} \tag{4}$$

Kwak and Choi [5] come up with another better estimation of MI between input features and output classes and proposed a greedy selection algorithm named MIFS-U, in which U stands for uniform information distribution. A change made in MIFS Eq. (4) to obtain MIFS-U. The method of MIFS-U differs from that of MIFS in the right-hand side term in Eq. (4) as shown in Eq. (5).

$$G_{\mathrm{MI}} = \max\left\{ I(c, f_i) - \beta \sum_{f_s} \frac{I(C, f_s)}{E(f_s)} I(f_i, f_s) \right\} \tag{5}$$

In 2005, another estimation of MIFS is proposed by Long and Ding [6] called min Redundancy—Max Relevance (mRMR) by replacing Eq. (4) with proposed Eq. (6)

$$G_{\mathrm{MI}} = \max\left\{ I(C, f_i) - \frac{1}{|S|} \sum_{f_s \varepsilon S,} I(f_i, f_s) \right\} \tag{6}$$

In 2009, Pablo et al. [7] proposed normalized mutual information feature selection (NMIFS) and are presented in Eq. (4).

$$G_{\mathrm{MI}} = \max\left\{ I(C, f_i) - \frac{1}{|S|} \sum_{f_s \varepsilon S, \, f_i \varepsilon F} \frac{I(f_i, f_s)}{\min\{E(f_i), E(f_S)\}} \right\} \tag{7}$$

In 2011, MMIFS [8] was proposed to extract the feature based on mutual information. In this, a variation to Eq. (4) is:

- Compute the $I(\mathrm{SU} f_i, C)$
- Choose f_i such which maximizes the value in step-a.

In 2014, Ambusaidi et al. [1] proposed flexible mutual information-based feature selection (FMIFS) and are presented in Eq. (8).

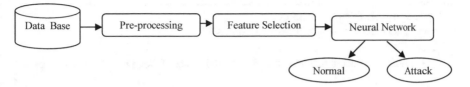

Fig. 1 Framework for the proposed model

$$G_{\mathrm{MI}} = \max\left\{ I(C, f_i) - \frac{1}{|S|} \sum_{f_s \,\varepsilon\, S,\ f_i \varepsilon F} \frac{I(f_i, f_s)}{I(C, f_i)} \right\} \tag{8}$$

In 2016, Hoque et al. [14] proposed a fuzzy mutual information-based feature selection (FMIFS-ND) to extract the features from a dataset. Some other [2, 10, 11, 13, 15] feature selection methods are also proposed based on filter-based methods to extract the features in different areas.

We proposed a new feature selection simplified mutual information-based feature selection (SMIFS).

3 Proposed Model

The framework of the proposed mutual information-based multilayer neural network for IDS is shown in Fig. 1. The model consists of four phases, and in the first phase the data is collected, in the second phase the preprocessing is done, in the third phase the relevant features are selected by the proposed SMIFS algorithm, and in the final phase the multilayer neural network classification the trained on the selected features.

3.1 Data Collection

The type of data source and the place where the data is collected are two main factors in the construction and effectiveness of the IDS. In this proposed model, we have used KDD CUP 99 dataset for several experiments.

3.2 Data Preprocessing

Due to the collection of data from heterogeneous sources, data may contain noisy, irrelevant, and not in standard form to build the model. Such data must be converted into standard form.

Input: A dataset D with n features.

Output: Selected subset of Informative features SF.

begin

    1. Initialize selected feature subset SF= Φ and let set of features of a given dataset
        F= $\{f_1, f_2, ..., f_n\}$.

    2. For each feature $f_i \varepsilon$ F and target class label C, calculate MI(C, fi).

    3. Select the feature f_i which is a max (MI(C, f_i . Then, Set F=F- $\{f_i\}$ and
        SF= $\{f_i\}$.

    4. Repeat the until |SF|=k(a Greedy Selection)

        a. calculate MI(f_i, f_s) for each $f_i \varepsilon$ F and recently selected feature $f_s \varepsilon$ SF.

        b. Evaluate C_{MI} = max $\{$ MI(C, f_i) - MI(f_i, f_s) $\}$ (9)

        c. Select the feature f_i which gives C_{MI}. Set F= F- $\{f_i\}$ and S=S U $\{f_i\}$

    5. Output the subset of k- features containing selected features -SF

end.

Fig. 2 Simplified mutual information-based feature selection

3.2.1 Data Transforming

The input data for the model may contain different types of values—integer, numeric, symbolic, and so on. The KDD cup 99 dataset contains integer, numeric (serror-rate, rerror-rate, Dst-host-count, etc.), and symbolic (service type, flag, protocol) features. The data symbolic and numeric data are converted into integer data to calculate the mutual information.

3.3 Simplified Mutual Information-Based Feature Selection (SMIFS)

It is the heart of the constructed model. The number of features affects the performance of the neural network. To train the multilayer neural network on high-dimensional data, it consumes more time and also requires high computational resources. It also reduces the intrusion detection accuracy. Dimensionality reduction is one of the main tasks in building any type classification model. For this, we proposed a filter-based mutual information to select the optimized informative features. In the SMIFS, the first feature is extracted from the given set of features which gives $\max\{MI(C, f_i)\}$. The next feature is selected based on a greedy approach, and this procedure is continued till the threshold is met. The detailed SMIFS is presented in Fig. 2.

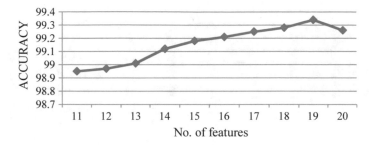

Fig. 3 Accuracy the proposed model with different number of features

3.4 Multilayer Neural Network

Multilayer neural network is a supervised classification algorithm. It reads the given dataset with target class labels, analyses the data instances, and constructs an optimal network to classify the data either attack or normal. Neural network model is constructed with selected features to classify data instances into attack or normal. For this, we have used tenfold cross-validation method. We have built the model with several experiments with different dimensions of the dataset, starting from 11 features to 22 features to evaluate the detection rate by extracting the features with proposed SMIFS, and it is observed that the neural network model performance is increased as a number of features are increased. With 19 features, it reaches maximum detection rate from and then the performance is decreased. So, we concluded that the number of best and important features for the model is 19.

4 Experimental Results

The proposed model is implemented in Intel I3 processor with 8 GB RAM using R tool. Currently, there are only few standard datasets available for IDS. From these KDD cup 99, NSL—KDD dataset and Kyoto 2015 have been commonly used to evaluate the performance of the constructed models. According to the literature, it was observed that the majority of the experiments are done on the KDD—CUP 99 dataset. In our experiments, we have used complete KDD CUP 99 dataset. The number of features extracted using mutual information is tested with multilayer neural network model. The accuracy of the model is increasing as the number of selected feature by SMIFS is increasing. After 19 features, the accuracy is decreasing as so the maximum number of selected features is 19 (Figs. 3 and 4).

The results of proposed model SMIFS + NN are compared with other existing models. The existing techniques used various feature selection methods such as principal component analysis (PCA), genetic algorithm (GA), mutual information (MI) to extract the features on different datasets with various sizes. The selected features are

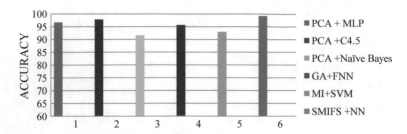

Fig. 4 Performance comparison with other models

then classified with multilayer perceptron (MLP), Naïve Bayes, feedforward neural network (FNN), support vector machine (SVM), RBF-based neural network, simple neural network. The existing proposals selected more features, whereas our proposal SMIFS + NN selected less number of features but the accuracy is comparatively equal.

5 Conclusion

The proposed intrusion detection system is developed using mutual information and multilayer neural network. The simplified mutual information-based feature selection algorithm selects the next features by considering only the recently selected feature. The multilayer neural network is trained on the selected features, and it produced 99.34 percentage of accuracy when the selected feature count is reached 19. It is also reduced the computational resources and time to train the model. Previous work shows that some more investigations are required in mutual information to find optimum value of number of selected features.

References

1. Mohammed A. Ambusaidi, Priyadarsi nanda: "Building an intrusion detection system using a filter-based feature selection algorithm"- IEEE transactions on Computers (2014).
2. N. Hoque, D.K. Bhattacharyya, J. K. Kalita: "MIFS-ND: A mutual information based feature selection" - Expert systems with Applications-Volume 64, Issue 14, (2014).
3. Fatemeh Amiri, Mohammad Rezael Yousefi, Caro Lucus, Azadeh Shakery, Nasser Yazdani: Mutual information based feature selection for intrusion detection systems - Journal of Network and Computer applications- Elsevier (2011).
4. Roberto Battati: Using mutual information for selecting features in supervised neural net learning -IEEE transactions on neural network volume 5(1994).
5. N. Kwak and C. H. Choi: Input feature selection for Classification problems- IEEE Transactions of Neural Networks, Vol 13(1)-(2001).

6. H. Peng, F. Long and C. Ding- Feature selection based on mutual information criteria of max-dependency, max-relevance and min-redundancy- IEEE transaction on pattern analysis and Machine intelligence-vol 27(2005).
7. P. A. Estevez, M. Tesmer, Claudio A. Perez and J. M. Zurada-Normalized Mutual Information Feature Selection- IEEE transactions on Neural Networks, vol 20(2)- (2009).
8. F. Amiri, d. Rezaei Yousefi, Caro Lucus, A. Shekery ans N. Yazdani- Mutual Information based feature selection for Intrusion detection systems - Journal of Network and Computer Application 34 (2011).
9. Dr. Saurabh Mukherjee, Neelam Sharma: Intrusion Detection using Naive Bayes Classifier with Feature Reduction -Procidia Technology-Vol-4 119–128: Elsevier(2012).
10. Fleuret F- Fast binary feature selection with conditional mutual information - Journal of Machine learning research 2004: 1531-55.
11. S. Cang, H. Yu, -Mutual information based feature selection for classification problems- Decision support systems-54(1), (2012) 691–698.
12. F. Amiri, M. Rezei Yousefi, C. Lucas, A. Shekery, N. Yazdani: Mutual Information -based feature selection for intrusion detection systems, Journal of Computer Applications 34 (2011) 1184–1199.
13. Ambusaidi, Mohammed A.; He, Xiangjian; Nanda, Priyadarsi; Tan, Zhiyuan: Building an intrusion detection system using a filter-based feature selection algorithm- IEEE transactions on Computers (2016).
14. N. Hoque, A. Ahmed, K. Bhattacharyya, K. Kalita: A Fuzzy Mutual Information- based Feature Selection Method for Classification - Fuzzy Information and Engineering- Elsevier (2016).
15. Mohammed A. Ambusaidi: Using Mutual Information for Feature Selection in a Network Intrusion Detection System-Proceedings of the Third International Conference on Digital Security and Forensics (DigitalSec), Kuala Lumpur, Malaysia, 2016.

Improving Student Academic Performance Using an Attribute Selection Algorithm

K. Anvesh, M. Srilatha, T. Raghunadha Reddy, M. Gopi Chand
and G. Divya Jyothi

Abstract The aim of this paper is to predict the student academic performance using data mining techniques. Data mining is the process of extracting data and prediction. Prediction of student academic performance helps the students to take decision about their progress in academics. It can help not only the current students but also the future students to take decision. In this work, kaggle student dataset was used to predict the student performance. Different properties of the collected data were investigated and developed a classification hypothesis in order to apply data mining algorithms. A new attribute selection algorithm was proposed to identify the best attributes to increase the efficiency of student performance prediction. In this work, a machine learning tool called WEKA tool develop by the University of New Zealand was used for testing different algorithms on the data. The experimental results are validated against test data, and interesting co-relations are observed. The experimentation carried out with Naïve Bayes, IBK, bagging, simple logistic, logistic, and random forest classifiers. The prediction results of student academic performance promising than most of the research work happened in educational data mining.

Keywords Student academic performance · Attribute selection algorithm · Naïve Bayes algorithm · Random forest algorithm · WEKA tool · Classification model

K. Anvesh (✉) · T. Raghunadha Reddy · M. Gopi Chand
Department of IT, Vardhaman College of Engineering, Hyderabad, India
e-mail: anveshitse@gmail.com

T. Raghunadha Reddy
e-mail: raghu.sas@gmail.com

M. Gopi Chand
e-mail: gopi_merugu@yahoo.com

M. Srilatha
Department of CSE, VR Siddhartha Engineering College, Vijayawada, India
e-mail: srilatha.manam@gmail.com

G. D. Jyothi
Department of CSE, MLR Institute of Technology, Hyderabad, Telangana, India
e-mail: divyag.1605@gmail.com

© Springer Nature Singapore Pte Ltd. 2019   539
R. S. Bapi et al. (eds.), *First International Conference on Artificial Intelligence and Cognitive Computing* , Advances in Intelligent Systems and Computing 815,
https://doi.org/10.1007/978-981-13-1580-0_52

1 Introduction

The data mining techniques are used in different research domains to extract hidden patterns in a database. In engineering educational domain, several researchers used various data mining techniques to improve the student academic performance, to increase the placements of students in companies, to improve the percentages of students and in many more applications. In this work, we concentrated on predicting the student academic performance by using data mining techniques.

The student database contains a set of student records which is a combination of different attributes values. In general, the data mining techniques are used to generate a model by using existing known data and this model is used to predict the class label of a new anonymous record. The classification algorithms in data mining used all the attributes that were described the student record to prepare classification model. In general, some of the attributes are not having discriminative power to predict the class label of new record. In this work, a new algorithm is proposed to find the best attribute to improve the prediction accuracy power of the classifiers.

This paper is organized in six sections. The literature survey of educational data mining is explained in Sect. 2. The characteristics of dataset, classification procedure, and evaluation measures were described in Sect. 3. The proposed attribute selection algorithm was elaborated in Sect. 4. The experimental results of attribute selection algorithm were presented in Sect. 5. Section 6 concludes this paper.

2 Literature Review

Getaneh Berie Tarekegn et al. applied [1] data mining techniques to predict the performance of students in placements of different departments. They experimented on student database which contains the entrance examination results. Three algorithms such as random forest, J48, and Naïve Bayes were used for generating the prediction model. It was observed that the random forest algorithms performed well for predicting whether students will be placed or not when compared with other two algorithms.

In [2], they predicted the student achievements in academics using neural networks and decision trees. They collected the dataset of 1600 students with 22 features joined in a university in Thailand between 2001 and 2011 years. They observed that decision tree classifier obtained good accuracy than neural networks. The attributes in the database are student ID, gender, age, status of student (single, married, and divorced), educational background, continent, qualification, father occupation, mother occupation, scholarship status, on-campus residence, department, pre-university English, nativity language, grade point average, cumulative grade point average, number of credits, highest cumulative grade point average of subject, extracurricular subject, extra study hour, number of work hours, number of activity hours. It was also observed that the student groups who had good English proficiency skills obtained good performance in academics.

Khan [3] used educational data mining techniques for prediction of scholarship of students. They analyzed student data with ID3 and J48 decision tree algorithms for predicting winning chances of scholarship by converting decision tree into IF-THEN rules. The features—the rank or position in a class, grades in a semester, minimum and maximum number of credit hours taken, achievements and extracurricular activities—were used to predict the scholarship winning chances. It was observed that ID3 achieved good accuracy when compared with J48 even though J48 generates a smaller tree and faster in classification than ID3.

Agrawal [4] experimented with four classification algorithms such as random forest, J48, Naïve Bayes, and multilayer perception and seven attribute evaluation techniques such as Filtered Attribute Eval, Chi-Squared Attribute Eval, Info Gain Attribute Eval, Gain Ratio Attribute Eval, OneR Attribute Eval, Relief Attribute Eval, and Symmetric Uncert Attribute Eval.

It was observed that the J48 algorithm with no attribute evaluator method performance was good when compared the performance of J48 alone. It was also observed that the random forest algorithm perform well when combined with Relief Attribute Eval, Naïve Bayes algorithm performance is good when combined with Filtered Attribute Eval, Symmetric Uncert Attribute Eval, and finally the combination of multilayer perception algorithm and Chi-Squared Attribute Eval obtained good performance for prediction of student performance in exams.

Siri [5] predicts the dropouts of students at University of Genoa in the academic year 2008–09 using artificial neural networks. They used the database of 810 students registered for healthcare professions degree and collected the information through telephone conversation with students who were not registered for successive years, administrative information relevant to student careers and statistical data collected through ad hoc survey.

Ahmad [6] used the student database of July 2006 to July 2014 which contains the student academic records, demographics information, and information related to family background. They experimented with Naïve Bayes, decision tree, and rule-based classification algorithms to predict the student performance. It was found that the rule-based classifier obtained the highest accuracy of 71.3% for predicting student academic performance among three classifiers.

Koutina [7] used machine learning techniques to predict the performance of postgraduate students. They used the postgraduate students database of Ionian University Informatics. They experimented with six classifiers such as J48 decision tree classifier, K-nearest neighbor classifiers (1-NN, 3-NN, 5-NN), Naïve Bayes, JRIP, random forest, and support vector machines. It was observed that Naïve Bayes and 1-NN classifiers along with feature selection algorithm obtained good accuracy results for predicting student performance.

Table 1 Characteristics of student dataset

Feature	Number of students
Total students	480
Male	305
Female	175
From Kuwait	179
From Jordan	172
From Palestine	28
From Iraq	22
From Lebanon	17
From Tunis	12
From Saudi Arabia	11
From Egypt	9
From Syria	7
From USA, Iran, and Libya	6
From Morocco	4
From Venezuela	1
Number of students exceeded 7 absent days	191
Number of students absence under 7 days	289

3 Corpus Characteristics and Evaluation Measures

3.1 Corpus Characteristics

The student dataset consists of 480 student records with 16 attributes [8]. The attributes were categorized into three classes such as demographic attributes, academic background attributes, and behavioral attributes. The attributes description is represented in Table 2. The dataset contains 175 females and 305 males. The dataset includes two educational semesters' student data wherein 245 student records from first semester and 235 records from second semester. The characteristics of student dataset are represented in Table 1.

3.2 Evaluation Measures

Most of the approaches for educational data mining used recall, precision, F1-score, and accuracy measure to evaluate the efficiency of their model for predicting the student academic performance. In this work, the accuracy measure is used to test the efficiency of our approach for student academic performance prediction. The accuracy is the ratio of the number of student records correctly predicted their performance to total number of student records in the test corpus.

3.3 Classification Procedure

Student performance prediction is a complex problem, and few probabilistic classifiers were used to predict the performance of a student by analyzing the dataset of students. In order to implement our solution, we first need to construct a corpus. In order to do this, we need to collect data from various sources. For our work, we choose kaggle dataset. In this work, tenfold cross-validation was used to validate our classifiers. Finally, we experiment on and compare our classifiers with various parameters and arguments.

Classification is a problem of identifying which category a new input belongs in. An algorithm that implements a classification is called a classifier. There are many classification algorithms like Naïve Bayes multinomial, random forest, decision trees, bagging, support vector machines, and neural networks used for classification. In this work, the experimentation carried out with Naïve Bayes multinomial, simple logistic, logistic, IBK, bagging, and random forest classifiers (Table 2).

Classifiers generally follow two steps. First, a model is created by using training data using an algorithm. Second, test data was used to test the accuracy of model. In testing part, we test our model with the data that we already have a known category or group. Success of the classifier depends on the type of the data to be classified. Thus, it is very important to examine the characteristics of the data and understand the details of it. Classifiers usually need to be modified and specialized according to data.

In this work, K-fold cross-validation is used to evaluate the classifiers performance. K-fold cross-validation implemented with K iterations. On every iteration, K-1 sample was used for training and one sample is used for testing. This procedure is repeated K number of times. With this method, we ensure that all data is used for both training and testing. In our work, we used tenfold cross-validation.

As mentioned above, the classification method used various features to predict the student performance. Features can be numeric or nominal. We search for effective features that will predict the student performance effectively. As we discussed above, we used tenfold cross-validation. For each fold of the validation, we construct the training vectors and test vectors. We pass the training vectors to classifier, and the classifier will create a model by iterating all the training vectors. This classifier model will be an input to the tester as well as testing vectors. The tester class will test the testing vector with the model and find a success result. The average of all folds will be the final success of our classification methodology.

Table 2 Description of attributes in a student dataset

S. No.	Attribute	Type of attribute	Description	Possible values
1	Gender	Nominal	Student's gender	'Male' or 'Female'
2	Nationality	Nominal	Student's nationality	'Kuwait', 'Lebanon', 'Egypt', 'Saudi Arabia', 'USA', 'Jordan', 'Venezuela', 'Iran', 'Tunis', 'Morocco', 'Syria', 'Palestine', 'Iraq', 'Libya'
3	Place of birth	Nominal	Student's place of birth	'Kuwait', 'Lebanon', 'Egypt', 'Saudi Arabia', 'USA', 'Jordan', 'Venezuela', 'Iran', 'Tunis', 'Morocco', 'Syria', 'Palestine', 'Iraq', 'Libya'
4	Educational stages	Nominal	Educational level student belongs	'Lower level', 'Middle School', 'High School'
5	Grade levels	Nominal	Grade student belongs	'G-01', 'G-02', 'G-03', 'G-04', 'G-05', 'G-06', 'G-07', 'G-08', 'G-09', 'G-10', 'G-11', 'G-12',
6	Section ID	Nominal	Classroom student belongs	'A', 'B', 'C'
7	Topic	Nominal	Course topic	'English', 'Spanish', 'French', 'Arabic', 'IT', 'Math', 'Chemistry', 'Biology', 'Science', 'History', 'Quran', 'Geology'
8	Semester	Nominal	School year semester	'First', 'Second'

(continued)

Table 2 (continued)

S. No.	Attribute	Type of attribute	Description	Possible values
9	Parent responsible for student	Nominal	Parent responsible for student	'Mom', 'father'
10	Raised hand	Numeric	How many times the student raises his/her hand on classroom	0–100
11	Visited resources	Numeric	How many times the student visits a course content	0–100
12	Viewing announcements	Numeric	How many times the student checks the new announcements	0–100
13	Discussion groups	Numeric	How many times the student participate on discussion groups	0–100
14	Parent answering survey	Nominal	Parent answered the surveys which are provided from school or not	'Yes', 'No'
15	Parent school satisfaction	Nominal	The degree of parent satisfaction from school	'Yes', 'No'
16	Student absence days	Nominal	The number of absence days for each student	above-7, under-7

4 Attribute Selection Algorithm

In this work, a new attribute selection algorithm was proposed to identify the best attributes for classification model generation. This algorithm assigns weights to the attributes based on the values of the attributes. The attribute selection algorithm as follows.

$C1$ is a first category class, $\{I_{11}, I_{21}, ..., I_{x1}\}$ is a set of Instances in a first category class

$C2$ is second category class, $\{I_{12}, I_{22}, ..., I_{y2}\}$ is a set of instances in second category class

$C3$ is third category class, $\{I_{13}, I_{23}, ..., I_{z3}\}$ is a set of instances in third category class

D is a dataset, $\{I_1, I_2, ..., I_n\}$ is a set of instances in a dataset, where $n = x + y + z$

Instance is a representation of student information with features $\{A_1, A_2, ..., A_m\}$ is set of attributes in a student dataset initialize weight $[A]=0$ for every attribute in a student dataset for $p=1$ to n do

select an instance I_{ran} randomly from dataset D

find the nearest instance in first category class (I_{c1}), nearest instance in second category class (I_{c2}), and nearest instance in third category class (I_{c3}) for selected instance (I_{ran}) using the following distance measure equation

$$\text{Dist}(I_a, I_b) \frac{m}{j1} \left| A_{Iaj} A_{Ibj} \right|$$

for each attribute A_i, calculate the weight of the attribute

$$W(A_i) \frac{n}{i1} \text{Dist}(A_i, I_{ran}, I_{c1}) \ \text{Dist}(A_i, I_{ran}, I_{c2}) \ \text{Dist}(A_i, I_{ran}, I_{c3})$$

end for
end for
To normalize the attribute weight values for each attribute, A_i do

$$W[A_i] = W[A_i]/n$$

end for
For categorical attributes, the distance between attribute values

dist(A_i, I_a, I_b)	= 0,	if	$I_a[A_i] = I_b[A_i]$
	= 1,	if	$I_b[A_i] \neq I_b[Ai]$

The positive weight value of an attribute indicates that the attribute is relevant for prediction.

The negative weight value of an attribute indicates that the attribute is irrelevant for prediction.

The next section shows the experimental weights of 16 attributes that were used in kaggle dataset.

4.1 Experimental Results of ASA

The weighted results of ASA were presented in Table 3. The ASA assigns positive weights to 11 attributes and negative weights to 6 attributes.

Table 3 Weights of attribute values

0.0171246	A11	0.0051206	A5	−0.0001304	A7
0.0091525	A10	0.004868	A9	−0.0002346	A8
0.006378	A13	0.0043249	A14	−0.0007688	A3
0.005766	A16	0.004112	A15	−0.000772	A2
0.0045896	A4	0.0040716	A1	−0.0007895	A6
0.005287	A12				

Table 4 Accuracies of student academic performance prediction

S. No.	Classifiers	Accuracy of student performance prediction	
		16 attributes	11 attributes
1	Naïve Bayes multinomial	78.02	82.65
2	Simple logistic	76.34	81.71
3	Logistic	74.84	80.54
4	IBK	79.46	82.21
5	Bagging	80.57	85.32
6	Random forest	83.13	88.89

5 Experimental Results

The experimentation is performed with six classifiers and with 16 attributes and 11 attributes. The accuracies of student performance prediction are high when experimented with 11 attributes compared to 16 attributes. The random forest classifier obtained good accuracy for student performance prediction (Table 4).

6 Conclusions

In this work, the experimentation carried out with a new attribute selection algorithm to identify the best attributes for generating classification model. The random forest classifier obtained 88.39% accuracy for predicting student performance by using the attributes identified by the attribute selection algorithm.

In the future, further rigorous study to match between demographic data and academic data will lead to much determining factors in order to predict the student performance.

References

1. Getaneh Berie Tarekegn, Vuda Sreenivasarao, " Application of Data Mining Techniques to Predict Students Placement into Departments ", in International Journal of Research Studies in Computer Science and Engineering (IJRSCSE), Volume 3, Issue 2, 2016, PP 10–14.
2. Pimpa Cheewaprakobkit, " Predicting Student Academic Achievement by Using the Decision Tree and Neural Network Techniques", in catalyst, Volume 12, No. 2, 2015, pp 34–43.
3. Irfan Ajmal Khan and Jin Tak Choi, "An Application of Educational Data Mining (EDM) Technique for Scholarship Prediction", in International Journal of Software Engineering and Its Applications, Vol. 8, No. 12 (2014), pp. 31–42.
4. Richa Shambhulal Agrawal, 2Mitula H. Pandya, "Data Mining With Neural Networks to Predict Students Academic Achievements", in International Journal of Computer Science And Technology, Vol. 7, Issue 2, April - June 2016, PP. 100–103.
5. Anna Siri, "Predicting Students' Dropout at University Using Artificial Neural Networks", in ITALIAN JOURNAL OF SOCIOLOGY OF EDUCATION, 7 (2), 2015, PP 224–247.
6. Fadhilah Ahmad*, Nur Hafieza Ismail and Azwa Abdul Aziz, "The Prediction of Students' Academic Performance Using Classification Data Mining Techniques", Applied Mathematical Sciences, Vol. 9, 2015, no. 129, 6415–6426.
7. Maria Koutina, Katia Lida Kermanidis, "Predicting Postgraduate Students' Performance Using Machine Learning Techniques", Artificial Intelligence Applications and Innovations, pp 159–168.
8. https://www.kaggle.com/aljarah/xAPI-Edu-Data.

Brain Tumor Detection in MR Imaging Using DW-MTM Filter and Region-Growing Segmentation Approach

Bobbillapati Suneetha and A. Jhansi Rani

Abstract Brain tumor analysis is most challenging and emerging exploration area in medical image processing. For appropriate regimen of brain tumor, early detection and scrutiny are essential. To provide better detection of tumor without affecting a normal tissue is a very difficult process. So as to amend the downsides, we propose another novel technique for brain tumor detection through magnetic resonance imaging (MRI). It is a commonly processed method for providing high-quality imaging. It provides higher details about the soft tissue of human anatomy. In this proposed method, MR image is preprocessed by optimized kernel possibilistic C-means algorithm. Then, the image is enhanced by adaptive DW-MTM filter. It helps to neglect the unwanted noise from the MRI. Finally, the image is segmented by region-growing algorithm. The segmented image is utilized for the detection and diagnoses of brain tumor in earlier stage.

Keywords MR imaging · Optimized kernel possibilistic C-means algorithm ·
Adaptive double window modified trimmed mean filter
Region-growing algorithm

1 Introduction

Brain tumor is an irregular augmentation of cells esoteric the skull. Regularly, the tumor will expand since the cells of the brain, blood nerves, and vessels ascent up out of the brain. Benign (non-cancerous) and malignant (cancerous) are the two sorts of tumors [1]. By producing irritation, applying stress on the parts of brain and expanding heaviness inside the skull tumors can destruct the ordinary brain cells.

B. Suneetha (✉)
ECE Department, Acharya Nagarajuna University, Guntur, Andhra Pradesh, India
e-mail: suneetha.bobbillapati@gmail.com

A. Jhansi Rani
ECE Department, V. R. Siddhartha Engineering College, Vijayawada, Andhra Pradesh, India
e-mail: jhansi9rani@gmail.com

© Springer Nature Singapore Pte Ltd. 2019   549
R. S. Bapi et al. (eds.), *First International Conference on Artificial Intelligence and Cognitive Computing* , Advances in Intelligent Systems and Computing 815,
https://doi.org/10.1007/978-981-13-1580-0_53

Three normal sorts of tumors are: (1) benign; (2) premalignant; and (3) malignant (cancer must be malignant). A benign tumor is a tumor that does not augment in an unrefined way. Benign tumor does not ready its neighboring ordinary tissues and additionally does not expand to non-nearby tissues. Moles are the trifling case of benign tumors. Precancerous phase of tumor is otherwise called premalignant tumor. It is treated as an infection. Premalignant tumor may prompt cancer on the off chance that it is not appropriately handled. Malignant tumor is the sort of tumor that deteriorates with the advancement of time and at last outcomes in the passing of a man. In therapeutic term, malignant is for the most part used to show a genuine continuing ailment. The depiction of cancer is ordinarily expressed in terms of malignant tumor [2]. MRI is the comprehensively utilized method for tumor determination. Tumor discovery and grouping are exceptionally costly. To recognize the tissues and the sickness of brain cancer, MRI is a development strategy. X-ray contributes the diverse data about the unmistakable structures of the body. Contrasted with X-ray, computed tomography (CT) outputs and ultrasound MRI are the best systems for giving higher-quality images. X-ray innovation has a magnetic field and prepares beats of radio wave vitality. It is utilized to catch structures and organs inside a body. Besides, physically the amount of the resultant information is so quite investigated [3]. To accomplish noise diminishment and image enhancement, more strategies are utilized to enhance the image quality. To decide the tumor in the image, some morphological operations are executed. In view of a few suppositions, the morphological operations are appointed. The size and state of the tumor are figured from the image. At that point, tumor is mapped on to the first grayscale picture with two 55 intensities to make obvious the tumor in the picture. [4]. Introducing displaying stage, more algorithms are used in image-based tumor discovery. The algorithms are utilized to distinguish edges and to recognize shapes while others can identify that different elements have been to a great degree dynamic exploration region in the shrewd medicinal group [5].

The particle swarm optimization (PSO) should not be executed in the arena of brain image exploration. A decent survey of this optimization procedure is concentrated on and watched that PSO procedures can be employed in brain elucidation. [6] These strategies lessen administrator communication via mechanizing a few parts of applying the low-level operations, for example, limit choice, histogram examination, grouping [7]. To beat the disadvantages, the proposed approach uses the optimized kernel possibilistic C-means algorithm for successful pre-preparing. At that point, feature enhancement is finished by adaptive DW-MTM filter. At last, the segmentation is ended by region-growing algorithm. Region-based strategies search out clusters of pixels that piece some extent of comparability. In Sect. 2, related works are described. Proposed work is explained in detail in Sect. 3. Section 4 incorporates with experimental results. Finally, Sect. 5 declares the conclusion part.

2 Related Work

Handling a picture is a convoluted assignment. Prior to any picture prepared, it is critical to evacuate any undesirable relics it might hold. At exactly that point can the picture be prepared effectively. Handling a medicinal picture includes two principle steps. The first is the pre-handling of the picture. This includes performing operations like noise decrease and separating so the picture is appropriate for the subsequent stage. The second step is to perform division and morphological operations. These decide the size and the area of the tumor. channel and Center Weighted Median channel. The execution of above channels is measured and evaluated. At last, the best channel of focus weighted center channel is perceived and utilized for MR cerebrum picture change. It utilized MRI. Venugopala Krishnan et al. implemented a filtering algorithm for MRI images. Noise is one of the snags in programmed image understanding, and noise decreasing is essential to enhance the after effects of this procedure [8]. X-ray images when caught as a rule have Gaussian noise and salt-and-pepper noise. To expel this noise, filtering algorithms are presented. The outcomes are investigated and assessed.

3 Proposed System

In the proposed system, brain tumor detection is carried out by the following steps. Initial step is obtaining MRI input. Preprocessing is handled in the second step. MRI feature enhancement is the third process. Finally, segmentation process is explained. **Here, we considered 457 images out of 238 that are considered from UCI machine learning repository and remaining 210 images from Ramakrishna Scan Centre, Guntur, and remaining 9 from** "https://www.ncbi.nlm.nih.gov/pm c/?term=16685825%5BPMID%5D&report=imagesdocsum". **All images undergo preprocessing and image enhancement and are applied to proposed method.**

1. **Obtaining MRI input**

In this proposed method, magnetic resonance image is taken as an input. MR images are most broadly used in image processing owing to its high-quality image. By utilizing the magnetic resonance image, user is able to extract the necessary information required for the detection and analysis of brain tumor. Also, MRI uses low radiation compared to other methods. So MRI image is given as an input for the process.

2. **Preprocessing**

After getting the input, preprocessing is carried out in the next step. Optimized kernel possibilistic C-means algorithm is employed in the projected system for preprocessing. In preprocessing step, the image is grouped into a cluster.

In preprocessing stage, possibilistic C-means uses the Euclidean aloofness to estimate a fuzzy membership u_{ik} by using the following equation

$$u_{ik} = \left[1 + \left(\frac{D_{ik}^2}{\eta_i}\right)^{\frac{1}{m-1}}\right]^{-1}, \forall i, k \tag{I}$$

and

$$v_i = \frac{\sum_{k=1}^{n} u_{ik}^m x_k}{\sum_{k=1}^{n} u_{ik}^m}, \forall i \tag{1}$$

Here, v_i is the cluster center or prototype of u_i. Though in real world, to deal with more practical problem, the Euclidean distance is not complex enough so that we use kernel techniques to determine the distance $D_{ik} = \|P_k - V_i\|$.

Let Φ be a nonlinear mapping to some feature space F, and the input space X is mapped into a novel big dimensional feature space F.

$$P = (a_1, \ldots, a_m) \rightarrow \Phi(P) = (\phi(P_1), \ldots, \phi(P_N)) \tag{2}$$

Then, $\|P_k - V_i\|$ is mapped into space F.

$$\|P_k - V_i\| \rightarrow \|\phi(P_k) - \phi(V_i)\| \tag{3}$$

If F is very high dimensional, it will be impractical to compute Eq. (3).

$$K(a_i, a_j) = < \phi(a_i).\phi(a_j) > \tag{4}$$

By utilizing a nonlinear mapping, scalar product computation in input space is converted into kernels computation

$$\{a_i.a_j\} \rightarrow \{\phi(a_i).\phi(a_j)\} = K(a_i, a_j)$$

$$K(a_i, a_j) = \exp\left(-\frac{\|a_i - a_j\|^2}{2\sigma^2}\right) \tag{5}$$

Then, the subsequent equation is acquired.

$$\|\phi(P_k) - \phi(V_i)\|^2 = < [\phi(P_k) - \phi(V_i)].[\phi(P_k) - \phi(V_i)] > \tag{6}$$

Then, the objective function (1) in big dimensional article space F is converted as the following

$$J_m(U, V) = \sum_{k=1}^{n} \sum_{i=1}^{c} u_{ij}^m \|\phi(P_k) - \phi(V_i)\|^2 + \sum_{i=1}^{c} \eta_i \sum_{i=1}^{n} (1 - u_{ij})^m \tag{7}$$

To reduce Eq. (7) as

$$(u_{ij})^{m-1} \|\phi(P_k) - \phi(V_i)\|^2 - \eta_i(1 - u_{ij})^{m-1} = 0 \tag{8}$$

$$\sum_{k=1}^{n} u_{ij}^m (\phi(P_k) - \phi(V_i)) = 0 \tag{9}$$

By computing Eqs. (8) and (9), we propose the following equations

$$u_{ik} = \left[1 + \left(\frac{2 - 2K(a_k, v_i)}{\eta_i}\right)^{\frac{1}{m-1}}\right]^{-1}, \forall i, k \tag{10}$$

$$\phi(V_i) = \frac{\sum_{k=1}^{n} u_{ij}^m \phi(P_k)}{\sum_{k=1}^{n} u_{ij}^m}, \forall i \tag{11}$$

The Eq. (11) cannot be directly calculated. After that, both sides of the equations are right multiplied by $\phi(a_j)^T$ and the Eq. (11) is revised as

$$K(a_k, v_i) = \frac{\sum_{k=1}^{n} u_{ik}^m K(a_k, v_i)}{\sum_{k=1}^{n} u_{ik}^m}, \forall i, j \tag{12}$$

Likewise, Eq. (I) is amended by kernel approaches to work out the η_i in high-dimensional feature cosmos F.

$$\eta_i = \beta \frac{\sum_{k=1}^{n} u_{ik}^m (2 - 2K(a_k, v_i))}{\sum_{k=1}^{n} u_{ik}^m}, \beta > 0 \tag{13}$$

Usually, β is chosen to be 1

By utilizing optimized kernel possibilistic C-means procedure, the given MRI input image is preprocessed and grouped into clusters. Then, this preprocessed picture is provided as an input for feature enrichment process.

MRI Feature Enhancement

Feature enhancement process is handled after completing the preprocessing. Adaptive double window modified trimmed mean filter is cast off for feature enhancement process. In this process, noise is eliminated from the magnetic resonance image (MRI).

Double window modified trimmed mean filter

DW-MTMF was particularly implemented for grayscale pictures. The abstraction of DW-MTM is used in the case of multi-channel pictures. The filter encompasses two phases: In the first one, the borderline median $a_m = [a_{mR}\ a_{mG}\ a_{mB}]^T$ of the vectors in a 3×3 neighborhood is calculated; here, P_{mR}, P_{mG}, and P_{mB} are the separable medians of the R and G and B modules of the vectors in the area. In the subsequent step, pixels xi in a bigger $(2Q+1) \times (2Q+1)$ window are tested. Given pixel located at a, b within the image, a median filter (MED $\{g(a, b)\}$) is calculated within an n by n local area surrounding the part a, b. The median value computed from this filter is

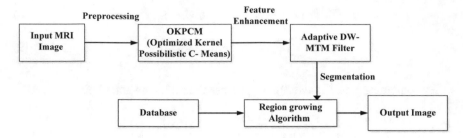

Fig. 1 Enhanced architecture of proposed work

used to evaluate the mean value of the n by n local area. Then, a large-sized window surrounding the pixel at location a, b of size $q\,a\,q$ is used to find out the mean worth. In figuring the man value in the $q \times q$ window, only pixels within the gray level range of

$$\text{MED}[g(a, b)] - c \text{ to } \text{MED}[g(a, b)] + c$$

It is used to disregarding any outliers from the mean calculation. The crop of the DW-MTM filter is the $q \times q$ mean filter. The value of c is chosen as a function of the noise standard deviation as $c = K$. σ_n Adaptive DW-MTM filter removes the Gaussian noise present in the magnetic resonance image. In the next phase, enhanced image is considered as an input. This enhancement process is helpful for the further segmentation process (Fig. 1).

Segmentation

The enhanced information is fragmented by region-growing algorithm. By and large image segmentation depends on the detachment of the image into locales. Segmentation is prepared on the foundation of proportional characteristics. Semblances are quarantined out into gatherings. Extraction of imperative components from the image is the fundamental objective of segmentation. This will serve to effectively see the data. In the arena of medical imaging, brain tumor segmentation from MRI images is an engaging yet forcing process.

Region-growing technique

Region-growing technique is a simple segmentation technique; it implicates the choice of preliminary seed points. This way to deal with division looks at adjoining pixels of introductory seed focuses and decides if the pixel nationals ought to be added to the locale. The procedure is iterated on, in an indistinguishable way from general information bunching calculations. Let $g(i, j)$ demonstrate the pixel being taken care of, the seed. The district developing framework works in two phases: an underlying stride that goes for deciding the area coarsely and a moment remedial stage. The underlying stride of area developing is according to the accompanying:

For each eight-related neighbor (k, l) of the seed, the Euclidean separation concerning the seed is ascertained.

$$d = \| g(m, n) - g(o, p) \|$$

Utilizing region-growing system, the region of interest is isolated from the magnetic resonance image. Finally, brain tumor is effectively detected from the segmented image.

4 Experimental Results

Here, a new algorithm based on clustering and enhancement method with optimized kernel possibilistic C-means (OKPCM) for preprocessing is proposed (Figs. 2, 3, 4, 5, and 6).

Here, we considered 457 images out of 238 that are considered from UCI machine learning repository and remaining 210 images from Ramakrishna Scan Centre, Guntur, and remaining 9 from "https://www.ncbi.nlm.nih.gov/pmc/?ter m=16685825%5BPMID%5D&report=imagesdocsum". **Some of the images considered in Figs.** 2, 3, 5. Our proposed method consists of the following stages. Initially, our method uses OKPCM for clustering the image pixels method. For optimization of OKPCM, particle swarm optimization is used for cluster center ini-

Fig. 2 Original MRI image of brain

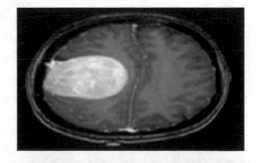

Fig. 3 Image with Gaussian noise

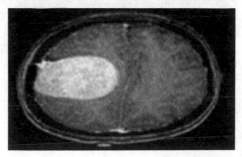

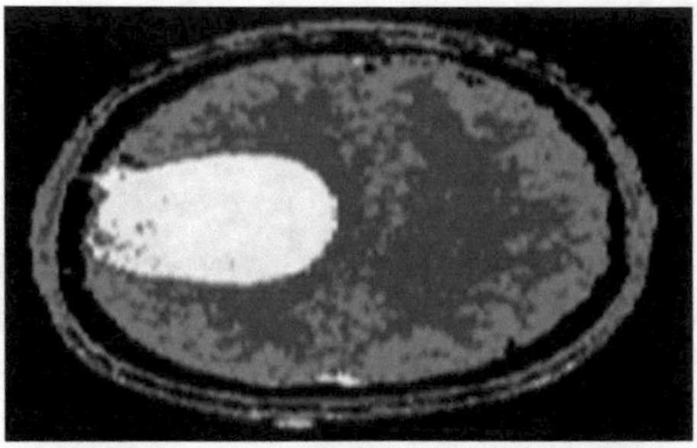

Fig. 4 Preprocessed image
using optimized kernel
possibilistic C-means
algorithm

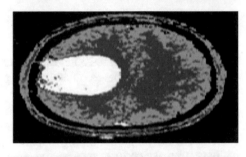

Fig. 5 Enhanced image
using adaptive DW-MTM
filter

Fig. 6 Segmented image
using region-growth
algorithm 3

tialization. Then, the new filtering technique, adaptive DW-MTM filter, is used to enhance the image of brain tumor which can efficiently remove the outliers such as salt-and-pepper noise in the calculation of the mean estimate. Figure 2 represents the original image of tumor affected brain, and image with Gaussian noise is shown in Fig. 3. The image is preprocessed by optimized kernel possibilistic C-means algorithm. OKPCM provides a better improved initial contour for extraction of brain. It is used to initialize the center of cluster as shown in Fig. 4. After that, segmentation is executed by using region-growth algorithm for segmenting the region of tumor detected in brain. The proposed system for the segmentation of images is implemented in the working policy of MATLAB with the following system configuration. This method is implemented in MATLAB R2014a using Intel i5 environments under a personal computer with 2.99 GHz CPU, 2 GB RAM with Windows 8 system. In this proposed methodology, we have used brain MRI database and the experimental results are compared with existing method.

Fig. 7 Accuracy
comparisons of different
clustering methods

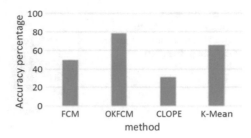

After preprocessing, image will be enhanced by using adaptive DW-MTM filter as shown in Fig. 5. Adaptive DW-MTM filter is used to overcome the difficulties of using the other filters in the presence of impulsive noise by using the median estimator to estimate the local mean.

5 Performance Analysis

In this section, we perform some experiments to compare the performance of KPCM and FCM algorithm with brain dataset. The entire process is done by using MATLAB. Clustering results using FCM, OKPCM, CLOPE, and K-means algorithms are shown in Fig. 7. The KPCM is used to initialize the memberships. In this way, the before observed drawbacks of the FCM can be prevented. And OKFCM performs better than remaining methods; it gives more accuracy in clustering (Figs. 8, 9, and 10).

Table 1 gives the corresponding accuracy value between OKPCM, FCM, CLOPE, and K-means techniques.

Fig. 8 Elapsed time of
various clustering algorithms

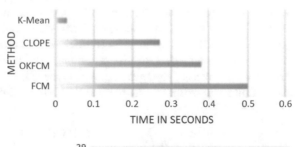

Fig. 9 MSE comparisons of
different filters in brain
tumor detection

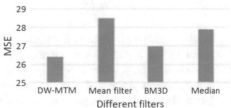

Fig. 10 Accuracy comparisons of different segmentation techniques in brain tumor detection

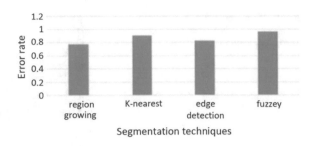

Table 1 Accuracy comparison

Segmentation technique	Accuracy
OKPCM	78.9
FCM	49.625
CLOPE	30.84
K-means	65.94

It shows that KFCM has much better powerful and precise than other methods. In this case, KPCM and FCM both of them describe the basic texture of the dataset, but KFCM is better than FCM. When PCM is processed, the results are entirely changed. The prototypes got by PCM are nearly identical. However, that case in not appear in KPCM. K-means is better, but in this brain tumor detection scenario OKPCM gives far better accuracy than all other methods.

Figure 8 shows the elapsed time of various clustering algorithms of OKPCM, FCM, CLOPE, and K-means in the scenario of brain tumor image processing; here, K-means has fast processing but less accuracy and OKFCM has moderate elapsed time and gives more accuracy. The attainment of the projected system is gauged by paralleling its enhancement results with old-fashioned scheme which uses the adaptive DW-MTM filter-based tumor noise filtering practice. Fig. 9 represents the assessment graph of the adaptive DW-MTM filter, mean filter, BM3D filter, and median filter depending on mean square error value, and Fig. 5 represents comparison graph between DW-MTM filter, mean filter, BM3D filter, and median filter based on PSNR, which is employed to give a compression quality of image. PSNR is easily described by means of the mean squared error (MSE). Consider noise-free image I and its noise approximation K, MSE and PSNR are defined as:

$$\text{MSE} = \frac{1}{n} \sum_{i=1}^{n} (x_i - \tilde{x}_i)^2 \quad \text{PSNR} = 10 \cdot \log_{10}\left(\frac{\max_i^2}{\text{MSE}}\right)$$

Based on this equation, filtering techniques are compared as shown in Table 2. Consider image 5 and 11 represent error rate and accuracy comparison between region-growing algorithm, KNN algorithm, edge detection and fuzzy algorithms for segmentation process. Error Rate is defined as, Following Table 3 Represents the Error Rate and Accuracy Comparison between Region-growing algorithm and

Table 2 Filtering technique comparison

Filtering technique	PSNR	MSE
Adaptive DW-MTM filter	29.98	26.4
Mean filter	28.17	28.5
BM3D	28.68	26.9
Median	28.98	27.9

Fig. 11 Error rate comparisons of different segmentation techniques in brain tumor detection

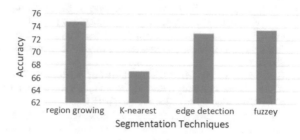

Table 3 Segmentation technique

Technique	Accuracy	Error rate
Region-growing algorithm	74.8	0.78
K-nearest neighboring algorithm	67	0.91
Edge detection	73	0.83
Fuzzy	73.6	0.97

KNN algorithm. The classification presentations such as mean square error (MSE), PSNR, and accuracy are also relatively better than the orthodox method. From these evaluates, we can aspire that the proposed procedures turn out to be the greatest varieties for brain tumor mining in brain MR image.

6 Conclusion

The real objective of this paper is to identify brain tumor utilizing medical imaging systems. In this proposed strategy, magnetic resonance image (MRI) is considered as a contribution because of its top-notch picture. At that point, MRI information is preprocessed by optimized kernel possibilistic C-means algorithm. After that, the picture is enhanced by adaptive DW-MTM filter. At last, the enhanced picture is sectioned by region-growing algorithm. Locale developing techniques can give the first pictures which have clear edges with great segmentation results. Using this enhanced work, brain tumor is accurately detected and diagnosed. It will provide the facility of early tumor detection to the user. The proposed method identifies the brain tumor with 94.5% accuracy.

References

1. Kimmi, Pandey, Shardendu, "Image Processing Techniques for the Enhancement Of Brain Tumor Patterns" International Journal of Advanced Research in Electrical, Electronics and Instrumentation Engineering Vol. 2, Issue 4, April 2013.
2. Roopali, Ladhake "A Review on Brain Tumor Detection Using Segmentation And Threshold Operations" International Journal of Computer Science and Information Technologies, Vol. 5 (1), 2014.
3. Saini, Mohinder, "Brain Tumor Detection in Medical Imaging Using Matlab" International Research Journal of Engineering and Technology (IRJET) Volume: 02 Issue: 02 | May-2015.
4. Patil, Bhalchandra "Brain Tumour Extraction from MRI Images Using MATLAB" International Journal of Electronics, Communication & Soft Computing Science and Engineering, Volume 2, 2011.
5. Geoffrey S. Young, MD "Advanced MRI of Adult Brain Tumors" 2007.
6. Abdel-Maksoud, Elmogy, Rashid, "Brain tumor segmentation based on a hybridclustering technique" Egyptian Informatics Journal (2015).
7. Rajendran,Dhanasekaran "Fuzzy Clustering and Deformable Model for Tumor Segmentation on MRI Brain Image: A Combined Approach" International Conference on Communication Technology and System Design 2011.
8. Anisha, Venugopala Krishnan "Comparison of Various Filters for Noise Removal in MRI Brain Image" International Conference on Futuristic Trends in Computing and Communication (ICFTCC-2015).

A Hybrid Biometric Identification and Authentication System with Retinal Verification Using AWN Classifier for Enhancing Security

B. M. S. Rani and A. Jhansi Rani

Abstract Biometrics deals with perceiving a man or confirming the authentication of physiological or behavioral attributes. To ensure the genuine nearness of a component in contrast to a fake self-made simulated or reproduced trial is a noteworthy issue in biometric confirmation, which require the advancement of novel and proficient security measures. So that we requires an automatic efficient model which can cut out the irregularities and fake access endeavors before matching and decision making. In this paper, here a novel programming-based fake discovery strategy is utilized as a part of retinal check to distinguish distinctive sorts of. The motivation behind the proposed framework is to enhance the security of biometric acknowledgment systems. In this proposed novel cross-breed, adaptive weighted neighbor (AWN) classifier is the procedure to order the information retinal picture relies upon the highlights extraction and coordinating the highlights with the prepared highlights. Initially, the captured image is pre-processed by middle separating procedure and Gaussian filter. What is more, upgrade algorithm used for achieving the complexity of vasculature particularly in the thin vessels and stumpy vasculature differentiate vessels. The proposed model can be utilized as an underlying procedure in numerous security fundamental applications. This paper likewise proposes the algorithm for discovering the bifurcation point in the veins. It enables high security, good performance, and greater accuracy. Also, it provides better FAR, FRR and decreases the error rate.

B. M. S. Rani (✉)
ECE Department, Acharya Nagarjuna University, Guntur, Andhra Pradesh, India
e-mail: ranibms@gmail.com

B. M. S. Rani
ECE Department, Vignan's Nirula Institute of Technology and Science for Women, Guntur, Andhra Pradesh, India

A. Jhansi Rani
ECE Department, V. R. Siddhartha Engineering College, Vijayawada, Andhra Pradesh, India
e-mail: jhansi9rani@gmail.com

© Springer Nature Singapore Pte Ltd. 2019
R. S. Bapi et al. (eds.), *First International Conference on Artificial Intelligence and Cognitive Computing* , Advances in Intelligent Systems and Computing 815,
https://doi.org/10.1007/978-981-13-1580-0_54

Keywords Biometric system · Blood vessels · Vasculature detection and identification · Retinal images · Identification methods · Bifurcation point · Authentication · Biometric template

1 Introduction

Conventional verification frameworks require the client play out the undertaking of retaining various passwords, individual recognizable proof numbers (PIN), pass-express, and additionally replies to mystery questions like, "what is your mom's last name by birth?", and so on so as to get to different databases and systems. All the more frequently, it turns out to be practically difficult to the different designs because of case affectability, prerequisite of alphanumeric content, and the need to change passwords or pass-expresses occasionally to keep from unplanned trade-off or robbery. Numerous clients pick passwords to be a piece of their names, telephone numbers, or something which can be speculated. Also, to deal with the hard errand of recalling such a significant number of passwords, individuals have a tendency to keep in touch with them in files, and obvious places, for example, work area schedules, which uncovered odds of security infringement [1]. Biometric validation comes in play to manage these difficulties with customary secret word frameworks. Conceivably, biometric frameworks can be utilized in all applications that need validation system, thus in all applications that today utilize passwords, PINs, ID cards, or the like. To date unique biometrics have been investigated and utilized, for example, unique mark, hand geometry, confront, scent, voice, ear, step.

Veins designs are diverse for each eye (even indistinguishable twins have unmistakable examples) [2]. Albeit retinal examples put forth change in defense of sicknesses with the end goal that minimized safe effectiveness disorder, glaucoma, and diabetes, the retina remains essentially stable around a people lifetime [3]. The execution of retina-based biometric frameworks is analyzed as the best in term of precision in performance [4]. More than that, human retina has not yet been manufactured and the retina of dead people rots too quickly to be in any way used to cheat a retinal sweep [5]. Along these lines, with the exception of the way that retina examine is regularly thought to be badly arranged and meddlesome so it is hard to increase general acknowledgment by the end clients, the retina is one of the body parts that fulfills superbly all the alluring issues for a biometric framework [6]. Milestone-based strategies for vessel bifurcation and hybrid focus as highlight focuses for retinal enlistment [7]. Another retinal enrollment technique in light of area of optic plate is executed for retina confirmation [8]. A cross-relationship coefficient-based retinal recognizable proof was finished. They initially enrolled the information picture and after that coordinating is finished by corresponding the vascular example.

In this view, we display another auxiliary component for machine-based retinal picture confirmation framework. By contrasting and other procedure, the proposed AWNRF classifier gives better outcome. The possibility of enhancement algorithm and classification process utilized as a part of the proposed framework to enhance the vasculature differentiate. Especially, it focuses on the thin and low difference vessels and concentrates more number of highlights to enhance the precision of grouping. Finally, improvement algorithm, arrangement of altered algorithm, and paired morphological recreation is utilized to extricate the highlights, for example, edge among bifurcation and width of retina and hybrid purpose of retina.

2 Proposed Methodology

In this proposed technique, three algorithms are accomplished for vein ID and confirmation. The proposed technique portrays the approach of enhancement algorithm and classification process is utilized to accomplish the differentiation of vasculature especially in thin vessels and low difference vessels and concentrates the more number of highlights to improve the precision of characterization. At that point, the improvement algorithm, grouping of altered algorithm and twofold morphological recreation is utilized concentrate the highlights, for example, edge among bifurcation and width of retina vessels retina.

The fundamental modules are pre-handling, feature extraction, classification, blood vessel estimation coordinating and decide the veins edge (bifurcation focuses) found in retinal veins. Underneath Fig. 1. Demonstrates that engineering for proposed technique.

2.1 AWN Classifier

To group the preparation tests and making a moderately uniform circulation of preparing tests an enhanced algorithm is utilized. At that point relies upon the preparation tests another parcel grouping-based AWN order algorithm is embraced to arrange the test tests. In the new algorithm, we alter a dynamic W' change in every cycle. For upgraded preparing tests, we have gotten another parcel grouping-based AWN characterization algorithm.

These algorithms have following four phases: focus determination, focus modification, group refinement, and AWN arrangement. As indicated by this algorithm, W is a preset weight of class, and we utilize a specific bunch focus in statement technique to decide the underlying point of convergence; that is, select W objects from the first archive information, each protest speaking to the underlying each group focus. And afterward whatever remains of each protest will be allocated to the bunch of biggest comparability as per the similitude estimate in all the underlying group focus. At that point concentrated on all the test reports, we can utilize the adaptive weighted neighbor technique to rename the document. This procedure is rehashed until the point when the sorts of all records are never again changed. In this calculation two parameters anticipated that would set up, first is the amount of gathering W, the other is W' used in the midst of each cycle of W-versatile neighbor. There are

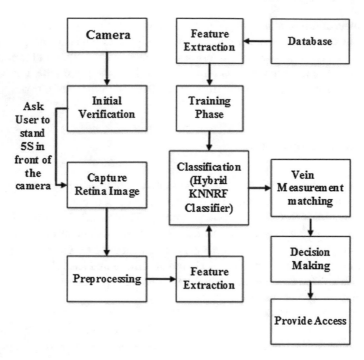

Fig. 1 Design for proposed system

three concentration assurance calculations: sporadic, buckshot, and fractionation. In this paper, we pick sporadic seed decision calculations since self-assertive is direct. Generally, it picks k concentrates indiscriminately from the data set as the basic centers. The choice of W' directly hurt the execution of portrayal, if the parameter is pretty much nothing, you cannot find enough chronicle to authentically gathering, else it would be find more neighbors in the far away groups, so the report would be doled out wrongly to additional distant bundles. We change a dynamic W' adjustment in each accentuation, setting the nearest neighbor parameter W' is the amount of reports contained in the humblest class size of the gathering incorporating one in the current.

2.2 RF Classifier

Arbitrary forests develop numerous grouping trees. Each tree is developed as takes after: On the off chance that N is the quantity of cases in the preparation set, at that point which tests N indiscriminately case–however, by supplanting it from the first information. This specimen will enable the preparation to set for developing the tree. In the event that there is an information variable M, at that point mM a number which is indicated in every hub, after that variable m is chosen at specific out of M and it splits these m when is utilized to isolate the hub. Consistent incentive for m is held amid the backwoods developing. Every last trees are enlarged to its greatest augmentation conceivable. No pruning happens.

2.3 Detection of Bifurcation Point

The particulars extraction is a most essential issue in extricating the bifurcation point. It is handled on skeleton picture by using a thought of intersection number (CN). For an eight neighboring pixels to the inside pixels, here 3×3 size pixel windows are utilized. In this window, if focal pixel is 1 and has precisely 3 one-esteem neighborhoods, at that point the focal pixel is a bifurcation point. For instance, if the focal pixel esteem is considered as 1 and additionally it has just single adjacent values, at that point the focal pixels are specified as an edge finishing point (edge point where it closes) that constitute vein design. That is, on the off chance that $CN(P) == 1$, it is an edge and if $CN(P) == 3$, then it is a bifurcation point. Each and every progression need to compute, so as to see either a pixel is a vessels bifurcation point nor bounded more than veins, put this window on the picture with the goal that the considered pixel is at the focal point of the window and discover the bifurcation point. In the event that the focal pixel does not lie on any vessel, it is not a bifurcation point. On the off chance that the middle pixel lies on the vessel then it is considered as a bifurcation point.

2.4 Detection of Angle

The distinctive strides of angle discovery algorithm for finding the edge between two lines of a bifurcation point is portrayed underneath. Check any retinal twofold picture from its upper left corner to right-base corner. While crossing, a little inward window of any size is considered. The double fundus picture is presently changed over into one-pixel width skeleton shape. The white pixels are recognized when each pixel in the images is navigated. At that point, check for the quantity of its encompassing neighboring pixels of white shading. If the encompassing white pixels are not as much as a limit esteem, overlook those white pixels. The chose white pixels are then made intense by shading their encompassing four pixels as white (same shading as the picture) for a superior view. If the encompassing pixels numbers are more than the limit (say 5–8), those white pixels are checked. Presently, the checked white pixels are shaded dark (same as foundation) to get a solitary pixel skeleton of the picture. In the event that any bifurcation point happens, at that point the internal window transverse as clockwise. In the event that any progress emerges from white to dark, at that point spare the focuses and name the point as locale 1, area 2 and 3 individually. Directly in every district finds the beginning stages of the line as the bifurcation area is done itself by utilizing this algorithm. Likewise the same is sequenced for different locales. At that point, the slants of the districts are computed. Considering the slant of the primary line as $m1$ and the incline of the second line as $m2$.

Calculates the angle between two lines by using formula as below

$$\tan \theta = \frac{m1 - m2}{1 + m1.m2}$$

$$\theta = \tan^{-1}\left(\frac{m1 - m2}{1 + m1.m2}\right)$$

3 Experimental Result

The point of this strategy is to discover the edge and veins width at that point confirm the retina veins, we consolidate it into an individual distinguishing proof framework. The strategy was connected on the DRIVE database, which is worked for estimation of retinal acknowledgment frameworks. DRIVE database comprises aggregate of 40 shading fundus picture. Each picture is caught by a picture size of 565×584 with 24 bits for every pixel which has been JPEG compacted. The outcome is created according to the distinctive strides of proposed strategy.

The recognizable proof framework without pre-preparing step guarantees just 98.6% because of the low quality of information retinal pictures which by and large endured because of low differentiation in dark level picture and brightening varieties which damages the key focuses extraction and coordinating. In spite of the fact that applying the best cap change as a pre-processing procedure, the recognizable proof rate has been enhanced from 98.6 to 99.57%, it stills lower than our proposed framework. Regardless of the possibility that the proposed framework enhance marginally the distinguishing proof rate contrasted with different frameworks, the most pick up is in term of usage time. Fig. 2 shows to the contribution of unique retinal, Fig. 3 shows to feature extracted picture, and Fig. 4 shows the bifurcation focuses in numbers and limits. It likewise shows the width of the vessel with specific vessel position.

By this paper, we presume that characterization exactness of retinal pictures is perfect and productive by utilizing proposed strategies as contrast with existing techniques. In this paper, concentrate on two strategies for retinal picture bifurcation structure has been presented to be specific width coordinating and bifurcation point coordinating of veins. We delineate how highlight focuses are removed from retinal pictures and additionally its coordinating systems. Moreover, includes are isolated utilizing details focuses which are organizes as width and point. Subsequent to applying the predefined informational index which gathers all extricated bifurcation and its directions. We gauged that this technique has better distinguishing proof accuracy

Fig. 2 Input image of retina

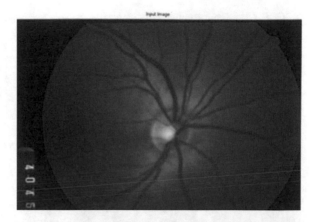

Fig. 3 Extracted image of retina

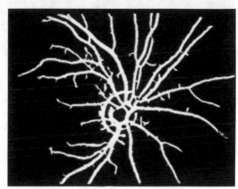

Fig. 4 Bifurcation point, breaking point and blood vessel width

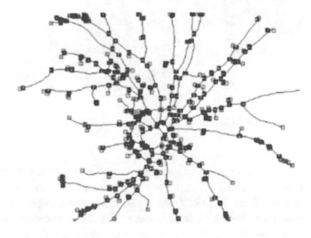

and low blunder. Our proposed method is giving higher accuracy when contrasted to the normal width measurement (Table 1).

Table 1 Classification output of proposed and existing technique

Classification output	Accuracy	FER	FAR	FRR	Sensitivity	Specificity	Precision	Recall
Proposed method	94.3687	73.25	65.02	92.88	95.98	18.265	8.11	1.65
Existing method	76.22	82.03	81.35	68.02	73.00	13.44	19.58	0.89

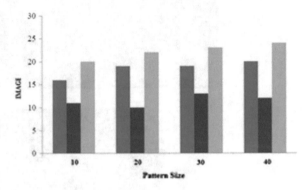

Fig. 5 Histogram for points detected from the training set

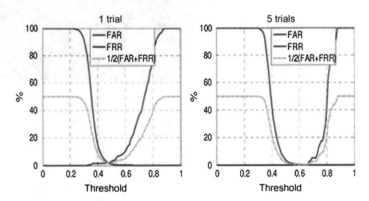

Fig. 6 Graphical comparison of FAR versus FRR

As we ascertain the inspecting rate of the proposed method techniques regarding the traverse time required to isolate the veins from the picture retina. We used info pictures from the DRIVE database to figure a normal preparing time for the proposed plan and similar pictures were utilized as contribution to the first programming-based algorithm. The execution of our approach fundamentally in light of the picture estimate should be investigated and the I/O data transfer capacity (Figs. 5 and 6).

The attainment of biometrics-based identification systems is also commonly depicted using FAR and FRR curves. The graph shows both the FAR and FRR curves are a characterization of the error between the FAR and the FRR as the threshold value is being changed. In Fig. 6, we display the FAR and FRR curve for DRIVE dataset.

In existing identification system, the rate of classification accuracy is 83.4%. In our proposed system, the retina verification method gave 93.6% accuracy of classification. Also, the rate of error is lower than compared to varied techniques.

4 Conclusion

In this paper, we proposed another procedure for machine-based alternative biometric retinal verification system in view of looking at the edges between the bifurcation points and discover finding the width of the veins utilizing a half and half AWNRF classifier for user identification and authentication. Our proposed algorithm has more precise outcomes and less demanding than contrast and different algorithms. The drive dataset is used for analyzing which is bigger than datasets in different papers. In this work, a critical issue in the design of a multibiometric system, namely template security is addressed. In future, the extracted template can be stored in sparse matrix representation for reducing memory wastage to store the template and security mechanisms are implemented for authenticating a user.

References

M. S. Kalyana Sundaram, K. Gokulakrishnan "Width Measurement from True Retinal Blood Vessels" *International Journal of Emerging Technology and Advanced Engineering*, Volume 4, Special Issue 2, April 2014.

Seto, Yoichi. "Retina recognition." *Encyclopedia of biometrics*, pp. 1321–1323, 2015.

Dehghani, Amin, Zeinab Ghassabi, Hamid Abrishami Moghddam, and Mohammad ShahramMoin. "Human recognition based on retinal images and using new similarity function." *EURASIP Journal on Image and Video Processing*, no. 1, pp. 1–10, 2013.

Panchal, Parth, Ronak Bhojani, and TejendraPanchal. "An Algorithm for Retinal Feature Extraction Using Hybrid Approach", *Procedia Computer Science* Vol.79, pp. 61–68, 2016.

Meng, Weizhi, Duncan S. Wong, Steven Furnell, and Jianying Zhou. "Surveying the development of biometric user authentication on machine phones." *Communications Surveys & Tutorials*, IEEE Vol.17, no. 3, pp. 1268–1293, 2013.

Ashokkumar, S., and K. K. Thyagharajan. "Retina biometric recognition in moving video stream using visible spectrum approach." *In Green Computing, Communication and Conservation of Energy (ICGCE)*, pp. 180–187, 2013.

Seyed Mehdi Lajevardi, Member, IEEE, Arathi Arakala, Stephen A. Davis, and Kathy J. Horadam "Retina Verification System Based on Biometric Graph Matching" *IEEE Transactions on Image Processing*, Vol. 22, No. 9, September 2013.

Tripti Rani Borah, Kandarpa Kumar Sarma, PranHari Talukdar "Retina and Fingerprint based Biometric Identification System" *Machine and Embedded Technology International Conference* (MECON - 2013).

Multistage Interconnection Networks in Reliability Shuffle Exchange Networks

Balarengadurai Chinnaiah

Abstract Multistage interconnection networks (MINs) are highly demand in switching process environment. MINs are well known for its cost efficiency which enables economic solution in communication and interconnection. Important features in interconnection topologies are improvement toward performance and increment of reliability. In general, the importance of the reliability will be proportionally increased with the increment of size as well as system complexity. In this paper, we proposed replicated shuffle exchange network and compared with Benes and shuffle exchange network (SEN) topology to investigate the behavior analysis of MINs reliability. Three types of behaviors, namely basic stage, extra stage, and replicated stage, have been compared in this paper. The results show that replicated network provides highest reliability performance among all topological measured.

Keywords Multistage interconnection network · Shuffle exchange network · Benes · Reliability performance · Terminal reliability

1 Introduction

A decade before, MINs have been applied in field such as parallel computing architecture. MINs are known for its cost efficiency which enables economic solution in communication and interconnection [1]. MINs were operated by multiple stages with the input and output switches. The switching element (SE) in MINs has the ability to transmit input either straight or cross-connection. The types of MINs can be differentiating by its topological representation, switching element connected, and the number of stages used in the network configuration [2]. A reliable interconnection network is a crucial factor to ensure overall system performance. The reliability performance is applied to measure the systems ability to convert information from input to output. Reliability itself can be known as the possibility of the network to work

B. Chinnaiah (✉)
Marri Laxman Reddy Institute of Technology and Management, Hyderabad, India
e-mail: cbalarengadurai@yahoo.com

© Springer Nature Singapore Pte Ltd. 2019　　　　　　　　　　　　　　571
R. S. Bapi et al. (eds.), *First International Conference on Artificial Intelligence and Cognitive Computing* , Advances in Intelligent Systems and Computing 815,
https://doi.org/10.1007/978-981-13-1580-0_55

effectively according to specific period and environment [3]. The different types of behaviours in MINs could affect the reliability performance of the network. Several multipath MINs have been developed that offer alternate path in the network; however, the challenge in accepting the new design to enhance the performance and reliability becomes a crucial task. Several performance measures have been implemented such as by adding the number of stages in the network and replicating the network topology [4]. The increment number of stage in the network, it is reliable to offer additional paths for source and destination pairs. It can increase the reliability to some extent. Alternatively, the replicating approach has been analysed to provide a better reliability compared to increasing stages approach. This approach offers the opportunity to decrease the number of stages in the network while maintaining the reliability performance [10]. Therefore, this paper will investigate the behavior of non-blocking MINs known as SEN and Benes network toward the reliability performance. The topology measure in this paper consists of shuffle exchange network (SEN), shuffle exchange network with additional stage (SEN+), replicated shuffle exchange network (RSEN), Benes network, extra-stage Benes network (EBN), and replicated Benes network (RBN). The performance measurement focused on the reliability performance called terminal reliability.

2   Background

The reliability of interconnection networks is the important measurement in system performance. Evaluation of the reliability were believed that each component was stated either completely worked or completely failed. Reliable operation in multistage interconnection networks depends on their topology, network configuration, and number of stages in the system. Performance improvement and reliability increasing are the two major attributes for multistage interconnection network topology. As the number of stages and system complexities increase, the reliability performance becomes an important issue. Therefore, the reliability performance has become a concern for researchers in the field of MINs. Mostly, the basic networks provide lower reliability performance as compared to modification networks. Therefore, the researcher initiates the new approach by adding the stage into the network to increase the reliability performance [9]. In this approach, a fixed number of stages were added to the network. With advancement in the multiprocessing field, an increasing focus needs to be placed on multipath network. Replicating is a method to create the redundant path by replicate the network by L layer with the advantages of an auxiliary link. The increasing number of layers in replicated MINs can improve the reliability compared to basic and additional stage network.

2.1 Basic Network

Shuffle exchange network (SEN) consists of a unique path for each input and output pairs [3]. In this network, all SEs are assumed as series connection. In this connection, the input for SEN can be transmitted either straight or cross-connection. The connections between stages are shuffle exchange. SEN consists of N input and output switches with n stage, where $n = \log_2 N$. SEN provides $(N/2) \log_2 N$ switching element in series [5]. Figure 1 represents the topological view of SEN.

Benes topology is built through an extension of inverse baseline. The first stage and the last stage of the inverse baseline are merged. This leads to $2n - 1$ stages for Benes networks. When switching is performed in the network, the Benes networks can be classified as non-blocking MINs [7]. Compared to other non-blocking MINs, Benes has the smallest complexity packet switched in the network. Figure 2 shows the topology for Benes network.

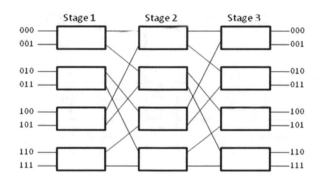

Fig. 1 Shuffle exchange network

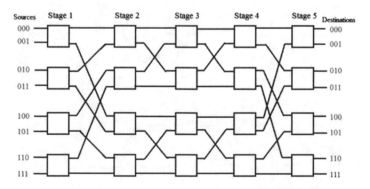

Fig. 2 Benes networks

2.2 Additional-Stage Network

Shuffle exchange network with additional stage (SEN+) implementing by adding one extra stage to the network. SEN+ has four stages and N input–output pairs. Referring to SEN+ topology, the increment of stages leads to the increment of the number of switches and links. It consists of $n = \log 2\, N + 1$ stages with $N/2$ switching element for each stage. The additional stage of SEN+ has enabled extra path connection to provide double paths for communication process for each source and destination pairs [8]. SEN+ topology is illustrated in Fig. 3. Extra-stage Benes network (EBN) was designed to construct redundancy the number of paths between each source and destination pairs as shown in Fig. 4. This network improves the path length for each source and destination. However, the additional-stage method is not efficient to be used in Benes network because this approach decreases the reliability in this network [6].

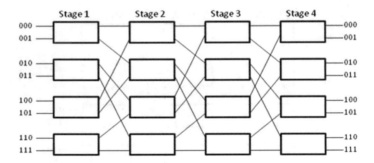

Fig. 3 Shuffle exchange networks with additional stage

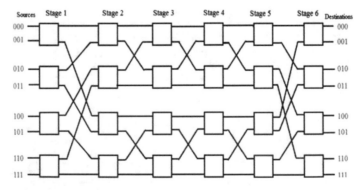

Fig. 4 Extra-stage Benes networks

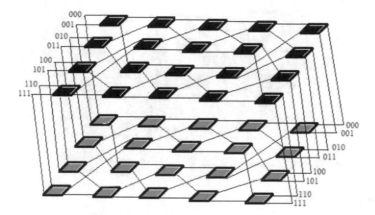

Fig. 5 Replicated Benes networks

2.3 *Replicated Network*

The design pattern of the replicated Benes network (RBN) is a modification from Benes network by replicating the network to create the redundant path [6]. The replicated Benes is arranged in L layer. Normally, for RBN topology, all switching elements are size 2×2. It has $[2(\log_2 N) - 1]$ number of stages. Replicated Benes network is shown in Fig. 5.

3 Proposed Topology Mechanism

3.1 *SENs Topology*

In this section, we proposed a new topology named the replicated shuffle exchange network by replicating the network on SEN's topology. Replicated SEN enlarges regular SEN by replicating the network equal to L times. The corresponding input and output was synchronously connected. Packets were received by the inputs of the network and distributed throughout the layers. Different to the MINs, the replicated MINs lead to out of order packet sequence due to the availability of multipath for each source and destination pair. The rationale for sending packets belonging to the same source is to avoid packet-order destruction. For the reliability purposes, the increment of the number of layers will lead to a reliability improvement in the network. In a view of terminal reliability, the system can be said as a parallel system, where the system consists of two series systems in parallel. Each series system comprises $\log_2 N$ of SE. Figure 6 shows the three-dimensional view of the replicated SEN with two layers of replication.

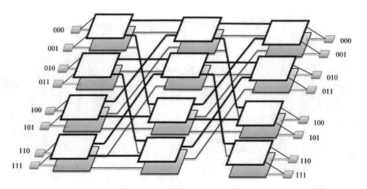

Fig. 6 Replicated shuffle exchange network

3.2 Reliability Measurement

Interconnection network topologies provide the ability to improved performance and the reliability in the network. The choice of interconnection network is dependent on a number of factors, such as the topology, routing algorithm, and communication properties of the network [11–14]. The reliability of an interconnection network is defined as measurement method to identify the capability of a system to convert information from the input to the output.

Terminal reliability can be described for the possibility of existing of at least one fault-free path connecting a selected inputs and outputs pairs [9]. The basic switching element for time-dependent reliability can be referred as α equal to (0.90–0.99). SEN consist of N input/output switches and $n = \log_2 N$ stage, where each stage has N/2 switching elements. The variable α can be defined as the possibility of a switch being operational. The terminal reliability for SEN topology is derived as follows:

$$SEN = \alpha^{\log_2 N} \tag{1}$$

According to Benes topology as shown in Fig. 2, the terminal reliability for 8 × 8 can be calculated by the following equation:

$$Benes = \alpha^2[1 - (1 - (\alpha^2(1 - (1 - \alpha)^2)))^2] \tag{2}$$

SEN+ has two paths MIN. It consists of $\log_2 N + 1$ stages, providing two paths for communication process for each source and destination pairs. The SEN+ with $N * N$ sizes is derived as follows:

$$SEN+ = \alpha^2[1 - (\alpha^{(\log_2 N - 1)2}] \tag{3}$$

Table 1 Terminal reliability comparison

TR(SE)	SEN	SEN+	Rep.SEN	Benes	Extra Benes	Rep.Benes
0.90	0.73	0.78	0.93	0.78	0.77	0.95
0.92	0.78	0.82	0.95	0.82	0.82	0.97
0.94	0.83	0.87	0.97	0.87	0.87	0.98
0.95	0.86	0.89	0.98	0.89	0.89	0.99
0.96	0.88	0.91	0.99	0.91	0.91	0.99
0.98	0.94	0.96	0.99	0.96	0.96	0.99
0.99	0.97	0.98	0.99	0.98	0.98	0.99

Similar to Benes network, for EBN the terminal reliability for this topology can be formulated based on Fig. 4. Therefore, we have the following equation to calculate the terminal reliability for EBN:

$$\text{Extra Stage Benes} = \alpha^2[1 - (1 - (\alpha^2(1 - (1 - \alpha)^2)^2))^2] \tag{4}$$

The time-dependent reliability of the basic switching element can be defined by α. By replicating the SEN topology, it will improve the reliability performance. Therefore, replicated SEN is derived as follows:

$$\text{R.SEN} = 1 - (1 - (\alpha^{\log_2 N}))^2 \tag{5}$$

In addition, the replicated Benes arranged L layer as shown in Fig. 5. Therefore, the terminal reliability for 8×8 replicated Benes can be calculated as follows:

$$\text{R.Benes} = 1 - (1 - \alpha^2[1 - (1 - (\alpha^2(1 - (1 - \alpha)^2)^2))^2])^2 \tag{6}$$

4 Result and Discussion

This paper focuses on the reliability measurement known as terminal reliability for the SEN and Benes network topologies. Table 1 shows the comparative analysis for MINs including the SEN, SEN+, replicated SEN, Benes, extra-stage Benes and replicated Benes. From Table 1, the overall reliability performance shows that the replicated behavior provides better performance compared to extra stage and basic network. The basic stage for both networks provides the lowest reliability performance in the network. However, for basic-stage comparison in Fig. 7, the results indicate that Benes network provides the highest reliability, since it provides more redundant path as compared to SEN topologies.

Extra-stage network comparison in Fig. 8 shows that SEN+ and extra-stage Benes networks provide an equal reliability performance. In contrast with SEN, the Benes network reliability does not essentially affect by adding the extra stage in the network.

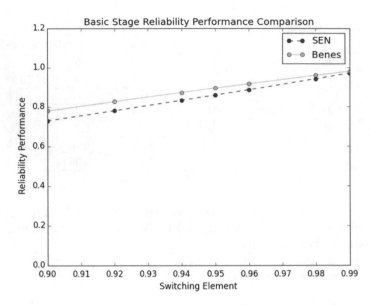

Fig. 7 Basic network reliability

The extra-stage network can increase the path length in the network; however, the approach is inefficient to be implemented in certain network and does not always lead to increment of reliability performance. Additional of extra stage in Benes network provides more complexity due to its configuration and redundant paths availability.

For replicated topology, the results indicate that terminal reliability for replicated Benes is superior compared to replicated SEN as shown in Fig. 9. The advantages of redundant paths in the replicated Benes enhance the reliability in the network. The result shows the reliability of the replicated Benes and replicated SEN is superior to extra-stage and basic-stage topology due to the network offers redundant paths for the network connectivity.

Figure 10 shows the additional stage has the highest switching element, and the replicated network has the lowest switching element used in both network. In general, the switch complexity for basic-stage network and replicated network is less than additional-stage network. It can be concluded that for each size comparison, replicated SEN and replicated Benes had a lowest SEs compared to additional-stage SEN and Benes topology. It can be seen that the switch complexity for basic-stage network is equal to replicated network. Despite the equal switch complexity in the network, the replicated network for both networks leads to highest reliability performance for SEN and Benes topology. The advantage of lesser SE in the network will reduce the implementation cost of the network.

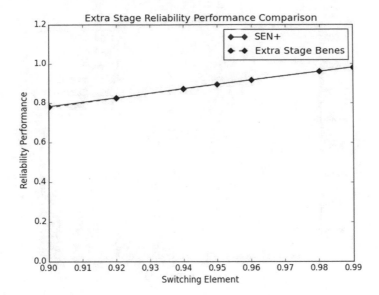

Fig. 8 Additional-stage network reliability comparison

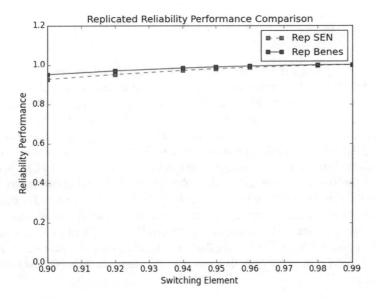

Fig. 9 Replicated network reliability comparison

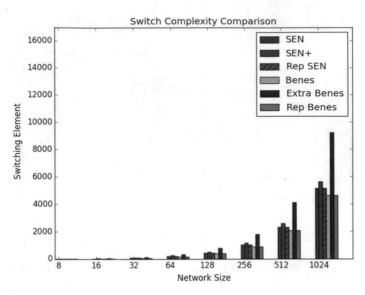

Fig. 10 Switch complexity comparison

5 Conclusion

The network topology, the number of stage and the types of network configurations are different for each interconnection network. The existed of additional paths in interconnection network resulting in the increment of reliability performance. For basic stage, additional stage, and replicated network, the result comparison indicates highest terminal reliability for Benes as compared to SEN topology. The SEN topology itself has lower reliability performance due to the disadvantages of lesser stage and lesser path to route the message in the network compared to Benes topology. However, for overall performance, the behavior of replicated network provides a higher reliability for SEN and Benes topology. The additional stage for both networks enhances the reliability performance to small extent. It can be summarized that the replicated network provides highest reliability compared to basic-stage and additional-stage network. We can see that the replicated behavior provides the availability of more redundant path that can lead the network with highest reliability performance.

References

1. Tutsch D and Hommel G.: Multilayer multistage interconnection networks, Proceedings of 2003 Design Analysis and Simulation of Distributed System. Orlando, 155–162, (2013).
2. Soni S, Dhaliwal A.S. and Jalota A.: Behavior Analysis of Omega Network Using Multi-Layer Multi-Stage Interconnection Network", International Journal of Engineering Research

and Applications. 4, 4, 127–130 (2014).

3. Yunus N.A.M, Othman M., Hanapi Z.M. and Kweh Y.L.: Reliability Review of Interconnection Networks, IETE Technical Review. 33, 6, 596–606, (2016).

4. Aulah N.S.: Reliability Analysis of MUX-DEMUX Replicated Multistage Interconnection Networks, Experimental Techniques. 19–22 (2016).

5. Bistouni F. and Jahanshahi M.: Analyzing the reliability of shuffle-exchange networks using reliability block diagrams, Reliability Engineering and System Safety. 132. 97–106, (2014).

6. Jahanshahi M. and Bistouni F.: Improving the reliability of Benes network for use in large-scale systems, Microelectronics Reliability, 55. 679–695 (2015).

7. Tutsch D.: Performance Analysis of Network Architecture, Springer-Verlag Berlin Heidelberg (2006).

8. Yunus N.A.M. and Othman M.: Reliability Evaluation for Shuffle Exchange Interconnection Network. Procedia Computer Science. 59. 162–170, (2015).

9. Rajkumar S. and Goyal N.K.: Reliability Analysis of Multistage Interconnection Networks. Quality and Reliability Engineering International. 32. 8. 3051–3065, (2015).

10. Bistouni F. and Jahanshahi M.: Pars Network: A multistage interconnection network with fault tolerance capability, Journal of Parallel Distributed and Computing.168–183, (2015).

11. Gunawan I.: Reliability Analysis of Shuffle-Exchange Systems, Reliability Engineering and System Safety, 93. 2. 271–276, (2008).

12. Gunawan I. and Fard N.S.: Terminal reliability assessment of gamma and extra-stage gamma networks. International Journal Quality Reliability Management. 29. 820–831, (2012).

13. Gunawan I. Sellapan P. and Lim C.S.: Extra Stage Cube Network Reliability Estimation Using Stratified Sampling Monte Carlo Method, Engineering e-Transaction.1. 13–18, (2006).

14. Moudi M., Othman M., Kweh Y.L., and Rahiman A.R.A. x-Folded TM.: An efficient topology for interconnection networks, Journal of Network and Computer Applications, 73. 27–34, (2016).

Swarming the High-Dimensional Datasets Using Ensemble Classification Algorithm

Thulasi Bikku, A. Peda Gopi and R. Laxmi Prasanna

Abstract In dealing the typical issues associated with a high-dimensional information search space, the conventional streamlining algorithms off limits a sensible course of action in light of the fact that the interest space increases exponentially with the performance issue, along these lines handling these issues using exact techniques are not helpful. Without a doubt, the comparing data has demonstrated its strength as an indispensable advantage for the business elements and legislative association to take incite and consummate choices by methods for surveying the relevant records. As the number of features (attributes) expands, the computational cost of running the acceptance errand develops exponentially. This curse of dimensionality influences supervised and in addition unsupervised learning algorithms. The characteristics inside the informational collection may likewise be unimportant to the undertaking being contemplated, hence influencing the unwavering quality of the results. There might be a relationship between qualities in the informational index that may influence the execution of the order. In this way, a novel methodology known as ensemble classification algorithm is proposed in the view of the feature selection. We show that our algorithm compares favorably to existing algorithms, thus providing state of the art performance. This algorithm is proposed to lessen the computational overheads, adaptability, and information unbalancing in the Big Data.

Keywords Clustering · Big data · Classification · Feature selection · Accuracy

Please note that the LNCS Editorial assumes that all authors have used the western naming convention, with given names preceding surnames. This determines the structure of the names in the running heads and the author index.

T. Bikku (✉) · A. P. Gopi · R. L. Prasanna
Vignan's Nirula Institute of Technology and Science for Women, Palakalur, Guntur, India
e-mail: thulasi.bikku@gmail.com

A. P. Gopi
e-mail: gopiarepalli2@gmail.com

R. L. Prasanna
e-mail: happy.prasanna44@gmail.com

© Springer Nature Singapore Pte Ltd. 2019
R. S. Bapi et al. (eds.), *First International Conference on Artificial Intelligence and Cognitive Computing* , Advances in Intelligent Systems and Computing 815,
https://doi.org/10.1007/978-981-13-1580-0_56

583

1 Introduction

Big data attempts to discuss what creates huge data, what estimations depict the magnitude and diverse characteristics of enormous information, and what gadgets and developments exist to saddle the limit of huge information. From corporate sector to metropolitan facilitators and scholastics, huge information is the subject of thought and, to some point, fear [1].

The unexpected growth of gigantic data has left badly prepared traditional data mining tools and techniques. Sometime recently, new creative upgrades at first showed up in particular and academic arrangements. Authors and specialists bounced to books and other electronic media for speedy and wide stream of their work on tremendous data. In like manner, one finds two or three books on Big data, including Big Data for Dummies [2], however, inadequate with regards to basic trades in academic preparations. One of the effective factors in information mining ventures relies upon choosing the correct algorithm for the inquiry close by. One of the more mainstream data mining capacities is clustering [3] the data sets and classification. For this examination, we have picked to utilize a few order calculations, as we will probably characterize our information into two labels, referred to as binary classification. Now a days there are diverse sorts of classification algorithms accessible. For the motivations behind this investigation, we have picked few algorithms like C4.5, a decision tree classification, k-nearest neighbor (K-NN), and support vector machine (SVM) calculations [4]. The big information can be portrayed as three V's to be specific: volume, variety, and velocity. *Volume*: It suggests the degree and quantity of data. Big data sizes are accounted for different terabytes, petabytes and zettabytes to store and process. Because of the huge capacity threshold values are constructed, allowing impressively more prominent datasets. In like manner, the sort of data, analyzed under variety, describes what is inferred by 'huge' [5]. *Variety*: It implies the basic heterogeneity in a dataset. Innovative advances allow firms to use distinctive sorts of structured, a combination of structured and unstructured known as semi-structured, and unstructured data [6]. Structured data, which constitutes only 5% of all present data, implies the even data found in spreadsheets or social databases. An unusual condition of combination, and analysing of enormous data, is not on a very basic level new. *Velocity*: The multiplication of electronic contraptions, for instance, mobile phones and sensors, has provoked an extraordinary rate of data creation and is driving a creating requirement for progressing examination and affirmation-based orchestrating [7].

2 Related Work

C. L. Philip Chen et al. (2014) examine the overview of big data on the data-intensive applications. It is presently bona fide that Big Data has drawn gigantic thought from specialists in information sciences, system, and chiefs in governments and tries [8].

As the rate of information improvement outperforms Moore's law toward the start of this new century, irrational data is making an uncommon burden to the people. Then again, there are so much potential and significantly important qualities concealed in the huge volumes of data. Another investigative perfect model is considered as data-intensive scientific discovery (DISD), generally called Big Data issues [9]. There are many fields, for example, open association, government security to the logical research, which experiences the enormous information issues. From one point of view, Big Data is incredibly critical to convey benefit in associations, likewise formative accomplishments in consistent controls, which accommodate a lot of opportunities to make staggering advances in various fields. Of course, Big Data also develops with various troubles, for instance, challenges in finding the relevant data, data accumulating, data examination, and data representation [10]. In Fig. 1, 50% of the enterprises think big data will help them in increasing operational efficiency, etc. Chakraborty et al. (2014) portrays on the best way to enhance the activity execution utilizing Map-reduce in the Hadoop Clusters [11]. In datasets of this present reality, there are numerous superfluous, excess or deceiving highlights that are pointless. In such cases, include determination of Feature Selection (FS) is used to defeat the curse of dimensionality issue. The motivation behind FS is to diminish the intricacy and increment the execution of a framework by choosing unmistakable highlights. At the end of the day, the objective of FS is to choose an applicable subset of relevant data, R highlights from an aggregate arrangement of information, I includes $(R < I)$ in a given dataset. Selection of attributes is one of the dynamic fields in numerous different territories, for example, information handling, information mining, design acknowledgment, machine learning, arrangement issues, PC vision. Dimension reduction method is applied in the first phase to reduce dimension of the datasets and in next phase of classifying the relevant data. Unler et al. introduced a half and half channel wrapper important attribute subset selection calculation model in view of particle swarm optimization, PSO for SVM characterization. The best component subset is chosen from an arrangement of low-level picture properties, including local, intensity, shading, and literary highlights, utilizing the hereditary calculation. This calculation has a superior in tackling different improvement issues and was inspected in different issues like channel displaying, unconstrained enhancement, model classifier, highlight determination such as attribute selection, arrangement, and multi-objective streamlining. To deal with swarming the information depends on k-means clustering, must be of high caliber as to an appropriately picked homogeneity measure. K-means is a partitioning based clustering procedure that endeavors to discover a client indicated a number of clusters based on their centroids. The significant disadvantages distinguished are there is a trouble in looking at nature of the clusters formed and delivered; initial partitions can bring about final clusters. It is extremely hard to anticipate the quantity of groups (k-value) and sensitive to scale. Another approach is proposed in light of the Gravitational Search Algorithm GSA as an enhancement procedure to upgrade both the optimal feature subset and SVM parameters at the same time. Two kinds of GSA are utilized as a part of the type of a remarkable algorithm: the continuous (real-valued) version and discrete (binary-valued) version [12]. The continuous-valued variant is utilized to enhance the best

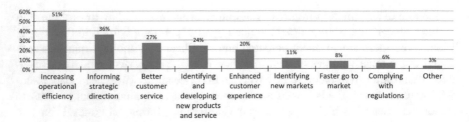

Fig. 1 Survey chart big data opportunities

SVM demonstrate parameters, while the discrete rendition is utilized to look through the optimal feature subset. Rashedi et al. have built up the binary variant of the first GSA for parallel advancement issues in which refreshing position intends to switch in the vicinity of 0 and 1 as opposed to adjacent. The issues to be dissued in twofold space instead of real space. So in this paper, we proposed another calculation which can beat the computational issues and enhance the performance and accuracy.

3 Proposed Algorithm

Our examination point is to viably speak to the information regarding the decrease in computational overheads, adaptability, and information adjusting. The database might be DBMS or RDBMS. The enormous information can be ordered into three information structure to be specific organized information, unstructured, and semi-organized information. An advanced methodology is proposed in the field of information stored in large volumes in light of the law of gravity. In this proposed calculation, specialists are considered as items and their execution is estimated by their masses. The items are pulled into each other because of their gravitational power and their development toward their heavier masses. In the proposed algorithm named as elector-based algorithm for high-dimensional datasets (EHD), every one mass has four particulars: position, latency mass, dynamic gravitational mass, and static gravitational mass. Every one mass presents an answer, and the figuring is investigated by really changing the gravitational and latency masses. By disappointment of time, we expect that masses be pulled in by the heaviest mass. This mass will demonstrate a perfect course of action in the search space. The proposed algorithmic (EHD) steps are given: Here, $t = 0$ means that it is not the past, not the future but the present state. $B(t) = $ B0 epsilon-alpha*t/T;

The number of iterations defined by T;
Create a random initial population;
Consider a system with N agents.

$$Y_i = (y_{i1} \ldots y_{id} \ldots y_{in}) \quad \text{for} \quad i = 1, 2 \ldots, N;$$

where Y_{id} presents the position of the agents of the ith agent in the dth dimension.

$$\text{Mass}_{\text{active } i} = \text{Mass}_{\text{passive } i} = \text{Mass}_{ii} = \text{Mass}_i, \ i = 1, 2, \ldots, N;$$

The following calculations are done until the criterion dissatisfies:
Obtain the total force that exerts on the ith solution.
Force = gravitational constant (B) [active mass$_i$* passive mass$_j$]/square of the distance between the two masses.
Total force acting on agent i is given as:
Forceid(t) = summation of best j values of all the groups are calculated by the best and worst values of j.
All agents in the group are evaluated, and best and worst agents are initialized.

$$\text{Mass}_i(t) = [\text{fit}_i(t) - \text{worst}(t)]/[\text{best}(t) - \text{worst}(t)];$$

$$\text{best}(t) = \text{minimum of } j[\text{fit} j(t)], \text{ where } j = 1, 2, \ldots, n;$$

$$\text{worst}(t) = \text{maximum of } j[\text{fit} j(t)];$$

It can also be noted that for maximization problem, we can change the above equation as: best (t) = maximum of $j[\text{fit} j(t)]$;

$$\text{worst}(t) = \text{minimum of } j[\text{fit} j(t)];$$

In this way, all operators apply the power as time goes on, K_{best} is decreased specifically, and toward the end, there will be one force applying on the others. Finally, K_{best} is the arrangement of first K operators with the best fitness esteem and greatest mass.
K_{best} is the function of time with the initial value K_o at the beginning and decreasing with time.
K_{best} is the capacity of time with the underlying quality K_o at the start and diminishing with time.
Calculate the acceleration of the ith solution.
According to Law II of Newton:
Acceleration = force/mass of the object;
According to the proposed algorithm (EHD):
Acceleration$_i$d(t) = summation   of j at   best   [gravitational   constant (B)*[mass$j$/distance between agents + epsilon (for initial value)]*displacement between agents;
Compute the new velocity of the ith solution; the velocity of an object is the rate of progress of its position as for an edge of reference and is an element of time.
Velocityid $(t + 1)$ = random$_i$*velocity$_i$d(t) + accelaration$_i$d(t);
Where random$_i$ can be from 1 to n

Obtain the new position of the ith solution

$Y_i(t+1) = \text{randomi*velocityid}(t+1)$;

Now consider two distinct random solutions from the Y values.

Calculate the elector operation is

$\text{Electort}(i,j)\ (\Omega) = Yj + \text{random}(0,1)*(Y(t)r1 - Y(t)r2)$;

$Yi(t+1) = Y(t)i,j + \text{velocityid}(t+1)$, if $\text{random}(0,1) < \text{elector}$;

$Yi(t+1) = Yi + \text{random}\ (0,1)*[Y(t)r1 - Y(t)r2]$ otherwise

Where elector (Ω) is the rate at which it controls the probability of inheriting from the new position.

The advantages of the proposed algorithm are the heavy inertia mass moves slowly and thus search space becomes easier. Finding gravity of the data improves the accuracy of the search which reduces the time complexity. More precise search of the data leads to the scalability of the data. Data balancing is maintained due to the inertia and mass calculations.

4 Experimental Analysis

In this proposed model, medical documents are collected from each data source as structured or unstructured format. In this system, a novel gene-based document's relationships are analyzed using MapReduce framework to discover the textual patterns from a large collection of medical document sets. This segment gives the outcomes, accuracy, and the proposed algorithm's performance, with reference of k-mean and GSA algorithm. The proposed method is assessed as far as precision, error rate and compared with other algorithms. The four diverse datasets are collected and utilized for the proposed model for result analysis and the significant execution of the framework is given underneath.

Average separation index measures the greatness of gaps between any two clusters in a segment, by anticipating the information in a couple of groups into a one-dimensional space in which they have the most extreme partition. Table 1 shows the comparison between different algorithms and the proposed EHD model, which had given good results.

The Separation Index is measured for various calculations with various datasets of different algorithms. Hence, the results prove that the proposed algorithm is efficient than the conventional algorithms such as k-means and GSA.

Entropy is the difference between the original label of the class and the predicted class label. If the entropy of the cluster is low indicates it as the better cluster. The entropy increases when ground truth of objects in the cluster additionally expands. The more noteworthy entropy implies that the grouping is not great. The proposed algorithm shows that it is having less entropy than other traditional algorithms. Table 2 shows the comparison on different datasets based on entropy between different algorithms and the proposed EHD model, which had given good results.

The k-means algorithm is broadly utilized for clustering vast arrangements of information. However, the standard algorithm does not generally ensure great out-

Table 1 Comparing different instances based on average separation index

Data set	Average separation index (with five instances)			Average separation index (with ten instances)		
	K-means	GSA	Proposed method (EHD)	K-means	GSA	Proposed method (EHD)
Abalone	0.25722	0.250	0.1798	0.280546	0.24801	0.176145
Brest cancer	0.25101	0.2492	0.1761	0.278084	0.25612	0.173817
Forest fire	0.21245	0.201	0.1744	0.276339	0.23789	0.172168
Iris	0.26159	0.254	0.1789	0.280957	0.2187	0.176533

Table 2 Comparing different algorithms on different datasets based on entropy

Dataset	Average cluster entropy		
	K-means	GSA	Proposed method (EHD)
Abalone	0.618	0.52375	0.2128
Brest cancer	0.629	0.52755	0.2548
Forest fire	0.632	0.5304	0.2884
Iris	0.641	0.53895	0.3392

Table 3 Comparing different algorithms on different datasets based on accuracy

Data set	Accuracy%		
	K-means	GSA	Proposed method (EHD)
Abalone	65.27	93.33	95.63330165
Brest cancer	64.89	91.23	94.39843083
Forest fire	63.68	91.32	94.98560788
Iris	65.71	93.74	94.33791411

comes as the accuracy, which means nearness to original evaluation of the final groups, relies upon the choice of initial centroids. Table 3 shows the comparison on different datasets based on accuracy between different algorithms and the proposed EHD model, which had given good results.

Sensitivity measures the extent of positives that are effectively recognized. All experiments, regardless of how carefully arranged and executed, have some level of mistake or vulnerability or uncertainty. The accuracy can be evaluated by computing the percentage of the error, which can be calculated when the true value is known. Though the percent error is an absolute value, it can be expressed with magnitude to indicate the direction of error from true value. Table 4 shows the comparison based on true positive, error, and outlier between different algorithms and the proposed EHD model, which had given good results.

Table 4 Comparing different algorithms on based on sensitivity and error rate

Algorithm	True positive (sensitivity)	Error (%)	Outlier (%)
K-means	72.921	27.91	16.2
GSA	82.415	22.64	12.4
Proposed method (EHD)	84.624	19.85	9.92

Table 5 Accuracy measured at different values of elector (Ω)

Dataset	Distinct elector (Ω)		
	$\Omega = 0.1$	$\Omega = 0.2$	$\Omega = 0.3$
Abalone	94.225	95.101	95.63330165
Brest cancer	94.2849	94.2585	94.39843083
Forest fire	93.1258	93.5265	94.98560788
Iris	94.1232	94.2459	94.33791411

The computational complexity nature of the typical calculation is frightfully high due to reassigning the data points at various circumstances, amid each iteration of the loop. The proposed algorithm computed at different elector values and the accuracy computed for different datasets are shown in Table 5.

5 Conclusion

High-dimensional document clustering and classification is one of the essential machine learning models for the knowledge extraction process of the real-time user recommended systems. As the amount of information in the high-dimensional data repositories increases, many organizations are facing the unprecedented issues of how to process the available huge volumes of data efficiently. This model is used as a user recommended system on larger document sets using the map reduduce of Hadoop framework. Experimental results show that the proposed algorithm (EHD) has given better results compared to traditional document clustering and classification models. The computational complexity of the Mapper phase is O(nlogn) and Reducer phase is O(logn). In future, this work can be extended to protein clustering and classification using the Hadoop framework in the biomedical datasets.

References

1. Kruger, Andries F. Machine learning, data mining, and the World Wide Web: design of special-purpose search engines. Diss. Stellenbosch: Stellenbosch University, 2003.
2. Hurwitz, Judith, et al. Big data for dummies. John Wiley & Sons, 2013.

3. Wu, A. H., et al. "Soy intake and breast cancer risk in Singapore Chinese Health Study." British journal of cancer99.1 (2008): 196–200.
4. Wu, Xindong, et al. "Top 10 algorithms in data mining." Knowledge and information systems 14.1 (2008): 1–37.
5. Zikopoulos, Paul, and Chris Eaton. Understanding big data: Analytics for enterprise class hadoop and streaming data. McGraw-Hill Osborne Media, 2011.
6. Wang, Lei, et al. "Bigdatabench: A big data benchmark suite from internet services." High Performance Computer Architecture (HPCA), 2014 IEEE 20th International Symposium on. IEEE, 2014.
7. Jagadish, H. V., et al. "Big data and its technical challenges." Communications of the ACM 57.7 (2014): 86–94.
8. Chen, CL Philip, and Chun-Yang Zhang. "Data-intensive applications, challenges, techniques and technologies: A survey on Big Data." Information Sciences 275 (2014): 314–347.
9. Tolle, Kristin M., D. Stewart W. Tansley, and Anthony JG Hey. "The fourth paradigm: Data-intensive scientific discovery [point of view]." Proceedings of the IEEE 99.8 (2011): 1334–1337.
10. Kaisler, Stephen, et al. "Big data: Issues and challenges moving forward." System Sciences (HICSS), 2013 46th Hawaii International Conference on. IEEE, 2013.
11. Chakraborty, Suryadip. Data Aggregation in Healthcare Applications and BIGDATA set in a FOG based Cloud System. Diss. University of Cincinnati, 2016.
12. Sarafrazi, Soroor, and Hossein Nezamabadi-pour. "A New Class of Hybrid Algorithms Based on Gravitational Search Algorithms: Proposal and Empirical Comparison".

Review on Autonomous Vehicle Challenges

Supriya B. Sarkar and B. Chandra Mohan

Abstract In the growing of government participation, many countries like USA, UK, Japan attention the production of autonomous car. Google is the first full autonomous car got the license for driving. Autonomous car is a vehicle which is moving from source to destination without human intervention. The technologies embedded into the vehicle like camera, radar, lidar for sensing the feature of environment behavior. To improve the quality of safety, to improve the congestion, and to reduce the accident due to happen by human wrong behavior, autonomous vehicle plays a crucial role. In this paper proposed, many challenges which are come across in autonomous car are described by author. The challenges are sensing the vehicle environment, GPS system must be reliable, avoid obstacle, and multi-sensor fusion must be accurate, path tracking, curb detection. In this paper evaluated, the approach and technologies used overcome the above challenging issues for autonomous car on the road. For detection, tracking, shortest distance, smooth driving in this parameters are solved by different algorithm to get optimal and efficient driving.

Keywords Sensor · Autonomous vehicle · Perception algorithm

1 Introduction

In 1950, first research takes place and began with skill-based driving model. Since 1986, autonomous vehicle concept started by various different country like Japan, USA, Europe, UK. Sensor functionality used for solving the problem tunnel, wet road. Then in 1998, urban traffic problem discussed in autonomous vehicle for highway scenarios. In 2005, author described only some parameter such as localization, mapping, navigation and control. In 2009, Google car started self-driving car and

S. B. Sarkar (✉) · B. C. Mohan
SCOPE, VIT Vellore, Vellore 632014, India
e-mail: supriya.sarkar@rediffmail.com

B. C. Mohan
e-mail: dr.abc@outlook.com

© Springer Nature Singapore Pte Ltd. 2019
R. S. Bapi et al. (eds.), *First International Conference on Artificial Intelligence and Cognitive Computing* , Advances in Intelligent Systems and Computing 815,
https://doi.org/10.1007/978-981-13-1580-0_57

Fig. 1 Autonomous car

it allows to detect pedestrian, cyclist, roadwork, and lane in all direction. Advance driver assistant controller is used to control the movement of full automatic vehicle defined by Joel et al. [1].

The definition of autonomous driver is driving a vehicle from one place to another place without human controller. Automated vehicle is classified into five levels such as level 0 based on no automation, level 1 based on assisted automation, level 2 based on partial automation, level 3 based on high automation, and level 4 based on full automation means no driver is expected discussed by Nishith [2].

The companies estimate their driverless future car such as Tesla motor, Google, BMW, Jaguar, Nissan, Toyota, etc. Autonomous vehicle incorporated in three dimensions of performance: economic, social, environmental. In economic, insurance are provided to the road accident or if any component damage then manufacture company provide the insurance claim defined by Daniel and Kara [3]. In social, human can utilize their driving time in other activities also due to human wrong decision accident happened like drink and drive so it is also reduced. In environmental, the fuel emission increases up to 50%, reduces carbon emission up to 90%, road accident reduces up to 90% and it will help to maintain green in environment proposed by Umit et al. [4].

Around 1.25 million people were killed every year and millions of people injured due to wrong driving. So, human behavior is very important characteristic in road safety and increases the collision risk also explained by Lynn et al. [5]. Figure 1 shows many technologies equipped into vehicle like sensor, radar, GPS, lidar, and camera to detect the pedestrian, lane, and tunnel. Autonomous vehicle is safer for pedestrian behavior. The ACC controller used to maintain the speed and brake of the vehicle.

The vehicle infrastructure network works to collect all the data from environment and send to the entire vehicle. The collisions occur and then break the vehicle to prevent the accident. If any break is needed to control, then alert is given to driver. The main important goal of autonomous car is to reduce accident, traffic, parking space, pollution emission, etc. In urban area perception, capability must be reliable for autonomous vehicle [5]. Every system having some advantage and disadvantages defined by Kransnigi [6]:

Advantages of autonomous vehicle are:

1. Improvements of safety on the road without driver.
2. Independent mobility for elder, younger, and disable.
3. Improve the road congestion and driver utilized that time in other activities.
4. Less congestion so improve the fuel efficiency and parking space.
   Disadvantages of autonomous vehicles are:

   1. Cost of car is increased.
   2. More unemployable due to driverless.
   3. Automobile insurance provides by manufacture industry.
      Challenging issue of self-driving vehicles is explained by Umit et al. [4]:
      1. Sensing capability of the vehicle environment;
      2. Perception of unknown behavior;
      3. Insure safety performance of the vehicle;
      4. Traffic control capabilities;
      5. Lack of good software;
      6. Lack of road ahead.

2 Literature Review on Research Challenges

See Tables 1 and 2 .

Table 1 Comparison between autonomous and non autonomous car defined by Todd [7]

Parameters	Autonomous car	Non autonomous car
Driverless	Driverless means no human intervention	Control by driver
Avoid driving drunk	Avoid driving drunk	Driving drunk
Safety from accident	Increase safety from accident	Less safety due to human poor judgment
Independent mobility	Independent mobility for younger, elder, and disable	No mobility for younger and disable
Fuel efficient	Improve fuel efficient	Fuel utilization more
Utilize driving time	Driving time utilized for other activities or take rest	Driver concentrates on driving

Table 2 Comparison table of literature survey based on autonomous vehicle technology and functionality

Title of paper	Author name with published year	Methodology	Description
Technology need for autonomous car [8]	Claudiu et al. (2016)	Hierarchical control structure such as strategic level, tactical level and low level control	This system based on AC design consists on solving unexpected driving problem. All the action had been taken in the form of states. Locomotion problem which was solved by learning technique because it was improved the capability of the system and also managed more complex locomotion problem
Curb detection and tracking [9]	Jamein et al. (2011)	Hough transformation is used for driving such as road detection, obstacle detection, curb and lane detection	The laser range finder technology used for detecting correct curb, firstly crub detection through crub geometric shape and detect the position, second crub estimation and tracking using particles filters. Extracted the feature from the assumption and calculated similarity between points by pattern recognition algorithm. The crub detection technology is a fast and robust but not used to control the steering of vehicle

(continued)

Table 2 (continued)

Title of paper	Author name with published year	Methodology	Description
Facial recognition and geo location services [10]	Mohhamad et al. (2016)	Virla Jones algorithm is implemented to identify face with stages such as Hear feature selection, creating integral image	These paper proposed stages correctly identified the front view of face and real time result. As every system has some corn and porn. In this system face with head portion is not detected accurate result
IoT technology to drive the automatic industry [11, 6]	Supriya et al. (2016) Kransnigi et al. (2016)	RFID, WIFI, 5G, and DSRC, v2v communication	Author also proposed the market forecast to connect autonomous vehicle. It also described 5 phases for car evaluation path
Visual perception system in autonomous vehicle [12]	Weijing et al. (2017)	ADAS(Advanced Driver Assistant system) provided simple and partial feature of low level of autonomy	At the level of autonomous car require to sense signalization, location, moving pedestrian decided by human making decision. The visual perception is one of the critical technology, if there is no correct perception then decision will be taken by vehicle and it is not safe
Platooning autonomous mobile robot [13]	Satoshi et al. (2013)	Adaptive cruise control system used to define the relationship between velocity	In this system bottleneck for lane crossing and inter vehicle communication based on ACC and create a small cluster of platoon. In this paper author proposed to solve the jam or congestion method for moving efficiently to reduce the travel time of the robot

(continued)

Table 2 (continued)

Title of paper	Author name with published year	Methodology	Description
A Multi sensor fusion for moving object detection [14]	Hyunggi et al. (2014)	It is based on classification algorithm	For intelligent vehicle system detection and tracking of object is play vital role. By using radar vehicle obstacle detection is not get qualities result. To improve this approach author proposed multisensory feature
Computer vision for autonomous vehicle [1, 15]	Joel et al. (2015) Chandra et al. (2008)	Support vector machine (SVM) used to the marginal line to classify simple and efficient image either 2D or 3D	For Dynamic Environment the PPS i.e. pedestrian protection system detect the presence of stationary and moving people around the vehicle in this way author proposed may state of art on different issues like recognition, reconstruction, tracking and end to end learning
Path tracking in autonomous vehicle [16]	Matthew et al. (2016)	Path planning and path tracking	In this paper author discussed about two part of vehicle modeling: first accurate capture vehicle dynamically second reduce modeling error. Nominal path is very precise sequence from higher level path planner of the vehicle behavior and environment
Strategic and operational concern [17]	Neda et al. (2017)	Greedy heuristic algorithm and the SVOR (Shared vehicle ownership and Ridership) used to form the cluster and cover all essential trip through spatio- temporal proximity	The Autonomous vehicle is providing more safety and mobility and environment benefits. It reduces the vehicle to perform task which based on daily for household. One solution given by autonomous vehicle is shared ownership with others

3 Challenges

3.1 System for Safety and Autonomous Behavior:

DARPA: The operation which performed by autonomous car is car following, lane keeping/lane changing, emergency stopping, traffic control capabilities, obstacle avoidance. So, it is expected to sense all the operation by self-sensing with respect to some coordinates, maps, speed and acceleration explained by Rodrigo et al. [18]. In 2005, DARPA involved in this technology and got some challenges such as sensing the vehicle environment, control over an unknown course, insure the safety performance of the vehicle, testing and validation. [4] proposed the automatic path which is detected by A* algorithm. This algorithm used to find the optimized the path with distance travelled, slope grade, negative bias. After sensing the obstacle, vehicle must be stopped automatically to assure safe performance.

3.2 Generation of Roadway Map for Autonomous Cars

Localization algorithm also used for the same concept. There are two requirements for roadmap, first geometric requirement which is for find out the quality of the map. Second, implementation requirement based on data structure to reduce data storage define by Chandra and Baskaran [19]. In proposed system, author [20] defined three steps: data acquisitions, sensor data fusion, road modeling. The data generated from data acquisition which equipped into vehicle, i.e. through RTK GPS To overcome this problem mathematical road model algorithm used for perception, vehicle localization, path planning through B-spline curve and get optimal smoothing process. This system based on 2D coordinates only not supported to 3D coordinates and map database management system for autonomous vehicle.

3.3 Traffic Light Detection and Recognition

While driving autonomous car on road, appropriate traffic action is required in urban environment. There are many types of traffic light such as vertical, horizontal, circle with color. In conventional system, traffic light used with fixed camera but for autonomous car more cameras are used for capturing image from front view or rear view. In proposed system, author [21] explained the new system which has been detected and recognized traffic light in real time. The environment perception consists of camera and radar for detecting lane, obstacle, traffic light signal. The decision making module based on data fusion sensor and the automatic control module based on steer control and speed control operation. Some of the featured detected under traffic light system such as preprocessing, detection, recognition defined by Pawel and Inga [22].

3.4 Ecall System for Autonomous Car Accident Detection

Author [23] proposed autonomous car accident detection are made by ecall system which is used to save the life and respond emergency. For this reason, European country built new regulation that is called ecall technology. Ecall technology is equipped in each car (semi or automatic) by 2015. The characteristic of MSD is message identified either call type or test call through vehicle direction. The sensor is equipped into the car that detects the accident and sends the alert message to admin. The Retro fill ecall module is equipped in the application of SAE vehicle to detect crash and detect emergency alert.

3.5 Human Dynamic-Based Driver Model for Autonomous Car

The autonomous vehicle moved road in the near future widely. The behavior of humans influences their driving. The important goal of autonomous vehicle is to reduce accident, collision, and eliminate human error and poor judgment. It is determined by attitude, subjective norms, perceived behavior and depends on age and gender explained by Li et al. [24].

3.6 Autonomous Driving System for Unknown Environment

While the autonomous vehicle drive which are considered all situation like speed, traffic signal, crosswalk, parking space etc. Some algorithms are very important like prediction for feature extraction, color-based detector for traffic information and sensor for obstacle detector discussed by Alfred et al. [25]. The local map building modules access the image detection, obstacle detection. The process for handling this module is based on three factors: computing power, network traffic, and memory usage of the module. For detecting traffic information, author [26] proposed an approached called unified map. It converts the traffic environment into imaginary obstacle.

3.7 Perception of Autonomous Vehicle: Relationship with Road Users

Now, full autonomous car used to drive by self without any driver. The number of advantages while used AC like reduce the collision, improve the road safety than conventional way. In this [5] proposed autonomous car interact with people and percept the route of people where they want to travel. Around 1.25 million people were killed every year and millions of people injured due to wrong driving. However, human behavior acted as important role in road safety and collision risk also.

3.8 Optimal Driving Path of a Car Using Modified Constrained Distance Transformation

As some method used for finding smooth path finding but only absence of obstacle. The optimal path based on the characteristics such as size, orientation of car, obstacle of arbitrary shape, smoothness of timing, and direction of cost. [27] proposed the static and dynamic properties for autonomous car. The static car defined shape and orientation but in dynamic properties described smoothness and direction. The chamfer is optimal algorithm which is to get numeric computation instead of floating point. The algorithm defines if the value is infinity, then the algorithm does not find feasible path from source to destination.

3.9 Dynamic Bottleneck Congestion

The speed of autonomous car is fast than normal car. Autonomous car is used for transportation application. There are some effects by congestion with three channels such as the resulting increase in capacity, value of time (vot) and implementation. [28] proposed only effect of bottleneck not any safety or transportation system. The mathematical consist two problems one is routing and other is on demand car sharing. To solve this model problem, find out the minimum number of autonomous vehicle used in a cluster with optimal routing status and also increase the car sharing request.

4 Conclusion

As conventional vehicle compares with autonomous vehicle, it is provided better safety from accident. Autonomous car basically used for transportation purpose. The number of steps processed in autonomous vehicle is vehicles' infrastructure network, lane detection, driver alert, camera, collision prevention. To described the number of challenges from DARPA grand which have been occurred in autonomous car on offline road. By using different algorithm need to overcome the challenges. Also, describe the Internet of Things applied to autonomous car as one of the applications for smart cities. In urban environment, 90% safety provides than non-autonomous car from accident. While driving car path tracking and planning done by discrete transformation method. Through unified map representation, avoid the obstacle occurred due to driving. For avoiding the accident, human dynamic behavior is playing major role also their age and gender. Multi-sensor fusion used for detects the pedestrian, lane detection but not traffic signal with classification algorithm. In this way integrate most of the challenges occurred in autonomous car in this paper.

References

1. Joel, J., Fatma G.u., Aseem B., Andreas, G., "Computer Vision for Autonomous Vehicles: Problems, Datasets and State-of-the-Art", Project Report, pp. 1–64 (2015)
2. Nishith D., "Preparing For a Driverless Future", pp. 1–29(2016)
3. Daniel J. F., Kara K., "Preparing a nation for autonomous vehicles: opportunities, barriers and policy recommendations,"Elsevier, pp. 167–181, (2015)
4. Umit O., Christoph S., Keith R., "Systems for Safety and Autonomous Behavior in Cars: The DARPA Grand Challenge Experience", proceeding in IEEE,vol.95.No.2. (2007)
5. Lynn M., Hulse H.X., Edwin R.G., "Perceptions of autonomous vehicles: Relationships with road users, risk, gender and age", Elsevier, vol. 102, pp. 1–13, (2018)
6. K. Kransnigi, E.H.," Use of IoT Technology to Drive the Automotive Industry from Connected to Full Autonomous Vehicles",Elsevier,pp. 269–274, (2016)
7. Todd L.,"Autonomous Vehicle Implementation Predictions Implications for Transport Planning" Victoria Transport Policy Institute,pp. 1–24, (2017)
8. Claudiu P., Csaba A.,"Issue about Autonomous cars,"11th IEEE International Symposium on Applied Computational Intelligence and Informatics,pp. 1–5, (2016)
9. Jamein B., Junyoung s.,M. chan r., sung hoon k.,"Autonomous driving through CURB detection and tracking",The IEEE 8th International Conference on Ubiquitous Robots and Ambient Intelligence (URAI 2011), pp. 273–278, (2011)
10. Mohhamad F., Anita T.,"Autonomous car system using facial recognition and geolocation services", 6th IEEE international conference-cloud system and Big data engg., pp. 417–420, (2016)
11. Supriya S., Chandra M.B.,"Review on recent Research in data mining based on IoT", International journal of control theory and application, vol. 9 no. 42, pp. 501–5011, (2016)
12. Weijing S., Mohamed B. A., Xin L., Huafeng Y.," Algorithm and hardware implementation for visual perception system in autonomous vehicle: A survey",Integration the VLSI Journal Elsevier, pp. 1–9, (2017)
13. Satoshi H.," Reactive Clustering Method for Platooning Autonomous Mobile Robots",2013 IFAC Intelligent Autonomous Vehicles the International Federation of Automatic Control,pp. 152–157, (2013)
14. Hyunggi cho, Young-Woo seo, B.V.K. Vijaya Kumar, Ragunathan (Raj) Rajkumar,"A mutisensor Fusion System for moving object detection and Tracking in Urban Driving Environment", 2014 IEEE International conference on Robotics and Automation(ICRA), pp. 1836–1843, (2014)
15. Chandra M., B., Sandeep, R. and Sridharan, D. "A Data Mining approach for Predicting Reliable Path for Congestion Free Routing using Self-Motivated Neural Network", The Ninth ACIS International Conference on Software Engineering; Artificial Intelligence; Networking and

Parallel/Distributed Computing, Thailand (Awarded as Best Paper), Vol.no. 149, pp. 237–246, (2008)

16. Matthew B., Joseph F., Stephen E., J. Christian G.," Safe driving envelopes for path tracking in autonomous vehicles", Elsevier, pp. 1–10, (2016)

17. Neda M., R. Jayakrishnan, "Autonomous or driver-less vehicles: Implementation strategies and operational concerns", Transportation Research Elsevier, pp. 179–194, (2017)

18. Rodrigo B., Thierry F., Michel P.," Achievable Safety of Driverless Ground Vehicles", IEEE 10th Intl. Conf. on Control, Automation, Robotics and Vision,pp. 515–521, (2008)

19. Chandra M. B. and Baskaran, R. "Reliable Barrier-free Services in Next Generation Networks", International Conference on Advances in Power Electronics and Instrumentation Engineering, Communications in Computer and Information Science (Vol. 148), Springer-Verlag Berlin Heidelberg, pp. 79–82, (2011)

20. Kichun J., Myoungho S.," Generation of a Precise Roadway Map for Autonomous Cars", IEEE Transactions on intelligent transportation systems, vol. 15, No. 3,pp. 925–938, (2014)

21. Guo Mu, Zhang X., Li Deyi, Z. Tianlei, An L.,"Traffic light detection and recognition for autonomous vehicles", The Journal of China Universities of Posts and Telecommunications Elsevier, pp. 50–56, (2015)

22. Pawel G., Inga R.,"Traffic models for self-driving connected cars", 6th Transport Research Arena Elsevier, pp. 2207–2216, (2016)

23. Attila B., Attila G., Olivér K., Hunor S., Balázs I., Judit K., Zsolt S., Péter H., Gábor H.," A Review on Current eCall Systems for Autonomous Car Accident Detection",40th IEEE conference,pp. 1–8, (2017)

24. Lin Li, Yanheng Liu, Jian Wang, Weiwen Deng, Heekuck Oh, "Human dynamics based driver model for autonomous car",IET Intelligent transport system, ISSN 1751–956,pp. 1–9, (2015)

25. Alfred D., Anand P., Newlin R., Seungmin R.,"Big Autonomous Vehicular Data Classifications: Towards procuring Intelligence in ITS",Vehicle communication Elsevier,pp. 1–11, (2012)

26. Inwook S., Seunghak S., Byungtae A., David H. S., and In-So K.,"An Autonomous Driving System for Unknown Environments Using a Unified Map", IEEE transactions on intelligent transportation systems, pp. 1–15, (2015)

27. Soo-Chang P., Ji-Hwei H.,"Finding the Optimal Driving Path of a Car Using the Modified Constrained Distance Transformation", IEEE Transactions on robotics and automation, vol. 14, NO. 5, pp. 663–671, (1998)

28. Vincent A.C., van den B., Erik T. V.," Autonomous cars and dynamic bottleneck congestion: The effects on capacity, value of time and preference heterogeneity", Elsevier, pp. 43–60, (2016)

Comparative Study of Automatic Urban Building Extraction Methods from Remote Sensing Data

V. S. S. N. Gopala Krishna Pendyala, Hemantha Kumar Kalluri, V. Raghu Venkataraman and C. V. Rao

Abstract Building foot prints and building count information in urban areas are very much essential for planning and monitoring developmental activities, efficient natural resource utilization, and provision of civic facilities by governments. Remote sensing data such as satellite/aerial imagery in association with digital elevation model is widely used for automatic extraction of building information. Many researchers have developed different methods for maximizing the detection percentage with minimum errors. A comparative study of different methods available in the literature is presented in this paper by analyzing the primary data sets, derived data sets, and their usage in the automated and semiautomated extraction methods. It is found that the success of the method for automatic building detection in urban areas primarily depends on using combination of high-resolution image data with digital elevation model.

Keywords Urban buildings · Automatic extraction · High-resolution satellite images · DEM · Clustering

1 Introduction

Attribute information of buildings, such as the size, height, the building type, and number of buildings in a cluster, is important for many applications [1]. This is an essential input for civic facility implementation by governments, planning infrastructure development, and auditing of natural resource utilization. This data is utilized for

V. S. S. N. G. K. Pendyala (✉) · V. R. Venkataraman · C. V. Rao
National Remote Sensing Centre, Hyderabad, Telangana, India
e-mail: gopalakrishna_pvssn@nrsc.gov.in

H. K. Kalluri
Department of Computer Science and Engineering,
Vignan's Foundation for Science Technology and Research,
Vadlamudi, Andhra Pradesh, India
e-mail: hemanth_mtech2003@yahoo.com

© Springer Nature Singapore Pte Ltd. 2019
R. S. Bapi et al. (eds.), *First International Conference on Artificial Intelligence and Cognitive Computing* , Advances in Intelligent Systems and Computing 815,
https://doi.org/10.1007/978-981-13-1580-0_58

works, such as implementation of smart city concept, town planning, microclimate investigation, population estimation, and site selection of cell tower transmitter [1, 2], to name a few.

Urban building footprints extraction by automatic methods is an active and popular research topic [3] for the past 20 years. The manual method of collecting this data involves extensive labor, time, and cost. Fully automated methods cannot be implemented in a single step [4, 5], especially for complex urban areas [5].

Many researchers used aerial imageries, satellite imageries, digital surface model (DSM), and digital terrain model (DTM) derived from airborne LiDAR (Light Detection and Ranging) and photogrammetry techniques to automatically extract the building information. The global research methods work on larger areas, concentrating on extracting many buildings at a single stretch, and provide the footprint information focusing mainly on quantitative statistics such as number of units and area of each unit. On the other hand, algorithms for detailed building information extraction focus on local information of each building with high level of detail (LOD) including modelling the undulations in the rooftops and other parametric details.

Once the building information is available, they can be clustered into the groups of buildings such as industrial, residential, planned, unplanned, slum settlements. City cluster development for efficient urban corridors and megacity planning [6] for inclusive economic growth, industrial concentration by clustering analysis [7], prediction of building energy efficiency measures [8, 9], 3D digital city applications [10], and aggregation and visualization [11] for virtual tours are some of the popular applications using this cluster.

2 Literature Review

In the effort to analyze the approaches adopted for automatic urban building extraction, literature study is taken up. The main objective of this study is to understand the type of data sets used by various researchers, the algorithms applied, and the analysis of the observed results to find out efficient and robust approach.

2.1 Data Sets

Aerial photographs of high-resolution, satellite data from mono- to multispectral ranging from 0.1 to 1.0 m resolutions, and airborne LiDAR generated DSM having various resolutions and point densities are some of the primary data sets used for this purpose. Fusion of more than one data set is also attempted to improve the detection efficiency. Table 1 lists the data sets used by various researchers:

Table 1 Data sets used for urban building extraction

Data set	Resolution	Author reference
Mono/fused satellite images	0.5–1.0 m	[1, 4, 10, 12–16]
LiDAR DSM	0.15 m height accuracy with 1–2 m posting	[2, 10, 15, 17, 18]
Aerial images	0.09–0.20 m	[2, 17–20]
Google Earth images	1–10 m	[21–23]

2.2 Preprocessing of Data

The primary data set was either directly used or preprocessed to generate the secondary information which helps in isolating the unwanted features from the image. The preprocessing consists of generating the derivative products such as photogrammetric DSM, DTM, normalized DSM (nDSM), and normalized difference vegetation index (NDVI).

DSM consists of all the height information including that of vegetation and man-made objects. DTM is the modified DSM by removing all vegetation and man-made objects and essentially consists of only ground information.

nDSM: It is obtained from DSM and DTM height differences as shown in (1), consisting of information about vegetation, buildings, and other man-made objects [16].

$$nDSM = DSM - DTM \tag{1}$$

NDVI: It is an index value comparing between the values of the near-infrared (NIR) and red image bands (RIB) of color image as shown in (2), widely used for the detection of the vegetated areas [3, 18, 24].

$$NDVI = \frac{(NIR - RIB)}{(NIR + RIB)} \tag{2}$$

The secondary information such as shadow information [4, 21], NDVI analysis [18, 24], DSM generation by photogrammetric method [5, 14, 16], normalized DSM (nDSM) [16, 18] is utilized for further maximizing the possibility of detection of buildings.

2.3 Methods

In the effort to detect and extract the building information, researchers have taken various approaches.

The local extraction methods were developed to estimate the height and shape of the roof for individual buildings [1, 25, 26]. One of the main algorithms used was PIVOT (Perspective Interpretation of Vanishing points for Objects in Three dimensions) [1, 25].

It is observed that on the global scale, a wide variety of methods were used. The most important parameter for selecting a particular algorithm is based on the characteristics of the data sets used. The importance was given to number of buildings detected rather than the individual size and modelling of each building. Fully automated and semiautomated approaches were adopted to achieve this objective.

Fully automated methods. Traditional methods of segmentation and classification [4, 15, 18, 20, 24, 27, 28] are applied in the fully automated approaches. The strength of building detection for this approach lies in the resolution of the data set used. The high-resolution satellite image is supplemented with other data such as shadow information [4], fusion of methods [18], multi-resolution image information [24], photogrammetry DSM [27], and LiDAR DSM [28].

The general approach consists of applying sequence of steps systematically such as data preparation, segmentation, classification, feature extraction, and selection. The data preparation uses shadow or NDVI information to mask out the unwanted information. Segmentation is carried out by applying region growing method on images and elevation-based threshold [15, 18, 24, 28] on height information. Edge detection [20] is used for building delineation.

In addition, methods like maximum likelihood algorithm [22, 29] and line-corner analysis method [25] are also applied. Advanced segmentation methods are also suggested such as chess board segmentation and multi-resolution segmentation to improve detection [24] for automatic detection of buildings.

Semiautomated methods. Semiautomated methods rely on training of data sets for building detection using techniques such as fuzzy and neural networks [2, 5, 12, 23], stacked sparse auto-encoder with optimized structures [17], stereo pair disparity index [19], region-based segmentation [21], and marked point process (MPP) [30].

3 Discussion

It is necessary to compare the methods on a common scale. Various researchers used the quantitative parameters: the detection percentage (DP) and quality percentage (QP) [4, 5, 15–17, 25]. This popular method makes use of the spatial comparison between the number of buildings actually present on the ground and the number of buildings detected by the algorithm.

Detection percentage: This metric measures the percentage of buildings in the manual results which were actually detected by the automated system.

$$\text{Detection Percentage (DP)} = 100\frac{\text{TP}}{\text{TP} + \text{FN}} \tag{3}$$

Table 2 Comparison of building extraction methods

Data	Method	DP (3)	QP (4)	Author
High-resolution images with LiDAR DSM	Thresholding	96.5	86.6	[15]
High-resolution images with LiDAR DSM	Object-based classification	95.1	81.7	[15]
LiDAR DSM with optical images	Neural network	94.2	79.8	[17]
High-resolution images	Line-corner analysis method	84.4	55.1	[25]
High-resolution images with DSM	Maximum likelihood classifier	82.6	–	[27]
Fusion of Images and nDSM	Object-based automatic extraction	81.5	74.4	[16]
High-resolution images	Object-based approach with shadow information	74.6	69	[4]

Quality percentage: This metric summarizes overall system performance. Any false positives or true negatives are reflected in this score and will lower the quality percentage.

$$\text{Quality Percentage (QP)} = 100\frac{TP}{TP + FN + FP} \tag{4}$$

where true positive (TP) represents the number of correctly extracted buildings, false positive (FP) represents the number of buildings detected by the automatic approach that is not truly present on the ground, and false negative (FN) represents undetected buildings actually present on the ground.

The comparison of the detection percentage (3) and quality percentage (4) for the global-level automatic urban building extraction methods on which study is conducted is listed in Table 2.

4   Conclusions

After thorough analysis of the various methods adopted by the previous researchers, it is concluded that the better detection results for automatic building detection in urban areas are achieved using high-resolution image data supplemented with height information by using DEM generated from optical data/DEM derived from LiDAR data.

References

1. Weidner, R.: An approach to Building Extraction From Digital Surface Models. International Archives of Photogrammetry and Remote Sensing, Vol XXXI, Part B3 (1996).
2. Yuhan, Huang., Li, Zhuo., Haiyan, Tao., Qingli, Shi., Kai, Liu.: A Novel Building Type Classification Scheme Based on Integrated LiDAR and High-Resolution Images. Remote Sensing (2017).
3. Mahak, Khurana., Vaishali, Wadhwa.: Automatic Building Detection Using Modified Grab Cut Algorithm from High Resolution Satellite Image. International Journal of Advanced Research in Computer and Communication Engineering Vol. 4, Issue 8 (2015).
4. Benarchid, O., Raissouni, N., El Adib, S., Abbous, A., Azyat, A., Ben Achhab, N., Lahraoua, M., Chahboun, A.: Building Extraction using Object-Based Classification and Shadow Information in Very High Resolution Multispectral Images, a Case Study: Tetuan, Morocco. Canadian Journal on Image Processing and Computer Vision Vol. 4 No. 1 (2013).
5. Bittnera, K., Cuia, S., Reinartz, P.: Building Extraction from Remote Sensing Data Using Fully Convolution Networks. ISPRS Workshop, Hannover (2017).
6. City Cluster Development: Triggering Inclusive Economic Growth. ADB Urban Innovations Internet Resource. https://www.adb.org/sites/default/files/publication/27883 (2018).
7. Density Based Satial Clustering- Identifying industrial clusters in the UK: Methodology Report by Department for Business, Energy and Industrial Strategy. https://www.gov.uk/ (2017).
8. Philipp, Geyer., Arno, Schlueter.: Performance-Based Clustering for Building Stock Management at Regional Level. Internet resource. https://lirias.kuleuven.be/handle/123456789/56294 4 (2016).
9. David, Hsu.: Comparison of integrated clustering methods for accurate and stable prediction of building energy consumption data. Applied Energy 160, 153–16 (2015).
10. Suhaibah, Azri., Alias, Abdul Rahman., Uznir, Ujang., François, Anton., Darka, Mioc.: 3D City Models. Encyclopedia of GIS, Editors: Shashi Shekhar, Hui Xiong, Xun Zhou, Internet resource, https://doi.org/10.1007/978-3-319-17885-1_100002 (2017).
11. Liqiang, Zhang., Chen, Dong.: A spatial cognition-based urban building clustering approach and its applications. International Journal of Geographical Information Science (2012).
12. Poonam, S. Tiwari., Hina, Pande., Badri Narayan, Nanda.: Building Footprint Extraction from IKONOS imagery based on multi-scale object oriented fuzzy classification for urban disaster management. Symposium of ISPRS Commission IV, Goa (2006).
13. Karuna, Sitaram Kirwale., Seema, S Kawathekar.: Feature Extraction of Build-up area using Morphological Image Processing. International Journal of Emerging Trends & Technology in Computer Science (IJETTCS), Vol. 4, Issue 5(1) (2015).
14. Tian, J., Krau, T., d'Angelo, P: Automatic Rooftop Extraction in Stereo Imagery Using Distance and Building Shape Regularised Level Set Evolution. ISPRS Workshop, Hannover (2017).
15. Txomin, Hermosilla., Luis, A Ruiz., Jorge A Recio., Javier, Estornell.: Evaluation of Automatic Building Detection Approaches Combining High Resolution Images and LiDAR Data. Remote Sensing (2011).
16. Sefercik, U G., Karakis, S., Atalay, C., Yigit, I., Gokmen, U.: Novel fusion approach on automatic object extraction from spatial data: case study Worldview-2 and TOPO5000. Geocarto International (2017).
17. Yiming, Yan., Zhichao, Tan., Nan, Su., and Chunhui, Zhao.: Building Extraction Based on an Optimized Stacked Sparse Auto encoder of Structure and Training Samples Using LIDAR DSM and Optical Images. Sensors (2017).

18. Demir, N, Poli, D, Baltsavias, E.: Extraction of Buildings and Trees Using Image and LiDAR data", The International Archives of the Photogrammetry, Remote Sensing and Spatial Information Sciences. Vol. XXXVII. Part B4. Beijing (2008).
19. Feifei, Peng., Jianya, Gong., Le, Wang., Huayi, Wu., Pengcheng, Liu.: A New Stereo Pair Disparity Index (SPDI) for Detecting Built-Up Areas from High-Resolution Stereo Imagery. Remote Sensing (2017).
20. Jiang, N., Zhang, J X., Li, H T., Lin, X G.: Semi-Automatic Building Extraction from High Resolution Imagery Based on Segmentation. International Workshop on Earth Observation and Remote Sensing Applications, Beijing (2008).
21. Gregoris, Liasis., Stavros, Stavrou.: Building Extraction in Satellite Images using Active Contours and Colour Feature. International Journal of Remote Sensing, 37:5, 1127–1153 (2016).
22. Parivallal, R., Nagarajan, B., Nirmala, Devi.: Object Identification Method using Maximum Likelihood Algorithm from Google Map. International Journal of Research in computer applications and Robotics, Vol. 2 Issue. 2, Pg.: 153–159 (2014).
23. Pascal, Kaiser., Jan, Dirk Wegner., Aur´elien, Lucchi., Martin, Jaggi., Thomas, Hofmann., Konrad Schindler.: Learning Aerial Image Segmentation from Online Maps. IEEE Transactions on Geoscience and Remote Sensing, Vol. 55, Issue: 11 (2017).
24. Małgorzata, Verőné Wojtaszek.: Opportunities of object-based image analysis for detecting urban environment. presentation at IGIT, Internet resource, www.geo.info.hu (2015).
25. McGlone, J.C., Shufelt, J.A.: Projective and Object Space Geometry for Monocular Building Extraction. Proceedings of Computer Vision and Pattern Recognition, 21–23; pp. 54–61 (1994).
26. Jeffery, A. shufflet.: Exploiting Photogrammetric Methods for Building Extraction in Aerial Images. The International Archives of the Photogrammetry and Remote Sensing Vol. XXXI, Part B6, Voenna (1996).
27. Ibrahim, F. Shaker., Amr, Abd-Elrahman., Ahmed, K. Abdel-Gawad., Mohamed, A. Sherief.: Building Extraction from High Resolution Space Images in High Density Residential Areas in the Great Cairo Region. Remote Sensing (2011).
28. Huan, Ni., Xiangguo, Lin., Jixian, Zhang.: Classification of ALS Point Cloud with Improved Point Cloud Segmentation and Random Forests. Remote Sensing (2017).
29. Jenifer, Grace Giftlin. C., Jenicka, S.: A Survey Paper on Buildings Extraction from Remotely Sensed Images. International Conference on Intelligent Computing Techniques (ICICT), Tirunelveli (2017).
30. Mathieu, Brédif., Olivier, Tournaire., Bruno, Vallet., Nicolas, Champion.: Extracting polygonal building footprints from digital surface models: A fully-automatic global optimization framework. ISPRS Journal of Photogrammetry and Remote Sensing, 77, pp. 57–65 (2013).

Difference Expansion based Near Reversible Data Hiding Scheme

Madhu Oruganti, Ch. Sabitha, Bharathi Ghosh, K. Neeraja
and Koona Hemanath

Abstract Reversibility is being widely explored in these days due to the emerging applications in remote sensing, medical imaging, and military communication. The new paradigm of near reversible data embedding has to be explored further. Recently, very few data embedding techniques have been presented in spatial and frequency domains. This paper presents a near reversible technique in spatial domain which partially uses the difference expansion technique for hiding the secret data into the cover content.

Keywords Near reversible · Visual quality · Data embedding
Difference expansion · Capacity · Bit error rate

M. Oruganti (✉)
Department of Electronics and Communication Engineering, Sreenidhi Institute of Science and Technology, Hyderabad, Telangana, India
e-mail: oruganti.madhu@gmail.com

Ch. Sabitha · B. Ghosh
Imaging Research Center (1304), Department of Computer Science and Engineering, Vardhaman College of Engineering, Shamshabad 501218, India
e-mail: sabithakiran.ch@gmail.com

B. Ghosh
e-mail: bharathighosh2511@gmail.com

K. Neeraja
Department of Computer Science and Engineering, MLR Institute of Technology, Hyderabad, Telangana, India
e-mail: kneeraja123@gmail.com

K. Hemanath
Department of Information Technology, MLR Institute of Technology, Hyderabad, Telangana, India
e-mail: hemanath.it@gmail.com

© Springer Nature Singapore Pte Ltd. 2019   613
R. S. Bapi et al. (eds.), *First International Conference on Artificial Intelligence and Cognitive Computing* , Advances in Intelligent Systems and Computing 815,
https://doi.org/10.1007/978-981-13-1580-0_59

1 Introduction

Reversibility has become a rising issue nowadays in the field of watermarking and steganography [6, 7]. Reversibility may be classified as full reversible and near reversible. Full reversible schemes are widely explored in the literature, whereas near reversible schemes are not explored much in the literature. There is a need of exploring the near reversible schemes more, due to their applications in remote sensing, medical image authentication, and military communication.

Near reversible watermarking techniques can be classified into spatial domain and frequency domain techniques. In spatial domain techniques, for the first time, Barni et al. [1] have first discussed the concept of near lossless embedding paradigm for protecting the copyright of remote sensing media. Later, Tang et al. [2] and Zhang et al. [3] have presented their schemes in spatial domain. Tang et al. [2] have given a robust near reversible data hiding technique using histogram projection. And Zang et al. [3] have presented LSB replacement technique-based near reversible schemes.

In frequency domain schemes, Ahmed et al. [4] and Gujjunoori et al. [5] have contributed their near reversible schemes. Ahmed et al. [4] have given a semi-reversible watermarking scheme for medical image authentication, whereas Gujjunoori et al. [5] have presented their near reversible technique for MPEG-4 video in DCT domain.

This paper presents a simple near reversible technique which partially depends on the difference expansion techniques.

2 Proposed Scheme

Here, the data embedding and extraction techniques are presented. The proposed technique is presented by partially using the principle of difference expansion technique. A random watermark is embedded into the cover image based on the availability of the pixel pairs of interest.

In Algorithm 1, the proposed data embedding technique is presented.

```
1.  Algorithm 1:Data Embedding Technique:
2.  Input : Gray scale image, A, of size m × n
3.  Output: Embedded Image
4.  evencount=0
5.  for a ← 1 to m do
6.  for  b ← 1 to n do
7.  if  b mod 2 == 0 then
8.  x=A[a,b]
9.  y=A[a,b+1]
10.     if  (x%2==0  and  y%2==0)   or  (x%2!=0  and
    x!=255 and y%2==0) then
11.     evencount=evencount+1
12.     end
13.     end
14.     end
15.     end

16.     Generate  random  watermark,W,
    of size=evencount*2.
17.     Initialize mark_im=[0,0,0…,0]ₘ*ₙ,k=0
18.     for a ← 1 to m do
19.     for  b ← 1 to n do
20.     if  b mod 2 == 0 then
21.     x=A[a,b]
22.     y=A[a,b+1]
23.     l=⌊(x+y)/2⌋
24.     h=x-y
25.     if  (x%2==0  and  y%2==0)   or  (x%2!=0  and
    x!=255 and y%2==0) then
26.     x1=l+⌊h/2⌋+W[k]
27.     y1=l-⌊h/2⌋+W[k+1]
28.     k=k+2
29.     mark_im[a,b]=x1
30.     mark_im[a,b+1]=y1
31.     end
32.     else  then
33.     mark_im[a,b]=x
34.     mark_im[a,b+1]=y
35.     end
36.     end
37.     end
38.     end

39.     Out put the mark_im Embedded Image
```

The data embedding and extraction are inverse to each other. The data extraction procedure is presented in Algorithm 2.

1. **Algorithm 2: Data Extraction Technique**

2. **Input** : Embedded Image, B , of size m × n, with hidden watermark W.

3. **Output:**   Restored   image,   A   and Extracted Watermark as **'Ex'**.

4. // Image Restoration

5. Ex=[ ]

6. A [m][n] = [0, 0, 0, . . . , 0]ₘₓₙ //Initialization of A

7. p=0

8. **for** a ← 1 to m **do**

9. **for** b ← 1 to n **do**

10.   **if** b mod 2 == 0 **then**

11.   x=B(a,b])

12.   y=B(a,b+1)

13.   if p<=|W| ∧ x ≠255 ∧ y ≠ 255 then

14.   b1=x mod 2

15.   b2=y mod2

16.   Ex □ b1 // where   □, is   appending the value of b1 to Ex

17.   Ex □ b2

18.   p=p+2

19.   if b1=0 ∧ b2=0: then

20.   A[a,b]=x; A[a,b+1]=y

21.   **end**

22.   **elif** b1=0 ∧ b2=1 **then**

23.   A [a, b] = x; A [a, b + 1] = y+1

24.   **end**

25.   **elif** b1=1 ∧ b2=0 **then**

26.   A [a, b] = x; A [a, b + 1] = y;

27.   **end**

28.   **elif** b1=1 ∧ b2=1 **then**

29.   A [a, b] = x; A [a, b + 1] = y+1

30.   **end**

31.   **else then**

32.   A [a, b] = x; A [a, b + 1] = y;

33.   **end**

34.   **end**

35.   **end**

36.   **end**

3   Experimental Results

We present our experimental results by considering the 512 * 512 TIFF formatted images which are collected from the dataset [8]. There is no near reversible scheme which can be compared, which is similar to the proposed scheme. We demonstrate the results in terms of PSNR, capacity, and BER matrices.

The embedding capacity is defined as the total number of bits that can be embedded into a cover image of size $m * n$ (here 512 * 512). The metric bit error rate (BER) gives the error rate at which the extracted bits exhibit. That is, the rate of wrongly

Table 1 Performance in terms of capacity, PSNR, and BER

Test image	Capacity	PSNR	Bit error rate (BER)
Lena	130732	54.17	0.50
Eline	128458	54.24	0.48
Boat	126630	54.30	0.48
Airplane	113484	51.02	0.43
Aerial	131338	54.15	0.49

Fig. 1 Relative analysis of PSNR and BER

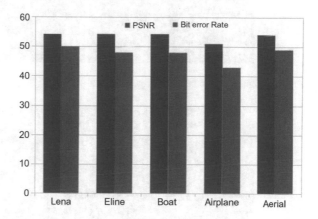

extracted bits is measured using BER metric. The PSNR gives the visual quality of the distorted image after performing the near reversible data embedding operation.

From the experimental results as shown in Table 1, we can observe that the proposed technique has higher visual quality in terms of PSNR. Here, PSNR >50 dB implies that the embedded image is almost seamlessly modified when compared to the original image. However, the BER achieved is around 50% for most of the test images. The relative analysis of PSNR and BER is shown in Fig. 1. From Figs. 2 and 3, we can observe that the original image and the embedded images are looking same.

4 Conclusion

As near reversibility of data hiding has emerging applications in remote sensing and medical image authentication, we presented a near reversible scheme in spatial domain. The proposed results demonstrate that a higher visual quality is achieved, by being able to maintain at least 50% of the bit error rate.

Fig. 2 Original and embedded images of Lena

Fig. 3 Original and embedded images of boat

References

1. Barni M, Bartolini F, Cappellini V, Magli E, Olmo G. (2002) Near-lossless digital watermarking for copyright protection of remote sensing images. In: IEEE international geoscience and remote sensing symposium, 2002. IGARSS '02., vol 3, pp 1447–1449.
2. Tang Y-L, Huang H-T (2007) Robust near-reversible data embedding using histogram projection. In: In IIH-MSP 2007, vol 02, pp 453–456.
3. Zhang B, Xin Y, Niu X-X, Yuan K-G, Jiang H-B (2010) A near reversible image watermarking algorithm. In: International conference on machine learning and cybernetics (ICMLC), vol 6, pp 2824–2828, July 2010.
4. Ahmed F, Moskowitz IS (2006) A semi-reversible watermark for medical image authentication. In: 1st Transdisciplinary conference on distributed diagnosis and home healthcare. D2H2, pp 59–62.
5. Gujjunoori S., Amberker B. (2013) A DCT Based Near Reversible Data Embedding Scheme for MPEG-4 Video. In: Proceedings of the Fourth International Conference on Signal and Image Processing 2012 (ICSIP 2012). Lecture Notes in Electrical Engineering, vol 221. PP. 69–79, Springer, India.

6. Sagar Gujjunoori, B.B. Amberker, DCT based reversible data embedding for MPEG-4 video using HVS characteristics, In Journal of Information Security and Applications, Volume 18, Issue 4, 2013, Pages: 157–166.
7. Lee, Chin-Feng and Weng, Chi-Yao and Chen, Kai-Chin, An efficient reversible data hiding with reduplicated exploiting modification direction using image interpolation and edge detection, Multimedia Tools and Applications, vol: 76 (7), 2017, pages: 9993–10016.
8. http://sipi.usc.edu/database/database.php?volume=misc.

A New Method for Edge Detection Under Hazy Environment in Computer Vision

Chatakunta Praveen Kumar, Koona Hemanath
and B. Surya Narayana Murthy

Abstract   Images captured in outdoor areas are typically degraded in quality by its turbid medium in the nature such as haze, fog, and smoke. The absorption and scattering of light on such kind of images effects the quality of the image. The degraded images will loss the contrast and color artifacts from the original image. Edge detection is another challenging issue on such kinds of degraded images. There are several research works are under progress to reduce the haze exists in the image. Despite the fact that haze exclusion systems will diminish the haze exists in the image, the results obtained from these techniques are having a penalty in the natural look of the image. In this work, we proposed a successful mechanism to find the edges for hazy images. As a first step, dark channel prior (DCP) is utilized to diminish the undesirable haze in the captured image. The statistics shows that this method effectively works for the images taken in an outdoor hazy environment. The key observation of this method is that at least one color channel is having a minimum intensity value in a local patch. The results of our work show the decent outcomes compared with other contrast adjustment strategies. Besides, we have applied Sobel edge mechanism to discover the edges of the resultant image.

Keywords   Image enhancement · Hazy images · DCP · Edge detection

C. P. Kumar
Department of Computer Science Engineering, Institute of Aeronautical Engineering,
Dundigal, Hyderabad, India
e-mail: 526.praveen@gmail.com

K. Hemanath (✉)
Department of Information Technology, MLR Institute of Technology, Hyderabad, India
e-mail: hemanath.it@gmail.com

B. S. N. Murthy
Department of Computer Science Engineering, Dravidian University, Kuppam,
Andhra Pradesh, India
e-mail: suryanarayanamurthy.b@gmail.com

© Springer Nature Singapore Pte Ltd. 2019   621
R. S. Bapi et al. (eds.), *First International Conference on Artificial Intelligence
and Cognitive Computing* , Advances in Intelligent Systems and Computing 815,
https://doi.org/10.1007/978-981-13-1580-0_60

1   Introduction

Image enhancement is considered as one of the most basic systems in the image processing domain. The essential point of enhancing the image denotes that upgrading the quality of the image in the sense of contrast, brightness, etc. In other words, to expand the visual appearance of an image contrasted with the captured image. It is vital to adjust the pixel values to change the quality of the image. The objective of image enhancement is the resultant image should have a good quality, i.e., natural look in which objects should be noticeable contrasted with the captured image. It is essential in numerous applications, for example, locating the edges, detection, and recognizing the objects in an image [1, 2].

The images collected in outdoor places having haze that obscures scenes decrease visibility and change the hues in the image. The quality degradation is annoying problem for the photographers while taking the picture. It is additionally a dangerous threat to reliability in numerous systems, for example, outdoor surveillance, human and face recognition, and aerial imaging. In this sense, it is crucial to reduce the blurriness or haziness exist in the image in several computer vision applications.

Images of outdoor areas are typically degraded in quality by its turbid medium in the atmosphere such as haze, fog. There is a variance in the original look of the image from the captured image by the hazy environment. The irradiance received by the camera from the point where the scene is captured is attenuated, and atmospheric air light will effect on the original scene [3]. The images taken from such environment will effect in the contrast and color of the original image as shown in Fig. 1.

Fig. 1 Images from hazy environment

The primary challenge exists in the uncertainty of the issue. Cloudiness constricts the light reflected from the scenes and further mixes it with some added substance sunlight in the environment. The objective of murkiness evacuation is to recuperate the reflected light (i.e., the scene hues) from the mixed light. This issue is numerically questionable. This issue is numerically questionable: there are an infinite number of arrangements given the mixed light. How might we know which arrangement is valid? We have to answer this inquiry in cloudiness expulsion. Equivocalness is a typical test for some PC vision issues. As far as arithmetic, uncertainty is on the grounds that the quantity of conditions is littler than the quantity of questions. The techniques in PC vision to comprehend the uncertainty can generally be arranged into two procedures. The first one is to get more known factors, e.g., some murkiness evacuation calculations catch various pictures of a similar scene under different settings (like polarizers). Be that as it may, it is difficult to acquire additional pictures by and by. The second methodology is to force additional limitations utilizing some information or presumptions known already, in particular, a few 'priors.' In general, a prior can be some statistical/physical properties, rules, or heuristic assumptions.

In the first technique, we achieve a faster speed by solving a large kernel linear system. This discovery is against conventional theories but we can prove its validity theoretically and experimentally. The second technique is a novel edge-aware filter. We find this filter superior to previous techniques in various edge-aware applications including haze removal. Thus, we advance the state of the art in a broader area.

There exists several methods to remove haze using the polarization methods [4, 5] with varying degree of polarization. There are several constraints in [6–8] to obtain the results from multiple images under various environmental conditions. We require some depth information from the depth-based methods to obtain the result [8, 9].

The remainder of this paper is organized as follows. The background work on the mathematical equation of haze image is discussed in Sect. 2. In Sect. 3, the foundational concepts of DCP, depth map estimation, and Sobel edge detection are discussed. The experiments we have done are discussed in Sect. 4. Finally, Sect. 5 depicts conclusion and further work in future.

2 Background Work

In computer vision, the model generally used to portray the formation of hazy image is shown below:

$$I(x) = J(x)t(x) + A(1 - t(x)) \tag{1}$$

Here 'I' denotes the intensity value which is observed, 'J' denotes the scene radiance, 'A' represents atmospheric light, and t is the transmission medium. The physical capturing of the Eq. 1 is shown in Fig. 2. The main intension of removing the haze denotes that recovering the 'J' and 'A' from 'I'. The terms used here $J(x)t(x)$

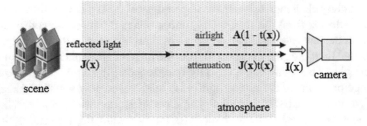

Fig. 2 Structure of physical haze imaging model

shown in (1) denote the direct attenuation. $A(1-t(x))$ denotes the air light. In the homogeneous atmosphere, the transmission t is given as:

$$t(x) = e^{-\beta d(x)} \tag{2}$$

In RGB color space, the terms shown in Eq. 1, such as vectors A, $I(x)$ and $J(x)$, are co-planar. The transmission ratio is computed as:

$$t(\mathbf{x}) = \frac{\|A - I(x)\|}{\|A - J(x)\|} = \frac{\|A^C - I^C(x)\|}{\|A^C - J^C(x)\|} \tag{3}$$

3 Related Study

In this session, we will discuss few more details about how to remove the haze exist in the image using dark channel prior (DCP), how to estimate the depth map and how to apply the Sobel edge detection algorithms.

3.1 Dark Channel Prior

The challenging issue in haze removal is due to the ambiguity, basically there is merely a single image as input. The dark channel prior is a technique used to remove the haze present in the outdoor images [10]. The key observation in this method is most of the local patches in the images (those do not contain haze) having few pixels are having very low intensities in any of the color channel; for example, in the RGB image, either R or G or B value will be very low value; i.e., it is almost 0. In general for any image J, the dark channel will be:

$$\mathbf{J}^{\text{dark}}(x) = \min_{c \in \{r,g,b\}} \left(\min_{y \in \Omega(x)} (J^C(y)) \right) = 0 \tag{3}$$

Eq. (3) is almost similar to image matting equation, denote by refined $t(x)$. Rewriting $t(x)$ and $t^1(x)$ in their vector form as t and t^1, we minimize the following cost function:

$$E(t) = t^{\mathrm{T}} L t + \lambda\, (t - \hat{t})^{\mathrm{T}} (t - \hat{t}). \tag{4}$$

Here L denotes the Laplacian matrix and λ denotes a regularization parameter, the former term denotes the smooth term, and latter denotes the data term.

This method works on the concept of identifying the dark pixels, which have minimum value in at least in any one of the channel of the RGB image excluding the sky area. It is effective and giving the accurate results among the other techniques recently developed. It has four stages starting with estimating the light and transmission map followed by refining and reconstructing the image, and it considers both contrast improvement and color restoration.

The working procedure of DCP is shown in Fig. 3. Initially, the hazy image captured from the outdoor is considered as input and it will be given for computing the dark channels. We can estimate the transmission function and atmospheric light. Color restoration algorithm is applied to restore the color. We can quantitatively compare the result with the other conventional techniques (Fig. 4).

Consider the image shown in Fig. 5. We have taken two input images: one is an outdoor image covering the buildings and roads, and another image which contains some text over the banner. We cannot clearly say the text present in that if there is more haze. We can dehaze and can get the natural quality; the result of the dark channel prior is shown in Fig. 5b, d for the input images Fig. 5a, c.

3.2 Depth Map Estimation

Consider a scene point position x at distance $d(x)$ from the observer. Here d is called as depth of the scene and transmission t is related to the depth t.

$$t(x) = \exp\left(-\int_{0}^{d(x)} b(z)\mathrm{d}z\right) \tag{5}$$

Here b is called scattering coefficient, and integral 0 to $d(x)$ is a line between a scene point and the human who is capturing the image. When we consider the scattering coefficient, we can rewrite the above equation as:

$$t(x) = \exp(-b(d(x))) \tag{6}$$

Here $d(x) = -\frac{\ln(x)}{b}$; we can estimate the depth by unknown scale.

Fig. 3 Working procedure
of DCP

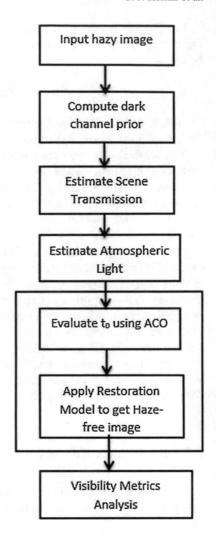

3.3 Sobel Operator

In image processing, Sobel operator is mainly used in the edge detection techniques. Actually, it is a derivative operator, to compute the gradients of the image. In all the points of image, it computes the gradient vector or norm of the corresponding vector. In other words, it is a convolution operator to compute the edges in both horizontal and vertical directions, and it is less expensive in computations than the other edge detection algorithms, and the results of this are relatively crude, specifically for high-frequency variation in the original image. The edge detection strategies exist in the literature follow the assumption the, an edge is considered the discontinuity in the neighborhood pixels [10]. This will help us to find the gradients of the image. We

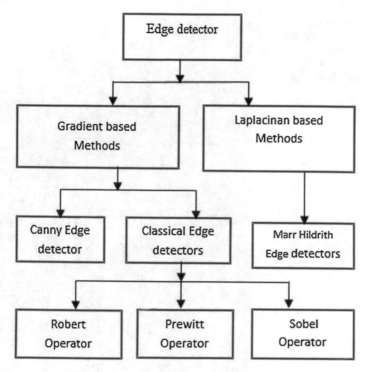

Fig. 4 Types of edge detectors

can find the derivatives of the intensity values in the complete image to find the maximum derivative to identify the edge in a theoretical manner. The gradient is simply a vector, whose intensity values are changed in the X and Y directions [11].

Consequently, we can found the gradients of the images with the following formula:

$$\frac{\partial f(x, y)}{\partial x} = \Delta x = \frac{f(x + dx, y) - f(x, y)}{\partial x} \tag{7}$$

$$\frac{\partial f(x, y)}{\partial x} = \Delta y = \frac{f(x, y + dy) - f(x, y)}{\partial y} \tag{8}$$

Here the ∂x and ∂y measures the distance in both the directions in axis; in general, ∂x and ∂y in terms of the number of pixels between the two given points. $\partial x = \partial y = 1$ is the point at which pixel coordinates are (I, j) denoted as:

$$\Delta x = f(i + 1, j) - f(i, j) \tag{9}$$

$$\Delta y = f(i, j + 1) - f(i, j) \tag{10}$$

Fig. 5 a Hazy outdoor
image **b** recovered haze free
image **c** hazy image
containing names **d**
recovered haze-free image

In addition, we can also apply the other edge detection algorithms. We have listed
out few of the categories of conventional edge detection mechanisms to apply as a
replacement for the Sobel edge detection algorithm. The list is shown in Fig. 4.

From the above equation, magnitude and the direction of the gradient (θ) can be
computed as:

$$\text{Magnitude} = \sqrt{\Delta x^2 + \Delta^2 y} \tag{11}$$

$$\theta = \tan^{-1} \frac{\Delta y}{\Delta x} \tag{12}$$

The derivative masks which convolve with image are:

$$\Delta x = \begin{bmatrix} -1 & 1 \\ 0 & 0 \end{bmatrix}$$

$$\Delta y = \begin{bmatrix} -1 & 0 \\ 1 & 0 \end{bmatrix}$$

Similarly, there are various masks are presented to find the edges from the given image [12, 13]. Robert's edge operator will consider the following masks:

$$\Delta x = \begin{bmatrix} 0 & 1 \\ -1 & 0 \end{bmatrix}$$

$$\Delta y = \begin{bmatrix} 1 & 0 \\ 0 & -1 \end{bmatrix}$$

The Sobel operator mask is considered as:

$$\Delta x = \begin{bmatrix} -1 & 0 & 1 \\ -2 & 0 & 2 \\ -1 & 0 & 1 \end{bmatrix}$$

The algorithm of Sobel edge detection algorithm is shown in Fig. 6. Initially, the gradients are computed in both direction, and we apply Sobel edge detection algorithm on the gradient to compute the magnitude and the angle of the direction of the edge.

We have applied the Sobel operator on the grayscale image, and the edge detection result can be observed in Fig. 7.

Fig. 6 Sobel edge detection algorithm

> **Algorithm: Sobel**
> **Input:** Two Dimensional image
> **Output:** Two Dimensional image having only edges
> **Step 1**: Read the input image
> **Step 2:** Apply mask in both directions such as Gx, Gy to the input image
> **Step 3**: Apply Sobel edge detection algorithm and the gradient
> **Step 4**: Masks manipulation of Gx, Gy separately on the input image
> **Step 5:** Results combined to find the absolute magnitude of the gradient. $|G|=\sqrt{Gx^2 + Gy^2}$
> **Step 6:** the absolute magnitude is the output edges

Fig. 7 **a** Input image **b** combined result of Sobel operator in X and Y directions

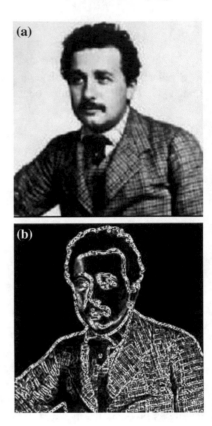

(a)

(b)

4 Results and Discussion

This section shows our works on the dark channel prior and the edge detection algorithms. Our work has done on OpenCV environment using Python. First, we have applied the DCP to the hazy image. Figure 8a is the input image; we can observe that how the image is appeared. It is completely covered with white pixel values in the front end. Figure 8b is the grayscale image converted from the input image. Figure 8c is the Sobel edge detection result. It is clear that most of the edges are not clear in the result; hence, most of the input image is covered with haze. Hence, we applied first DCP to the input image, and later, we applied the Sobel edge detection on the dehazed image. Figure 9a is the dehazed image, and Fig. 9b is the grayscale image, and Fig. 9c is the Sobel edge detection result. Our subjective comparison shows that the edge detection is better than the one which is applied directly without applying DCP.

There are several methods which exist in the literature to remove the haze present in the image. Fattal [14] and Tan [15] have done their research work to dehaze the

Fig. 8 **a** Hazy image **b** grayscale image converted (a) **c** results of Sobel

Fig. 9 **a** Dehazed image using DCP **b** grayscale image of (a) **c** result of Sobel

image. Figure 10a is the input image, and Fig. 10b is the result of Fattal, and Fig. 10c is the result of Tan, and Fig. 10d is the dark channel prior result.

Fig. 10 **a** Input image **b**
Fattel result **c** Tan result **d**
DCP result

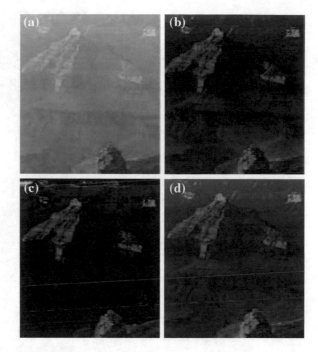

In addition, we have shown the depth map by taking the input image with different
haze levels. Figure 11a, c, e shows the various levels of input hazy images, and
Fig. 11b, d, f is their compute depth maps, respectively.

4.1 Quality Measurements

There exists numerous measures for assessing the quality of the gray images in
objective passion. Such kind of measures is crucial in the comparative study to
prove the proposed result is better than the existing one. Additionally, the time also
considered for few cases. The popular methods include entropy, MSE, and PSNR.

The mean square error can be computed as:

$$\mathrm{MSE} = \frac{1}{\mathrm{NXM}} \sum_{i=0}^{N-1} \sum_{j=0}^{M-1} [X(i,\ j) - Y(i,\ j)]^2$$

The minimum-valued MSE and maximized PSNR are considered for good results.
Entropy computes the available information which exists in the image, and maximum value of this is considered and indicates that the input image contains enough
information except in noisy conditions. The following is the formula for getting the
AIC.

Fig. 11 Depth estimation results for different haze levels

$$\text{AIC} = -\sum_{k=0}^{L-1}(k)\log P(k)$$

5 Conclusion

This paper contributes detection of edges present in the hazy image. Firstly, we have used the DCP and a haze removal technique to diminish the haze exists in the image. This new method gives best results compared to other haze removal techniques, and we have shown few subjective experiments on this work and also computed the depth map of the image. We omitted the running time in this work. Secondly, we have used SE detection for detecting sharp boundaries or edges from the dehazed image. The subjective comparison shows that the results are far better than applying SE alone. We have compared our work with second-order derivate such as Laplacian. Robert's, canny, and other techniques should be compared to get the efficient results, and this will be considered as our future work.

References

1. Rafael C, Gonzalez, Richard E, Woods & Steven L. Eddins, 2004, 'Digital Image Processing Using MATLAB', Pearson Prentice Hall.
2. S. Lau, 1994, "Global image enhancement using local information," Electronics Letters, vol. 30, pp. 122–123.
3. H. Koschmieder, "Theorie der HorizontalenSichtweite," Beitr. Phys. Freien Atm., vol. 12, pp. 171–181, 1924.
4. Y.Y. Schechner, S.G. Narasimhan, and S.K. Nayar, "Instant Dehazing of Images Using Polarization," Proc. IEEE Conf. Computer Vision and Pattern Recognition, vol. 1, pp. 325–332, 2001.
5. S. Shwartz, E. Namer, and Y.Y. Schechner, "Blind Haze Separation," Proc. IEEE Conf. Computer Vision and Pattern Recognition, vol. 2, pp. 1984–1991, 2006.
6. S.G. Narasimhan and S.K. Nayar, "Chromatic Framework for Vision in Bad Weather," Proc. IEEE Conf. Computer Vision and Pattern Recognition, vol. 1, pp. 598–605, June 2000.
7. S.K. Nayar and S.G. Narasimhan, "Vision in Bad Weather," Proc. Seventh IEEE Int'l Conf. Computer Vision, vol. 2, pp. 820–827, 1999.
8. S.G. Narasimhan and S.K. Nayar, "Contrast Restoration of Weather Degraded Images," IEEE Trans. Pattern Analysis and Machine Intelligence, vol. 25, no. 6, pp. 713–724, June 2003.
9. J. Kopf, B. Neubert, B. Chen, M. Cohen, D. Cohen-Or, O. Deussen, M. Uyttendaele, and D. Lischinski, "Deep Photo: Model-Based Photograph Enhancement and Viewing," ACM Trans. Graphics, vol. 27, no. 5, pp. 116:1–116:10, 2008.
10. Agaian, S. S., Baran, T. A., & Panetta, K. A. (2003). Transform-based image compression by noise reduction and spatial modification using Boolean minimization. IEEE Workshop on Statistical Signal Processing, 28 Sept–1 Oct, pp. 226–229.
11. Baker, S., & Nayar, S. K. (1996). Pattern rejection. Proceedings of IEEE Conference Computer Vision and Pattern Recognition, 544–549.
12. Chang-Huang, C. (2002). Edge detection based on class ratio. 152, sec.3, Peishen Rd., Shenkeng, Taipei, 22202, Taiwan, R.O.C.
13. John F. Canny, A Computational Approach to Edge Detection. IEEE Transactions on Pattern Analysis and Machine Intelligence, Vol. PAMI 8, No. 6, November, 1986.
14. R. Fattal, "Single Image Dehazing," Proc. ACM SIGGRAPH '08, 2008.
15. R. Tan, "Visibility in Bad Weather from a Single Image," Proc. IEEE Conf. Computer Vision and Pattern Recognition, June 2008.

iBeacon-Based Smart Attendance Monitoring and Management System

Suresh Limkar, Shubham Jain, Shraddha Kannurkar, Shweta Kale, Siddesh Garsund and Swarada Deshpande

Abstract The problem of fake attendance (proxy) is growing day by day. There are many technologies used to develop the attendance system like RFID, biometrics, bar code. Because all these technologies have been used, many students are able to mark the fake attendance (proxy) and some systems are very time consuming. There is a need to develop the riskless and shielded attendance monitoring and management system which is our main motto. The technology which we are going to use in the proposed system is "**iBeacon Technology**" which is mostly used for indoor positioning system and proximity detection and also beneficial in removing the problem of fake attendance. The newer technology which is found out to interact with hardware is iBeacon. An iBeacon works on Bluetooth Low Energy technology which is used to send a signal in a definite format. The iBeacon devices give accurate results in an indoor system in spite of the influencing factors of radio waves. In iBeacon devices, we can adjust the range of beacon up to which the device can transmit. The range of iBeacon is up to 200 m, which can vary according to the type of iBeacon used. An infinite number of devices can be connected to the iBeacon. The main advantage of the proposed system is to overcome the human intervention and follow the concept of paperless smart work. The accuracy of our proposed system is very high, i.e., 100%, and removes human intervention to almost zero.

S. Limkar (✉) · S. Jain · S. Kannurkar · S. Kale · S. Garsund · S. Deshpande
Department of Computer Engineering, AISSMS IOIT, Pune 411001, India
e-mail: sureshlimkar@gmail.com

S. Jain
e-mail: shubhamujain@gmail.com

S. Kannurkar
e-mail: shraddhak0697@gmail.com

S. Kale
e-mail: shwetakale29@gmail.com

S. Garsund
e-mail: siddesh.garsund@gmail.com

S. Deshpande
e-mail: swaradadeshpande2096@gmail.com

© Springer Nature Singapore Pte Ltd. 2019
R. S. Bapi et al. (eds.), *First International Conference on Artificial Intelligence and Cognitive Computing* , Advances in Intelligent Systems and Computing 815,
https://doi.org/10.1007/978-981-13-1580-0_61

Keywords iBeacon · Micro-localization · Tracking · Bluetooth low energy
Indoor location · Proximity detection · RSSI

1 Introduction

A brief introduction of wireless technologies has been discussed below:

1.1 RFID

The RFID reader and RFID tag consist of a wireless technology in them which has
the electromagnetic waves to collect the stored information automatically. The RFID
is especially recognized as the Radio Frequency Identification technology which is
uniquely identified and traced as well. The most commonly used RFID tags are:
active tag, passive tag, and semi-passive tag.

1.2 iBeacon (BLE-Based)

The wireless personal area network technology which is designed as well as marketed by Bluetooth Special Interest Group is Bluetooth Low Energy (also known
as Bluetooth LE and marketed by Bluetooth Smart). It is used in various innovative
applications like fitness, beacons, health care, security features, and home entertainment industries. The very newer technology developed by Apple is iBeacon which is
being used in the devices and operating system since 2013. It is based on Bluetooth
Low Energy (BLE). iBeacon functions as an indoor positioning system that allows
businesses to advertise their presence to nearby smartphones. With iBeacon's set
up, businesses can send messages to potential customers (such as special offers or
goods) when they walk past an iBeacon. RSSI is basically a unit of measurement
which is used to get a signal from an access point or a router. It is basically used in
video networks.

1.3 Motivation and Objectives

Students' attendance is becoming crucial analysis facet in ongoing education methodology in phrontistery as well as in colleges. The attendance monitoring system which
is conventionally used has numerous flaws. The most hectic job is circulating the
attendance page to an enormous quantity of students in the class daily. The additional
time is wasted by an unnecessary roll call. It can be also possible that the fake atten-

dance is registered by the student on daily passed attendance sheet. All the relevant attendance records are vanished if these documents are misplaced by the staff and it becomes very tedious to prepare all these documents correctly. These documents can be easily manipulated because the record of the attendance is maintained on the papers. The most terrible and sluggish task is that the staff has to perform plenty of computation while evaluating the attendance.

As a substitute to the conventional attendance monitoring system, a smart system is developed which can record the presence of pupils and staff without much human intervention involvement. Such systems are developed with the technologies like biometrics, bar code, and NFC [1]. To find out the identity of user, the enrolled students need to scan RFID tag which is mounted on the identification card and also verify by scanning the fingerprint. Since the fake attendance can be registered by students, the fingerprint scanner is incorporated in technique to prevent proxy which is being recorded by pupils. An attendance of particular student is cataloged in the database automatically, when the fingerprint and unique RFID tag UID are compared along with the data stored in information bank for specific student. An average presence of every student can also be calculated by the system automatically; i.e., the attendance reports can be generated.

Generally, the student attendance monitoring system is developed using RFID, bar code, biometric, etc., but these systems consist of human intervention. Hence in our proposed system, we have used iBeacon Technology. Basically, iBeacon is a Bluetooth Low Energy device; hence, the technology used is Bluetooth only but with some more modifications and effectiveness. Traditional attendance system is used to track the attendance of a particular person, and it is very time consuming and often leads to human errors. That is why we are using iBeacon which is an indoor tracking, a variant of wireless tracking. The main advantage of the proposed system is to overcome the human intervention and follow the concept of paperless work, and number of devices can be connected to the iBeacon. In our proposed system, the iBeacon of all the student will get directly connected to the central device as soon as they enter the class, i.e., through iBeacon, and the attendance of every student will be marked. For detection of fake attendance, there will be a bidirectional people counter placed at the door which will show the count of the student entered in the class. As soon as staff deactivates the device, no one else would be able to mark attendance again. Then, the collected data will be stored on cloud for further analysis.

2 Related Work

Jeongyeup Paek et al. [1] proposed "Geometric Adjustment Scheme for Indoor Location-Based Mobile Applications." The system proposed a simple class attendance checking application by performing a simple form of geometric adjustments to compensate for the natural variations in beacon signal strength readings. The advantage of this system is it checks the attendance automatically. The disadvantage of this system is that it is not resilient and robust to dynamic RSSI values and it only checks the attendance.

Germ'an H. Flores et al. [2] proposed "An iBeacon Training App for Indoor Fingerprinting." They have introduced a novel approach to streamline the process of collecting training data for the fingerprinting indoor localization technique. The training data can be collected from a Bluetooth-enabled iOS client device and requires no complex graphical maps since we only need the x-y location where the iBeacon training data was collected. The disadvantage is the app is not universally compatible.

Md Sayedul Aman et al. [3] proposed "Reliability Evaluation of iBeacon for Micro-Localization." This paper shows how the RSSI values from iBeacon can be taken and used to present the micro-localization. The mobile device used to get the desired output was R-pi3. The advantage is that the algorithm used is computationally inexpensive, and thus it can be employed in devices based on IoT. The disadvantage is it may not be totally accurate.

Zhenghua Chen et al. [4] proposed "Smartphone Inertial Sensor-Based Indoor Localization and Tracking with iBeacon Corrections." The paper proposed an indoor localization system which is assisted by an iBeacon and in which in-built smartphone inertial sensors are used. The advantage of the system compared to Wi-Fi is that the iBeacon Technology works on battery and is much more power efficient. The disadvantage of system is that the optimal location of each iBeacon is not known, and they have assumed the location.

Mr. Sehul A. Thakkar et al. [5] proposed "An iBeacon: Newly Emerged Technology for Positioning and Tracking in Indoor Place." The needs of the user for the particular indoor area can be fulfilled by "iBeacon Technology" which is for positioning and tracking. It can be used to locate and track any valuable asset, where it is actually located and where it moves from one place to another place.

Rahul More et al. [6] proposed a RFID-based student attendance monitoring system. Initially, students' partial attendance is marked as student swaps the I-Card on RFID reader. Then, teacher login for the lecture with lecture ID and OTP is given to students by orally or written, after entering this OTP into their android mobile, and the student attendance is marked completely. The advantage of this system is to solve the attendance problem and get the information of students at any time, and limitation of the system is that it requires human intervention and it is also time consuming.

Srinidhi MB et al. [7] proposed a "Web-enabled secured system for attendance monitoring and real-time location tracking using Biometric and Radio Frequency Identification (RFID) technology." The regularity of the child will be informed to parents through the SMSs. The main advantage is that the graph is generated which gives the information about the regularity of student. But the disadvantage of this system is that it requires the hardware devices like RFID, biometric which increases the cost.

Vishal Bhalla et al. [8] has proposed Bluetooth-Based Attendance Management System. In the proposed system, the staff's mobile is used for taking the attendance. Bluetooth connection is used for querying the student's mobile by staff's mobile which is having application software installed on it and over the transmission of

Table 1 Comparison of technologies

Parameters	Technologies			
	RFID [7]	Bar code	NFC	Bluetooth [8]
Range	Minimum distance over 1 m or 3 feet	Long-range bar code scanner that enables them to read 1D, 2D bar code up to 50 feet	Approx. up to greater than 20 cm	Approx. 100 m or 320 feet for class 1
Frequency	125 or 135 MHz	–	2483.5 MHz	13.56 MHz
Cost	499	250	599	4090
Data rate	4008 kbps	–	24 mbits	42 kbps
IEEE standard	802.15	–	802.15	802.15.1

media access control (MAC) addresses of student's mobile to the staff's mobile, and this confirms the presence of the particular student. The problem in the proposed system is student's phone is required for attendance.

Josphineleela. R et al. [9] proposed "An Efficient Automatic Attendance System using Fingerprint Reconstruction Technique." This system can be used for both student and staff. In this system, the fingerprint is taken as an input for attendance management. In this system, novel fingerprint reconstruction algorithm is used. This new algorithm reconstructs the phase image from minutiae. The disadvantage of this system is that student has to stand in the queue and has to wait for his turn to mark attendance.

A comparison in Table 1 shows the earlier implemented techniques require human interventions and they are time consuming. We have proposed a far much better system for attendance monitoring and management which is accurate and faster.

3 Proposed Work

3.1 Methodology

The elements used in proposed system are bidirectional people counter, student, the central device (server), staff member, cloud, iBeacon device. Bidirectional counter will be set at the doorstep. The value of counter will be incremented by one every time student enters the classroom. Now, the student's I-Card will be consisting of iBeacons which will get connected to the server immediately and the attendance will be marked. As the staff's device and server are connected to the cloud, the notification regarding the counter and the attendance of the student will be sent to staff member and also the student will only receive the attendance notification. If any of the proxy attendance has been marked, it will be known by matching the values between the counter and the server and staff member can discard the fake

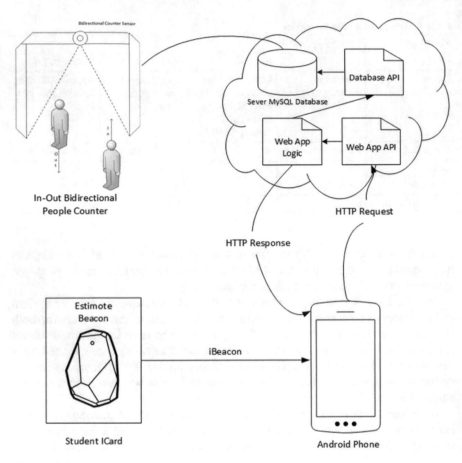

Fig. 1 Architectural diagram

attendance immediately. According to the data saved on cloud, attendance report will be generated; i.e., average students' attendance is calculated, and we can also generate the student's defaulter list (Fig. 1).

3.2 Algorithm

1. **Algorithm Student_Attendance ()**
2. $\alpha \leftarrow 0$ // α=counter
3. s ←the number of students whose attendance is to be marked
4. while($\alpha <= s$) do
5. if(s goes from out to in)
6. α ++;
7. end
8. else if (s goes from in to out)
9. α --;
10. end
11. end

// Map the estimote beacon on the student's id card to the central server
// Fetch the UUID of each beacon i.e. get the received signals (alpha)

12. if($\delta > (\alpha+1)$)   // δ = Received signal of iBeacon
13. Fake attendance is detected and immediate alert to staff through the cloud server;
14. end
15. else
16.   Push notification to s that attendance has been marked;
17. end
18. end

3.3 Hardware Specifications

3.3.1 iBeacon

It is based on Bluetooth Low Energy (BLE). iBeacon [10] functions as an indoor positioning system that allows businesses to advertise their presence to nearby smartphones.

3.3.2 Counter Sensor

It is bidirectional counter sensor [11] which records the precise number of people which are going in and out of the door.

3.3.3 Android Phone/iPhone (Central Device)

Android devices [12] with Bluetooth 4.0 and Android 4.3 and later [Samsung Galaxy S4 Mini, HTC Butterfly (aka Droid DNA)].

Apple iOS devices [10] with Bluetooth 4.0 [iPhone 5/5s/6 and later, iPad-3rd generation and later, iPad Mini-1st generation and later, iPod Touch (5th generation and later)].

3.4 Advantages

1. Removing the human intervention.
2. Development of reliable and highly accurate attendance monitoring system.
3. To overcome the problem of fake attendance (proxy).
4. Stick to work on paperless concept.

3.5 Limitations

The proposed system carries certain limitations which says that fake attendance (proxy) can only be identified, instead of discarding automatically.

4 Result Analysis

As you can see the below result analysis [2], the students' attendance is automatically marked as soon as they present in the class with their iBeacon-enabled identity card and push notification is sent with the help of the central server in response to it, i.e., present; otherwise, they are marked as absent. On comparing the number of received signals of iBeacon and counter value, if both the values are mismatched, then we can conclude that fake attendance has been marked (Table 2).

5 Conclusion and Future Work

In this paper, we proposed the smart attendance system which will totally overcome the problem of fake attendance (proxy) which was the main motto of our research. It will also minimize the human intervention to almost zero. Implementation of our proposed system will follow the paperless approach as there is no use of paper anywhere for attendance management leading to the green environment. The accuracy of the proposed system is very high, i.e., 100%. We can say that our proposed system

Table 2 Data set

Roll No	Name	iBeacon UUID	Attendance
F76001	Shraddha Kannurkar	C9561B04-DE7A-4A3C-83EB-E35AE76EE72B	Present
F76002	Shubham Jain	76ABEA19-9A95-47DB-BC05-8BCDA3970D50	Present
F76003	Shweta Kale	CEF2A40C-DE33-465F-BE40-CE7A88A1D18B	Absent
F76004	Parag Sanyashiv	BEF2A40C-DE33-465F-BE40-NE7A88A1D18B	Absent
F76005	Anushree Gawande	Z9561B04-DE7A-4A3C-83EB-R35AE76EE72B	Present

is more efficient than the existing systems which are based on various technologies such as RFID, bar code, biometric, wireless. Furthermore, based on the observations and results obtained in this work, our future work is to develop the system which will not only detect the fake attendance but it will also remove the fake attendance automatically.

6 Declaration

We have taken necessary permission to use the data set/images/other materials from respective parties. We, authors, undertake to take full responsibility if any issue arises in future.

References

1. Jeongyeup Paek, JeongGil Ko and Hyungsik Shin, "A Measurement Study of BLE iBeacon and Geometric Adjustment Scheme for Indoor Location-Based Mobile Applications", Hindawi Publishing Corporation Mobile Information Systems Volume 2016, Article ID 8367638, 13 pages.
2. Germán H. Flores, Thomas D. Griffin, Divyesh Jadav, "An iBeacon Training App for Indoor Fingerprinting", 2017 5th IEEE International Conference on Mobile Cloud Computing, Services, and Engineering, 978-1-5090-6325-3/17 $31.00 © 2017 IEEE.
3. Md Sayedul Aman, Haowen Jiang, Cuyler Quint, Kumar Yelamarthi, Ahmed Abdelgawad, "Reliability Evaluation of iBeacon for Micro Localization", 978-1-5090-1496-5/16/$31.00 © 2016 IEEE.

4. Zhenghua Chen, Qingchang Zhu, and Yeng Chai Soh, "Smartphone Inertial Sensor Based Indoor Localization and Tracking with iBeacon Corrections", https://doi.org/10.1109/tii.2016.2579265, IEEE Transactions on Industrial Informatics.
5. Mr. Sehul A. Thakkar, Mr. Sunil Patel, Mr. Brijesh Kamani, "iBeacon: Newly Emerged Technology for Positioning and Tracking in Indoor Place", IJARCCE, Vol. 5, Issue 3, March 2016.
6. Rahul More, Kiran Patel, Rhutika Tavasalkar, Yogita Khandagale, Prof. A. R. Uttarkar, "Student Attendance system and Monitoring using RFID and Processing", IERJ, Volume 2 Issue 2 Page 402–406, 2016, ISSN 2395-1621.
7. Srinidhi MB, Romil Roy, "A Web Enabled Secured System for Attendance Monitoring and Real Time Location Tracking Using Biometric and Radio Frequency Identification (RFID) Technology", 2015 International Conference on Computer Communication and Informatics (ICCCI-2015), Jan. 08–10, 2015, Coimbatore, INDIA.
8. Vishal Bhalla, Tapodhan Singla, Ankit Gahlot, Vijay Gupta, "Bluetooth Based Attendance Management System", IJIET-2013, Vol. 3 Issue 1 October 2013, ISSN: 2319-1058.
9. Josphineleela. R, Dr. M. Ramakrishan, "An Efficient Automatic System Using Fingerprint Reconstruction Technique", IJCSIS, Vol. 10, No. 3, March 2012.
10. https://community.estimote.com/hc/en-us/articles/204092986-Technical-specification-of-Estimote-Beacons-and-Stickers.
11. http://www.infoplus.com.sg/images/people_counter.pdf.
12. https://www.cisco.com/c/dam/en/us/solutions/collateral/enterprise-networks/connected-mobile-experiences/ibeacon_faq.pdf.

Fault Identification in Power Lines Using GSM and IoT Technology

**Bhavana Godavarthi, Vasumathi Devi Majety, Y. Mrudula
and Paparao Nalajala**

Abstract This chapter focuses on today's civil constructions that have been developed using closed wiring, which has resulted in the identification of fault locations and their subsequent rectification becoming a major task. There about a pulse-type reverberation extending shortcoming locator need been formed which may be suitableness to placing maintained faults once phone also control transmission lines an measure of these instruments bring been on organization to a couple very much percentage occasion when they bring been used to find a totally assortment about control power line faults for huge declines in direction pauses and economies in line watching. In front of the improvement from requesting new circlet it must a chance to be tried still guarantee that the whole development fact are right to transpositions. In order to avoid transmission losses and mismatch in telephone lines, this has to be tested. Has resut a new automatic fault locator has been developed which works with same standards and which automatically records the area from claiming managed alternately transient energy line faults.

Keywords Microcontroller · IoT · Relay · Oil sensor · Temperature sensor
Display unit

B. Godavarthi (✉) · P. Nalajala
Department of Electronics and Communication, Institute of Aeronautical Engineering,
Hyderabad, India
e-mail: bhavana.bhanu402@gmail.com

P. Nalajala
e-mail: nprece@gmail.com

V. D. Majety
Department of Computer Science,
Vignan Nirula Institute of Technology and Science for Women, Guntur, India
e-mail: mvasu_devi@rediffmail.com

Y. Mrudula
Department of Electronics and Communication Engineering, Farah Institute of Technology,
Hyderabad, India
e-mail: mrudulamudu@gmail.com

© Springer Nature Singapore Pte Ltd. 2019   647
R. S. Bapi et al. (eds.), *First International Conference on Artificial Intelligence
and Cognitive Computing* , Advances in Intelligent Systems and Computing 815,
https://doi.org/10.1007/978-981-13-1580-0_62

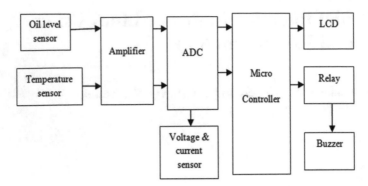

Fig. 1 System block diagram

1 Introduction

Many organizations which manufacturing and transmit electric power, such as EPC, need to be able identify faults on their transmission lines. Due to recent and dependable innovations, it is now straightforward to find such faults. In the past, identifying the correct areas was not always easy and was often achieved by the use of specialized groups charged with inspecting transmissions lines for faults in a lengthy and drawn-out process.

As a result it is important to minimize the amount of time outages occur over, and it is also necessary to increase the reaction time of the specialists involved in identifying the areas where faults are located. Furthermore [1, 2] for both the overhead and underground power transmission system network, utilizing both existing shortcoming pointer engineering for more business engineering will fast What's more exact pin side of the point issue ID number Furthermore fathoming those issue for just hours (Fig. 1).

The suggested worth of effort may be restricted to spot offering with line Furthermore ground faults done both overhead Also underground transmission lines for those implication [3] will control room the place the power line fault struck them.

2 Implementation

An electrical issue is that of varying voltages. Under ordinary operating states the voltages What's more ebbs and flows convey typical values coming about sheltered operational states. Be that as any deviation previously, voltage or current effects unforeseen secondary ebbs and flows Furthermore harms those gears Also units. Along these lines issue identification Also examination is necessary with select or configuration suitableness switchgear equipment, electromechanical relays, circuit

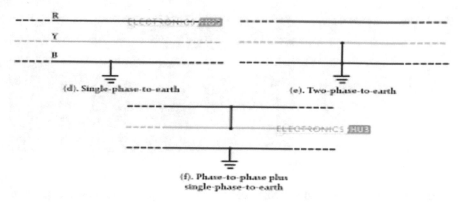

Fig. 2 Symmetrical fault lines

breakers, and different security units. There are two main types of faults in the electrical power network, and these are symmetrical faults and asymmetrical faults [1].

2.1 Symmetrical Faults

Symmetrical faults on the other hand balanced for complete structures. Symmetrical faults are two sorts specifically offering will transport (L–L–L) and agreement on ground (L–L–L–G). Only 2–5% of system faults require help symmetrical issues. In the event of such faults occurring, the system stays in balance despite the effects of the faults on the power network (Fig. 2).

2.2 Asymmetrical Faults

Asymmetrical faults would normal faults struck them furthermore would about three sorts to be specific transport to ground (L–G), line will line (L–L–L) Furthermore transport with ground (LL–G) faults. Accordance on ground faults will be ordinarily happened faults A maximum of around 70% of faults are asymmetric faults. 20% faults would two fold accordance with faults struck t by ground states. The remaining 10% of faults require aid line with accordance faults struck them by ecological states [4, 5]. These have support similarly known as unbalanced faults since their off chance reasons unbalance in the schema. Unbalance of the skeleton intimates that that impedance qualities might dissimilar on each stage making unbalance present once stream in the phases. These need aid a greater amount troublesome should

Fig. 3 Unsymmetrical fault
lines

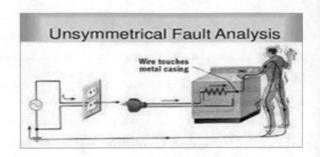

examine What's more need aid conveyed by for every period support comparative
should three period adjusted faults [5, 6] (Fig. 3).

LM324 can be arranged for four independent, secondary gains; internally recurrence adjusted operational amplifiers which were outlined particularly with work
from a single energy supply in an extensive variety about voltages. Operation starting with part force supplies will be also workable and the low energy supply present
channel may be free of the extent of the control supply voltage.

Requisition regions incorporate transducer amplifiers, dc addition obstructs
What's more every last one of customary op amp circuits which Notwithstanding
could a chance to be a greater amount effortlessly actualized done solitary energy supply frameworks. For example, LM324 arrangement can make specifically worked off
of the standard +5 V control supply voltage which will be utilized within advanced
frameworks. Furthermore will effectively give acceptable those needed interface
hardware without requiring the extra ±15 V power supply.

The status of these switches is transmitted utilizing a global system for mobile
communications (GSM) modem and received by the portable. The microcontroller
continuously monitors the status of these switches beginning with those that control
supply and then performs a comparison. The GSM modem is used to send the status
of the data via a cellphone. The controller will send the data using those commands.
Eventually Tom's perusing utilizing the during commands the controller will send the
majority of the data. A 16 × 2 liquid crystal display (LCD) screen on the controller
side will show the status of the powerlines.

This task employs 5 V, 500 mA and 12 V, 500 mA energy supplies. 7805 Furthermore 7812 three terminal voltage controllers are utilized to regulate voltage. Span
kind full wave rectifier can be used to correct the ac yield for auxiliary about 230/12 V
step-down transformer.

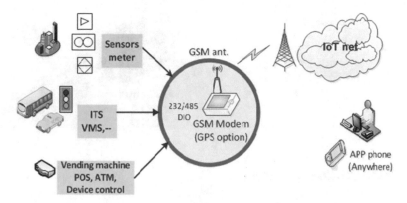

Fig. 4 IoT with GSM

Fig. 5 Oil sensor

3 Implementation

3.1 Internet of Things

Stretched out scope GSM IoT (EC-GSM-IoT) is a low force totally zone innovation that need been institutionalized toward 3GPP for utilize on authorized range. In light of GPRS, EC-GSM-IoT is intended similarly as a secondary capacity, in length range, low vitality and low multifaceted nature cell division framework with help those IoT [7–9] (Fig. 4).

3.2 Oil Level Sensors

Industrial applications and the automobile industry require a large number of oil sensors. These sensors range from simple float-type indicators to complex laser particle counters. Oil sensors analyze the level of oil with their mode of operation depending on the type of sensor used (Fig. 5).

Fig. 6 Temperature sensor

3.3 Temperature Sensors

Temperature sensors sense environmental conditions and monitor changes in temperature in a wide range of solids, liquids and gases. A temperature sensor requires to be physically connected with an object through electrical terminals (Fig. 6).

3.4 Amplifier

A speaker is an electronic device that stretches those voltage, current, alternately vitality of a banner. Intensifiers are used Concerning illustration An and only remote correspondences its more broadcasting, Also over sound supplies of arranged sorts. They camwood be orchestrated Concerning illustration Possibly weak banner enhancers or force speakers.

3.5 Voltage and Current Sensors

Current sensors are devices that identify electric current in a wire and produce a notification related to the size of the current present. This can be displayed notification for simple voltage levels even computer also using IoT. This can then be used to show the current via an ammeter, or can be omitted in order to encourage investigation into an information securing framework, or can be utilized with the end goal of control.

3.6 Liquid Crystal Display (LCD)

An LCD indicates the temperature of the planned component. CMOS improvement makes those contraption flawless to provision close Toward held, helpful and other battery bearing with low control usage [6].

Fig. 7 Relay

3.7 Buzzer

A ringer or beeper can be used as a sound signalling device. This can be mechanical, electromechanical or piezoelectric but usually a buzzer is used. These can be incorporated into alert devices or can be used to provide confirmation of a client entering a mouse click or a keystroke.

3.8 Relay

A transfer relay is an electrical switch that opens and shuts under the control of an alternate electrical circuit. The switch may be operated by an electromagnet to open or end you quit offering on that one or A large number sets about contact (Fig. 7).

AT+CMGF, SMS Format
This command controls the presentation format of short messages from the modem.

Command	Possible response
AT+CMGF=?	+CMGF:(list of supported <mode>s)
AT+CMGF?	+CMGF:<mode>
AT+CMGF=<mode>	<mode>:0 PDU mode

Obtaining messages and perusing commands
AT+CNMI, new message signs should tedium. This summon selects the procedure, how accepting from claiming new messages starting with the organize is shown of the DTE The point when DTE may be dynamic. Further majority of the data cam wood be found for GSM 03.38 [10, 11]. Unsolicited result codes can be buffered in the modem. When the buffer is full, indications can be discarded.

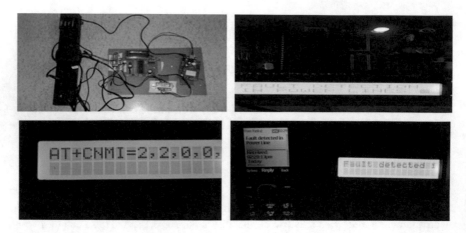

Fig. 8 Prototype hardware system

4 Hardware Results

See Fig. 8.

5 Applications

5.1 Access Control Devices

Access control devices communicate with servers and security staff through SMS services provided by GSM mobiles. This technology can be used for fingerprint access control as well as time attendance products. A high degree of security and reliability can be achieved.

5.2 Transaction Terminals

Electronic data capture (EDC) machines and point of sale (POS) terminals can use SMS messaging to confirm transactions from central servers.

6 Conclusion

It need been made toward incorporating highlights for every last one of gear segments used. Closeness about each module need been examined out and set meticulously along these lines including of the best working of the unit. Furthermore, using exceedingly moved IC's Furthermore with the support about Creating advancement those assignment need been adequately executed.

References

1. Senger, E. C., Manassero, G., Goldemberg, C. and Pellini, E. L. (2005), Automated Fault Location System for Primary Distribution Networks, IEEE Transactions on Power Delivery, pp. 1332–1340.
2. Das R., Sachdev, M. S. and Sidhu, T. S. (2000), "A Fault Locator for Radial Sub Transmission and Distribution Lines", In Proceedings of IEEE PES Summer Meeting, pp. 443–448.
3. Zhu, J., Lubkerman, D. L. and Girgis, A. A. (1997), Automated Fault Location and Diagnosis on Electric Power Distribution Feeders, IEEE Trans on Power Delivery, pp. 801–809. [13] Anon, Microcontroller (2009b).
4. Al-Shaher, M., Sabra, M. M. and Saleh, A. S. (2003), Fault Location in Multi-Ring Distribution Network using Artificial Neural Network, Electric Power Systems Research. pp. 87–92.
5. Wang, C., Nouri, H. and Davies, T. S. (2000), A Mathematical Approach for Identification of Fault Sections on the Radial Distribution Systems, 10 Mediterranean Electrotechnical Conference (MELECON), pp. 882–886.
6. PaparaoNalajala, BhavanaGodavarth, "Morse code generator by using microcontroller with alphanumeric keypad", International Conference on Electrical Electronics, and Optimization Techniques (ICEEOT), IEEE xplore 24th Nov-2016. pp. 762–766.
7. Bhavana Godavarthi, Paparao Nalajala, Vasavi Ganapuram, "Design and Implementation of Vehicle Navigation System in Urban Environments using Internet of Things (IoT)", IOP Conf. Series: Materials Science and Engineering, Volume 225, Sept. 2017.
8. PaparaoNalajala, D Hemanth Kumar, "Intelligent detection of Explosives using Wireless sensor Network and internet of things (IoT)", International journal of control theory and applications, Volume 9, Number 42, pp. 391–397. December-2016.
9. Paparao Nalajala, S Bhagya Lakshmi, "A Secured IoT Based Advanced Health Care System for Medical Field using Sensor Network", *International Journal of Engineering & Technology, 7 (2.20) (2018) 105–108.*
10. Bhavana Godavarthi, Paparao Nalajala, "Wireless Sensors Based Data Acquisition System using Smart Mobile Application," Internet of things, "International Journal of Advanced Trends in Computer Science and Engineering" Vol. 5 No. 1, pp. 25–29 Jan 2016.
11. Paparao Nalajala, Bhavana G, "Provide safety in school childrens vehicle in urban environments using navigation system", International journal of applied engineering research, Volume 12, Number 13, pp. 3850–3856, July-2017.

Design of a Smart Mobile Case Framework Based on the Internet of Things

Paparao Nalajala, P. Sambasiva Rao, Y. Sangeetha, Ootla Balaji
and K. Navya

Abstract The present research portray the design of building a smart case ready to follow its client making its way when cleared out. After a survey of the present way adherent gadgets accessible for performing distinctive assignments, we stretched out our approach that expects to build up a case that performs errands on its manmade brainpower when guided. Specifically, the approach depends on a control framework ready to execute hindrance shirking and focus on its Following conduct. Additionally, a relative area gadget in view of a flag producer (Set on the target single person) and an directional sensor. The folder case is created to be moderate for another innovation that can prompt new applications to help people further. Displayed a proposed Method to accomplish human after conduct as an initial move toward the advancement of a wise folder the smart case moving along with man. The system for monitoring, tracking, and controlling the smart case is based on the internet of things (IoT). This new system receives real-time information about the smart case and is able to upload the data to a monitoring center through the client/server architecture.

Keywords IR sensor · IoT · Server · Smartphone · Motors · GPS · LDR

P. Nalajala (✉) · O. Balaji
Department of Electronics and Communication Engineering,
Institute of Aeronautical Engineering, Hyderabad, India
e-mail: nprece@gmail.com

P. Sambasiva Rao
Department of Computer Science Engineering, GITAM University,
Visakhapatnam, India
e-mail: sambasiva.phd@gmail.com

Y. Sangeetha
Department of Information Technology, VR Siddhartha Engineering College,
Vijayawada, India
e-mail: sangeetha18.yalamanchili@gmail.com

K. Navya
Department of Computer Science and Engineering,
MLR Institute of Technology, Hyderabad, India
e-mail: navyareddy.karangula@gmail.com

© Springer Nature Singapore Pte Ltd. 2019
R. S. Bapi et al. (eds.), *First International Conference on Artificial Intelligence and Cognitive Computing* , Advances in Intelligent Systems and Computing 815,
https://doi.org/10.1007/978-981-13-1580-0_63

1 Introduction

My working definition of a self-sufficient case is self-ruling on the off chance that it has the computational assets both as far as equipment and software. Other than real time obstruction from a human operator, to assess how it is physically inserted in the earth to register most ideal activities limited by a few limitations to perceive and move if necessary, to accomplish an arrangement of objectives. As per this working definition, a case's capacity to gauge its present state (how it is physically inserted in the environment) is a fundamental segment of independence. At that point, it needs to have sufficient computational assets available to its to make a move inside limits, to see the earth progressively if necessary to accomplish a given objective.

Advanced mobile case integrates for your advanced mobile telephone should fulfill all assignments that no different suite could do it might move vertically alternately once a level plane undoubtedly. You might place more things or maybe another bag over Smart situation when it is voyaging self-rulingly in even mode. Keen case explores huge group and can perceive and stay away from objects as required.

The Ability to make good framework implies that smart case [1, 2] is at the front line of AI and self-governing development innovation. The Follow Me framework has been employed in sensors that identify and avoid obstructions, such as individuals and furniture. It utilizes the smart case application to follow you and to better pinpoint its location on a consistent basis. The self-driving innovation is perhaps the most talked about innovation in the car industry at present and many organizations are creating independent driving highlights to be included to their current generation of automobiles. The inspiration driving this innovation aims to enhance auto security and productivity. Sebastian Thrun, the principal designer of Google's driverless automobile venture wrote on his blog that the objective of creating a driverless vehicle is to "help anticipate car crashes, free up individuals' chance and diminish carbon outflows," (see Fig. 1).

2 Literature Review

The principal occasion of present day apply autonomy Indicated up to 1948. The point when George create Also Joe Engle Berger made a mechanical arm. The two men later produced the grade mechanical engineering association for 1956. Done 1979, those Stanford truck robot adequately crossed and situate loaded remain for no human help. It destroyed this by using a polaroid camera to take photographs and then upload them to a PC for examination. The robot might try toward an rate for around person meter each 15–20 min What's more quickly from that point onwards it might afterward take photographs furthermore reassess nature's domain. It completed this eventually Tom's perusing utilizing a introduced PC will make the vital figuring's.

This PC is practically indistinguishable twin of the introduce microcontrollers. Those fundamental real harsh driverless vehicle might have been the terregater, which

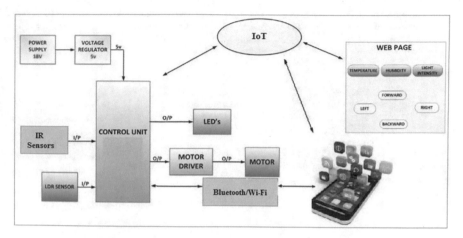

Fig. 1 Structural design of the system

might have been committed for 1984 to those essential parts about road mapping and mine examination. This was the first occasion where an autonomous robot was manufactured to undertake a particular mechanical task.

The idea of an free from outside control following robot is a relatively new innovation. Robots have been around for quite a long time, but one that can basically think for itself had not been considered until recent improvements and innovations in microcontroller technology. Recent advances in Wi-Fi and Bluetooth technology have likewise contributed to advances in remote control and information exchange. As a result of these developments, interest in this field has exploded, particularly in view of the following statement made by the US military: "The U.S. Congress has ordered that by the year 2015, 33% of ground battle vehicles will be unmanned, and the DOD is currently building up a huge number of unmanned frameworks that it plans to quickly field". Additional significance is being placed on the space meanderers used by NASA. These new traveller, particularly the ones currently on Mars, must have a self-sufficient capacity enabling them to adjust to the terrain.

3 Proposed System

The practical analysis system technique (PAST), is a compelling strategy used to decide the fundamental tasks a framework is required to perform. Perusing the PAST outline in the figure from left to right and to shows how the relationship is identified with the smart case. Read from appropriate to left and base to top shows the relationship is identified with the person. Through the use of this technique, it was considerably less demanding to making new plans, and see how the different parts would work with each other (see Fig. 2).

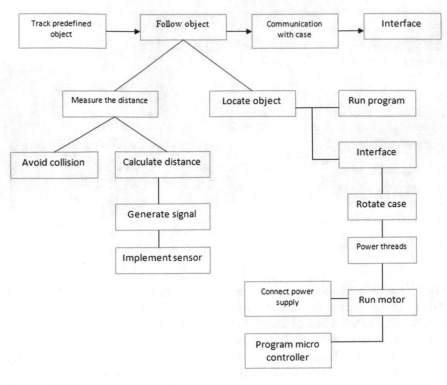

Fig. 2 PAST diagram

4 Implementation

The smart case is composed of two design elements

1. Mechanical Design
2. Circuit Design.

    1. Mechanical Design

This incorporates those fittings outline of the body of evidence that is engine and wheel placement, muscle to setup. Situation utilization two mechanical technology rigging engine and wheel to the development [3] which help it to move forward, exited or correct. The event utilization two engine and wheel in the over side Furthermore person freewheeling ball will be put toward the front which aides it on spare development those sensor are set such-and-such they cam wood disguise. the greatest zone before the case and can be skilled to distinguish a deterrent either impediment is little or enormous.

2. Circuit Design

The circuit configuration is comprised of two sections

(a) Control board
(b) Sensor

(a) Control board

This contains the basic driver circuit of the robot. It contains the microcontroller and the engine driver.

4.1 Systems Operation

Similarly as the robot is traded ON, it initially checks whichever start banner may be gotten alternately not, if not afterward those system counter won't try of the accompanying area it will sits tight for a comparative deliver until those perspective that it get a negative banner. At that point the robot continuously checks for any obstacle that might be in the way. If there is no obstruction present then the robot will continue in a straight line. In the event that an obstruction is present will change the moving point and controller and send information to motor to drive can move various directions [4].

In the unlikely event that an obstruction is found on both left and right sides, the controller sends an instruction to the engine driver to move the two engines in switch heading. Deterrent maintaining a strategic distance from method is extremely helpful, in actuality, this strategy can utilize Similarly as a dream cinch from claiming outwardly impeded people Eventually Tom's perusing evolving the IR sensor Toward An changing sensor, which is with respect to sort of microwave sensor whose identifying range may be secondary and the yield from claiming this sensor contrast Previously, Likewise shown by the address position transforms. This procedure makes visually impaired individuals ready to explore the impediment effortlessly by setting three vibratos in Left, correct and the point of convergence of a cinch named as dream cinch and makes outwardly impeded people primed to stroll anywhere. Through prevention keeping far starting with robot temperature/weight sensors might a chance to be included will screen the air conditions around. This is valuable in places where the earth isn't reasonable for people.

4.2 Vehicle Control Component

When plotting those vehicle control instrument flying particular case must know around the best payload for that vehicle in perspective for which those parts a chance to be gotten.

There are two courses Eventually Tom's perusing which those vehicle might a chance to be regulated.

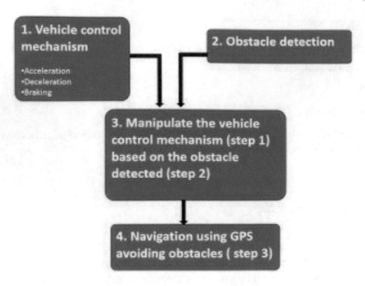

Fig. 3 Obstacle detection design

1. Fixed speed control
2. Variable speed control.

With fixed speed control, those vehicle a chance to be whichever moved (with an enduring speed) or ceased. A straightforward hand-off rationale controller is sufficient to play out this sort of control. You could whichever build your identity or hand-off built controller to your vehicle or disaster will be imminent you might buy a exchange shield on control those vehicle engines starting with those more modest scale controller.

4.3   Obstacle Detection

An infrared sensor may be an electronic contraption that produces recollecting the genuine goal on recognize several of parts of the world. An IR sensor cam wood assesses those shimmer of a Contrast Furthermore besides perceives the improvement. These sorts for sensors measures essentially infrared radiation, as opposed to discharge it that is known as similarly as a disengaged IR sensor [5, 6]. As manage in those infrared range, each a standout amongst those articles transmits some sort warm radiations. These sorts about radiations might indistinct will our eyes that could an opportunity will make recognized at an infrared sensor. The creator may be main a IR headed (Light Emitting Diode) and the locator may be fundamentally an IR pho-todiode which may be delicate with IR light from claiming an undefined wavelength from that discharged Eventually Tom's perusing the IR headed (see Fig. 3).

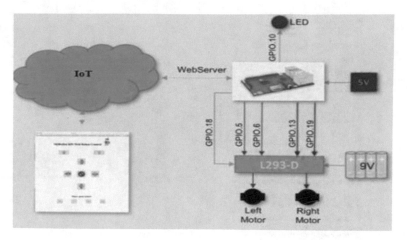

Fig. 4 Communication of information with the IoT

4.4 GPS

A global positioning system (GPS) will be a space based radio route skeleton committed up from claiming no short of what 24 satellites. It will be an around the world course satellite schema that provides for relocation and the duration of the time information to An GPS authority anywhere ahead alternately near those earth the place there will be an unhindered viewable pathway to no fewer than four GPS satellites [7, 8]. With discover 2-D spot (scope and longitude) and track development, a GPS beneficiary must be blasted with respect to of the banner for no fewer than three satellites. The GPS skeleton doesn't require those customer on transmit At whatever information. Also it meets expectations autonomously from claiming any telephonic or web gathering. Nonetheless these developments could overhaul the comfort of the GPS situating information [9].

4.5 Internet of Things

IoT can get to its mechanical limitations, for example, preparing, stockpiling and vitality by utilizing the boundless assets of the cloud. The cloud can likewise stretch out its assistance to manage certifiable elements in a more powerful manner by the utilization of IoT case can be controlled by web which makes it IoT [10, 11]. Utilizing Web IOPi system the web interface can control the case. The interface comprises a few catches to control the case like left, ideal, forward and so forth utilizing web persistent live video gushing of home condition is coordinated to the remote client by web (see Fig. 4).

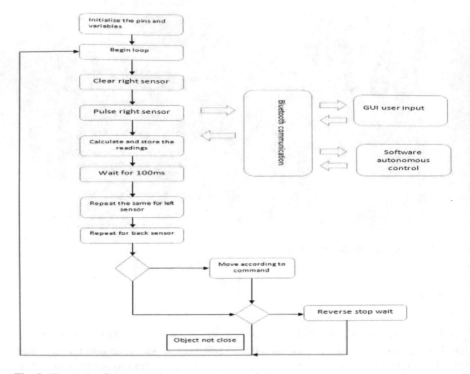

Fig. 5 Flowchart of system design

5 Flow Chart

See Fig. 5.

6 Results

In this paper, we have outlined an IoT-based smart case with innovative features which can be controlled through a remote webpage. Figure 6 shows a model of the completed IoT-based smart case. The client can control the development of the smart case like left, appropriate, forward and so forth with the controls gave in page. The smart case is fitted with sensors, notably IR, to check temperature and identify any obstacles separately. These sensor can be reset and their output viewed via a webpage (see Fig. 6).

Fig. 6 Hardware results of the system

7 Conclusion

The smart case is an inventive lightweight case that makes life simpler. Carrying luggage is a fundamental problem facing every single traveler. Here we aim to solve the problem of carrying luggage and furthermore provide improved security and new features appropriate to modern life. In this venture we built up a new minimal effort human after innovation to help ease buyer item execution, so the general creation cost of a programmed client following case will be less. The inbuilt power bank can give adequate power and in the meantime share energy to clients devices like advanced cell, workstations and so on.

8 Future Work

In the future, we intend to incorporate some new features, such as the ability to climb stairs and improved security features for women. These additional highlights will make the case all the more attractive and easy to use.

References

1. César Nuñez, Alberto García, RaimundoOnetto, Daniel Alonzo, SabriTosunoglu," Electronic Luggage Follower", *Florida Conference on Recent Advances in Robotics, FCRAR 2010—Jacksonville, Florida, May 20–21, 2010.*
2. Sebin J Olickal1, Amal Yohannan2, Manu Ajayan2, Anjana Alias," SMART BAG", International Research Journal of Engineering and Technology, Vol 4, Issue 4, 2017.

3. Keerthi.S. Nair, AnuBabu Joseph, Jinu Isaac Kuruvilla "Design of a low cost human following porter robot at airports" IJACTE, ISSN (Print): 2319–2526, Volume-3, Issue-2, 2014.
4. Kwan-HoonKim; Jun-Uk Chu; Yun-Jung Lee "Steeringby-Tether and Modular Architecture for HumanFollowing Robot" SICE-ICASE, 2006. International Joint Conference Digital Object Identifier: 10.1109/SICE.2006.315704 Publication Year: 2006 , Page(s): 340–343.
5. Bhavana Godavarthi, Mohammad khadir, Paparao Nalajala," Biomedical sensor based remote monitoring system field of medical and health care", Journal of advanced research in dynamical and control systems, Vol.9, Issue.4, pp 210–219.
6. Paparao Nalajala, P Madhuri, G Bhavana," RFID based security for exam paper leakage using electromagnetic lock system", International journal of pure and applied Mathematics, Vol. 117 No. 20, pp 845–852, ISSN: 1314-3395.
7. E. A. Topp and H. I. Christensen, "Tracking for following and passing persons," in Proc. IEEE/RSJ Int. Conf. Intell. Robots Syst., Edmonton, AB, Canada, 2005, pp. 2321–2327.
8. Paparao Nalajala, Bhavana Godavarthi," Provide Safety in School Children's Vehicle in Urban Environments using Navigation system", International Journal of Applied Engineering Research, ISSN 0973-4562 Volume 12, Number 13 (2017) pp. 3850–3856.
9. Paparao N, G Bhavana," RTOS Based Image Recognition & Location Finder Using GPS, GSM and Open CV", *International Advanced Research Journal in Science, Engineering and Technology*, Vol. 2, No. 12, pp. 85–88, Dec 2015.
10. Bhavana Godavarthi, Paparao Nalajala, V Ganapuram,"Design and implementation of vehicle navigation system in urban environments using internet of things (IoT)", IOP Conference Series: Materials Science and Engineering, Vol. 225 (1), 012262, Sep- 2017.
11. Bhavana Godavarthi, Paparao Nalajala," Wireless Sensors Based Data Acquisition System using Smart Mobile Application," Internet of things, "*International Journal of Advanced Trends in Computer Science and Engineering*" Vol. 5 No.1, pp. 25–29 Jan 2016.

Analysis and Optimization of FIR Filters Using Parallel Processing and Pipelining

Paparao Nalajala, C. Devi Supraja, N. Ratna Deepthika, Bhavana Godavarthi and K. Pushpa Rani

Abstract Streamlining of limited drive reaction filters may be a standout among the significant tests in the field of altogether vast-scale mix about circuits starting with those precise starting. In the FIR filters, the multipliers have a tendency with help the vast majority of the delay accompanied eventually Tom's perusing the adders. Hence, these adders and multipliers and their positioning in the filters decide the speed, area, and power of the filters. There are various techniques to optimize the FIR filters, namely pipelining, parallel processing, retiming, unfolding. In this project, an attempt to compare four of these techniques by applying them to three-tap FIR filters is done. After applying these techniques, the critical path of the concerned filter is analyzed for delay (primarily), as the project is primarily concerned about optimization of delay in the filters. The use of retiming, pipelining, and parallel processing reduces. Those basic ways of the concerned channel may be investigated to delay (primarily), Concerning illustration the task may be principally worried around streamlining from claiming delay in the filters. The utilization for retiming, pipelining What's more parallel preparing diminishes those basic way considerably, along these lines upgrading that execution of the channel. The outcomes need aid compared and the delay execution from claiming these filters for which those over specified strategies would connect might have been found should be unrivaled in examination of the typical FIR filters.

Keywords FIR · Retiming · Pipelining · Parallel processing · Critical path

P. Nalajala (✉) · C. D. Supraja · N. R. Deepthika · B. Godavarthi
Department of Electronics and Communication Engineering, Institute of Aeronautical
Engineering, Hyderabad, India
e-mail: nprece@gmail.com

B. Godavarthi
e-mail: bhavana.bhanu402@gmail.com

K. P. Rani
Department of Computer Science and Engineering, MLR Institute of Technology, Hyderabad,
India
e-mail: rani536@gmail.com

© Springer Nature Singapore Pte Ltd. 2019 667
R. S. Bapi et al. (eds.), *First International Conference on Artificial Intelligence
and Cognitive Computing* , Advances in Intelligent Systems and Computing 815,
https://doi.org/10.1007/978-981-13-1580-0_64

1 Introduction

Digital filters with high performance are most important in digital signal processing systems . The speed of the filter depends on potentialities of hardware stage employ and computational construction of the code [1]. In order to find smaller significant path delay in pipeline processing, a lengthy significant way is busted into the level of small, faster operation within the registers so that we can get higher-order operating frequency and higher throughput. In order to modify transfer function of IIR filter, the register can advise in a feedback loop delay alternately. Hence, IIR filter is pipelined, whereas conserves its inventive transfer function when the computation is reconstructed as look-ahead filter [2, 3], and the proposed system optimizes the system rate and minimizes the rate of hardware and software. Exclusive of sacrifice the design of unique filters the essential design builds filters with minimum delay. FIR filters are limited to linear phase, stability and few finite accuracy errors and well execution which have a most important disservice of higher arrange it require IIR with parallel performance. In order to meet the requirement of extra hardware, a design of fabricating filter with arithmetic operations, area usage, and power consumption is designed. Hence, the major aim of digital filters designing is to minimize or reduce the parameters [4].

1.1 Finite Impulse Response Filter

The below equation for FIR can define the relation between the input to output signal.

$$y[n] = b0\, x[n] + b1\, x[n-1] + \cdots + bN\, x[n-N]$$

where $X[n]$ is the enter signal, $Y[n]$ is the yield sign. Furthermore, bi may be those channel coefficients. N may be known as those channel order; an Nth-order channel needs $(N + 1)$ terms on the right-hand side; these would usually allude to likewise taps [5].

1.2 Pipelining

Pipelining is an execution method in which numerous instructions overlap in implementation. Today, Pipelining is a key to assembly processors speedy. The general execution time for all individual instruction is not changed by pipelining. Pipelining does not increase speed instruction execution time. Anyway it can quicken system execution occasion when toward expanding those number about guidelines completed for every unit chance [6, 7].

2 Existing Model

A number of the exhibit day advanced indicator preparing frameworks use best the transposed immediate type (TDF) for FIR Filters. This produces finer comes about As far as delay execution in examination of the usage for regulate manifestation FIR filters [7, 8]. However, speeder outputs What's more quicker transforming need generally been wanted particularly in the secondary limit hardware which find requisition clinched alongside indicator transforming similar to in the fields for satellite communication, radar communication, media development, air movement control and so forth.

3 Proposed Model

In the suggested model, the utilization of parallel processed, pipelined TDF FIR may be proposed concerning illustration that these have unrivaled execution as far as delay and the yield examining rate. That basic way is significantly reduced due to the application of pipelining, and the sampling rate is increased due to pipelining as the number of stages producing the output increases depending on the number of stages the FIR has been pipelined to. Hence, sample amount of outputs are available at faster rate due to usage of these techniques.

4 Methodology

4.1 Pipelining

The water pipe: keep on transfer water lacking waiting the water in the pipe to be out Whichever expand the clock speed alternately diminish the energy utilization toward same pace for a DSP framework (Fig. 1).

Fig. 1 Pipelining model

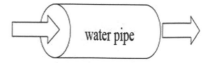

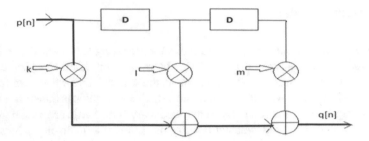

Fig. 2 Critical path of direct form FIR

4.2 Parallel Processing

Case 1: think about a 3-tap FIR filter: $y(n) = ax(n) + bx(n-1) + cx(n-$ way may be set Toward 1 increase Also 2 include times. Thus, the "sample period" 1 is provided for by:

$$T_{sample} \geq T_M + 2T_A$$

$$f_{sample} \leq \frac{1}{T_M + 2T_A}$$

4.3 Critical Path

Critical path is the highest computation moment in time among all the available paths which have zero delay elements.

For example:
The red-lined path is the critical path of it (Fig. 2).

5 Implementation

5.1 FIR Filters

The current project is a comparison of three types of FIR filters. These mentioned types differ in the techniques that are applied to them. The filters are:

1. Direct form three-tap FIR filter
2. Transposed three-tap FIR filter
3. Parallel processed, pipelined three-tap FIR filter.

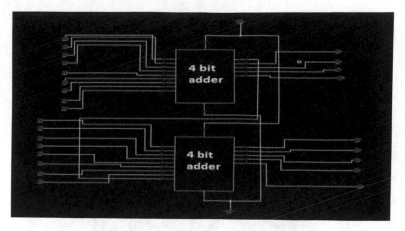

Fig. 3 Schematic of 8-bit adder

But before going to that a brief explanation of the basic structural elements in the FIR filters is given.

5.2 8-Bit Adder

This is the circuit designed to add eight bits each from two different input sequences and gives a 16-bit output. It is made using eight full adders (Fig. 3).

5.3 8-Bit Multiplier

In this unit, the task of multiplying eight bits is performed. Two inputs with eight bits each are fed as input, and it produces a 16-bit output. It is made of half adders and full adders (Fig. 4).

5.4 8-Bit Register

The 8-bit register is generally used as a shift register. To achieve this, it is made of eight D flip-flops (Fig. 5).

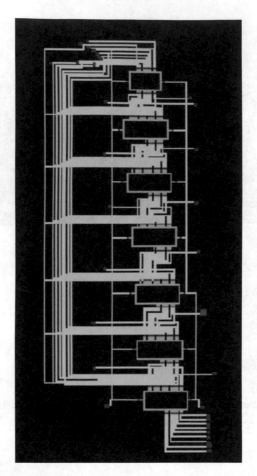

Fig. 4 Schematic of 8 bit multiplier

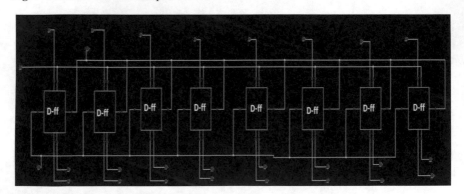

Fig. 5 Schematic 8-bit register

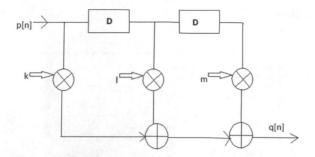

Fig. 6 Direct form FIR filter

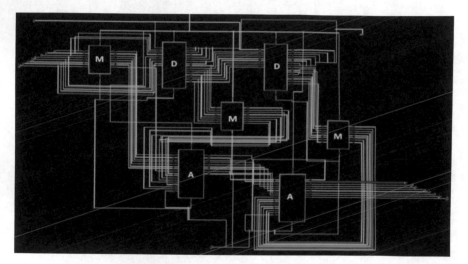

Fig. 7 Schematic of direct form FIR filter

5.5 *Direct Form Three-Tap Filter*

Description: This is a conventional and very basic three-tap filter. It has two delay elements, three multipliers, and two adders (Figs. 6 and 7).

5.6 *Parallel Processed, Pipelined Three-Tap FIR Filter*

Description: This FIR has a three-level implementation of the previously mentioned transposed FIR, thus increasing the sampling rate threefolds.

Critical path: The critical path gets reduced to $T_M + T_A$ (Fig. 8).

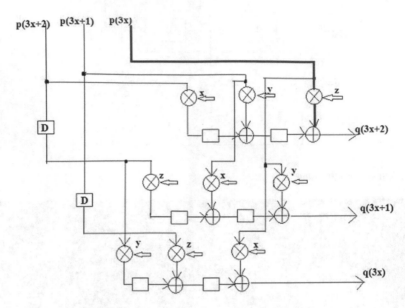

Fig. 8 Critical path of parallel processed, pipelined three-tap FIR filter

6 Result Analysis

6.1 Direct Form FIR

See Fig. 9.

Fig. 9 Output of direct form FIR

6.2   Transposed FIR

See Fig. 10.

6.3   Parallel Processed, Pipelined FIR

See Figs. 11, 12, 13 and 14.

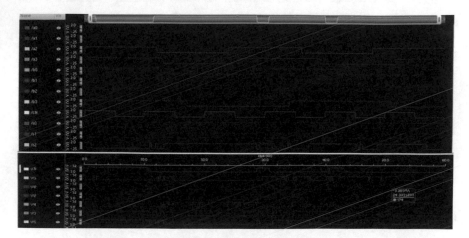

Fig. 10 .

Fig. 11 Outputs of parallel processed, pipelined FIR

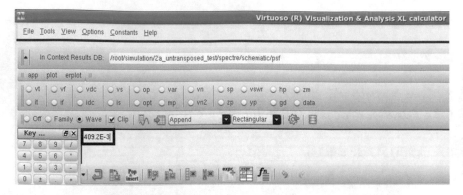

Fig. 12 Power of parallel processed, pipelined FIR with $V_{dc} = 1.7$

Fig. 13 Power of parallel processed, pipelined FIR with $V_{dc} = 1.5$

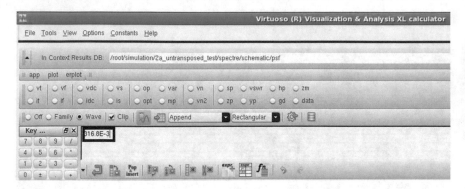

Fig. 14 Power of parallel processed, pipelined FIR with $V_{dc} = 1.3$

Table 1 Comparison of various filters at various V_{dc} voltages

Filter type	V_{dc} (in volts)	Delay (in sec)	Power (watts)	Speed-power product (nS-W)
Direct form FIR	1.8	0.915	0.548	0.502
Direct form FIR	1.7	0.976	0.518	0.506
Direct form FIR	1.6	1.053	0.382	0.402
Direct form FIR	1.5	1.149	0.358	0.411
Direct form FIR	1.4	1.274	0.335	0.427
Direct form FIR	1.3	1.44	0.311	0.449
Transposed FIR	1.8	0.555	1.347	0.747
Transposed FIR	1.7	0.528	1.297	0.685
Transposed FIR	1.6	0.680	1.221	0.831
Transposed FIR	1.5	0.687	1.145	0.787
Transposed FIR	1.4	0.747	1.069	0.814
Transposed FIR	1.3	0.837	0.99	0.831
Parallel process, pipelined FIR	1.8	0.437	0.615	0.269
Parallel process, pipelined FIR	1.7	0.464	0.409	0.189
Parallel process, pipelined FIR	1.6	0.498	0.386	0.192
Parallel process, pipelined FIR	1.5	0.541	0.362	0.196
Parallel process, pipelined FIR	1.4	0.597	0.339	0.202
Parallel process, pipelined FIR	1.3	0.671	0.316	0.212

6.4   Analysis and Comparison of the Output

See Table 1.

7   Conclusion

In the majority of the existing fill in those streamlining of the product-accumulation segment for TDF FIR filters will be disregarded. However in this paper that incredulous way for FIR channel may be made by that result amassing area with the assistance of parallel preparing Also pipeline strategies which could undoubtedly decrease that extra delay. Thus that parallel preparing and pipeline systems need aid executed Previously, discriminating way delay from claiming TDF FIR filters with make an item aggregation segment which diminishes those delay way from claiming filters.

References

1. C.-L. Su, Y.-T. Hwang, "Distributed Arithmetic Based Recursive Digital Filter Design for High Throughput Applications", 1996 IEEE SOUTHEASTCON (SECON), pp. 447–450, 1996-Apr.
2. Y.-T. Hwang, C.-L. Su, "A New Design Approach and VLSI Implementations of Recursive Digital Filters", 1996 IEEE International Symposium on Circuits and Systems (ISCAS), vol. IV, pp. 304–307, 1996-May.
3. C.-L. Su, Y.-T. Hwang, "Distributed Arithmetic-Based Architectures for High Speed IIR Filter Design", 1996 IEEE International Conference on Parallel and Distributed Systems (ICPADS), pp. 156–161, 1996-June.
4. K. K. Parhi, D. G. Messerschmitt, "Pipeline interleaving and parallelism in recursive digital Filters Part I: Pipelining using scattered look-ahead and decomposition", IEEE Trans. ASSP, vol. 37, pp. 1099–1117, July 1989.
5. UmmadisettyNagamani, VydehiMerusomayajula, PaparaoNalajala, Bhavana Godavarthi, "Design of Fault Tolerant ALU using Triple Modular Redundancy and Clock Gating", Journal of Advanced research in dynamical and control system, Vol 9, Sp-17, 2017.
6. K. K. Parhi, D. G. Messerschmitt, "Pipeline interleaving and parallelism in recursive digital filters–Part II: Pipelining incremental block filtering", IEEE Trans. Acoust. Speech Signal Processing, vol. 37, pp. 1118–1134, July 1989.
7. S. A. White, "Applications of distributed arithmetic to digital processing: A tutorial review", IEEE ASSP, vol. 16, pp. 4–19, July 1989.
8. Y. Harata, Y. Nakamura, H. Nagase, M. Takigawa, N. Takagi, "A high speed multiplier using a redundant binary adder tree", IEEE Journal of Solid-State Circuits, vol. 22, no. 1, pp. 18–34, Feb 19.

A Powerful Artificial Intelligence-Based Authentication Mechanism of Retina Template Using Sparse Matrix Representation with High Security

B. M. S. Rani, A. Jhansi Rani and M. Divya sree

Abstract Retinal picture examination has increased adequate significance in the exploration field because of the need for infection distinguishing proof systems. Anomaly recognition utilizing these strategies is exceedingly unpredictable since these infections influence the human eye slowly. Ordinary malady distinguishing proof methods from retinal pictures are generally subject to manual mediation. A biometric format is an advanced portrayal of the special highlights that have been extricated from a biometric test and is put away in a biometric database. These layouts are then utilized as a part of the biometric verification and recognizable proof process. Maintaining the security of the biometric formats is of most extreme significance as any assault on the biometric layouts can prompt a disappointment of the biometric framework. The real test in planning a biometric security conspires that it ought to have the capacity to deal with intra-user fluctuation in the procured biometric identifiers. A powerful biometric format assurance conspire which is fit for dealing with intra-user inconstancy ought to preferably have diversity, revocability, security, performance. In this work, mechanized picture characterization frameworks with these AI procedures are created for down-to-earth applications. The picture order framework is partitioned into numerous modules, for example, highlight extraction, include choice and sickness distinguishing proof utilizing the AI-based classifiers and sparse matrix representation is used for retina template storing which uses less memory, and the template is compared with the users for accurate authentication.

B. M. S. Rani (✉) · M. Divya sree
ECE Department, Acharya Nagarjuna University, Guntur, Andhra Pradesh, India
e-mail: ranibms@gmail.com

M. Divya sree
e-mail: divyasree.mikkili@gmail.com

B. M. S. Rani
ECE Department, Vignan's Nirula Institute of Technology
and Science for Women, Guntur, Andhra Pradesh, India

A. Jhansi Rani
ECE Department, V. R. Siddhartha Engineering College, Vijayawada, Andhra Pradesh, India
e-mail: jhansi9rani@gmail.com

© Springer Nature Singapore Pte Ltd. 2019
R. S. Bapi et al. (eds.), *First International Conference on Artificial Intelligence and Cognitive Computing* , Advances in Intelligent Systems and Computing 815,
https://doi.org/10.1007/978-981-13-1580-0_65

Keywords Identification techniques · Artificial intelligence · Template
identification · Classifiers · Sparse matrix

1 Introduction

Human retinal pictures have picked up an essential part in the identification and
determination of many eye sicknesses for ophthalmologists. A few maladies, for
example, glaucoma, diabetic retinopathy, and macular degeneration, are intense for
they can prompt visual impairment on the off chance that they are not recognized in
time with flawlessness. Consequently, retinal picture investigation has been a testing
research range that plans to give methods to aid the early discovery and conclusion
of many eye infections. Regular procedures depend on manual perception which
is exceedingly inclined to mistake. Subsequently, the prerequisite of computerized
advancements for sickness recognizable proof is fundamentally high.

At first, the picture arrangement framework is tried with the regular procedures.
These systems are for the most part non-AI procedures. Furthermore, the retinal
pictures are grouped utilizing the customary AI methods. Both ANN and fluffy procedures are executed as delegates of AI frameworks. Thirdly, half-breed frameworks,
for example, the blend of ANN and GA, ANN and PSO, and so forth are produced
for the picture arrangement framework. At last, an execution upgraded, altered ANN
is created and tried with the anomalous pictures.

2 Related Work

Aliaa et al. [1] have displayed a near report between different complexity improvement strategies for retinal pictures. These methods are actualized on freely accessible
databases, and the outcomes are classified. These outcomes investigate the benefits
and bad marks of the different complexity improvement methods. Bob et al. [2]
have utilized the change-based systems for edge upgrade in low difference pictures.
The yield pictures depend on the non-straight capacity which joins the impacts of
commotion.

Mairal et al. [3] have proposed a space information-based vein improvement system in shading retinal pictures. An adjustment factor is obtained from the evaluated
debasement and utilized as a part of this work to limit the difference and radiance
variety in retinal pictures. A model-based vessel upgrade method is proposed by John
et al. [4]. The foundation concealment measure, smoother "vesselness" measure, and
the reactions at intersections are better than the ordinary techniques.

Alsaade et al. [5] have utilized the textural highlights in view of wavelet coefficients for retinal picture characterization. The relative homogeneity of confined
regions of the retinal pictures is portrayed by the wavelet coefficients. A broad list of
capabilities is obtained by interpreting, scaling, and pivoting the textural highlights.

Fig. 1 Sequence of
operations for authenticating
users

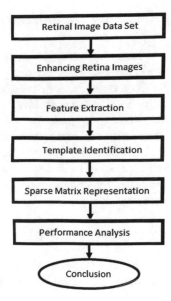

Be that as it may, the nature of the trial comes about detailed in this paper is low. Gaussian sifting-based picture highlights are utilized by Anwar et al. [6].

Zhang et al. [7] have utilized the unsupervised shading thresholding strategy for optic circle identification. Yellow shading and little size of the OD are utilized as critical highlights for exudates ID in this work. These highlights have plainly uncovered that the shading, size, and state of the OD are distinctive for ordinary and irregular pictures. Bunching calculation-based OD location is performed by Qu et al. [8]. The removed OD is utilized to separate the ordinary and obsessive eyes. Trial comes about have uncovered the prevalent idea of the proposed approach.

3 Proposed Methodology

The proposed mechanism considers the retinal images, and image recognition performs various operations on it which is illustrated in below figure. Two retinal databases called DRIVE and STARE are publicly available. When the set of retinal images undergoes these operations, then the person is said to be successfully authenticated and they can be allowed to make use of any application (Fig. 1).

3.1 Image Enhancement

The improvement step is a required errand in computerized picture grouping framework. In light of the securing procedure, all the time these pictures demonstrate vital lighting varieties, poor complexity, and commotion [3]. This issue may genuinely influence the demonstrative procedure and its result. Pre-preparing calculations are actualized to upgrade the first picture with the goal that it can expand the odds for achievement of consequent advances [7].

The distinction between the first d_0 and a blurred picture of it, $p * d_0$, is generally corresponding to its Laplacian [1, 11]. With a specific end goal to formalize this far-extending comment, we need to expect that p is spatially focused, and we should rescale p as

$$P_m(x) = 1/(m)\, p\big(X/\big(m^{1/2}\big)\big) \text{ with } m \to 0$$

Give us a chance to signify by $X = (x,y)$ a state of the plane and expect that d_0 is F_3 around X. Accept advance that p is a positive outspread bit fulfilling

$$\int \big(1 + |x|^2 + |x|^3\big) + |x|^3\big)\,\mathrm{d}x\ <\ \infty \text{ also, } \int x^2 p(x)\mathrm{d}x = \int y^2 p(y)\mathrm{d}y = 2.$$

Under these suppositions and utilizing a Taylor development of d_0 around x, it is a simple exercise to demonstrate that

$$\frac{p_m * d_0(x) - d_0(x)}{m} \to \Delta d_0(x)$$

As $m \to 0$, we can rewrite this equation as

$$P_m * d_0(x) - d_0(x) = h\Delta d_0(x) + o(m)$$

3.2 Retinal Template Matching

This methodology portrays a technique for affirmation of individuals in perspective of retinal cases. Since cases may encounter translational or rotational evacuations, it is essential to change the photographs to be composed. So the reference point area technique [6] is used to recognize the vein bifurcation concentrates honestly [12]. The illustrations investigated could have a substitute number of centers for a comparable individual, which is a direct result of different conditions of light and presentation [7] of the photograph in the acquiring system. Scaling is practically steady for all photographs as a result of eye closeness to the camera. Furthermore, unrests are greatly slight as the eye presentation while defying the camera is on a very basic level the same as. By then organizing of the bifurcation raises did.

Stage 1: Let the two configurations to be facilitated as T_1 and T_2 and subregions as S_1 and S_2.

Stage 2: Initialize the total number of composed concentrations in a configuration "TMP" to be 0.

Stage 3: For each subarea S_1 in T_1 and S_2 in T_2, perform:

(i) Initialize the total number of composed concentrations in a subarea "MP" as 0.

(ii) For each bifurcation point B_1 in S_1, perform:

- Find the BP_2 in S_2 and the eight neighbor subregions of S_2 which has minimum partition, D_{min} with B_1.
- If $D_{min} \leq D_{thresh}$, expel cutoff and B_2 is not authoritatively organized, by then $MP = MP + 1$.
- Mark B_2 as composed.

(iii) Find $TMP = TMP + MP$.

Stage 4: Bifurcation point organizing rate is

$P = (2 * TMP/(TB_1 + TB_2)) * 100$ (7) where TB_1 is the total number of intersection point centers in TP_1 and TB_2 is the total number of meeting centers in T_2.

Stage 5: Degree of matching is given by

Most outrageous {Template Matching (T_1, T_2), Template Matching (T_2, T_1)}.

As far as possible (D_{thresh}) addresses the most outrageous balance by which a comparative intersection point on different designs can be removed. It is used as a piece of demand to consider the quality mishap and discontinuities arrived in the midst of the vessel extraction process that prompts separation of feature centers by a couple of pixels.

The proposed procedure manhandles the way that the round structure of the OD is its most perceiving feature. Regardless of the way that the Green channel gives most prominent distinction, it is watched that the round system of the OD is observably separate out in the Red channel picture or the Gray power picture [10], heading to less requesting taking care of. The strategy depends in transit that the striking characteristics, for instance, the round condition of the OD and in addition the OD run filling in as meeting district of a screw up of nerves and the vein frameworks, remain solid over all subpictures that contain OD at their center.

It is acknowledged that all subpictures containing OD at within, lie in a lone subspace, and that any given subpicture, that contains OD at its center, can be imparted as an immediate blend of the positive cases among the readiness pictures [8]. The pixel estimations of each of the subpictures are associated to shape start vectors $\{v_1, v_2, ..., v_n\}$ in the dictionary while similar-sized insufficient grids with a singular nonzero part for each of the pixel zones $\{r_1, r_2, ..., r_m\}$ are taken as negative planning tests in the vocabulary cross section, A. For ensured test picture whose subpicture y is considered, 11-minimization-based edge work is used for OD recognizable proof as outlined underneath.

For a given *block-size parameter b*, partitions the $n \times n$.

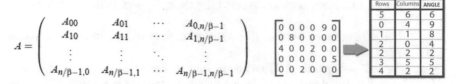

$$A = \begin{pmatrix} A_{00} & A_{01} & \cdots & A_{0,n/\beta-1} \\ A_{10} & A_{11} & \cdots & A_{1,n/\beta-1} \\ \vdots & \vdots & \ddots & \vdots \\ A_{n/\beta-1,0} & A_{n/\beta-1,1} & \cdots & A_{n/\beta-1,n/\beta-1} \end{pmatrix}$$

$$\begin{bmatrix} 0 & 0 & 0 & 0 & 9 & 0 \\ 0 & 8 & 0 & 0 & 0 & 0 \\ 4 & 0 & 0 & 2 & 0 & 0 \\ 0 & 0 & 0 & 0 & 0 & 5 \\ 0 & 0 & 2 & 0 & 0 & 0 \end{bmatrix} \Rightarrow$$

Rows	Columns	ANGLE
5	6	6
0	4	9
1	1	8
2	0	4
2	2	2
3	5	5
4	2	2

Fig. 2 Sparse matrix representation of bifurcation points

Matrix A into $n2/b2$ equal-sized $b \times b$ square *blocks* 4 (Fig. 2).

where the square $Ai\,j$ is the $b \times b$ submatrix of A containing components falling in lines ib, $ib + 1$, ..., $(I + 1)b - 1$ and segments jb, $jb + 1$, ..., $(j + 1)b - 1$ of A. The represented esteems with "x" are the edges figured from the retina layout and put away in the matrix. For effortlessness of introduction, we should accept that b is a correct energy of 2 and that it separates n; unwinding these suppositions is straight forward [5]. The computed bifurcation focuses at the enlistment time are spoken to utilizing meager network and they are contrasted and the recently figured inadequate frameworks at the season of verification as the time intricacy for scanty lattice examination [6] is low and the client is confirmed. This component gives a solid security and validation systems for different sorts of uses.

4 Sparse Matrix Representation

1. Input a variety of n subpictures, of size say $n_1 \times n_2$ with OD at the middle. Connect the pixel esteems [2] to shape the premise vectors $\{v_1, v_2, ..., v_n\}$. For every pixel area, embed an inadequate grid [9] of size $n_1 \times n_2$ with a solitary nonzero component. Link these inadequate grid passages to shape premise vectors $\{r_1, r_2, ..., r_m\}$, where $m = n_1 \times n_2$. Presently shape the word reference grid A $= [v_1, v_2, ..., vnr_1, ... r_m]$.
2. Normalize the segments of A to have unit l2 standard.
3. For a given test picture, rehash stages 4,5 for all subimages y of size $n_1 \times n_2$.
4. Solve the l1-minimization issue, for a picked estimation of ϱ : (11) : $\times$ 1 = argmin $\|x\|1$ s.t $\|Ax - y\|2 \_ \varrho$.
5. Compute the order file of the subpicture.
6. The pinnacle estimation of order file compares to the subpicture with OD at the middle.

5 Result Analysis

Two retinal databases called DRIVE [13] and STARE [13] are publicly available.

In this drive database is used which consists of 40 fundus images. The set of 40 images has been divided into training and a test set, both containing 20 images. The probabilities of finding a section in every quantum were taken care of from our database. In this little dataset, just 3¡5% of quanta had nonzero probabilities as were not able utilize this stream in the model direct. Hence, we input each of the quantization triplets in Table into the theoretical model what's more, utilized the ordinary likelihood transports the entropy of a retina design at different sort out edges s. The outcomes are given in below figure. They demonstrate a checked change in entropy at better quantization triplets. A superior quantization in position will fabulously redesign the greatest retina outline entropy conceivable, however an accommodating confinement to this is how much arrangement and enrollment accuracy can be refined. The method played out their arranging tests at $1 \pm$ affirmation for position at an encourage edge of 6 ¡ 7 highlights. We show that such a quantization on position may essentially permit a conventional of two incorporates into like way between trial of a comparable retina.

Regardless, if game plan and enlistment development improve adequately that position can be changed in accordance with $1 \pm$ precision (Figs. 3, 4, 5, 6 and 7; Table 1).

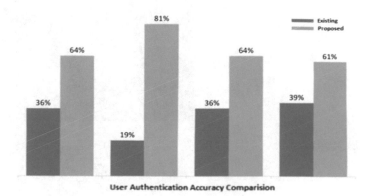

Fig. 3 Comparison of accuracy authentication rates in existing and proposed methods

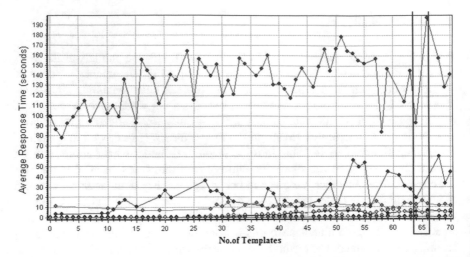

Fig. 4 Comparison of authentication level of existing and proposed mechanisms

Fig. 5 Retinal fundus (input image)

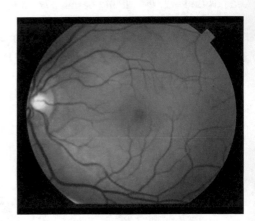

Fig. 6 Retina template

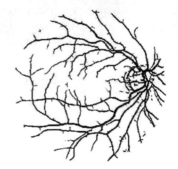

Fig. 7 Bifurcation points

Table 1 Analysis of parameters in existing and proposed methods

Method	Existing method-1 (using Fourier–Mellin function)	Existing method-2 (using blood vessel curvature function)	Proposed method
Average recognition rate with different orientation without noise	100%	100%	100%
Average recognition rate of noisy images	96%	97%	100%
Recognition rate of whole database	98%	98.5%	100%
Average recognition time	5.86 s	4.63 min	3.34 s

6 Conclusion

The strategy proposed here has ended up being effectual for the recognizing verification in perspective of retinal pictures. In the wake of being dealt with by the ICGF, the execution is strikingly improved, yet in the meantime has space for overhauls. In addition smoothed by the iterated spatial anisotropic smooth strategy, the EER fulfills 0. In addition, from the point of view of the normal keypoints examination, as plot in above Figure, the displayed ICGF extended the ordinary number of stable keypoints around 90 times by enhancing the inconspicuous components of the photographs. We have perceived that position is the standard supporter of retina plan entropy. This handles the most ideal retina outline entropy is obliged just by the precision of the enrollment and plan of intra-retina tests and the limitations showed by picture managing. In this paper, sparse matrix representation is used for storing the retina template angles, then the user retina template for authentication is again stored in a sparse matrix, and these two are compared for authentication which provides a strong authentication method enhancing the security.

Declaration We have taken necessary permission to use the dataset /images /other materials from respective parties. We, authors undertake to take full responsibility if any issue arising in future.

References

1. Aliaa Abdel-Haleim Abdel-Razik Youssif, Atef Zaki Ghalwash, and Amr Ahmed Sabry Abdel-Rahman Ghoneim, "Optic disc detection from normalized digital fundus images by means of a vessels direction matched filter," *IEEE Transactions on Medical Imaging*, vol. 27, no. 1, pp. 11–18, "Jan" 2008.
2. Bob Zhang and Fakhri Karray, "Optic disc detection by multi-scale gaussian filtering with scale production and a vessels directional matched filter," *Medical Biometrics Lecture Notes in Computer Science*, vol. 6165, pp. 173–180, 2010.
3. J. Mairal, F. Bach, J. Ponce, and G. Sapiro, "Online learning for matrix factorization and sparse coding," *Journal of Machine Learning Research,*, vol. 11, pp. 19–60, 2010.
4. John R. Vacca. Biometric Technologies and Veri_cation Systems. ElsevierInc., Burlington, MA, USA, 2007.
5. Alsaade, F & Zahrani, M 2009, _Enhancement of Multimodal Biometric Verification using a Combination of Fusion Methods', 5th International Conference: Sciences of Electronic, Technologies of in Formation and Telecommunications (SETIT 2009), pp. 1–5.
6. Anwar, F, Rahman, A & Azad, S 2009, Multibiometric Systems Based Verification Technique', European Journal of Scientific Research ISSN 1450-216X vol.34 No.2, pp. 260–270.
7. Zhang, L, Zhang, D & Zhu, H 2010, _Online Finger-Knuckle-Print Verification for Personal Authentication', Biometrics Research Center, Department of Computing, The Hong Kong Polytechnic University.
8. ZhongQu & Zheng-yong Wang 2010, _Research on Preprocessing of Palmprint Image Based on Adaptive Threshold and Euclidian Distance', Sixth International Conference on Natural Computation (ICNC 2010), pp. 153–159.
9. Zenga, Z & Huang, P 2011, _Palmprint Recognition using Gabor feature-based Two-directional Two-dimensional Linear Discriminant Analysis', International Conference on Electronic & Mechanical Engineering and Information Technology, IEEE, vol. 34, pp. 1917 –1921.
10. Yan, Y & Zhang, YJ 2008, _Multimodal Biometrics Fusion using Correlation Filter Bank', Proceedings of 19th International Conference on Pattern Recognition, pp. 1–4.
11. VijayaKumari, V & Suriyanarayanan, N 2008, _Performance Measure of Local Operators in Fingerprint Detectio', Academic Open Internet Journal, vol. 23, pp. 1–7.
12. John Daugman,‖How Iris works‖ IEEE Transaction on circuit and systems for Video Technology, VOL.14, No.1, January 2004.
13. http://www.ces.clemson.edu/~ahoover/stare/.

Big Data Sentiment Analysis Using Distributed Computing Approach

K. Rajendra Prasad

Abstract Big data refers to three properties, namely, volume, variety, and velocity. Big data is used because of its huge storage capacity, high processing power, and availability of data. Big data is used now due to its availability of powerful multi-core processors, possibility of low latency by distributed computing, partitioning—aggregating—isolating resources in any size and hot swapping dynamically, affordable storage, and computing with minimal man power with the help of cloud deployment models. Sentiment analysis is one of the data mining techniques that is used for measuring users sentiments through techniques like natural language processing (NLP), computational linguistics, digital technology, artificial intelligence, and text analysis. These techniques are used for identifying, extracting, and analyzing subjective information from multiple Web sources. The sentiment analysis explores the contextual divergence of the information/data. Every day, different varieties of data are generating in volumes with high velocity, typically in multiples of 1024 bytes, which means from a petabyte to exabyte. Users around the globe are sharing/communicating/exchanging huge bytes of data through the medium of different sources like social sites, e-commerce sites, file sharing, database repositories, and secondary storage devices. These huge bytes of data may be structured, unstructured, and semi-structured. Analyzing such vast voluminous and veracity of data plays a crucial role in knowing customer behavior and thoughts. Data analytics also known as data analysis refers to qualitative and quantitative techniques and processes used to enhance productivity and business gain. Data is extracted and categorized to identify and analyze behavioral data and patterns, and techniques vary according to organizational requirements. During sentiment analysis process, the data analyst is able to extract a best solution based on sentiment variation and fantasy analysis of a user. This solution could be used for enhancement of business/organization/product/or services, etc. In this work, we are analyzing social data using sentiment analysis, which checks the attitude of user interests, using distributed computing approach, to result in an effective solution.

K. R. Prasad (✉)
Department of Computer Science and Engineering, Institute of Aeronautical Engineering, Hyderabad, India
e-mail: krprgm@gmail.com

© Springer Nature Singapore Pte Ltd. 2019   689
R. S. Bapi et al. (eds.), *First International Conference on Artificial Intelligence and Cognitive Computing* , Advances in Intelligent Systems and Computing 815,
https://doi.org/10.1007/978-981-13-1580-0_66

Keywords Sentiment analysis · Distributed computing · Cloud deployment models · Computational linguistics · Text analysis · Sentiment variation

1   Literature Study

In this era, data is available in everything and everywhere. Data is available in many forms as number images, videos, and text. As data continues to grow, there is need to organize and manage it [1].

Collecting such large amount of data would just be waste of time, effort, and storage space. If data cannot be organized in a proper way, then it is required to rearrange for proper use. The need to sort, organize, analyze, and offer data in a logical use and systematic manner leads to the discussion of a word called Big data [2].

Large amount with veracity of data is known as Big data. The evolution of big data started in 1940s [3].

Datafication is a process of capturing or collecting big data. Big data is data field, so that it can be used productively. Big data cannot be made useful, by simply organizing it [4]. Rather the usefulness of data lies in determining what we can do with data. To extract the meaningful information from big data, we need:

- Optimal processing power
- Analytical capability
- Skill.

Big data is usually unstructured and qualitative in nature. Big data is a growing buzzword in every sector of human existence, like education, health, science, technology, defense, lifestyle, business. [5].

Big data may be anything from petabyte (1 PB = 1000 TB) to an exabyte (1 EB = 1000 TB) of data. This huge volume, variety, velocity, and veracity of data may be structured, unstructured, semi-structured, or heterogeneous in nature. The data can be collected from different sources like:

- Social data
- Machine data
- Transactional data
- World Wide Web.

Since the data is huge, it is to be structured in a manner, such that data becomes easy to study, analyze, and derive conclusions from it. Structuring data helps in understanding user behavior, requirements, and preferences to make personalized recommendations for every individual. The data is obtained mainly from following types of sources:

- Internal sources
- External sources.

The internal and external sources of data may be structured, unstructured, or semi-structured data [6].

The structured data is defined as data that has defined a repeating pattern, which makes it easier for any program to sort, read, and process the data. The unstructured data is set of data that might or might not have any repeating or logical patterns.

The semi-structured data is a form of structured data that contains tags or markup elements, in order to separate elements and generate hierarchies of records and fields in the given data [7]. The semi-structured data is also known as schema-less or self-describing structured data.

These types of data are available in huge volume and variety with rapid velocity and veracity [8]. The volume is the amount of data generated by organizations or individuals. The variety of data defines varied or different formats of data. The velocity data describes the rate at which data is generated, captured, and shared. The veracity of data defines the uncertainty of data, i.e., is whether the obtained data is correct or not [9].

Once the raw data is in hand, the next phase is analysis. Analysis is a process of:

1. Trying to solve a problem
2. Finding the required data to obtain required answer
3. Analyzing the data
4. Interpreting the results to suggest a recommendation.

In big data, the analytics is defined in two ways:

- Decision-oriented analysis—Analysts use the results of the analysis in the process of making business decisions.
- Action-oriented analysis—used when a quick response or an action is expected to a critical situation.

The following are the different phases of the analytical process:

1. Business understanding
2. Data collection
3. Data preparation
4. Data modeling
5. Data evaluation
6. Deployment.

Big data analyses have additional characteristics besides the four V's—volume, velocity, variety, and veracity—that make it different from traditional analysis [10].

- It can be programmatic—Big data analysts must use code to handle raw data. Analysts can use the code to manipulate or explore the data. This is useful when we have large volumes of data.
- It can be data driven—A hypothesis-driven approach is used by several analysts to develop a premise and gather data to see if premise is correct or not. However, big data uses a large volume of data to drive the analysis.
- It can be a lot of attributes—Big data uses thousands of attributes and millions of observations to conduct analysis.

• It can be iterative—Big data analysts use iterations on models that provide more computational power. These models allow the analysts to do iterations to the required level of analysis. Because of this, new applications are designed and developed to analyze requirements and time frames to complete the analysis.

Many organizations or individuals depend on the feedback of their services or products, in order to enhance their business or organization [11]. The feedback is normally one's review or opinion about a product or service. Users normally share their opinions through social sites as a platform. Therefore, the well-known important text mining component called "Sentiment analysis" comes into reality.

Sentiment analysis is also called opinion mining. Sentiment analysis involves the analysis of facts and opinions [12]. Sentiment analysis can also be applied in automated scoring system and rating applications to provide scores and rating to public companies, public opinions, sentiments, attitudes, appraisals, and evaluations [13].

Sentiment analysis measures user's sentiments through techniques like natural language processing (NLP), computational linguistics, digital technology, artificial intelligence, and text analysis. These techniques are used for identifying, extracting, and analyzing subjective information from multiple Web sources. The sentiment analysis explores the contextual divergence of the information/data [14].

The following parameters may be applied to classify the given text, in the process of sentiment analysis:

1.   Polarity ('+', '−', '±')
2.   Emotional states ('☺', '☹', '☺')
3.   Scaling system or numeric values
4.   Subjectivity or objectivity
5.   Features based on key entities.

During sentiment analysis process, the data analyst is able to extract a best solution based on sentiment variation and fantasy analysis of a user. This solution could be used for enhancement of business/organization/product/or services, etc.

Earlier, organizations or individuals were using a set of structured data, to analyze their business trends, market, product, and services. The resultant prediction analysis may be or may not be accurate. The reason was fewer test data sets and amount of structured data.

Nowadays, huge amount of data lies in unstructured format. It is very hard to analyze the unstructured data for predictions and recommendations, for business/organization growth. This unstructured data contains huge information, which needs to be understood and analyzed.

Many organizations are now transforming toward the big data for the facility of 4V's [15]. The organizations are looking for the user opinions, sentiments, and attitudes toward their services and products. We cannot achieve an accurate analysis with fewer structured data set.

Thus, sentiment analysis in collaboration with big data will work for analysis of facts, opinions, or sentiments for providing predictions/recommendations/solution to a problem, to organizations/individuals for their growth.

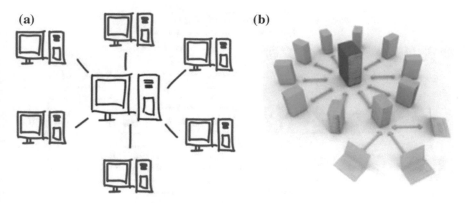

Fig. 1 Various Big data computing models

Now, we have the big data and text mining component called sentiment analysis for extraction of meaningful facts and opinions from unstructured data. We have to now focus on how to process such huge volume of unstructured data [16].

As learnt earlier, if data is on standalone file system and of small amount, then we cannot get accurate and efficient opinions. If we have huge amount of data, then the standalone file system cannot accommodate, and the accuracy, speed, storage, and performance hinders. Therefore, we approach for the distributed computing for big data.

Distributed computing approach provides huge storage facility (in multiples of petabytes) and good performance in the form of speed and accuracy. In distributed computing, multiple computing resources are connected in a network and computing tasks are shared across the multiple resources [17].

This distribution of tasks increases the speed, as well as the efficiency of the system. Because of the reason, the distributed computing system is considered as faster and efficient than traditional methods of computing. It is also more suitable to process huge amounts of data in a limited time.

Distributed computing system provides a level of parallelism, by sharing of tasks across the multiple resources. One of the distributed and parallel computing technologies is Hadoop, which is used to process the big data (Fig. 1).

Next, this paper will discuss on which sentiment analysis classification technique will be used for analyzing the big data for sentiment. Before that, let us see the sentiment analysis classification diagram (Fig. 2).

From the above sentiment analysis classification diagram, this paper will focus on the sentence-level-based approach [18]. The sentence-level-based approach falls under text view technique. The sentence-level sentiment classification is applied to individual sentences in a document or short sentences.

Sentence-level sentiment classification has two tasks, such as subjectivity classification and sentiment classification. Subjectivity classification identifies the sentence as subjective or objective [19]. Sentiment classification classifies the subjective infor-

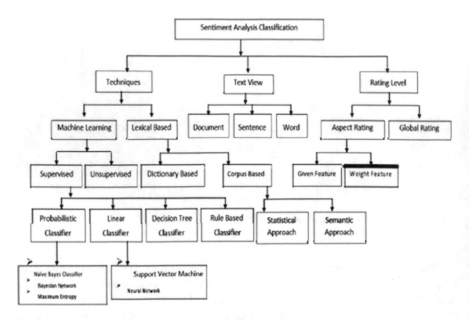

Fig. 2 Sentiment analysis classification diagram

mation based on polarity such as positive or negative or neutral category. Usually, a simple sentence conveys single opinion about an entity. Sentence level expresses overall opinions of each sentence [20]. In sentence-level classification, there are different aspects of a sentence that are indicative of its weight. The words in sentence, its link/correlation to other sentences, and its location in the sentence are important aspects that are used for assessing the weight of a sentence.

2   Existing System

- Topic learning method—Clustering the users based on retweet network and user's interest is mined as prior knowledge.
- Probability model-based sentiment analysis used to analyze text-based data.
- Aspect-based approach using a morphological sentence pattern model.
- Correlating the tweets posted by users from a specific location to understand the sentiments.
- Dictionary of emotions used in various language dimensions.
- Supervised learning model, using training set.
- Lexicon-based approach is used to classify the text according to polarity.
- Depending on the proximity of the words with adjectives, adverbs, and negations or adversative sentences in order to analyze overall polarity.

3 Proposed System

Today, Web users are reliant on the social networking sites, at most of times, in search of information, to know the business analysis and developments, product services, and subject.

The social networking sites hold the unstructured data and need a deep analysis of sentiments that user posts in the blog.

Sentiment analysis, when applied to social media channels, can identify spikes in sentiment, thereby allowing user to identify potential product advocates or social media influencers. As discussed earlier, sentiment analysis when applied to social media channels yields to identify probable negative gears that are emerging online regarding business, thereby allowing user to be proactive in dealing with it.

Twitter is a widespread social networking site that enables Web users to communicate and exchange information, etc. Twitter has become very popular and has grown rapidly. An increasing number of people are willing to post their opinions on Twitter, which is now considered a valuable online source for opinions.

As a result, sentiment analysis on Twitter is a rapid and effective way of appraising public opinion for business marketing or social studies. Twitter produces millions of bytes of data, which is used for business or social purpose. Analyzing data from these social networking Web sites is one of the new buzzwords for many business strategies, election campaigns, world health issues, technical concepts, inventions, and entertainment, etc.

This work utilizes Twitter repository data for sentiment analysis. Twitter repository contains huge data and easy to access. Sentiment analysis is a process of figuring and stratifying an assessment of a person given in a piece of text, in order to identify ones' thinking toward a specific topic, product, subject, or services. Twitter data is collected for analysis via Twitter API.

The proposed approach is based on topic origin method that includes text similarity and computing interactions between conversations of users and applies sentiment analysis to classify the text into positive, negative, or neutral opinion. We can also use machine learning technique, which is very accurate. But the machine learning technique takes too much time performing sentiment analysis. Machine learning is not trained initially, so that it is not efficient for handling of sentiment analysis over Big data.

4 Implementation Overview with Brief Case Study

This section focused briefly on the sentiment analysis and polarity classification of the collected tweets with respective to objective content similarity.

The programming technology by name SCALA is used for analyzing the sentiment variation between collected tweets.

Following are the overview of steps followed with respect to sentiment analysis computation:

A. Data Collection

Data collected from Twitter by using the Twitter API (Twitter 4j) is shown below. Twitter has created its own API for tweets retrieval. These Twitter API credentials are used in Scala code, by declaring a structure array object.

Then, a Twitter object is created. Using JSON/UBJSON encoder and a decoder in the native Scala language, convert a Scala data structure into JSON/UBJSON formatted strings, or decode a JSON/UBJSON file into Scala data structure.

This Scala data object holds the tweets, as and when tweets are posted, and count variable is incremented, with the length of a number of elements, for each retrieval of tweet from the Twitter. The count variable displays the count of the tweets retrieved in decimal notation.

Data Preprocessing and Extraction of Features

Data preprocessing is done using a utility method in Scala. Twitter object, which was created, stores tweets in structure array created from API response in JSON format. Then, sort the rows of a user table by follower count in descending order.

Sentiment Analysis and Polarity-wise Classified Data

In this work, the sentiment analysis is performed by using "Text View approach." This technique uses sentence-level approach. In this approach, the polarity of each sentence is calculated.

In this approach, classifier has to find objective sentences. The objective sentences contain true opinion words which help in determining the sentiment about the entity.

The net sentiment rate is calculated using an approximate formula below:

Sentiment rate $=$ ((sum of sentiments ≥ 0) $-$ (sum of sentiments < 0))$/$ length (sentir

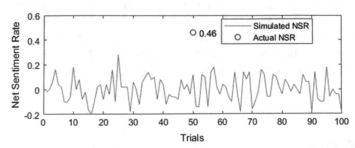

Task of polarity classification means the reviews collected are classified depending upon the opinions expressed as positive, negative, and neutral.

5 Conclusion

Customers are performing an important role in the changes of the business trends and services. They will actively participate, directly or indirectly in the business improvement strategies. Customers got demand for their opinions in the growth of the product services and brand services. It is fact that customers always feel the importance of their sentiments or opinions.

Therefore, today business organizations are taking at most attention while providing services to the customer. Every business organization is keenly observing the customer requirements, their choices toward the product. There are many methods, where organizations follow to get feedback from customers to understand the customer satisfaction and opinion.

Some methods are like as below:

1. Social engineering
2. Personal interview
3. Scheduling
4. Online or offline survey
5. Feedback forms.

All these methods, whichever is adopted, will result to know the customer experience and opinion.

Thus, there comes the concept of sentiment analysis, where organizations today are focusing and giving worth to the customer feedback.

Sentiment analysis is an emerging field with diversified applications. Many data scientists and analysts have made profound progress in the sentiment analysis, due to their demand and challenging tasks on natural language processing origins.

Business organizations are yielding fruitful results by analyzing the sentiments of the users. Companies are adopting this approach in order to know their market trends and services among the customers and as well as to know their success standpoint among the competitors.

Whereas the individuals (customers or users or consumers), wants to know the experiences or opinions of others before taking a decision on the product or brand service.

Sentiment analysis and opinion mining will direct an approach for forecasting the increasing necessity for product insights, customer requirements, the line of breach between product service constraints and customer requirements and bottlenecks that are aroused if any.

In the impending system, the sentiment analysis should focus to reduce the line of breach between the product service constraints and customer constraints. To attain this, it needs a sound knowledge in the objective line perspective bonded with human sentiment psychology with efficient intellectual methods and should be capable to provide an alternate solution, if customer does not satisfy with original solution. Thus, the natural language opinions are better understood and will lead to fill the gap between unstructured data and structured data.

The resultant focus in this paper was to capture polarity and variations of the sentiments captured from twitter data, based on level of objective data collected, by computing interactions between conversation of users.

References

1. R. Nugroho, D. Molla-Aliod, J. Yang, C. Paris, and S. Nepal, "Incorporating tweet relationships into topic derivation," in 2015 Conference of the Pacific Association for Computational Linguistics (PACLING 2015). Bali, Indonesia: PACLING, May 2015.
2. S. K. Bista, S. Nepal, and C. Paris, "Multifaceted visualisation of annotated social media data," in 2014 IEEE International Congress on Big Data (BigData Congress). Anchorage, Alaska, USA: IEEE, June 2014, pp. 699–706.
3. D. Lee and H. Seung, "Algorithms for non-negative matrix factorization," in Advances in Neural Information Processing Systems 13 (NIPS 2000), Denver, CO, USA, 2000, pp. 556–562.
4. D. Blei, A. Ng, and M. Jordan, "Latent dirichlet allocation," The Journal of Machine Learning research, vol. 3, pp. 993–1022, 2003.
5. T. Hofmann, "Probabilistic latent semantic indexing," in Proceedings of the 22nd Annual International ACM SIGIR conference on Research and Development in Information Retrieval. Berkeley, CA, USA: ACM, August 1999, pp. 50–57.
6. Y. Hu, A. John, F. Wang, and S. Kambhampati, "Et-lda: Joint topic modeling for aligning events and their twitter feedback." in AAAI Conference on Artificial Intelligence (AAAI 2012), vol. 12, Toronto, Ontario, Canada, July 2012, pp. 59–65.
7. X. Yan, J. Guo, S. Liu, X. Cheng, and Y. Wang, "Learning topics in short texts by non-negative matrix factorization on term correlation matrix," in Proceedings of the SIAM International Conference on Data Mining (SIAM 2013. San Diego, California, USA: SDM, July 2013.
8. K. Erk, "Vector space models of word meaning and phrase meaning: A survey," Language and Linguistics Compass, vol. 6, no. 10, pp. 635–653, 2012.
9. D. Ramage, S. T. Dumais, and D. J. Liebling, "Characterizing microblogs with topic models." The International AAAI Conference on Web and Social Media (ICWSM), vol. 10, pp. 130–137, May 2010.
10. J. Li, Z. Tai, R. Zhang, W. Yu, and L. Liu, "Online bursty event detection from microblog," in Utility and Cloud Computing (UCC), 2014 IEEE/ACM 7th International Conference on, Dec 2014, pp. 865–870.
11. J. Choo, C. Lee, C. K. Reddy, and H. Park, "Utopian: User-driven topic modeling based on interactive nonnegative matrix factorization," IEEE Transactions on Visualization and Computer Graphics, vol. 19, no. 12, pp. 1992–2001, 2013.
12. M. Albakour, C. Macdonald, I. Ounis et al., "On sparsity and drift for effective real-time filtering in microblogs," in Proceedings of the 22nd ACM International Conference on Information & Knowledge Management (CIKM 2013), October 2013, pp. 419–428.
13. J. Vosecky, D. Jiang, K. W.-T. Leung, K. Xing, and W. Ng, "Integrating social and auxiliary semantics for multifaceted topic modeling in twitter," ACM Transactions on Internet Technology (TOIT), vol. 14, no. 4, p. 27, 2014.
14. A. de Moor, "Conversations in context: a twitter case for social media systems design," in Proceedings of the 6th International Conference on Semantic Systems. New York, NY, USA: ACM, September 2010, p. 29.
15. R. Nugroho, J. Yang, Y. Zhong, C. Paris, and S. Nepal, "Deriving topics in twitter by exploiting tweet interactions," in Proceedings of the 4th IEEE International Congress on Big Data. New York, USA: IEEE Services Computing Community, July 2015.
16. G. Salton, Automatic Text Processing: The Transformation, Analysis, and Retrieval of Information by Computer. Addison-Wesley, 1989.

17. J. L. Fleiss, "Measuring nominal scale agreement among many raters." Psychological bulletin, vol. 76, no. 5, p. 378, 1971.
18. G. G. K. J. Richard Landis, "The measurement of observer agreement for categorical data," Biometrics, vol. 33, no. 1, pp. 159–174, 1977.
19. D. H. Von Seggern, CRC Standard Curves and Surfaces with Mathematica. CRC Press, 2006.
20. L. Yang, T. Sun, M. Zhang, and Q. Mei, "We know what@ you# tag: Does the dual role affect hashtag adoption?" in Proceedings of the 21st International Conference on World Wide Web (WWW 2012). Lyon, France: ACM, April 2012, pp. 261–270.

An Analysis of Digital Forensics in Cyber Security

D. Paul Joseph and Jasmine Norman

Abstract Digital forensics also called as computer forensics is a major field that
incorporates people regardless of their professions. Digital forensics includes various
forensic domains like network forensics, database forensics, mobile forensics, cloud
forensics, memory forensics, and data/disk forensics. Recent statistics and analytics
show the exponential growth of cyber threats and attacks and thus necessitate the
need for forensic experts and forensic researchers for automation process in the cyber
world. As digital forensics is directly related to data recovery and data carving, this
field struggles with the rapid increase in volume of data. In addition to that, day-
to-day increase of malware makes forensic field slacking. This paper provides the
users and researchers some information regarding forensics and its different domains,
anti-forensic techniques, and also an analysis of current status of forensics.

Keywords Anti-forensics · Cyber-attacks · Cyber targets · Cyber threats · Digital
forensics

1 Introduction

Digital forensics [1] is also called as Digital forensic science [2] is a branch of forensic
science that includes identify—search—seizure—preserving and investigation cycle
of digital data in the crime scenarios. Though the roots of this field were found in
early 1980s, the revolution of this field had started in mid 1990s with the invention
of multi-user, multi-tasking operating systems and wide area networks. In early
years, the forensic was confined to only unauthorized access of information, however
later extended to cyber-attacks, creation of malwares/viruses, financial frauds, child
pornography etc. With the rise in cyber threats and attacks, digital forensics has

D. Paul Joseph (✉) · J. Norman
School of Information Technology and Engineering, Vellore Institute of Technology, Vellore, India
e-mail: pauljoseph91@gmail.com

J. Norman
e-mail: jasmine@vit.ac.in

© Springer Nature Singapore Pte Ltd. 2019
R. S. Bapi et al. (eds.), *First International Conference on Artificial Intelligence
and Cognitive Computing* , Advances in Intelligent Systems and Computing 815,
https://doi.org/10.1007/978-981-13-1580-0_67

emerged as one of the key areas in the world of security. Recent KPMG cyber-crime [3] stated that 72% of companies in India faced cyber-attacks in 2016 followed by 63% of financial loss and 55% of sensual data stolen resulting in 49% of reputational damage. Survey per Symantec Corp [4] showed exponential growth in the rise of malwares to 430 million in 2016 from 2.3 million in 2009; that is, 1.1 million of malwares were created every day. The above issues sum up the one face of digital forensics and the other face includes data recovery or finding the lost data. The data may be either in raw format or multimedia format and it can include the hard drives, mobile phones, databases, GPS devices, IOT, and sophisticated electronic gadgets [5]. As the secure data storage and secure data retrieval mechanisms are advanced, data destroying or data shredding also became sophisticated, resulting the job of forensic experts more difficult. This paper concentrates and gives information on cyber threats, targets for cyber-attacks, results of that attack, anti-forensic techniques and eventually concludes with preventive measures of cyber-attacks from a user's perspective.

2 Background Work

Digital forensics is not solely assortment of multiple forensic disciplines; however, it is a combination of multiple subjects and techniques. For example, data storage and data recovery techniques contain data mining, machine intelligence, deep learning, algorithms and architectural framework techniques. Memory forensics [6] embodies kernel level debugging, hardware architectures knowledge, and mobile forensics which includes android programming knowledge, etc. As aforesaid earlier, it includes people from multiple professions like forensic experts, law enforcement agencies, attacker, victim, companies' courts. On whole, digital forensics incorporates multiple professions, technologies, and domains, and thus, the complexity is also multifold.

2.1 Digital Forensics Domain and Their Impact

As listed above, digital forensic domain consists of various domains like data forensics, cloud forensics, memory forensics, and android/mobile forensics. Though the digital forensics came into existence three decades ago, cloud forensics [7] and mobile forensics [8] were being into existence simply a decade past.

Computer/Disk Forensics
Disk forensics was started in the early 1990s; however, there has been tremendous research in this field in 2012 and 2014. Later, the graph fell down attributable due to the lack of proper data forensic tools. There have been small research gap in data forensics at that point as a result of the birth of big data [9] in forensic field. In the present year, still lot of research is going on in data forensics associated with big

Interest over time ❓

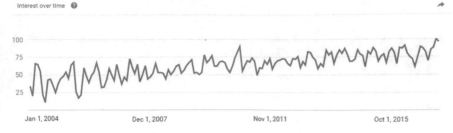

Fig. 1 Disk forensics in the time gap

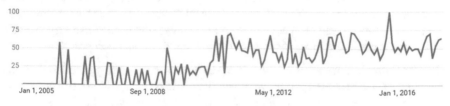

Fig. 2 Cloud forensics in time gap

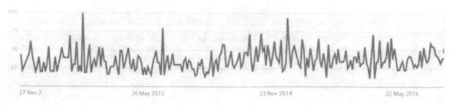

Fig. 3 Memory forensics in time gap

data either homogeneous or heterogeneous. Still there are lot of open problems in this field regarding automation and correlation of data [10]. The following diagrams represent the familiarity of research in those domains (Fig. 1).

Cloud Forensics

The term cloud forensics was introduced between 2005 and 2009, and initially, it had very less research scope as only a few cloud vendors were available at that time. Later in mid 13s, the research in this field raised to sky however with unresolvable questions at that time. Still this field faces difficulty in forensic area because it does not contain forensics as a service [7]. Secondly, forensic experts cannot gain access directly to the cloud servers (Fig. 2).

Memory Forensics

The domain memory forensic [11] is much acquainted with the forensic researchers. As this domain had started at the early birth of digital forensics, memory here refers to random access memory, read-only memory, un-separable memory (memory in mobile), flash devices, etc. A lot of research has been through this field, but still having some problems in live memory forensics (Fig. 3).

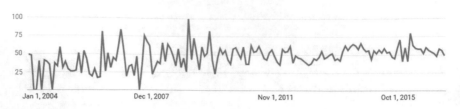

Fig. 4 Mobile forensics

In 2004, black analysis or death analysis is observed. That is, when system is in off stage or about to switch off, the experts used to take a snapshot of the memory and perform the analysis. But now many tools are available for live memory analysis [6] but these could not give efficient results, which is still an open problem.

Mobile/Android Forensics: Mobile forensics was started in the year 2000 with the invention of new Motorola phones which work on java mobile editions. But nowadays with the development of android and Windows operating systems, the day-to-day increase of mobile phones had been greatly increased. Since android phones serve as a mini computer, major work is done through the mobile itself. In the forensic stream, mobile forensics plays a vital role when compared to other domains. For the past two years, many android forensic tools [8] were developed and commercial tools were also available (Fig. 4).

But the major drawback [12] of this field is that the memory of the mobile and operating system cannot be separated, which is a difficult task for forensic examiners. The cause for the downfall of the graph is as many new mobiles with different architectures and new operating systems are being developed, the tools so far developed are not sufficient to gather the evidence from new devices.

2.2 Cyber World Crimes and Targets

Cyber-crime, synonymously called as computer crime, involves a computer and a network. The computer may be a target computer or may be an attacker's stand-alone computer. The network is simply not restricted to LAN, however extended to MAN, WAN, and SAN. Different types of cyber-crimes with their brief are as follows:

Cyber Terrorism: It is nothing but entering into security agencies of different countries, searching for the loopholes, and with the help of that loopholes, stealing the confidential matter regarding country's internal security. The persons who do these types of activities are referred to as cyber terrorists and their role is to steal the highly confidential matters like military plans and moves.

Cyber Warfare: Cyber warfare became a national concern as each country is performing their attacks on their enemies. For example, USA on Russia, China on USA, India on Pakistan and the list follows by many other countries. This mainly deals

Table 1 Cyber-attacks at cyber targets

S. No	Cyber-attack	Cyber target						Severity
		Desktop	Mobile	IOT	Server	SCADA	ERP	
1	Application-layer attack	•	•	•	•	•	•	Severe
2	SQL injection	•	–	–	•	•	•	Catastrophic
3	Spear phishing	•	•	–	•	•	–	Moderate
4	DDOS attack	–	–	–	•	•	•	Moderate (no data loss)
5	Malware/virus	•	•	•	•	•	•	High
6	Botnets	–	–	–	•	•	•	Moderate
7	Social engineering	•	•	–	•	–	–	Moderate

with hacking of nuclear plants, electricity distribution plants, air defense systems, government Web sites defacing, military agencies, etc.

Cyber Extortion: It is a cyber-attack within which the attackers send multitudes of requests to a Web server, thereby reducing the ability of that server in handling requests of other users. Then, these attackers demand some money from that server admins so as to stop that flooding attacks.

Financial Fraud: The most common cyber-attack is financial frauds. This happens so by unauthorized altering of data. This is observed mostly during online money transactions. This attack occurs by phishing mails, spam mail, and unauthorized second authentication factor from attackers.

Cyber Stalking: The name itself indicates that this attack could be a quite of either online or offline harassment of the persons. This attack includes defamation, depreciation, and false impeachments. This attack comes underneath criminal offense and has its own rules.

Identity Theft: This attack is also rising in an alarming way within which the assaulter uses the user's information like his name, mobile number, address, bank details and card numbers and does the financial transactions, which results in the great financial loss of the user.

Child Soliciting and Abuse: This attack can be seen at the time of children chatting in their personal chat rooms online. Some attackers use the chat rooms as baits and attract thousands of children into pornography field, which adversely affects the country itself.

Targets of Cyber World

Since there are multiple cyber-crimes [3], the targets also existed in multiple ways. The different targets of cyber attackers are as follows:

1. Desktop/Laptops/Mobiles/PDA
2. IOT
3. Servers (File Server, Web server, Email server)
4. SCADA systems and ERP systems (Fig. 5; Table 1).

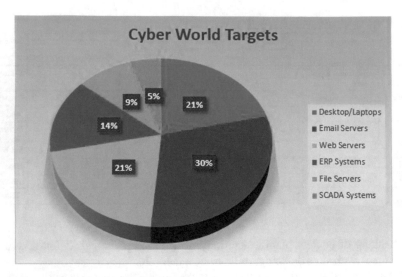

Fig. 5 Targets of cyber world with their impact [4]

3 Forensic and Anti-Forensic Techniques

The goal of the digital forensics is to recover, identify, and analyze the information so as to grasp who the offender is. By forensic process irrespective of domains, the goal is to recover the data that was either formatted, deleted, or shredded, and therefore, the goal of anti-forensics [13] is to delete the data without leaving any traces of log file, text file, or any temp file. So far for the last three decades, thousands of forensic tools were available in all the forensic domains. The anti-forensic tools were also developed which were a boon to attackers and criminals and ban to forensic experts as they struggle a lot to cope with that data. But still, only a few tools are available that supports all the domains. Examples of forensic tools are Encase, Caine Forensic tools, Oxygen Forensic Suite, etc.

Anti-forensic techniques in present scenario:

1. Overwriting metadata/Shredders [13]
2. Cryptographic techniques and Steganography techniques
3. Program Packers
4. Live CDs and Bootable Drives
5. Virtual systems
6. S.M.A.R.T technique in hard disks
7. Denial of kernel access
8. Altering of MAC (modified, access, control) and timestamps
9. Hiding in slack space and boot records
10. Using of encrypted and secured network protocols
11. Proxy and anonymous surf (for network attacks).

4 Discussion

As security enhances from day to day, thus were the security breaches. As the forensic tools are being developed, so are the anti-forensic tools. Though many security vendors come up with new software and technologies, there were not efficient tools available such that they cannot stop deadliest threats and attacks like ransomware, SQL injections, and spear phishing. Day to day the cyber-crime statistics reveals the alarming rise of attacks which makes many companies, professionals fear of falling in those traps. Though the forensic domain attracts many researchers, it could not yield up to its mark. Still android forensics and cloud forensics [14] stand as a major challenge for forensic examiners and researchers in implementing forensics as a service. The second major challenge for mobile forensics [15] is that there is a need for multi-tool that supports all operating systems and all architectures. Disk/data forensics challenges the researchers in correlating the massive amount of big data [10] and automated approaches in it.

5 Conclusion

So far in this paper, we discussed brief introduction of forensics in light of cyber world, types of cyber threats, cyber-attacks, current research scope of digital forensic domain, and various forensic and anti-forensic methods. The further research will be extended in implementing various professional tools in the environments of Windows and Linux operating systems with different critical test cases, i.e., testing the efficiency of forensic toolkits when used with anti-forensic techniques, performing the live analysis with test cases. The further work also includes the designing and implementing integrated forensic tool that supports all the platforms.

References

1. Raghavan, S. (2013). Digital forensic research: current state of the art. *CSI Transactions on ICT*, *1*(1), 91–114. https://doi.org/10.1007/s40012-012-0008-7.
2. Beebe, N. (2009). Digital forensic research: The good, the bad and the unaddressed. *Advances in Digital Forensics V*, 17–36. https://doi.org/10.1007/978-3-642-04155-6_2.
3. Cybercrime survey report. November (2017).
4. State, U. S., & Survey, C. (2017). Key findings from the 2015 US State of Cybercrime Survey of Cybercrime Survey, (July).
5. Stamm, M. C., Wu, M., Liu, K. J. R., Member, M. C. S., Fellow, M. I. N. W. U., & Fellow, K. J. R. A. Y. L. I. U. (2013). Information forensics: An overview of the first decade. *IEEE Access*, *1*, 167–200. https://doi.org/10.1109/ACCESS.2013.2260814.
6. Chan, E., & David, F. (2010). Forenscope: A Framework for Live Forensics, 307–316.
7. Pichan, A., Lazarescu, M., & Soh, S. T. (2015). Cloud forensics: Technical challenges, solutions and comparative analysis. *Digital Investigation*, *13*, 38–57. https://doi.org/10.1016/j.diin.2015.03.002.

8. Martinez, J. (2007). Mobile Forensics. *Technology*, *1*(1), 40. Retrieved from www.susteen. com.
9. Irons, A., & Lallie, H. S. (2014). Digital Forensics to Intelligent Forensics, 584–596. https://d oi.org/10.3390/fi6030584.
10. Mohammed, H., Clarke, N., & Li, F. (2016). AN AUTOMATED APPROACH FOR DIGITAL FORENSIC ANALYSIS OF HETEROGENEOUS BIG DATA. *Journal of Digital Forensics, Security and Law*, *11N2*, 1–16.
11. Vo, S., Freiling, F. C., Vömel, S., & Freiling, F. C. (2011). A survey of main memory acquisition and analysis techniques for the windows operating system. *Digital Investigation*, *8*(1), 3–22. https://doi.org/10.1016/j.diin.2011.06.002.
12. Abdallah, A., Alamin, M., Babiker, A., & Mustafa, N. (2015). A Survey on Mobile Forensic for Android Smartphones. *IOSR Journal of Computer Engineering*, *17*(1), 2278–661. https:// doi.org/10.9790/0661-17211519.
13. Garfinkel, S. (2007). Anti-Forensics: Techniques, Detection and Countermeasures. *2nd International Conference on I-Warfare and Security*, 77–84. https://doi.org/10.1.1.109.5063.
14. Manoj, S. K. A., & Bhaskari, D. L. (2016). Cloud Forensics-A Framework for Investigating Cyber Attacks in Cloud Environment. *Procedia Computer Science*, *85*(Cms), 149–154. https:// doi.org/10.1016/j.procs.2016.05.202.
15. Khan, S., Ahmad, E., Shiraz, M., Gani, A., Wahab, A. W. A., & Bagiwa, M. A. (2014). Forensic challenges in mobile cloud computing. *2014 International Conference on Computer, Communications, and Control Technology (I4CT)*, (I4ct), 343–347. https://doi.org/10.1109/I4 CT.2014.6914202.

Efficient Mining of Negative Association Rules Using Frequent Item Set Mining

E. Balakrishna, B. Rama and A. Nagaraju

Abstract Mining of negative association rules (NARs) is useful to find out the items from the database which are alternative to each other. Various algorithms and techniques have been implemented to find the NAR which are efficient and capable for finding the item sets, but producing many conditional FP-Trees. In this paper, we proposed an algorithm called frequent item set mining algorithm (FISM), used to find frequent patterns from databases for extracting NAR without generating conditional FP-Tree.

Keywords Improved FP-Tree · Support · Confidence

1 Introduction

In order to find frequent item sets, association rule mining (either positive association rules (PARs) mining method or NAR mining method) [1–5] is one of the significant data mining systems. NARs are very important and similar to PAR in which either the antecedent or the subsequent or both are negated. For example, for the rule A → B the negative rules are as follows: (a) $X \rightarrow \sim Y$, (b) $\sim X \rightarrow Y$, and (c) $\sim X \rightarrow \sim Y$. In order to discover the NAR, first we need to discover frequent item sets from the database. Next, from frequent item sets we can extract NAR which are valid. For example, in supermarket data, NAR is HotCoffee → ~ApplePie, which mean that people who

E. Balakrishna (✉)
CSE Department, Vaagdevi College of Engineering, Warangal, Telangana, India
e-mail: balakrishnakits@gmail.com

B. Rama
CSE Department, Kakatiya University, Warangal, Telangana, India
e-mail: rama.abbidi@gmail.com

A. Nagaraju
CSE Department, Central University of Rajasthan, Ajmer, Rajasthan, India
e-mail: nagaraju@curaj.ac.in

© Springer Nature Singapore Pte Ltd. 2019
R. S. Bapi et al. (eds.), *First International Conference on Artificial Intelligence and Cognitive Computing* , Advances in Intelligent Systems and Computing 815,
https://doi.org/10.1007/978-981-13-1580-0_68

purchase HC may not give importance AP. Such sort of NAR is needful for growing the sales in the supermarket.

2   Methodology

2.1   Basic Terminology

In this paper, first build improved FP-Tree [6], Sparse Table, and FISM algorithm. In the FISM algorithm, it uses improved FP-Tree and generates frequent item sets. By using these frequent item sets, find out the NAR. This paper is concentrated on generating NAR from frequent items.

2.2   Algorithm-I: Generating Improved FP-Tree

Improved FP-Tree consists of two main elements—the Tree and a Sparse Table. The procedure to generate IFP-Tree is as follows: (a) Find every transaction from database and sort the items in descending arrange based on the support of the items. (b) Make a root node of improved FP-Tree, which is at first "NULL". (c) For every transaction perform: (1) Items in each transaction is characterizing by $[M/N]$ wherever M-first item, N-remaining items. (2) If M is the most frequent item set then, (i) If a root has a direct child node say k such that k's item name $= m$'s item name, then increment the Occurrence of the item k and send the root from k to M denoted as X. (ii) For each item in N, perform the below steps until N turn into blank. For every latest item, make a node from root and increment the Occurrence of the new item say, X by 1. Send the root to latest node. If item, Ni has no distinct edge from the current root then Ni is go for as Sparse item. Add Ni to Sparse Table with Occurrence count by 1. (3) Add all items to Sparse Table with Occurrence 1, for every item in transaction $[M/N]$. (4) Compute the total number of Occurrences of every item in the Sparse Table.

2.3   Algorithm-II: Frequent Item Set Mining (FISM) Algorithm

We can find frequent item sets by making use of improved FP-Tree and Sparse Table, obtained in previous table. FISM algorithm finds frequent item sets by making using of improved FP-Tree and minimum_support (MS) (user-specified value) as inputs and by using the given rules:

(i) If the Occurrence of the item in the improved FP-Tree = MS of that item (user-specified threshold value), then Occurrence of frequent item set for that item = improved FP-Tree Occurrence. And the frequent items should be only those items, which contain greater or equal improved FP-Tree Occurrence at FP-Tree.

(ii) If the MS of the item > Occurrence of that item in the improved FP-Tree, then Occurrence for the frequent item sets = Total number of Occurrences of that item in the improved FP-Tree Occurrence + Sparse count.

(iii) If the MS of the item < Occurrence of that item in the improved FP-Tree, then Occurrence for frequent item set = Occurrence of that item in the improved FP-Tree.

2.4 Generating NARs

Now consider Certainty Factor (CF) [4]—is to determine of dissimilarity of the probability that B is in a transaction when only taking transactions with A. A rising CF means a decrease of the possibility that B is not in a transaction that A is in. Negative CFs have a similar interpretation.

$$CF(A \Rightarrow B) = [conff(A \Rightarrow B) - sup(B)]/sup(\sim B) \qquad (1)$$

3 Illustration: Generating Valid NAR for Sample Database

We take the below given database for building improved FP-Tree (Table 1).

Now take negative items and consider MS of item defined by the user is 2, i.e., 50% (Table 2).

Table 1 Database

ID	Items
T1	CoffeeEclair (CE), HotCoffee (HC)
T2	HotCoffee (HC), ApplePie (AP), CoffeeEclair (CE)
T3	HotCoffee (HC), ApplePie (AP)
T4	AlmonTwist (AT), HotCoffee (HC)
T5	OrangeJuice (OJ), VenillaEclair (VE), OperaCake (OC)
T6	VenillaEclair (VE), OperaCake (OC), LemonCookie (LC)
T7	OrangeJuice (OJ), AlmonTwist (AT)
T8	HotCoffee (HC), AlmonTwist (AT), ApplePie (AP)

3.1 Constructing Improved FP-Tree

First, we examine all the transactions and compute the number of Occurrences of every item, and we evaluate with the MS, if there are any items with less than MS than we reject those items from database. Next, we arrange the items in every transaction in descending sort based on the support and the item "LC" is set up to be infrequent (number of Occurrence of LC < MS). So, we reject "LC" from the database. Now, sort the items in descending order based on their support (Table 3).

3.1.1 Operation-1: {HC, CE}

Considering the above table, the first operation (transaction) is {HC, CE}. At initial, root of the improved FP-Tree is formed, and initially, the value is set to "NULL". Occurrence (Frequency) of any item in FP-Tree is characterized as {Item: IFP-Tree Occurrence}, such as {HC}. Occurrence of node "HC" is increased by 1 as it is a direct child node of root (as shown in the Fig. 1). Other items are in dissimilar nodes. The Sparse Table is empty since the transaction hold the most frequent item "HC" and the other item "CE" in transaction occurring for the first time. So, no item is regarded as Sparse here (Table 4).

Table 2 Transactional database—positive and negative items

ID	Original items	Augmented items (positive and negative items)
T1	CE, HC	CE, HC, ~AP, ~VE, ~AT, ~OC, ~OJ, ~LC
T2	HC, AP, CE	HC, AP, ~CE, ~VE, ~AT, ~OC, ~OJ, ~LC
T3	HC, AP	HC, AP, ~CE, ~VE, ~AT, ~OC, ~OJ, ~LC
T4	AT, HC	AT, HC, ~AP, ~VE, ~CE, ~OC, ~OJ, ~LC
T5	OJ, VE, OC	OJ, VE, OC, ~AP, ~VE, ~AT, ~HC, ~LC
T6	VE, OC, LC	VE, OC, LC, ~AP, ~HC, ~AT, ~CE, ~OJ
T7	OJ, AT	OJ, AT, ~AP, ~VE, ~CE, ~OC, ~HC, ~LE
T8	HC, AT, AP	HC, AT, AP, ~CE, ~VE, ~OC, ~OJ, ~LE

Table 3 Items after discarding infrequent items and in descending order

ID	Original items
T1	HC, CE
T2	HC, AP, CE
T3	HC, AP
T4	HC, AT
T5	VE, OC, OJ
T6	OC, VE
T7	AT, OJ
T8	HC, AT, AP

Fig. 1 Improved FP-Tree
after first operation

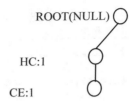

ROOT(NULL)

HC:1

CE:1

Fig. 2 Improved FP-Tree
after second operation

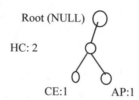

Root (NULL)

HC: 2

CE:1   AP:1

Fig. 3 Improved FP-Tree
after third operation

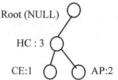

Root (NULL)

HC : 3

CE:1   AP:2

3.1.2 Operation-2: {HC, AP, CE}

Now consider with new item "AP" in second operation, the "Root" is changed to "HC" [from step (2) in Algorithm-I]. As a result, a new node is made from the node "HC" for the reason that this is the current root and its Occurrence is increased by 1. A new node "AP" is attached to node "HC" and its Occurrence is indicated by 1 as it is first time occurring in IFP-Tree. Next item is "CE". As here is no distinct edge from root "HC" to node "CE" (as shown in the Fig. 2), so the item CE is stored in Sparse (Table 5).

3.1.3 Operation-3: {HC, AP}

Most frequent item "HC" is added in this operation, so the root is still "HC", and all the other items are constructing (as shown in the Fig. 3) with respect to it (Table 6).

Table 4 Sparse Table for Fig. 1

Item name	Occurrence
–	–

Table 5 Sparse Table for Fig. 2

Item name	Occurrence
CE	1

Table 6 Sparse Table for Fig. 3

Item name	Occurrence
CE	1

Fig. 4 Improved FP-Tree after fourth operation

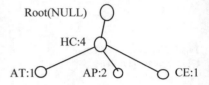

Table 7 Sparse Table for Fig. 4

Item name	Occurrence
CE	1

Fig. 5 Final improved FP-Tree after all operations

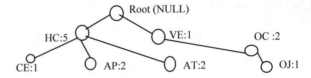

3.1.4 Operation-4: {HC, AT}

In this operation, new item "AT" comes out. So according to step-2 from Algorithm-I, create a new node from root and increment the Occurrence of the new item by 1(as shown in the Fig. 4). Sparse Table remains same (Table 7).

3.1.5 Final Operation: {HC, AT, AP}

Lastly, improved FP-Tree is produced (as shown in the Fig. 5). In this Tree, the number of nodes in the Tree is equal to the number of dissimilar items and Occurrence of each item is different from the number of Occurrence of Operation at "Table 2", because this Tree indicates, only the relationship along with the items (Tables 8 and 9).

3.2 Finding Frequent Item sets Using FISM

Now take the item "HC", and the Occurrence of item "HC" in the improved FP-Tree is "7", and minimum support defined by the user is 2 (i.e., 50%). So, we obtain frequent item set: {HC: 7}.

Table 8 Sparse Table for Fig. 5

Item name	Occurrence
CE	1
OC	1
VE	1
AT	1
OJ	1
AP	1

Table 9 Sparse Table count after all operations

Item name	Occurrence
CE	1
OC	1
VE	1
AT	1
OJ	1
AP	1

Table 10 Frequent item sets

HC	{HC: 2}
AP	{{AP: 2},{AP, HC: 2}}
AT	{{AT:2},{AT,HC: 2}}
CE	{{CEs: 2},{CE, HC: 2}}
OJ	{OJ: 2}
OC	{{OC: 2},{VE: 2}}
VE	{{VE: 2},{VE, OC: 2}}

Next we take item "AP", and the Occurrence of item "AP" in the improved FP-Tree is 2 that is equal to the user-specified value (i.e., 2). Based on the FISM, Occurrence of the frequent item set is also 2. For frequent item sets, we consider only those items which contain upper or equivalent FP-Tree. So, the frequent item sets for the item "AP" are: {AP: 2}; {AP, HC: 2}.

Now consider the item "AT", and the Occurrence of IFP-Tree is 2 that are equal to user specified value (i.e., 2). Based on the FISM, Occurrence of the frequent item set is also 2. For frequent item set, we consider only those items which contain upper or equivalent FP-Tree Occurrence. So, the frequent item sets are {AT: 2}; {AT, HC: 2} (Table 10).

3.3 Generating NAR

We can find NAR by making use of Eq. (1) as follows:

Consider minimum confidence (MC) threshold value is 50%. The NARs are as follows: **NAR-1**: AP → HC

$$Cf(\text{AP} \rightarrow \text{HC}) = \{\sup(\text{AP } U \text{ HC}) - \sup(\text{HC})\} / \sup(\sim\text{HC})$$
$$= \{(3/3) - (5/8)\} / (3/8) = 1 \text{ so } \textbf{NAR}-\textbf{1 is chosen}.$$

NAR-2: HC → AP

$$Cf(\text{HC} \rightarrow \text{AP}) = \{\sup(\text{HC } U \text{ AP}) - \sup(\text{AP})\} / \sup(\sim\text{AP})$$
$$= \{(3/5) - (3/8)\} / (5/8)$$
$$= 0.36 \text{ so } \textbf{NAR}-\textbf{2 is discarded}.$$

Finally, in this process, we found four valid NARs.

4 Experimental Results

For the database (Table 1), the given valid four NARs are produced: AP → HC, HC → AT, VE → OC, and OC → VE.

5 Conclusion and Future Work

The proposed novel approach produced useful and valid NAR. In the future, we will implement the NAR mining tools.

References

1. R. Agarwal, E. R. Srikant, "Fast Algorithms for Mining Association Rules in Large Databases," Proc. of the 20th International conference on very Large Databases, pp. 487–499, Santiago, Chile, 1994.
2. Wu X., Zhang C., Zhang S.: Mining both positive and negative association rules. In: Proc. of ICML (2002) 658–665.
3. X. Yuan, B.P. Buckles, Z. Yuan and J. Zhang, "Mining Negative Association Rules", Proc. Seventh Intl. Symposium on Computers and Communication, Italy, 2002, pp. 623–629.
4. F. Bezal, I. Blacno, M.A. Villa "A definition for fuzzy approximate dependencies" published in journal fuzzy sets and systems vol 149 issue 1 January 2005 page 105–129.
5. O. Daly and D. Taniar, "Exception rules mining based on negative association rules", lecture notes in computer science, vol. 3046, 2004 and pp. 543–552.
6. E. Balakrishna, B. Rama, A. Nagaraju, "Mining of NAR using Improved Frequent Pattern Tree", IEEE, INSPEC Accession Number: 15022153, https://doi.org/10.1109/iccct2.2014.7066748, ICCCT-2014.